Paleomagnetic Rotations and
Continental Deformation

NATO ASI Series

Advanced Science Institutes Series

A Series presenting the results of activities sponsored by the NATO Science Committee which aims at the dissemination of advanced scientific and technological knowledge, with a view to strengthening links between scientific communities.

The Series is published by an international board of publishers in conjunction with the NATO Scientific Affairs Division

A	Life Sciences	Plenum Publishing Corporation
B	Physics	London and New York
C	Mathematical and Physical Sciences	Kluwer Academic Publishers
		Dordrecht, Boston and London
D	Behavioural and Social Sciences	
E	Applied Sciences	
F	Computer and Systems Sciences	Springer-Verlag
G	Ecological Sciences	Berlin, Heidelberg, New York, London,
H	Cell Biology	Paris and Tokyo

Series C: Mathematical and Physical Sciences - Vol. 254

Paleomagnetic Rotations and Continental Deformation

edited by

Catherine Kissel

and

Carlo Laj

Centre des Faibles Radioactivités,
Laboratoire mixte CNRS/CEA, Gif-sur-Yvette, France

Kluwer Academic Publishers

Dordrecht / Boston / London

Published in cooperation with NATO Scientific Affairs Division

Proceedings of the NATO Advanced Research Workshop on
Paleomagnetic Rotations and Continental Deformation
Loutra Edipsou, Greece
May 8–13, 1988

Library of Congress Cataloging in Publication Data
Paleomagnetic rotations and continental deformation.

(NATO ASI series. Series C, Mathematical and physical
sciences ; vol. 254)
1. Earth--Crust--Congresses. 2. Paleomagnetism--
Congresses. 3. Continental drift--Congresses. I. Kissel,
Catherine. II. Laj, Carlo. III. Series: NATO ASI series.
Series C, Mathematical and physical sciences ; no. 254.
QE511.P335 1988   551.1'3   88-26648
ISBN-13:978-94-010-6878-9   e-ISBN-13:978-94-009-0869-7
DOI: 10.1007/978-94-009-0869-7

ISBN-13:978-94-010-6878-9

Published by Kluwer Academic Publishers,
P.O. Box 17, 3300 AA Dordrecht, The Netherlands.

Kluwer Academic Publishers incorporates the publishing programmes of
D. Reidel, Martinus Nijhoff, Dr W. Junk, and MTP Press.

Sold and distributed in the U.S.A. and Canada
by Kluwer Academic Publishers,
101 Philip Drive, Norwell, MA 02061, U.S.A.

In all other countries, sold and distributed
by Kluwer Academic Publishers Group,
P.O. Box 322, 3300 AH Dordrecht, The Netherlands.

This book contains the proceedings of a NATO Advanced Research Workshop held within the programme of activities of the NATO Special Programme on Global Transport Mechanisms in the Geo-Sciences running from 1983 to 1988 as part of the activities of the NATO Science Committee.

Other books previously published as a result of the activities of the Special Programme are

BUAT-MENARD, P. (Ed.) – *The Role of Air-Sea Exchange in Geochemical Cycling* (C185) 1986

CAZENAVE, A. (Ed.) – *Earth Rotation: Solved and Unsolved Problems* (C187) 1986

WILLEBRAND, J. and ANDERSON, D. L. T. (Eds.) – *Large-Scale Transport Processes in Oceans and Atmosphere* (C190) 1986

NICOLIS, C. and NICOLIS, G. (Eds.) – *Irreversible Phenomena and Dynamical Systems Analysis in Geosciences* (C192) 1986

PARSONS, I. (Ed.) – *Origins of Igneous Layering* (C196) 1987

LOPER, E. (Ed.) – *Structure and Dynamics of Partially Solidified Systems* (E125) 1987

VAUGHAN, R. A. (Ed.) – *Remote Sensing Applications in Meteorology and Climatology* (C201) 1987

BERGER, W. H. and LABEYRIE, L. D. (Eds.) – *Abrupt Climatic Change – Evidence and Implications* (C216) 1987

VISCONTI, G. and GARCIA, R. (Eds.) – *Transport Processes in the Middle Atmosphere* (C213) 1987

SIMMERS, I. (Ed.) – *Estimation of Natural Recharge of Groundwater* (C222) 1987

HELGESON, H. C. (Ed.) – *Chemical Transport in Metasomatic Processes* (C218) 1987

CUSTODIO, E., GURGUI, A. and LOBO FERREIRA, J. P. (Eds.) – *Groundwater Flow and Quality Modelling* (C224) 1987

ISAKSEN, I.S.A. (Ed.) – *Tropospheric Ozone* (C227) 1988

SCHLESINGER, M. E. (Ed.) – *Physically-Based Modelling and Simulation of Climate and Climatic Change* 2 vols. (C243) 1988

UNSWORTH, M. H. and FOWLER, D. (Eds.) – *Acid Deposition at High Elevation Sites* (C252) 1988

Table of Contents

PREFACE   xv

BLOCK ROTATIONS IN CONTINENTAL CRUST: EXAMPLES FROM WESTERN
NORTH AMERICA
   by M. E. BECK   1

      1. Introduction   1
      2. Definitions   3
      3. Types of rotation and their characteristics   3
      4. Discussion   12

THE KINEMATICS AND DYNAMICS OF DISTRIBUTED DEFORMATION
   by D. McKENZIE and J. JACKSON   17

      1. Introduction   17
      2. Fluid Dynamics   17
      3. Rotations about Vertical Axes in Deforming Zones   21
      4. Pinned or Floating Blocks?   26
      5. Kinematic Stability   28
      6. Conclusion   29

RELATIONS BETWEEN SEISMICITY AND PALEOMAGNETIC ROTATIONS IN
ZONES OF DISTRIBUTED CONTINENTAL DEFORMATION
   by J. JACKSON and D. McKENZIE   33

      1. Introduction   33
      2. A simple description of continuous deformation   34
      3. Seismicity and distributed deformation   36
      4. An example of distributed deformation achieved by faulting   38

THE DETECTION OF ROTATIONS BY SURVEYING TECHNIQUES
   by P. A. CROSS   43

      1. Introduction   43
      2. Terrestrial method   44
      3. Satellite techniques   45
      4. The surveying estimation process   56
      5. Assessement of quality   60
      6. Optimal design methods   63
      7. Conclusions   65

GEODETIC MEASUREMENTS OF CONTINENTAL DEFORMATIONS: PROJECTS
AND FIRST RESULTS
by E. GEISS, CH. REIGBER, and P. SCHWINTZER   69

    1.Introduction   69
    2. Methods   71
    3. Data   72
    4. Comparison with a Plate Tectonic Model   73
    5. Interpolation Techniques   78
    6. Conclusions   79

CONTINENTAL ROTATIONAL DEFORMATION: EXAMPLES FROM GREECE
by S. PAVLIDES   83

    1. Introduction   83
    2. Neotectonic rotational deformation in Chalkidiki Peninsula (Northern Greece).   85
    3. Paleomagnetic and structural evidence for recent deformation in the South
       Aegean Active Arc (Melos Island).   89
    4. Conclusion   91

CENOZOIC MAGMATISM, DEEP TECTONICS, AND CRUSTAL DEFORMATION IN
THE AEGEAN SEA
by G.A. PAPADOPOULOS   95

    1. Introduction   95
    2. The data   96
    3. Space-time distribution of the magmatism   96
    4. Petrochemistry   98
    5. Implications for the deep tectonics   99
    6. Discussion   105

A PATTERN OF BLOCK ROTATIONS IN CENTRAL AEGEA
by C. KISSEL, C. LAJ, A. POISSON, and K. SIMEAKIS.   115

    1. Introduction   115
    2. Geological setting.   116
    3. Paleomagnetic method   119
    4. Results and discussion.   120
    5. Conclusion   127

LATE CENOZOIC ROTATONS ALONG THE NORTH AEGEAN TROUGH FAULT
ZONE (GREECE); STRUCTURAL CONSTRAINTS
by C. SIMEAKIS, J.L. MERCIER, P. VERGELY and C. KISSEL   131

    1. Introduction   131
    2. Kinematics of the Late Cenozoic faults in the Aegean basins located on both
       sides of the North Aegean trough fault zone   133

3. Directions of the Early Cenozoic folds on both sides of the western termination
   of the North Aegean trough   137
4. Conclusions   139

SOME EXPERIMENTS ON BLOCK ROTATION IN THE BRITTLE UPPER CRUST
by P.R. COBBOLD, J.P. BRUN, P. DAVY, G. FIQUET, C. BASILE and D. GAPAIS   145

1. Introduction   145
2. Domino domains in Coulomb materials.   146
3. Pull-apart domino   150
4. Continental indentation.   150
5. Conclusions   153

LARGE RATES OF ROTATION IN CONTINENTAL LITHOSPHERE. UNDERGOING
DISTRIBUTED DEFORMATION.
by P. ENGLAND   157

1. Introduction   157
2. Deformation of the continental lithosphere   158
3. Discussion   163

STRAIN AND DISPLACEMENT IN THE BRITTLE FIELD
by P. CHOUKROUNE   165

1. Introduction   165
2. The local characteristics of strain in the brittle field.   166
3. The spatial integration of local strain data: the use of strain trajectories. 175
4. Conclusion.   178

REGIONAL DEFORMATION BY BLOCK TRANSLATION AND ROTATION
by Z. GARFUNKEL   181

1. Introduction   181
2. Deformation by block displacement and rotation: kinematics.   183
3. Pre- and post-faulting deformation   194
4. Faults as surfaces of weakness   196
5. Relation between shallow and deep crustal deformation   199
6. Concluding remarks   203

MECHANICS OF DISTRIBUTED FAULT AND BLOCK ROTATION
by A. NUR, H. RON and O. SCOTTI   209

1. Blocks and fault rotations   209
2. Field evidence   213
3. Material rotation vs. stress field rotation   221
4. Relevance to geodynamics   222
5. Relevance to earthquake prediction.   224

6. Conclusion   225

CRUSTAL ROTATION AND FAULT SLIP IN THE CONTINENTAL TRANSFORM
ZONE IN SOUTHERN CALIFORNIA
    by B. P. LUYENDYK   229

    1. Introduction   229
    2. Facts concerning deformation   230
    3. Amounts of dextral shear   235
    4. History of fault movements   238
    5. What has been learned?   242
    6. Questions and problems remaining   242

EVIDENCE FOR CONTEMPORARY BLOCK ROTATION IN STRIKE-SLIP
ENVIRONMENTS: EXAMPLES FROM THE SAN ANDREAS FAULT SYSTEM,
SOUTHERN CALIFORNIA
    by C. NICHOLSON and L. SEEBER   247

    1. Introduction   247
    2. Data and interpretations   250
    3. Discussion   272

THE IMPORTANCE OF MAGNETOSTRATIGRAPHY FOR STUDIES OF TECTONIC
ROTATIONS: EXAMPLES FROM THE MIO-PLIOCENE OF CALIFORNIA
    by K. L. VEROSUB and E. J. HOLM   281

    1. Introduction   281
    2. Ridge Basin   283
    3. Purisima Formation   287
    4. Conclusion   291

THE APPLICATION OF PALAEOMAGNETISM TO EXTENSIONAL TECTONICS: A
PALAEOMAGNETIC STUDY OF THE PARKER DISTRICT, BASIN AND RANGE
PROVINCE, ARIZONA.
    by P.DAGLEY and J.D.A. PIPER   293

    1. Introduction   293
    2. Geology and sampling   296
    3. Paleomagnetic results   299
    4. Discussion   305

MECHANISMS OF CENOZOIC TECTONIC ROTATION, PACIFIC NORTHWEST
CONVERGENT MARGIN, U.S.A.
    by R.E. WELLS   313

    1. Introduction   313
    2. Long-term dextral shear along the margin   317
    3. Basin-Range extension   321
    4. Discussion   323

ROTATION OF CENTRAL AND SOUTHERN ALASKA IN THE EARLY TERTIARY:
OROCLINAL BENDING BY MEGAKINKING?
by R. S. COE, B. R. GLOBERMAN, and G. A. THRUPP   327

    1. Introduction   327
    2. Paleomagnetic Evidence   328
    3. Discussion   332
    4. Conclusions   339

PALEOGEOGRAPHY AND ROTATIONS OF ARCTIC ALASKA - AN UNRESOLVED
PROBLEM.
by D. B. STONE   343

    1. Introduction   343
    2. Background   349
    3. Paleomagnetism   351
    4. Paleomagnetic data for Alaska   353
    5. Discussion   360

PALAEOMAGNETIC ESTIMATES OF ROTATIONS IN COMPRESSIONAL REGIMES
AND POTENTIAL DISCRIMINATION BETWEEN THIN-SKINNED AND DEEP
CRUSTAL DEFORMATION.
by E. McCLELLAND and A. M. McCAIG   365

    1. Introduction   365
    2. Rotations in thin-skinned thrusting; an example from SW Dyfed, Wales.   366
    3. Rotations in Basement thrust sheets; an example from the Axial zone, Spanish
        Pyrenees   370
    4. Conclusions   377

PALAEOMAGNETIC EVIDENCE FOR BLOCK ROTATIONS AND DISTRIBUTED
DEFORMATION OF THE IBERIAN-AFRICAN PLATE BOUNDARY
by M.L. OSETE, R. FREEMAN and R. VEGAS   381

    1. Introduction   381
    2. Palaeomagnetic Results   383
    3. Discussion   385

FAULT BLOCK ROTATIONS IN OPHIOLITES: RESULTS OF PALAEOMAGNETIC
STUDIES IN THE TROODOS COMPLEX, CYPRUS.
by S. ALLERTON   393

    1. Introduction   393
    2. Crustal Structure of the Troodos ophiolite   394
    3. Palaeomagnetic units   395
    4. Method of analysis   396

5. Extensional deformation   397
6. Strike-slip deformation   403
7. Conclusions   408

PALEOMAGNETISM IN SE ASIA: SINISTRAL SHEAR BETWEEN PHILIPPINE SEA
PLATE AND ASIA.
   by M. FULLER, R. HASTON and E. SCHMIDTKE.   411

   1. Introduction.   411
   2. Philippine Sea Plate.   411
   3. Luzon, northern Philippines.   418
   4. Tectonics of Philippines and Philippine Sea Plate.   428

PALAEOMAGNETIC CONSTRAINTS ON THE EARLY HISTORY OF THE MØRE-
TRØNDELAG FAULT ZONE, CENTRAL NORWAY
   by T.H. TORSVIK, B.A. STURT D.M. RAMSAY A. GRØNLIE, D. ROBERTS,
   M. SMETHURST, K. ATAKAN, R. BØE and H.J. WALDERHAUG   431

   1. Introduction   431
   2. Regional geology sampling and magnetic fabrics   433
   3. Paleomagnetic and rock-magnetic experiments   437
   4. Paleomagnetic reference data   444
   5. Interpretation of remanence data   444
   6. Discussion   449

PALEOMAGNETICALLY OBSERVED ROTATIONS ALONG THE HIKURANGI
MARGIN OF NEW ZEALAND
   by R.I.WALCOTT   459

   1. Introduction   459
   2. Plate reconstructions   461
   3. The Hikurangi margin   462
   4. Paleomagnetic studies on the Hikurangi margin   467
   5. Mechanics of rotation   469
   6. Conclusions   470

ROTATIONS ABOUT VERTICAL AXES IN PART OF THE NEW ZEALAND PLATE-
BOUNDARY ZONE, THEORY AND OBSERVATION
   by S. LAMB   473

   1. Introduction   473
   2. Floating block model   474
   3. New Zealand plate-boundary zone   479
   4. Northern Marlborough domain   481
   5. Southern Marlborough domain   485
   6. Development of Marlborough domains   486
   7. Conclusion   487

PALEOMAGNETIC ROTATIONS IN THE COASTAL AREAS OF ECUADOR
AND NORTHERN PERU
by C. LAJ, P. MITOUARD, P. ROPERCH, C. KISSEL, T. MOURIER, F. MEGARD   489

   1- Introduction.   489
   2. Choice of sites and sampling methods.   491
   3. The sampled regions   492
   4- Results   496
   5 - Tectonic implications of the data.   502
   6 - Discussion   504
   7 - Conclusions.   508

LIST OF PARTICIPANTS   513

Preface

One of the most interesting results obtained in the last two decades in the study of crustal deformation has been the recognition that large regions of continental crust undergo rotations about vertical axis during deformation. Proof of such rotations has come through the paleomagnetic studies, which reveal rotations when paleomagnetic declinations within the deforming region are compared with those found in coeval rocks in the stable regions outside the deforming zone.

Such rotations were first described in Oregon then in the North American Cordilleras and in Southern California and were a surprise to everyone. Even in California which, as a result of oil exploration, was among the best geologically explored regions in the world, no one could claim to have predicted that these rotations would be found. Rotations have subsequently been found in other areas of recent continental tectonic activity, notably in the Basin and Range province, New Zealand, the Andes, Greece and Western Turkey, so that they appear as an important feature of continental deformation.

These new paleomagnetic observations coincided with the development of new models of continental deformation which all originated from the recognition that continents, unlike the oceans, deform in a widespread, diffuse way not confined to narrow linear belts. Within the deforming zones the observed motions were described as those of a deforming fluid in which there is no strain discontinuity at all. The success of these continuum models in describing the long wavelength characteristics of continental deformation arises from the fact that only the very upper layer of the lithosphere deforms in a brittle fashion, so that it is reasonable that at large scale the deformation is controlled by the distributed flow in the lower crust and mantle lithosphere.

It is clear, however, that the deformation is discontinuous in the upper rigid crust, where paleomagnetic measurements are made and where faults may be observed. What fault geometry can then accomodate the creep below and rotations about vertical axis in the brittle crust? What constraints do the paleomagnetic rotations put on the kinematic evolution of this geometry? At what scale should these phenomena be described? What is the shape of the rotating blocks and how large do they get? Does their size put some constraints on the velocity of the movements? How deep do the rotations extend?

These questions, which need to be answered before the mechanisms of continental deformation can claim to be understood, have been the theme of the first NATO Advanced Research Workshop on "Paleomagnetic Rotations and Continental Deformation", held at Loutra Edipsou (Greece) , May 8-13, 1988.

This book arises from this workshop as a written version of the lectures given there. Its aim is to give a comprehensive up-to-date coverage of the subject, presented in a mode which unfortunately cannot entirely reflect the lively atmosphere of the workshop, the debated evening sessions, the field trip and the informal discussions held all along.

The first articles give a review of the topics fundamental to many studies, including the new satellite surveying techniques. A session of the workshop, dedicated to the Aegean Area, is then described. The following articles reports the results of experiments with analogic and numerical models, and then the results of studies in different tectonic situations, the Western Coast of North

America, the Pyrenees, the active margin of New Zealand, South East Asia and the Andean Cordillera. The final discussion, not reported here, stressed the importance and necessity of joint structural, paleomagnetic and modelling studies at different lengthscales in zones of continental deformation. The suggestion of having a second workshop on the same subject in two years time was also rather enthousiastically approved by all the participants.

The idea of organizing this workshop arose from discussions which we had in Gif-sur-Yvette with Dan McKenzie. We wish to thank him for this suggestion. We then benefitted from the active participation of Myrl Beck and James Jackson in the difficult task of selecting the participants and setting up a program. Jacques Mercier managed to find a few days in his crowed time-table to prepare, with the help of Spyros Pavlides, the field trip in Central Aegea which has been one of the highlights of the workshop. We wish to thank Iannis Karantzis, Mayor of Loutra Edipsou, for his wonderful Greek hospitality in the cultural center of the Municipality. At the hotel, Anthony Capolo has never considered us as clients but as guests: his friendly yet efficient organisation has greatly contributed to the overall success of the workshop. Marika Spyridakis took care of many secretarial problems and put much of her living personality in the life of the conference.

The workshop has been entirely funded by the Scientific Affairs Division of NATO, with a small contribution from the Centre des Faibles Radioactivités. We know of no other organisation which would have responded so quickly to our demand: barely two months after the proposal was submittted we received a positive answer and only a few minor formalities were needed to have the funding available in a most flexible way, well before the beginning of the workshop. Had it not been for this wonderful simplicity and efficiency, we are not sure that we would have succeeded in getting everything ready in time! For this support we wish to thank the Scientific Affairs Division, and particularly its Director, Dr. L. da Cunha.

Gif-sur-Yvette, July 1988.

Catherine Kissel and Carlo Laj

BLOCK ROTATIONS IN CONTINENTAL CRUST: EXAMPLES FROM WESTERN NORTH AMERICA

M. E. BECK Jr.
Department of Geology
Western Washington University
Bellingham, Washington 98225
U.S.A.

ABSTRACT. The North American Cordillera has many instances of rotated crustal blocks. The most important processes causing rotation seem to be:
1) Small-block rotations in discrete shear zones, caused by ductile deformation below the seismogenic layer.
2) Small-block rotations caused by distributed shear within zones of oblique subduction.
3) Rotation caused by northward transport along the western edge of North America.
4) Rotation by terrane transport within the Pacific basin. Other processes may be locally important:
5) Oroclinal bending.
6) Rotation during accretion.
7) Rotation during overthrust faulting.
Some conclusions:
1) Rotations in zones of oblique subduction are favored by a high angle of obliquity, a relatively steep subduction angle, and relatively high resistance to strike-slip faulting.
2) Slivers of continental crust undergoing coastwise transport may move as large coherent blocks for short distances but tend to fragment into smaller, independently-rotating blocks if transport continues.
3) Cordilleran rotations since the Middle Cretaceous have been very consistently clockwise, arguing for dextral shear. Late Jurassic and Early Cretaceous rotations may have been counterclockwise, suggesting sinistral shear.

1. Introduction

One of the earliest uses of paleomagnetism was to detect in situ rotations of large blocks of continental crust; e.g., the island of Newfoundland (relative to the North American craton), by Black (1964) and the bending of the two halves of Japan (on either side of the fossa magna) with respect to one another (Kawai et al., 1961). Paleomagnetism also was used to test for rotations in thrust sheets (e.g., Norris and Black, 1961) and for oroclinal bending associated with Appalachian syntaxes (Irving and Opdyke, 1965). Systematic rotations associated with an orogenic belt were first detected in the Alpine-Mediterranean area (summary by Zijderveld and Van der Voo, 1973), but the significance of this early work was obscured by the question of whether the rotated terranes had moved independently or as part of Africa. Carey's (1958) Bay of Biscay sphenochasm and rotation of the Iberian peninsula were studied using paleomagnetic declinations by several investigators (e.g., Watkins and Richardson, 1968). These and other early applications of paleomagnetism to problems of in situ rotation testify to the unmatched resolving power of the method; paleomagnetism can detect and measure rotations about vertical axes in cases where all

1

C. Kissel and C. Laj (eds.), Paleomagnetic Rotations and Continental Deformation, 1–16.
© *1989 by Kluwer Academic Publishers.*

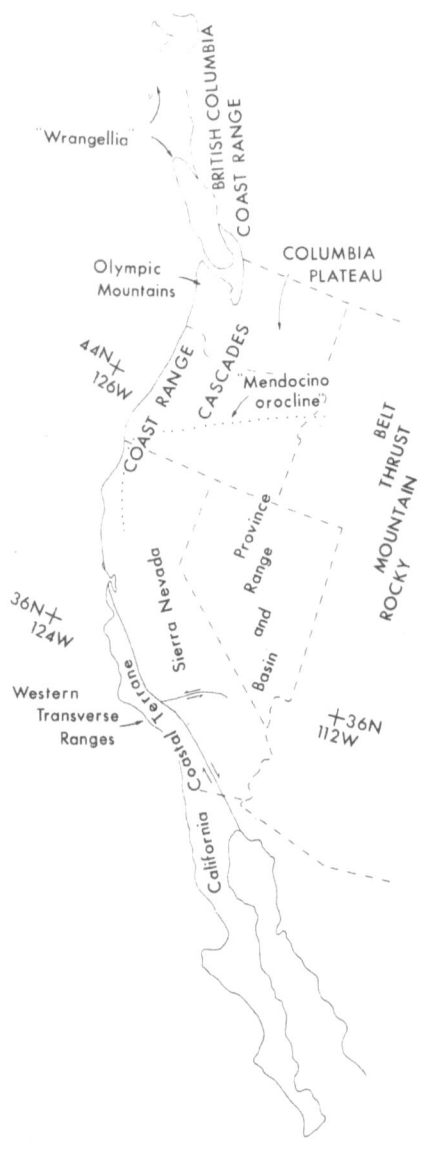

Figure 1. Index map showing location of main areas discussed in text

other methods fail.

Study of block rotations in the western Cordillera of North America began inadvertently with Cox (1957), who measured a highly suspicious easterly paleomagnetic declination in the Eocene Siletz Volcanics of western Oregon. Later Irving (1964, p. 249) pointed out that Cox's results could be explained (fortuitously, it now appears) by invoking the Mendocino orocline concept of Carey (1958). This exciting tectonic interpretation sent Watkins (1965) into the Columbia Plateau Miocene basalts in search of more evidence of oroclinal bending. Watkins concluded that some post-Miocene rotation was supported by his results, but this conclusion is no longer accepted (Gromme et al., 1986). Further interest in rotations followed discovery (Beck, 1976) that a remarkably consistent pattern of clockwise rotation existed in nearly every section of the western Cordillera, from Southern California to Alaska. For summaries of the extensive work done since that time see Gromme et al. (1986) and Beck (1988a, b).

This paper reviews possible mechanisms for tectonic rotation of blocks of continental crust, placing special emphasis on rotations driven by plate interactions. Examples cited are mostly from western North American (Figure 1); other papers in this volume provide additional evidence (see articles by Coe, Dagley and Piper; Luyendyk; Nicholson and Seeber; Stone; and Wells).

2. Definitions

"Rotation," in common with many other scientific terms borrowed from ordinary English, means quite different things to different people. In what follows the term rotation will be used as it is commonly defined in paleomagnetic studies: $R = D_O - D_X$, where D_O is the observed declination (determined in a paleomagnetic study) and D_X is the expected declination, calculated from the apparent polar wander (APW) path of the continent to which the rock unit is attached. (It also is possible to refer rotations to other reference frames, such as the present spin axis or paleomagnetic directions from other nearby units of the same age.) When in situ or nearly in situ rotation of a crustal fragment about a nearby vertical axis is implied, the term block rotation will be used. Euler rotation in what follows will imply rotation of some crustal element about a distant Euler pole; when the result of such movement is to translate a crustal fragment along a continental margin the term coastwise transport will be employed. All of the above can (and usually do) rotate the paleomagnetic vector. So equally can simple tilt ("rotation" about one or more local horizontal axes). In interpreting paleomagnetic results from orogenic belts it frequently happens that two or more forms of displacement have rotated the magnetic direction; the task often is to sort them out.

3. Types of Rotation and their Characteristics

In the western North American Cordillera rotations are found associated with the following tectonic situations:

1) Block rotations, with or without coastwise transport, in discrete shear zones (e.g., Southern California).

2) Block rotations, with little coastwise transport, occurring in zones of distributed shear and probably caused by oblique subduction (e.g., the Pacific Northwest).

3) Block rotations during accretion (perhaps exemplified by some Oregon-Washington Coast Range basalt exposures).

4

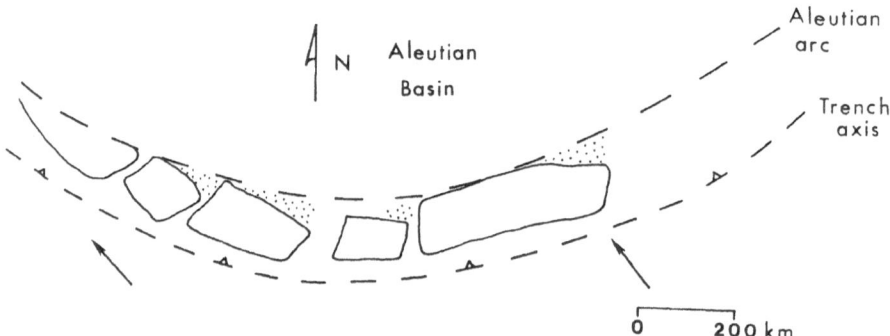

Figure 2. Diagram showing rotation of forearc blocks in Aleutian subduction zone, in response to oblique subduction. Dotted pattern indicates basins opening behind rotating blocks. Modified from Geist et al. (1988).

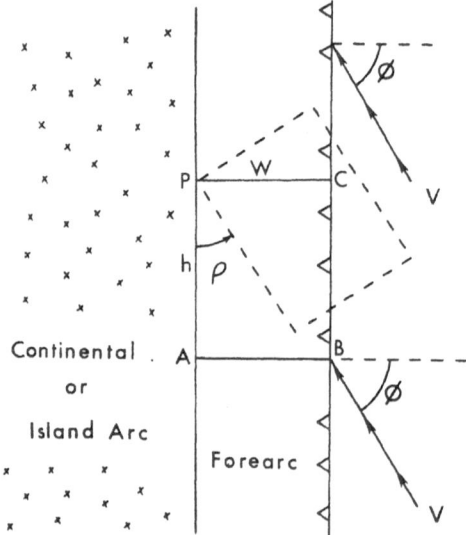

Figure 3. Cartoon illustrating rotation of forearc blocks in zone of oblique subduction. Block PABC has rotated through angle ρ because of traction applied at its base. Φ angle of obliquity; h, w are dimensions of block. Torque tending to rotate block given by $T = (cA/2) (w \sin \Phi - h \cos \Phi)$ where A is area of block and c is the shearing stress. The condition for rotation is that $\tan\Phi > h/w$.

4) Rotation (of the paleomagnetic declination) resulting from Euler rotation of exotic crustal fragments (e.g., the Wrangellia terrane).

5) Rotation by coastwise transport (Cretaceous plutons of California, Washington, and British Columbia).

6) Block rotation caused by oroclinal bending (the periphery of the Olympic uplift in northwest Washington; other examples possibly located in Alaska and coastal Oregon).

7) Block rotation during overthrust faulting (possible examples in the Rocky Mountain fold and thrust belt of Utah, Idaho and Wyoming).

Each of these kinds of rotation have distinctive characteristics:

3.1. SHEAR ZONES.

Block rotations found in the broad shear zone associated with the San Andreas transform in Southern California are the best examples of this kind of rotation (Kamerling and Luyendyk, 1985; Luyendyk et al., 1985; Luyendyk, this volume). As emphasized by Cox (1980) these are probably driven from below (beneath the rigid seismogenic layer) and are characterized by torques that are proportional to the horizontal rate of change of flow-velocity in the lower (ductile) part of the crust. The rotating rigid crustal fragment--termed a "scholle" by Dewey and Sengor (1979)--ought to be roughly 10-20 km in lateral dimensions, on average, which agrees quite well with the models of Luyendyk and co-workers. Some coastwise transport is inevitable in this kind of deformation but in the Southern California example it is small, except for the outboard Western Transverse Ranges terrane (Luyendyk and others, 1985; Beck, 1988 a, b).

3.2. DISTRIBUTED SHEAR RESULTING FROM OBLIQUE SUBDUCTION.

There are no discrete through-going shear zones associated with the Cascade and Olympic Mountains of Oregon and Washington, but dextral rotation is widely distributed there (Beck et al., 1986; Gromme et al., 1986). Rocks as young as Miocene and as old as mid-Cretaceous are involved (Magill et al., 1982). Magill et al. (1981) and Wells and Heller (1988) argue that a significant amount of this rotation is due to rigid-body block rotation of the Cascades and Coast Range, driven by differential extension inboard in the Basin and Range Province; this is discussed below under "oroclinal bending." However, much of it must be attributable to quasi in situ block rotations, probably caused by oblique subduction.

Figure 2 shows rotating blocks in the forearc of the Aleutian subduction zone, which is strongly oblique. Some elementary calculations and common sense indicate that, if such rotations are driven from below by oblique subduction, then the torque acting on a rotating block is a function of the shape of the block, the efficiency of coupling, and the angle of obliquity (Figure 3). This being the case, it follows that reducing the angle of obliquity by detaching a forearc sliver and translating it parallel to the arc (Fitch, 1972; Beck, 1986) should tend to suppress in situ block rotation, except locally along the detachment shear itself. In the Washington and Oregon Cascades and Coast Range the total amount of margin-parallel relative motion has been very small (no more than a few hundred kilometers since the early Tertiary), so oblique subduction (and block rotation) have prevailed for most of Tertiary time. However, during the Cretaceous it appears that sliver-detachment and northward transport of pieces of the continental margin were common throughout the Cordillera (e.g., Irving et al., 1985; Beck, 1986, 1988a, b). Under what circumstances are block rotations compatible with coastwise transport?

In Beck (1986) I considered parameters controlling whether or not oblique subduction will detach a sliver from the front of the overriding plate and translate it parallel to the plate margin. In the absence of a buttress preventing motion of the sliver it appears that the important parameters are the angle of obliquity (Φ), measured from the normal to the plate margin, the angle of dip on the subduction zone (Δ), and R, the ratio of resistance to slip on the vertical fault separating the sliver from the rest of the over-riding plate, to resistance to slip on the subduction surface itself. In this highly oversimplified model, high values of Φ and low values of Δ and R favor detachment.

Subsequently, Jarrard (1986) considered the same problem using parameters of actual subduction zones and made the important discovery that oblique convergence commonly is partitioned between oblique subduction and transport of a detached sliver. Thus in many instances of oblique convergence a sliver of overriding plate is detached and transported along the margin, but transport is at something less than maximum (this given by the margin-parallel component of relative motion). Clearly, if the sliver moves relative to the overriding plate at its maximum velocity it will experience no oblique shear whatsoever, and thus will have no tendency to fragment into rotating blocks. Moreover, any motion of the sliver will reduce the effective angle of obliquity and therefore the ability of oblique subduction to cause block rotations.

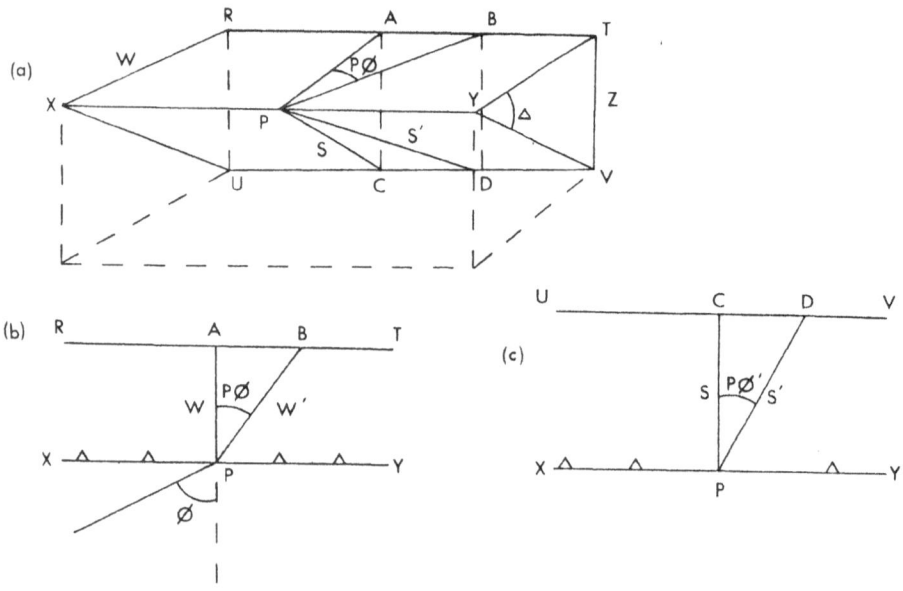

Figure 4. Cartoon of a subduction zone undergoing oblique subduction. XY is trench, XUVY is inclined subduction zone, and RUTV is a fault separating a forearc sliver (XYRUTV) from overriding plate. Angle of obliquity is Φ. In the absence of a buttress, oblique shear will detach the sliver and move it to the right, thus partitioning obliquity between motion on the fault and oblique subduction at a reduced angle (P Φ). P is a partitioning factor; P = 0 implies that the sliver is moving at its maximum rate (equal to the arc-parallel component of the relative motion); P = 1 for cases in which there is no sliver motion.

It is possible to extend the sort of analysis used in Beck (1986) to cover this new complication. Figure 4 shows a block diagram of a simplified subduction zone. Slanting plane XUVY is the subduction "fault" itself. Because subduction is oblique (at angle Φ as shown in plan view) a sliver is detached along a vertical fault (RUVT). If there is no buttress to the right of the diagram to prevent it, the sliver will move to the right, parallel to the margin. Because it is moving it will experience oblique shear at a reduced angle, PΦ where P is a partitioning factor (P varies from 1 for pure oblique subduction -- no detached sliver -- to 0 when the sliver moves at its maximum velocity). The angle Φ' is related to Φ by $\tan \Phi' = \cos \Delta \tan \Phi$.

Assuming that the partitioning factor will be that which yields minimum work per unit of convergence, it is possible to find a relationship between P as independent variable and the other convergence parameters defined above. This is:

$$\tan (P\Phi) = R \tan \Delta \left(\frac{1}{1-R^2 \sin^2 \Delta} \right)^{1/2} \qquad (1)$$

all variables as defined earlier.

Figure 5 illustrates this relationship. For a mature subduction zone the ratio of resistence to slip on the sliver-defining fault to resistence to slip on the subduction zone probably is fairly small, as indicated by the earthquake energy released (Jarrard, 1986). For R = 0.3 and a subduction angle of 20°, angles of obliquity of as little as about 10° will initiate displacement of a forearc sliver, and at only about 15° half the obliquity is partitioned into sliver motion. For a new subduction zone, presumably the resistance to slip on potential faults cutting the overriding lithosphere is higher (Beck, 1986); for R = 1 Figure 5 shows that an angle of obliquity of more than 20° is required to initiate sliver motion.

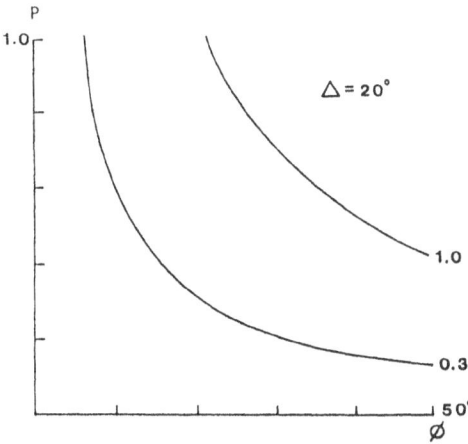

Figure 5. Solutions to equation (1) for a subduction angle (Δ) of 20° and two different values for R (ratio of resistance to slip on vertical fault RUTV to resistance to slip on subduction "fault" XUVY; see Figure 4). P and Φ as defined in Figure 4.

Figure 6 shows the relationship between the minimum angle of obliquity needed to initiate sliver-motion and R, for various values of Δ. Φ_{min} also is the effective angle of obliquity that a forearc will experience, for given values of R and Δ. This follows because, for $\Phi > \Phi_{min}$, sliver motion will ensue (in the absence of a buttress) and will maintain the effective value of obliquity at Φ_{min}. From Figure 3, the minimum angle needed for rotation must exceed arctan (w/h). Thus rotation by this method should be favored by relatively high resistance to slip along the fault cutting the overriding plate, by moderately steep subduction angles, and by the creation of thin, lath-shaped plates approximately normal to the subduction zone. Of course a buttress, by preventing or limiting motion of the sliver, also will promote in situ block rotation. Probably there was such a buttress aiding rotation in the Pacific Northwest during the early and middle Tertiary. Rotation of crustal blocks in Peru and Chile also may have been aided by such a mechanism (Beck, 1987).

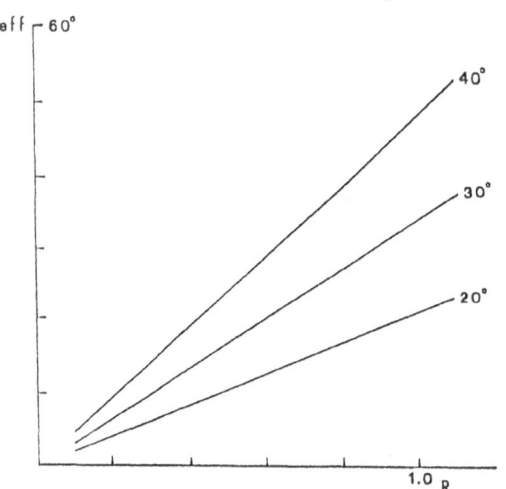

Figure 6. Minimum angle of obliquity (Φ_{min}) needed to initiate sliver motion for various values of subduction angle and R (defined in Figure 5). Φ_{min} also gives the maximum angle of obliquity that a moving sliver will experience, and thus helps determine its propensity to fragment and rotate. Analysis presupposes absence of a buttress preventing terrane motion. See text.

3.3 ROTATIONS DURING ACCRETION.

Whenever non-equidimensional crustal fragments are accreted to a continental margin ("offscraped") during subduction, or whenever subduction is markedly oblique, accreting crustal fragments may rotate (Figure 7); Roperch et al. (1987) give an example. This mechanism was one of several suggested by Simpson and Cox (1977) to account for clockwise rotation of accreted seamounts now lodged in the Coast Range of Oregon; however, Wells and Heller (1988) feel that rotation-accretion in the Coast Range has been minor. From a theoretical standpoint it is clear that this mechanism can cause very rapid rotation, and that the sense of rotation it causes can vary from place to place. Highly oblique convergence probably will produce rotation in a uniform sense, however.

3.4. EULER ROTATION.

Except under extremely unusual circumstances, transported, accreted crustal fragments will have paleomagnetic declinations and/or inclinations that are conspicuously discordant with respect to the expected paleomagnetic direction at their current location. The various rotated fragments of

Wrangellia (Jones et al., 1977) are well known examples. In principle it ought to be easy to reconstruct the relative plate motion responsible for transport of the exotic fragment, although the time of accretion often presents a problem. However, in North America, because convergence has been markedly oblique, rotations acquired before accretion are likely to have been "overprinted" by later block rotations caused by other mechanisms. Coastwise transport comprises a particular problem in terrane-trajectory analysis (e.g., Debiche et al., 1988) because it makes it difficult to know precisely where an exotic crustal fragment first docked. Beck (1988a) discusses some complications introduced by this kind of rotation.

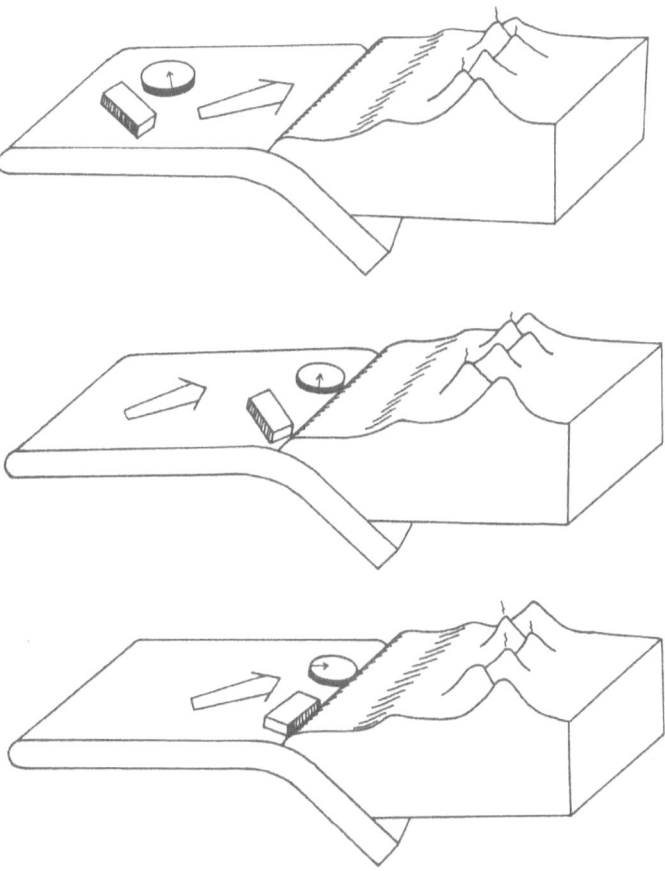

Figure 7. Cartoon illustrating rotation of "offscraped" crustal fragments at a zone of oblique convergence.

10

3.5. COASTWISE TRANSPORT.

This can be regarded as a special case of Euler rotation, and like Euler rotation generally will result in a discordant declination or inclination, usually both. Transport can be by transform faulting (as for instance the sliver of Mexico and California west of the San Andreas fault), or by behind-the-trench strike-slip faulting caused by oblique subduction (e.g., the forearc portion of Sumatra; Fitch, 1972). The paleomagnetic record for western North America indicates that one or both of these mechanisms were very active during the Late Cretaceous and early Tertiary, because most mid-Cretaceous paleomagnetic poles for the Cordillera are offset from the APW path in such a way as to suggest relative northward transport (Beck, 1988a, b).

Rotations of the first type considered in this paper (block rotations in discrete shear zones) ought to accompany coastwise transport. Of more interest here are rotations associated with the transport itself. Except under extremely unusual circumstances coastwise transport will rotate the magnetic declination (that is, move a crustal fragment into a new position where its declination is discordant). Because North America is convex toward the west, northward coastwise transport causes clockwise rotation.

Figure 8 illustrates the situation. Curve TR shows the amount of rotation (R) acquired by moving northward along the present continental margin, expressed as a function of distance traveled (given as P, the poleward transport; Beck, 1988a). Cretaceous rocks from the Sierra Nevada region lie on TR, hence have acquired their small rotation entirely by coastwise transport. Four of five directions for Cretaceous rocks from coastal California cluster above the line, perhaps indicating some rigid-block rotation in addition to coastwise transport. The most interesting group

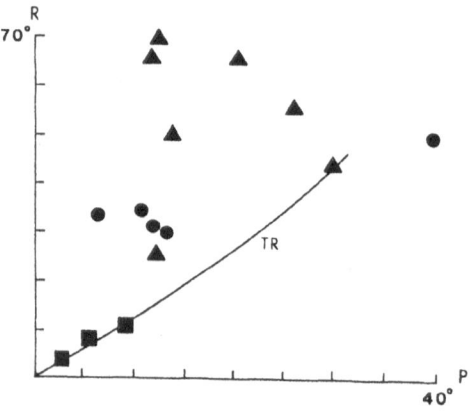

Figure 8. Rotation that would be acquired by a crustal fragment moving northward along the present western edge of North America. R, P are the rotation and poleward transport, respectively. $R = D_0 - D_x$, where D_0 is the observed declination and D_x is the expected declination. P is given by $L_x - L_0$, where L_x and L_0 are the expected and observed paleolatitudes, respectively. Symbols are actual values determined in Cordilleran paleomagnetic studies, all of Cretaceous age. Squares, Sierra Nevada region; circles, coastal California and Baja California; triangles, rocks from British Columbia and northwest Washington.

is the seven studies from British Columbia and Northwest Washington. These rocks seem to have moved farther (higher values of P) than rocks further south, and perhaps as a result they have acquired values of R that are markedly scattered. Evidently blocks from this region have behaved as small, independently rotating entities, and probably not as large, coherent terranes (e.g., the "Cordilleria" of Chamberlain and Lambert (1985) or the "Baja British Columbia" of Umhoefer (1987)). Note that all of the data plotted in Figure 8 lie on or above curve TR, implying local clockwise rotations. This consistency argues strongly that the scatter of points shown is not simply error; some unifying explanation is required. Figure 8 is discussed in more detail in Beck (1988a). Note that coastwise transport around the big topographic and tectonic bend northward from SE Alaska would produce counterclockwise rotation, as discussed by Coe (this volume).

3.6 OROCLINAL BENDING.

Since Carey (1955, 1958) introduced the orocline concept it has rarely been applied successfully, owing to the necessity of accounting for the very large horizontal strains that must accompany rotation of a large block of crust about a fixed pivot-point. In Carey's original model the pivoting crustal blocks had lateral dimensions of up to 1000 km or more and pivoted through angles of many tens of degrees. Thus displacements of the order of 500-700 km were required. Plate tectonics has eased the burden of accounting for such displacements where entire crustal blocks are concerned (e.g., the island of Japan, mentioned earlier), but the orocline concept remains difficult to apply in instances where the horizontal displacement must be accounted for within a continent or orogenic belt. An example of the latter type is Carey's (1958) Bolivian orocline, which has been supported recently by Kono et al. (1985), using paleomagnetic data. Isacks (1988) finds some geological support for the large crustal displacements this feature would require. An alternative and much simpler interpretation of the paleomagnetic evidence has been given by Beck (1987).

The North American Cordillera has several possible examples of rotation by oroclinal bending. By far the largest such feature is the "big-bend" of Alaska, entailing significant counterclockwise rotation. This is discussed by Coe (this volume). In the Pacific Northwest there are two "oroclinal" features. In the simplest case (Figure 9) underthrusting caused by oblique subduction appears to have oroclinally bent an external band of basalt exposures partly ringing the Olympic Mountains (Beck and Engebretson, 1982); observed crustal thickening in the Puget Lowland and dextral strike-slip faulting in southern Vancouver Island may have accompanied oroclinal bending.

Figure 9. A possible instance of oroclinal bending. Large arrow indicates relative motion between the Farallon and North American plates, as determined by Engebretson et al. (1985). Small arrows are observed paleomagnetic declinations for Late Eocene and Oligocene rocks. Stippled pattern is Crescent volcanics. Note how declinations bend around Olympics core.

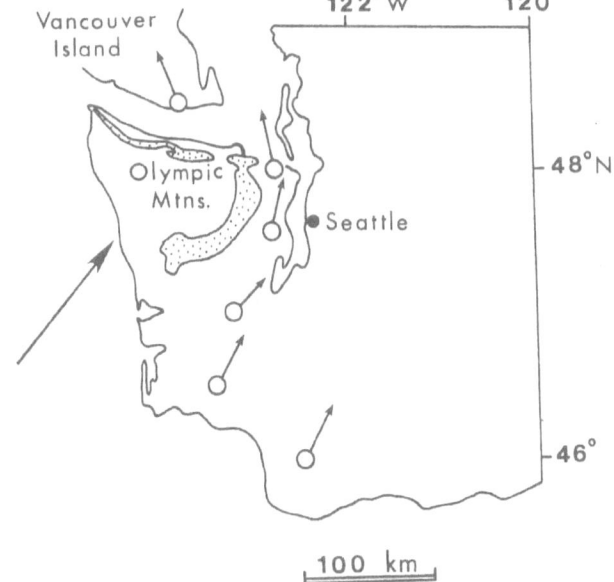

12

Wells (this volume) and Wells and Heller (1988) discuss another rotation process in the Pacific Northwest that is essentially oroclinal; rotation by differential extension in the Basin and Range province (Figure 10). This process has been thoroughly described by many authors (e.g., Simpson and Cox, 1977; Magill and others, 1982; Gromme et al., 1986; Wells and Heller, 1988) and is reviewed in this volume by Wells, so an extended discussion is not called for here.

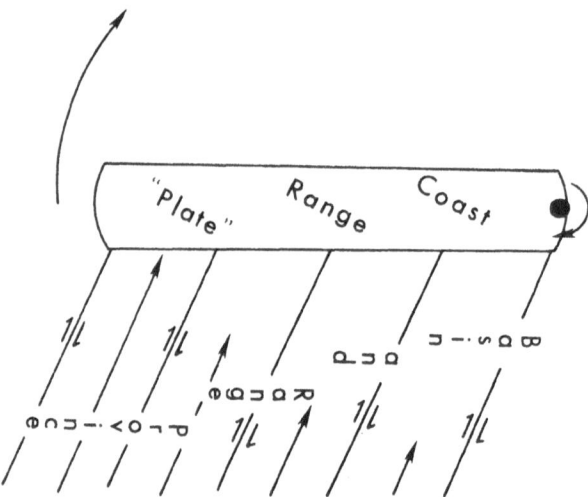

Figure 10. Cartoon illustrating clockwise rotation of coastal Oregon by differential extension in Basin and Range province, modified from Magill et al. (1981).

3.7. ROTATION DURING THRUSTING.

If a thrust sheet does not advance at a uniform velocity everywhere, rigid-body rotations may occur within it. Rotations are most likely to develop where advance of the thrust sheet is impeded by some sort of buttress. Van der Voo (1988) summarizes examples for the North American Cordillera and the Appalachian Mountains. The single well-described Cordilleran example known to me is from the Triassic of Wyoming where Grubbs and Van der Voo (1976) measured site-mean declinations from several thrust sheets that swung through roughly 90o in the same sense as bends in the horizontal trace of the thrust surface. As mentioned earlier a similar problem involving the Lewis thrust in the Canadian Rocky Mountains was investigated by Norris and Black (1961), with negative results. Perhaps the most prolific paleomagnetist of the 1960s and early 1970s, N.D. Watkins, owed his introduction to paleomagnetism to a desire to test the rotated-thrust-sheet hypothesis (N.D. Watkins, personal communication, about 1960).

4. Discussion

The most impressive thing about the pattern of rotation found in the western North American Cordillera is its consistency; of 60 studies on rocks of Cretaceous and younger age from the Cordillera south of Alaska tabulated in Beck (1988b), 44 show rotations that are significant at the 95% confidence level, and of those 39 are clockwise. In fact, the consistency is even greater than suggested by these numbers. For instance, six of the 16 studies that do not show rotation are from rocks of Neogene age, and many of the remainder are from a restricted area in Northwest Washington and southern British Columbia. Of the five counterclockwise-rotated areas, two are involved in the Olympics "orocline" mentioned earlier, and two more are from an area located well inboard of the continental margin, in western Arizona. Of rocks located within 200 km. of the continental margin and having ages in the 20-120 Ma range, nearly all are rotated clockwise. It seems to me that this overwhelming consistency calls for a unifying tectonic explanation (it has in fact seemed so to me for more than ten years; Beck, 1976). Because virtually all plate models for the eastern Pacific basin call for right-oblique convergence or dextral transform interaction along the leading edge of North America since mid-Cretaceous time, it also seems likely to me that dextral shear ultimately is responsible for most of the rotations. Several of the rotation mechanisms discussed above explicitly require shear to make them operate, and given a uniform sense of shear they inevitably produce a uniform sense of rotation. Dextral shear also is supported by an impressive array of paleomagnetic evidence indicating that outboard terranes within the Cordillera moved relatively northward during Late Cretaceous and early Tertiary time (Beck, 1986, 1988a, b; Irving et al., 1985). There is, of course, some contrary opinion (e.g., May and Butler, 1986; Price and Carmichael, 1986; Silver, 1987), mostly directed at the concept of large-scale northward transport. However, as presently formulated, most of these arguments rely too heavily on indirect geological inference or require too great a degree of coincidence (e.g., southwestward tilting of all Cretaceous plutons and remagnetized Cretaceous layered rocks the length of the Cordillera) to be very convincing.

Prior to about mid-Cretaceous time, and beginning some time in the late Jurassic, transport relative to North America seems to have been sinistral (Beck, 1987b, 1988a). Thus one might expect to find counterclockwise rotations associated with terranes containing rocks of pre-mid-Cretaceous age, perhaps overprinted by later clockwise rotation. Table 5 of Beck (1988b) contains 25 entries for rocks of Triassic and Jurassic age. Of these one has been so disturbed tectonically that a meaningful average declination cannot be determined, and four are so greatly rotated (over 100°) that it is uncertain whether they have rotated clockwise or counterclockwise (or even whether they originated in the northern or southern hemisphere). The remaining 20 are distributed almost evenly between those that are rotated counterclockwise (5), those that are rotated clockwise (8), and those that are statistically unrotated (7). This is in marked contrast to the pattern found in younger rocks, nearly all of which are rotated clockwise, and could be taken as evidence in support of two periods of rotation--counterclockwise prior to early Cretaceous time and clockwise thereafter--but obviously other explanations are possible. Finally, little is known about potential rotations in the western North American Cordillera earlier in the Mesozoic; plate modelling gives little if any reliable information about relative plate motions that far back in Earth's history (e.g., Engebretson et al., 1985), and the inventory of promising Cordilleran pre-Cretaceous rock units measured or waiting to be studied is, unfortunately, vanishingly small.

Acknowledgements Thanks are due to R. Burmester for many useful discussions. Diagrams drawn by D. Graham. Supported by NSF EAR 8718896.

References

Beck, M.E., Jr., Discordant paleomagnetic pole positions as evidence of regional shear in the western Cordillera of North America; *American Journal of Science*, v. **276**, p. 694-712, 1976.

Beck, M.E., Jr., Model for late Mesozoic--early Tertiary tectonics of coastal California and western Mexico and speculations on the origin of the San Andreas fault: *Tectonics*, v. **5**, p. 49-64, 1986.

Beck, M.E., Jr., Tectonic rotations on the leading edge of South America: The Bolivian orocline revisited: *Geology*, v. **15**, p. 806-808, 1987a.

Beck, M.E., Jr., North America APW and terrane transport, Late Jurassic--early Cretaceous: Tectonic consequences: *EOS Transactions of the American Geophysical Union*, v. **68**, p. 1254, 1987b.

Beck, M.E., Jr., Paleomagnetism of continental North America: Implications for displacement of crustal blocks within the western Cordillera, Baja California to British Columbia, in Pakiser, L.C. and W.D. Mooney, editors, Geophysical framework of the continental United States: *Geological Society of America Memoir*, in press 1988a.

Beck, M.E., Jr., Tectonic significance of paleomagnetic results for the western conterminous United States, in Burchfiel, B.C., P.W. Lipman, and M.L. Zoback, editors, The Cordilleran Orogen; Conterminous United States: Boulder, Colorado, Geological Society of America, *The Geology of North America*, v. **G3**, in press 1988b.

Beck, M.E., Jr., and D.C. Engebretson, Paleomagnetism of small basalt exposures in the west Puget Sound area, Washington, and speculations on the accretionary origin of the Olympic Mountains: *Journal of Geophysical Research*, v. **87**, p. 3755-3760, 1982.

Beck, M.E., Jr., R.F. Burmester, D.E. Craig, C.S. Gromme, and R.E. Wells, Paleomagnetism of middle Tertiary volcanic rocks from the Western Cascade Series, northern California: *Journal of Geophysical Research*, v. **91**, p. 8219-8230, 1986.

Black, R.F., Paleomagnetic support of the theory of rotation of the western part of the island of Newfoundland: *Nature*, v. **202**, p. 945-948, 1964.

Carey, S.W., The orocline concept in geotectonics: *Proceedings of the Royal Society of Tasmania*, v. **89**, p. 255-288, 1955.

Carey, S.W., The tectonic approach to continental drift, in Carey, S.W., editor, Continental Drift--A Symposium: Hobart, University of Tasmania Press, p. 177-358, 1958.

Chamberlain, V.E., and R. St. J. Lambert, Cordilleria, a newly defined Canadian microcontinent: *Nature*, v. **314**, 707-713, 1985.

Cox, A., Remanent magnetism of Lower to Middle Eocene basalt flows from Oregon: *Nature*, v. **179**, p. 685-686, 1957.

Cox, A., Rotation of microplates in western North America, in Strangway, D.W., editor, The Continental Crust and its Mineral Deposits: *Geological Association of Canada Special Paper* **20**, p. 305-321, 1980.

Debiche, M.G., A. Cox, and D.C. Engebretson, The motion of allochthonous terranes across the northern Pacific basin: *Geological Society of America Special Paper* **207**, 61 pp., 1987.

Dewey, J.F., and A.M.C. Sengor, Aegean and surrounding regions: complex multiplate and continuum tectonics in a convergent zone: *Geological Society of America Bulletin*, v. **90**, p. 84-92. 1979.

Engebretson, D.C., A. Cox, and R.G. Gordon, Relative motions between oceanic and continental plates in the Pacific basin: *Geological Society of America Special Paper* **206**, 59 pp., 1985.

Geist, E.L., J.R. Childs, and D.W. Scholl, The origin of summit basins of the Aleutian Ridge: Implications for block rotation of an arc massif: *Tectonics*, in press, 1988.

Gromme, C.S., M.E. Beck, Jr., R.E. Wells, and D. C. Engebretson, Paleomagnetism of the Tertiary Clarno Formation of central Oregon and its significance for the tectonic history of the Pacific Northwest: *Journal of Geophysical Research*, v. **91**, p. 14,089-14,103, 1986.

Grubbs, K.L., and R. Van der Voo, Structural deformation of the Idaho-Wyoming overthrust belt (U.S.A.), as determined by Triassic paleomagnetism: *Tectonophysics*, v. **33**, p. 321-336, 1976.

Fitch, T.J., Plate convergence, transcurrent faults, and internal deformation adjacent to Southeast Asia and the western Pacific: *Journal of Geophysical Research*, v. **77**, p. 4432-4461, 1972.

Irving, E., Paleomagnetism and its application to geological and geophysical problems: New York, John Wiley & Sons, Inc., 399 pp., 1964.

Irving, E., and N.D. Opdyke, The paleomagnetism of the Bloomsburg red beds and its possible application to the tectonic history of the Appalachians: *Geophysical Journal of the Royal Astronomical Society*, v. **9**, p. 153-167, 1965.

Irving, E., G.J. Woodsworth, P.J. Wynne, and A. Morrison, Paleomagnetic evidence for displacement from the south of the Coast Plutonic Complex, British Columbia: *Canadian Journal of Earth Science*, v. **22**, p. 584-598, 1985.

Isacks, B.L., Uplift of the central Andean plateau and bending of the Bolivian orocline: *Journal of Geophysical Research*, v. **93**, p. 3211-3231, 1988.

Jarrard, R.D., Terrane motion by strike-slip faulting of forearc slivers: *Geology*, v. **14**, p. 780-783, 1986.

Jones, D.L., N.J. Silberling, and J. Hillhouse, Wrangellia--A displaced terrane in northwestern North America: *Canadian Journal of Earth Science*, v. **14**, p. 2565-2577, 1977.

Kamerling, M.J., and B.P. Luyendyk, Paleomagnetism and Neogene tectonics of the northern Channel Islands, California: *Journal of Geophysical Research*, v. **90**, p. 12,485-12,502, 1985.

Kawai, N., H. Ito, and S. Kume, Deformation of the Japanese Islands as inferred from rock magnetism: *Geophysical Journal of the Royal Astronomical Society*, v. **6**, p. 124-130, 1961.

Kono, M., K. Heki, and Y. Hamano, Paleomagnetic study of the central Andes: Counterclockwise rotation of the Peruvian block: *Journal of Geodynamics*, v. **2**, p. 193-209, 1985.

Luyendyk, B.P., M.J. Kamerling, R.R. Terres, and J.S. Hornafius, Simple shear of Southern California during Neogene time suggested by paleomagnetic declinations: *Journal of Geophysical Research*, v. **90**, p. 12,454-12,466, 1985.

Magill, J., A. Cox, and R. Duncan, Tillamook Volcanic Series: Further evidence for tectonic rotation of the Oregon Coast Range: *Journal of Geophysical Research*, v. **86**, p. 2953-2970, 1981.

Magill, J., R.E. Wells, R.W. Simpson, and A. Cox, Post-12 m.y. rotation of southwestern Washington: *Journal of Geophysical Research*, v. **87**, p. 3761-3776, 1982.

May, S.R. and R.F. Butler, North America apparent polar wander: Implications for plate motions, paleogeography, and Cordilleran tectonics: *Journal of Geophysical Research*, v. **91**, p. 11,519-11,544, 1986.

Norris, D.K., and R.F. Black, Application of paleomagnetism to thrust mechanics: *Nature*, v. **192**, p. 933-935, 1961.

Price, R.A., and D.M. Carmichael, Geometric test for Late Cretaceous-Paleogene intracontinental faulting in the Canadian Cordillera: *Geology*, v. **14**, p. 468-471, 1986.

Roperch, P., F. Megard, C. Laj, L. Mourier, T.M. Clube, and C. Noblet, Rotated oceanic block in western Ecuador: *Geophysical Research Letters*, v. **5**, p. 558-561, 1987.

Silver, L.T., Southern California geology: A product of evolution in marginal environments: *Geological Society of America Abstracts with Program*, v. **18**, p. 184-185, 1986.

Simpson, R.W., and A. Cox, Paleomagnetic evidence for tectonic rotation of the Oregon Coast Range: *Geology*, v. **5**, p. 585-589, 1977.

Umhoefer, P.J., Northward translation of "Baja British Columbia" along the Late Cretaceous to Paleocene margin of western North America: *Tectonics*, v. **6**, p. 377-394, 1987.

16

Van der Voo, R., Paleomagnetism of continental North America: The craton, its margins, and the Appalachian belt, in Pakiser, L.C., and W.D. Mooney, editors, Geophysical framework of the Continental United States: *Geological Society of America Memoir*, in press, 1988.

Watkins, N.D., Paleomagnetism of the Columbia Plateaus: *Journal of Geophysical Research*, v. **70**, p. 1,379-1,406, 1965.

Watkins, N.D., and A. Richardson, Paleomagnetism of the Lisbon volcanics: *Geophysical Journal of the Royal Astronomical Society*, v. **15**, p. 287-304, 1968.

Wells, R.E., and P.J. Heller, The relative contribution of accretion, shear, and extension to Cenozoic tectonic rotation in the Pacific Northwest: *Geological Society of America Bulletin*, v. **100**, p. 325-338, 1988.

Zijderveld, J.D.A., and R. Van der Voo, Paleomagnetism in the Mediterranean area: Implications of continental drift for the Earth Sciences, v. 1, p. 133-161, 1973.

THE KINEMATICS AND DYNAMICS OF DISTRIBUTED DEFORMATION

D. McKENZIE and J. JACKSON
Department of Earth Sciences
Bullard Laboratories
Madingley Road
Cambridge CB3 0EZ
U K.

ABSTRACT. Distributed deformation within deforming zones must take up the motion between the plates on either side. The simplest model of this process consists of rigid blocks rotating about both vertical and horizontal axes, and is consistent with paleomagnetic and seismological observations within a deforming zone in central Greece. A similar zone about 1/10th the size of that in Greece joins the propagating and doomed rifts on the Galapagos Rift. The observed rotations in several such zones suggest that the rotation rate is controlled by the vorticity of the fluid in the zone.

1. Introduction

Rotation of rigid blocks or plates is commonly associated with tectonic processes. On the largest scale plate motions can be described by rotations about vertical axes only. On a smaller scale rapid rotation of smaller blocks is now known to be a common feature of deforming zones on the continents. The rotation axes in such regions are not in general vertical, since the deformation commonly changes the dip of the strata. But such rotations generally have a vertical component. From a fluid dynamical point of view large scale rotations with a vertical component are less easy to produce than are those with horizontal components. The required conditions are discussed in section 2. On a smaller scale it is the rotation about horizontal axes which requires explanation. An interesting feature of all finite rotations of rigid blocks is that they are kinematically unstable, and the geometry of the blocks must change with time. Two examples of such instabilities are discussed in section 5.

2. Fluid Dynamics

Any movement of a rigid body can be described by the velocity vector **u** of its centre of mass, together with an angular velocity vector Ω about an axis through the centre of mass. A fundamental quantity in fluid mechanics is the *vorticity* ω of the fluid, which is related to the fluid velocity **u** by

$$\omega = \nabla_{\wedge} \mathbf{u} \tag{1}$$

17

C. Kissel and C. Laj (eds.), Paleomagnetic Rotations and Continental Deformation, 17–31.
© *1989 by Kluwer Academic Publishers.*

One reason why vorticity is such a useful concept is that the angular velocity of a small rigid particle floating in a fluid is related to the vorticity (see for instance Tritton, 1977) by

$$\omega = 2 \ \Omega \tag{2}$$

This result is easily proved using Stokes's Theorem, and allows the rotation rate of any particle to be calculated if **u** is known. It is not straightforward to decide what the vorticity of a fluid is simply by looking at the flow. For instance if the velocity of a fluid is given by

$$\left(-\frac{V}{\rho}\sin \phi, \ \ \frac{V}{\rho}\cos \phi, \ \ 0 \right) \tag{3}$$

in cartesian coordinates, or

$$\left(0, 0, \frac{V}{\rho} \right) \tag{4}$$

in cylindrical polars, $\hat{z}$, $\hat{\rho}$ and $\hat{\phi}$, then

$$\omega = \nabla_\wedge \mathbf{u} \ = \ \frac{1}{\rho} \ \begin{vmatrix} \hat{z} & \hat{\rho} & \rho\hat{\phi} \\ \partial_z & \partial_\rho & \partial_\phi \\ 0 & 0 & V \end{vmatrix} \ = 0 \tag{5}$$

Therefore $\omega = 0$ everywhere except at $\rho = 0$. Yet the velocity, the velocity gradient and the strain rate $\dot{\varepsilon}_{xy}$ are non-zero everywhere.

The energy source for all movements within the Earth is gravitational: heavier material sinks and lighter material rises. Though the differences in density in fact arise from differences in temperature, the dynamics is controlled by the direction of gravity and is not concerned with how the density variations arise. The curl of the equations governing the conservation of momentum in a fluid whose Prandtl number Pr ($=v/\kappa$) is large, where v is the viscosity and κ is the thermal diffusivity of the fluid, is

$$\varepsilon_{ijk} \ \frac{\partial}{\partial x_j} \left[\frac{\partial}{\partial x_l} \eta' \left(\frac{\partial u_k}{\partial x_l} + \frac{\partial u_l}{\partial x_k} \right) \right] \ = \ - Ra \ \varepsilon_{ijk} \ \delta_{j3} \ \frac{\partial T}{\partial x_k} \tag{6}$$

where the summation over repeated indices is implied, ε_{ijk} is the alternating tensor, Ra is the Rayleigh number

$$Ra \ = \ \frac{g\alpha d^3 \Delta T}{\kappa v} \tag{7}$$

g is the acceleration due to gravity, α the thermal expansion coefficient, and ΔT the temperature difference across the fluid layer of depth d. The viscosity of the fluid is $v\eta'$ (x,y,z), where v is constant. Within the Earth the Prandtl number is very large ($\approx 10^{24}$), and therefore equation (6) applies. If the viscosity is constant the vertical component of equation (6) is

$$\nabla^2 \omega_z = 0 \qquad (8)$$

Therefore ω_z will be zero everywhere at all times unless it is maintained by the boundary conditions. If a small perturbation to η' is introduced which is a function of position, we can use perturbation theory to discover what form of viscosity variation will produce vertical vorticity. If the dimensionless viscosity is written $1 + \eta'_1 (x,y,z)$, where $\eta'_1 \ll 1$, and if second order terms like $\eta'_1 \omega_z$ are neglected, then the vertical component of equation (6) becomes

$$\nabla \cdot [\partial_y \eta'_1 (\nabla u_x + \partial_x u) - \partial_x \eta'_1 (\nabla u_y + \partial_y u)] - \nabla^2 \omega_z = 0 \qquad (9)$$

If $\eta'_1 = \eta'_1(z)$ this equation can be satisfied when $\omega_z = 0$, $u \neq 0$. But if η'_1 depends on x or y, the term in square brackets is non-zero and the equation cannot in general be satisfied if $\omega_z = 0$. In a spherical system, where gravity acts radially, the corresponding condition is that the radial component of the vorticity is zero if the viscosity is constant or varies only with radius.

These general results are important because they show that rotations about vertical axes will only occur when there are *horizontal* variations in material properties. It is the weakness of plate boundaries compared with plate interiors which causes ω_z to be non-zero. Futhermore ω_z is non-zero almost everywhere at the surface, not just on the plate boundaries. Since the largest scale of flow within the Earth generates vertical vorticity, it is to be expected that all small scale flows will also possess vertical vorticity, since they are maintained by the larger scale motions which alone can release gravitational energy stored in the thermal density differences.

The use of Euler's theorem to describe the movement of plates requires the vorticity of each plate to be parallel to its rotation axis everywhere. Convective models in which the viscosity is constant or a function of radius alone are unable to produce flows with *any* vertical vorticity. They are therefore unlikely to provide much understanding of the three dimensional dynamics of the Earth. When the relative movement between plates is distributed across a zone, the width of the zone is commonly small compared with the size of the two adjacent plates. The relative motion between the plates can then be approximately described by a constant velocity parallel to the Earth's surface. McKenzie and Jackson (1983) show that a simple model of distributed deformation across a plate boundary zone has a velocity u within the zone of width a given by

$$u = (Wy, -2Ty, 2Tz) \qquad (10)$$

where T and W, and therefore the velocity gradients, are constant and the zone is parallel to the xaxis. If T is not zero, shortening or extension occurs by vertical motions. Since $\nabla \cdot u = 0$, equation (10) describes an incompressible flow. The horizontal component of the velocity between the plates is $(Wa, -2Ta)$ and, from (1), the vorticity is given by

$$\omega = (0, 0, -W) \qquad (11)$$

Therefore only the constant vertical component of the vorticity is non-zero. This behaviour is a direct consequence of the weakness of plate boundaries, which allows the movement at the Earth's surface to be described by the movement of large rigid plates, separated by weak zones. It is the horizontal variations in rheological properties which generate the vertical vorticity of the plate motions, and these motions in turn generate strong vertical vorticity in the weak zones. It is therefore not surprising that large rotations about vertical axes occur in deforming zones. From a fluid dynamical point of view it is rotations about horizontal axes that are unexpected. As the expression (11) for ω shows, in McKenzie and Jackson's model ω_x and ω_y are both zero and the dip of all strata remains horizontal. But steep dips are a common feature of deforming zones, which therefore must possess non-zero horizontal vorticity. The way in which the necessary horizontal vorticity can be produced can also be understood using the equation (6) that governs momentum conservation. If the buoyancy forces within the zone can be neglected, the equations governing ω_x and ω_y are respectively

$$\nabla . [\partial_z \eta' (\nabla u_y + \partial_y \mathbf{u}) - \partial_y \eta' (\nabla u_z + \partial_z \mathbf{u})] - \nabla (\eta' \nabla \omega_x) = 0 \qquad (12)$$

$$\nabla . [\partial_x \eta' (\nabla u_z + \partial_z \mathbf{u}) - \partial_z \eta' (\nabla u_x + \partial_x \mathbf{u})] - \nabla (\eta' \nabla \omega_y) = 0 \qquad (13)$$

If a small perturbation $\eta'_1 (x,y,z)$ is made to the constant viscosity in McKenzie and Jackson's (1983,1986) model, then substitution of equation (10) into (12) and (13) gives

$$(W\partial_x - 8T\partial_y)\partial_z \eta'_1 = \nabla^2 \omega_x \qquad (14)$$

$$(4T\partial_x - W\partial_y)\partial_z \eta'_1 = \nabla^2 \omega_y \qquad (15)$$

The left hand sides of equations (14) and (15) are only non-zero if $\eta'_1 = \eta'_1 (x,z)$ or if $\eta'_1 = \eta'_1 (y,z)$. This condition is not satisfied if $\eta'_1 = \eta'_1 (x,y)$ only. In deforming zones rotations about horizontal axes are often produced by reactivation of existing parallel faults with normals in the direction $\hat{\mathbf{n}} = (n_1, n_2, n_3)$. Provided $\hat{\mathbf{n}}$ is not horizontal or vertical, the horizontal vorticity cannot be zero and the dip of the bedding is changed by the deformation. In this case the expressions are easily evaluated if the viscosity η'_1 only a function of the distance, s, in the normal direction. Then

$$\frac{\partial \eta'_1}{\partial x_i} = n_i d_s \eta'_1 \qquad (16)$$

$$\frac{\partial^2 \eta'_1}{\partial x_i \partial x_j} = n_i n_j d_s^2 \eta'_1 \qquad (17)$$

and equations (14) and (15) become

$$(Wn_1 - 8T n_2) n_3 d_s^2 \eta'_1 = \nabla^2 \omega_x \qquad (18)$$

and

$$(4Tn_1 - Wn_2) n_3 d_s^2 \eta'_1 = \nabla^2 \omega_y \qquad (19)$$

The vertical component of the vorticity has attracted great geological interest, because its existence was not foreseen and because the observed changes in paleomagnetic declination produced by such movements are so large. In fact it is the horizontal components of the vorticity which are of most fluid dynamical interest. Though rotations about horizontal axes are an obvious feature of all deforming zones, their dynamical importance has been somewhat overlooked. But, as the argument above shows, deformation distributed across plate boundary zones must produce vertical vorticity, but need not generate horizontal vorticity.

3. Rotations about Vertical Axes in Deforming Zones

In the last ten years paleomagnetic observations have revealed large and systematic rotations about vertical axes in western North America (Beck, 1976, Kamerling and Luyendyk, 1979, Coney et al., 1980, Beck 1980, Luyendyk et al., 1980, Jones et al. 1982), the Mediterranean (Ron et al, 1984, Kissel et al. 1985, 1986, Allerton, this volume), and New Zealand (Walcott, 1984, this volume). It is a basic belief of most physicists that the laws of physics are not affected by a rigid rotation. Such invariance leads directly to the law of conservation of angular momentum, and to the requirement that earthquake source mechanisms must be represented by a double, rather than a single, couple. The same invariance requires all rigid rotations to be unobservable: only the relative rotation of one particle or block with respect to another can be observed. The explanation of why large relative rotations about vertical axes were discovered using paleomagnetic observations, rather than by structural geologists, may be because distributed deformation commonly does not generate relative rotation between the blocks within the zone (see Figs. 2 and 3). Paleomagnetic observations make use of a reference frame fixed to the Earth's rotational axis. In this frame they are affected by ω_r and ω_θ, but not by ω_ϕ, a rotation about a polar axis. Structural geology can only detect rigid body rotations where the relative rotation changes. For reasons discussed below the structures on block boundaries that might reveal such changes are likely to be complicated and hard to interpret. Paleomagnetic observations are therefore likely to remain the main method of mapping such rotations.

There has been considerable discussion about the nature of the structures which accommodate the rotations about vertical axes. One of the earliest proposals (Beck, 1976) was concerned with distributed strike slip deformation, and suggested that circular blocks were rotating like ball bearings between two plates on either side of the deforming zone (Fig. 1). This proposal suffered from a number of difficulties. Circular rigid blocks have never been described from a region undergoing distributed deformation. The fault plane solutions that their motion would generate would be very characteristic and have also not been reported. Earthquake fault plane solutions show that distributed deformation is more common where a zone is being deformed by a combination of strike slip and extension (or thrusting). Sonder et al., (1986) have shown that this difference in behaviour is to be expected from thin plate theory. But the rotations in Fig. 1 conserve area and cannot describe a zone undergoing stretching or shortening.

An alternative scheme attempts to take up the distributed deformation on systems of strike slip faults (Freund, 1970, Garfunkel, 1974, Luyendyk et al., 1980, Ron et al., 1984, Fig. 2). From the point of view of structural geology this geometry is more plausible than is Beck's proposal, but still has geometric problems. If the deforming zone has to take up extension or shortening, two systems of strike slip faults are required, with different strikes.

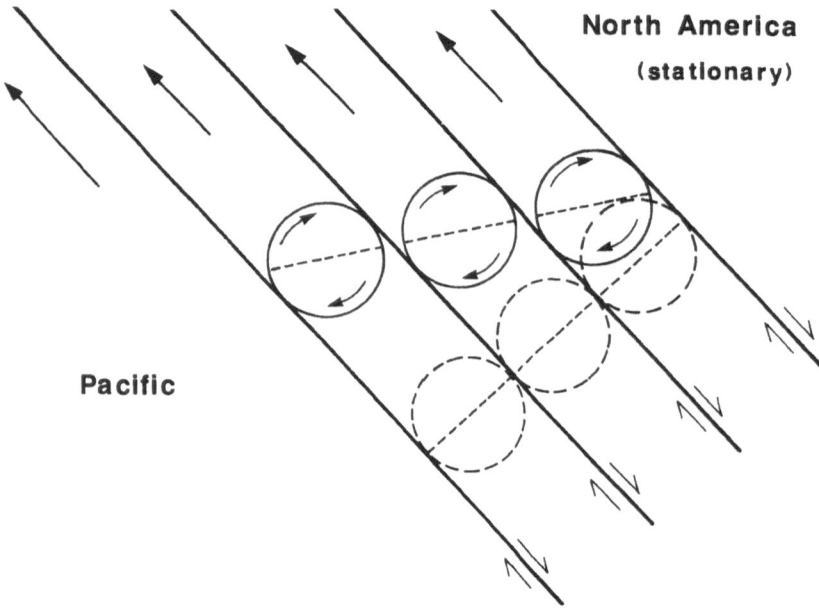

Fig. 1: Sketch illustrating the clockwise rotation of circular blocks within a distributed right-lateral shear zone. The heavy lines represent individual strike slip faults (Beck 1976).

Where the faults intersect they will lock each other, and therefore the simple model cannot describe such intersections. The other problem with the scheme in Fig. 2 results from the fact that the deformation conserves surface area, and is therefore a form of plane strain. Shortening (or extension) can only be accommodated by pushing (or pulling) material out of (or into) the ends of the deforming zone. The deformation within the zone then becomes rather complicated.

A simple scheme that avoids some of these difficulties involves movement on faults inclined to the vertical (McKenzie and Jackson, 1983, Fig. 3). The rotation of individual blocks relative to the plates on either side of the deforming zone must then be described by rotations about axes which are not vertical, but lie in the vertical plane containing the long axes of the blocks. Such rotations therefore involve both horizontal and vertical vorticity, and, unlike the motions in Figs. 1 and 2, can change the dip of strata. This scheme, shown in Fig. 3, is a generalisation of the familiar domino model of faulting to include three dimensional motions. It is the simplest model of faulting within a deforming zone which can accommodate strike slip and shortening (or extensional) movement without requiring movement of material along the strike of the zone. The motions in Fig. 3 differ from those involved in the generation of tension gashes. Their generation does not conserve volume, and can be understood in terms of two dimensional movements in a plane that is normal to the gashes. In contrast the movement on the faults in Fig. 3 involves only shear and conserves volume. Shortening (or extension) is taken up by movement of the free outer surface of the Earth, and leads to thickening (or

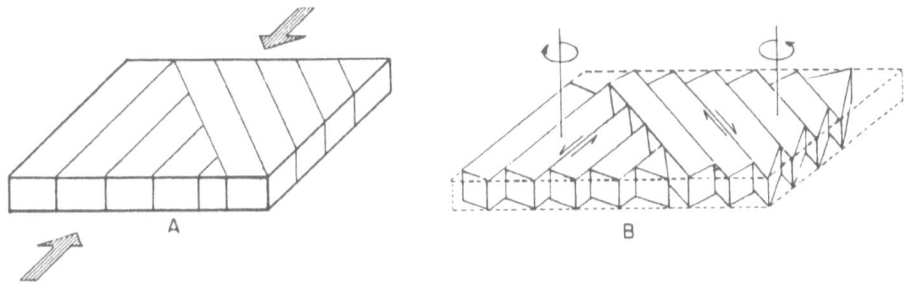

Fig. 2: Sketch illustrating the simultaneous activity of two systems of strike slip faults, (a) initial geometry, and (b) after deformation. The blocks bounded by left-lateral faults rotate clockwise, those bounded by right-lateral faults anticlockwise (Ron et al. 1984).

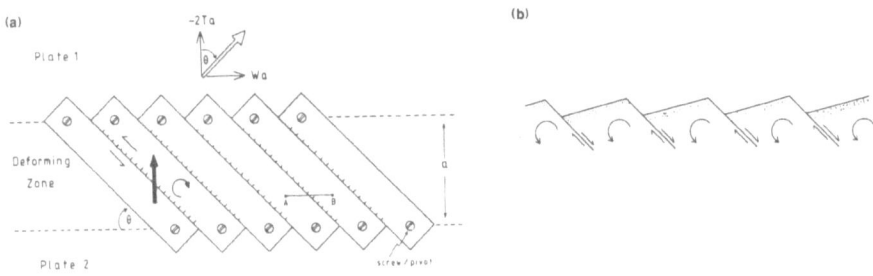

Fig. 3: Plan view, (a), and section, (b), of blocks bounded by normal faults. The motion of plate 1 with respect to plate 2 is shown by the large white arrow, and requires the zone to take up normal and right-lateral strike slip movement. The heavy black arrow shows the relative movement between individual blocks, and the movement on the normal faults bounding the blocks therefore has a left-lateral component, and the blocks rotate about both vertical and horizontal axes (McKenzie and Jackson 1983, 1986).

thinning) of the crust. McKenzie and Jackson (1983, 1986) discuss the relationship between the movement and rotation rate of the blocks within the zone to the strike slip and shortening (or extensional) movements between the plates on either side. The horizontal projection of the slip vectors on the faults bounding the blocks are constrained to be normal to the strike of the deforming zone. In the horizontal plane the angle θ between the faults and the strike of the zone and the angular velocity $\dot\theta$ of the blocks are constrained by the plate motions, but their values are model dependent.

Like all domino models, McKenzie and Jackson's (1983, 1986) scheme ignores the complicated motions which must occur below the lower boundary of the blocks. Because the Moho is often flat beneath regions where strong tilting is observed at the surface (see for instance Gans 1987), the horizontal vorticity of the flow must decrease with depth. How it does so is unclear, because there is at present no means of observing lower crustal flow. A similar problem with McKenzie and Jackson's scheme occurs where the deforming zone meets the rigid plates on either side. Here the vertical vorticity changes rapidly. However, unlike the motions at the base of the blocks, those at their ends can be studied using earthquake seismology and structural geology. Indeed *only* at the ends of the blocks is there deformation associated with their rotation. Little work has yet been done on the kinematics of such motions, probably because the problem has only recently been recognised and because such deformation produces few large earthquakes.

A question of considerable interest is the depth to which the observed surface motions extend. In the Aegean region the active faults bounding the rotating blocks are known to be roughly planar to a depth of about 10 km (Eyidoğan and Jackson, 1985), and aftershocks of large earthquakes in Greece are nearly all shallower than 10-12 km (Soufleris et al., 1982, King et al., 1985, Lyon-Caen et al., 1988). This depth is therefore likely to be the thickness of the rotating blocks, which probably have an aspect ratio in a vertical plane of 2 or 3.

At present there are few areas where the geometry of the distributed deformation is well known and suitable paleomagnetic observations have been carried out. One such area is central Greece (Fig. 4) where the NE-SW right lateral strike slip motion on the North Aegean Trough is taken up on a wide zone of normal faulting across central Greece. The slip vector changes from right lateral motion on vertical NE-SW planes beneath the North Aegean Trough, to approximately N-S extension with a small component of left lateral strike slip movement on WNW-ESE normal faults crossing Greece. This change in strike of the slip vector is a direct consequence of the block rotations, and was very puzzling until the kinematics were understood. The only structural indication of such rotations is the small left lateral component of the motion seen in fault plane solutions and as slickensides on the faults themselves (Mercier et al., 1979). The significance of this strike slip component was only undestood after the model in Fig. 3 had been proposed. In an area such as California, which has fewer large earthquakes and where the slickensides are less well preserved, it is more difficult to determine the fault geometries responsible for the rapid rotations. That central Greece is indeed undergoing rapid clockwise rotation has now been clearly demonstrated paleomagnetically (Kissel et al., 1985, 1986, and this volume). It is likely that similar fault geometries in Idaho (Scott et al., 1985, Crone and Machette, 1984, see McKenzie and Jackson, 1986) and New Zealand (Lamb, 1987, this volume) are also being formed in response to rapid rotations. The fault geometries responsible for the large rotations reported from many older circum-Pacific continental regions are as yet unclear.

Though the strikes of the faults and of the slip vectors in Figs. 3 and 4 are constrained by the relative motion of the plates on either side of the deforming zone, the faults involved can dip to the NE or to the SW. In fact all the large faults must dip the same way if they are closely spaced

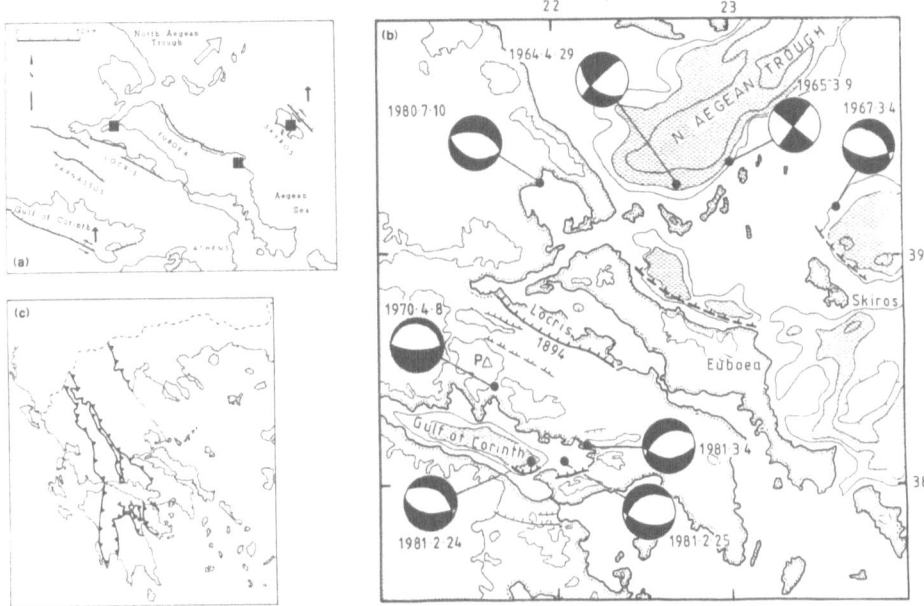

Fig. 4: The main zone of distributed normal faulting across central Greece, (a) and (b), takes up the motion shown by the large white arrow in (a). The major active normal faults are shown with tick marks on their downthrown sides. Those drawn in (a) are known or inferred to have moved in historic earthquakes. Slip vectors for two of the largest recent earthquakes, in the Gulf of Corinth, and near Skyros, are shown by black arrows in (a), and are obtained from the fault plane solutions in (b). Both show a small left lateral component of motion. Filled black squares in (a) mark the sites of paleomagnetic measurements where Kissel et al. (1986) found clockwise rotations (McKenzie and Jackson 1983, 1986).
(c) Sketch of major overthrusts in Greece, with triangles on their upthrust sides (Aubouin 1965).

and are not to intersect in the upper crust. The reason why they tend to dip to the NNE in Greece is probably because they are reactivating older thrusts which also dip to the NNE. These thrusts are shown in Fig. 4(c), taken from Aubouin (1965). But it is not generally possible to show that the thrust plane itself is reactivated as a normal fault, though the strike and dip directions of the two systems are similar.

Distributed deformation is not restricted to the continents. Transform zones joining propagating and doomed rifts (Hey et al., 1986) take up the strike slip motion by distributed deformation. The geometry has been studied in most detail on the Galapagos Rift, where the propagating rift is moving towards the triple junction with the East Pacific Rise. Though the major features in the deforming zone resemble those of central Greece, the entire zone is only about 30 km by 20 km, and the spacing between the faults 2 or 3 km (Fig. 5). Though there is as yet no direct evidence that the fault-bounded blocks are rotating, the geometry of the propagating region, the trend of the bathymetry, and the general similarity to regions like central Greece suggest that large rotations around vertical axes have occurred (McKenzie,1986). If the aspect ratio of the blocks is similar to those in Greece, their thickness should be only 1 or 2 km.

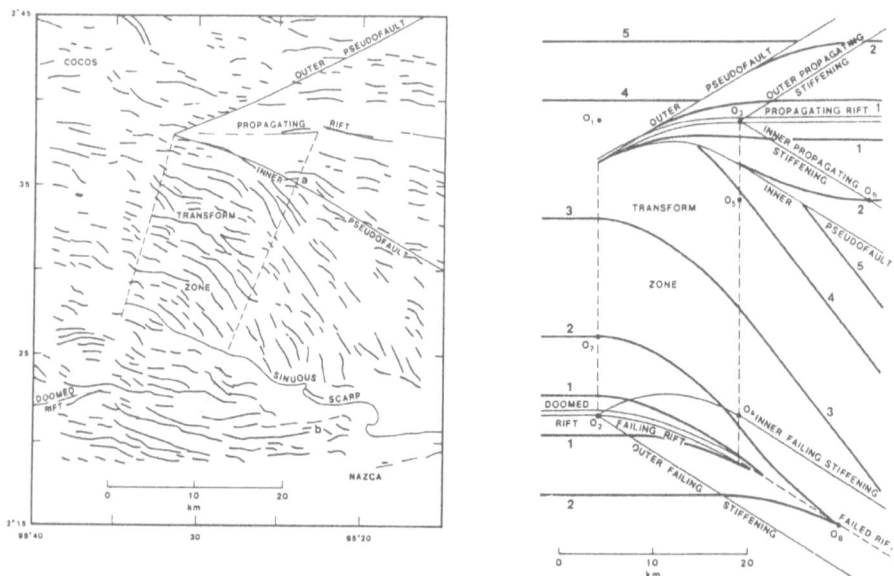

Fig. 5: The heavy lines in (a) show the trends of bathymetric features on the Galapagos Rift. The slip vector between the Nazca and Cocos plates is in the N-S direction, and the movement is distributed across the transform zone. Notice the difference in the size of the blocks compared with those in Fig. 4 (modified from Hey et al. 1986). (b) shows the calculated shapes of the magnetic anomalies, assuming the transform zone deforms by uniform simple shear (McKenzie, 1986).

4. Pinned or Floating Blocks?

If all dimensions of the blocks are small compared with the width of the zone, their vorticity is given by equation (11) and

$$\tan \theta_1 = - \frac{W}{4T} \tag{20}$$

$$\theta_1 = \frac{W}{2} \tag{21}$$

(McKenzie and Jackson, 1983), where the angle between the blocks and the zone boundary is θ_1. This model will be referred to as the 'floating block model'. Equations (20) and (21) are only exact if the blocks are circular, but will be good approximations if they are approximately equidimensional in plan view. If, however, the blocks are strongly elongated Lamb (1987) showed that their angular velocity $\dot{\theta}$ will vary as they rotate, and may even be zero for certain orientations. If, however, individual blocks span the deforming zone, and their ends are constrained to move with each plate, the angle θ_2 between the faults and the boundary of the deforming zone is given by

$$\tan \theta_2 = -\frac{W}{2T} \qquad (22)$$

and

$$\dot{\theta}_2 = W. \qquad (23)$$

This model of a deforming zone will be referred to as the `pinned block model'.

It is of considerable importance to discover whether the blocks are pinned or floating. To answer this question requires two different methods of estimating $\dot{\theta}$ or θ. Paleomagnetic observations provide one estimate, and in New Zealand Walcott (1984) has used geodetic observations to estimate W and to show that equation (21) provides a better description than does (23). A similar result has been found by Nelson and Jones (1987) from paleomagnetic measurements in the Las Vegas Shear Zone in the Basin and Range of the western U.S.A.. They used the geometry of the structures deformed by the motion to estimate θ. Laj et al., (personal communication) have carried out a similar study of the Cajamarca deflection in northern Peru, and, like Nelson and Jones (1987), they find that the paleomagnetic rotational angle is only about one half of the change of strike produced by the deformation. Therefore their observations are consistent with the floating, rather than with the pinned, block model. Brun et al., (this volume) have used sand box models of deforming zones to show that the rotation of markers spanning the zone is twice as great as that of the boundaries of deforming blocks. Their observations are therefore also consistent with control by vorticity.

The same result can be demonstrated for rotations in central Greece using a different argument. If the blocks are pinned, the slip vector between the plates on either side of the deforming zone must be at right angles to the block boundaries. If, however, the blocks are floating, the angle between the blocks and the zone boundaries, θ_1, is given by equation (20), and that between the normal to the zone and the slip vector between the plates, θ_2, by equation (22). Hence $\tan\theta_2 = 2\tan\theta_1$, or, if θ_1, $\theta_2 \ll 1$, $\theta_2 \approx 2\theta_1$. These relationships are illustrated in Fig. 6 for $\theta_2 = 45°$, when $\theta_1 = 26.6°$, and shows that the velocity vector between the plates on either side of the zone is not normal to the block boundaries when the blocks are floating. In the northern Aegean the mean strike of the slip vector from earthquakes of 1964.4.29, 1965.3.9, 1965.12.20, 1968.2.19, 1981.12.19, 1981.12.27 and 1982.1.18 near the North Aegean Trough (McKenzie, 1972, 1978, Jackson et al. 1982) is $41° \pm 9°$. Therefore, if the blocks are pinned, their boundary faults should have a strike of 131°. If, however, the blocks are floating, the strike of their boundaries depends on that of the zone. Taking the general trend of the zone to be E-W gives an expected strike of the faults of 110°. The average strike of the faulting in the Locris earthquake and that on the southern boundary of the Gulf of Corinth is about 117°, in somewhat better agreement with the estimate from the floating than the pinned model.

Another estimate of the strike of the zone can be obtained from the slip vectors of earthquakes. In both models these should be normal to the strike of the zone. However, some of the earthquakes for which fault plane solutions have been determined are from smaller earthquakes that

28

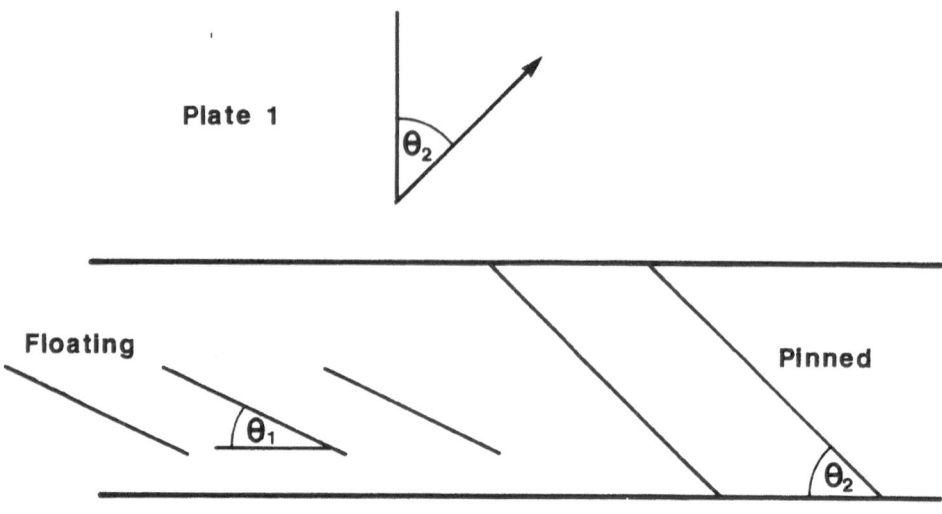

Plate 2

Fig. 6: Diagram to illustrate how the geometry of a zone of distributed deformation can be used to decide whether the blocks within it are floating or pinned to the margins of the zone. The slip vector between plate 1 and plate 2 is at right angles to the boundaries of the pinned blocks on the right, but makes an angle of 73.4° to those of the floating blocks on the left.

did not occur on major fault systems. The two largest shocks were on 1981.2.24 and 1967.3.4. The azimuth of the slip vectors are 350° (Kim et al., 1984) and 8° respectively. The expected strike is therefore 89°, in good agreement with the estimate from Fig. 4. The fault plane solutions also give the strike of the faults that slipped as 105° and 144° respectively. The average strike is therefore in reasonable agreement with the estimate of 117° from the geomorphology. In this area of central Greece the observations therefore agree somewhat better with the floating model than with one in which the blocks are pinned. These results also suggest that the floating blocks are approximately equidimensional (Lamb, 1987).

5. Kinematic Stability

McKenzie and Morgan (1969) showed that only certain geometries could occur at junctions between three moving plates, and they called these triple junctions stable. Other geometries are unstable, and immediately change to stable junctions as the plates move. But stable junctions can only exist when the motions are translations, an approximation which is obviously only valid on a plane. In practice this approximation is valid for large plates, because their poles are generally far from the triple junction, and only small movements of the junctions are of interest. McKenzie and Morgan showed that finite plate rotations could not occur about the instantaneous poles. The kinematic evolution of plates therefore requires changes in the position of their relative poles of rotation, and hence in the geometry of their plate boundaries. Mathematically this complication arises because infinitesimal rotations commute with each other, whereas finite rotations do not. This profound difference makes finite rotations much harder to analyse than instantaneous angular velocities. Finite rotations must be represented by non-commuting matrices or by quaternions, instead of by the pseudovectors used for instantaneous plate motions. The same problem arises for the same reason in the motions illustrated in Fig. 3. Finite rotations cannot be taken up by the block motions illustrated, and something must change. However the vorticity of the blocks is about two orders of magnitude greater than that of large plates. Therefore the complications resulting from the kinematic instability of the block rotations are likely to be more important than they are for major plate motions. As in the case of plate rotations, what changes is presumably governed by dynamic rather than kinematic considerations. If so the analysis of what happens will be difficult, since it will involve the stresses generated by the kinematic incompatibilities of finite rotations and the interaction of these stresses with existing weaknesses within the plate. To our knowledge no work has yet been done on these problems.

6. Conclusion

There are now a considerable number of geological studies and sand box experiments which suggest that the rotation rate of blocks within a deforming zone is controlled by the vertical vorticity. A central and difficult problem is then to understand the relationship between the instantaneous rotations responsible for the earthquakes and the finite rotations responsible for the changes in paleomagnetic declination.

Acknowledgement

The work described above forms part of a study of continental deformation supported by NERC. Contribution ES 1235.

References

Aubouin, J., 1965. *Geosynclines*. Amsterdam: Elsevier.
Beck, M.E., 1976. `Discordant paleomagnetic pole positions as evidence for regional shear in the western Cordillera of North America'. *Am. J. Sci.*, **276**, 694.
Beck, M.E., 1980. `Paleomagnetic record of plate-margin tectonic processes along the western edge of North America'. *J. geophys. Res.*, **85**, 7115.
Coney, P.J., Jones, D.L. and Monger, J.W.H., 1980. `Cordilleran suspect terranes'. *Nature*, **288**, 329.
Crone, A.J. and Machette, M.N., 1984. `Surface faulting accompanying the Borah Peak earthquake, central Idaho'. *Geology*, **12**, 664.
Eyidogan, H. and Jackson, J.A., 1985. `A seismological study of normal faulting in the Demirci, Alasehir and Gediz earthquakes of 1969-1970 in western Turkey: implications for the nature and geometry of deformation in the continental crust'. *Geophys. J. R. astr. Soc.*, **81**, 569.
Freund, R., 1970. `Rotation of strike slip faults in Sistan, southeast Iran'. *J. Geology*, **78**, 188.
Gans, P.B., 1987. `An open-system two-layer crustal stretching model for the eastern Great Basin'. *Tectonics*, **6**, 1.

Garfunkel, Z., 1974. `Model for the late Cenozoic tectonic history of the Mojave Desert, California, and for its relation to adjacent regions'. *Geol. Soc. Am. Bull.*, **85**, 1931.

Hey, R.N., Kleinrock, M.C., Miller, S.P., Atwater, T.M. and Searle, R.C., 1986. `Seabeam/Deep-tow investigation of an active oceanic propagating rift system, Galapagos 95.5°W '. *J. geophys. Res.* , **91**,3369.

Jackson, J.A., King, G. and Vita-Finzi, C., 1982. `The neotectonics of the Aegean: an alternative view'. *Earth planet. Sci. Lett.*, **61**, 303.

Jones, D.L., Cox, A., Coney, P. and Beck, M., 1982. `The growth of western North America', Sci. Am. November, 50.

Kamerling, M.J. and Luyendyk, B.P., 1979. `Tectonic rotation of the Santa Monica Mountains region, western Transverse Ranges, California, suggested by paleomagnetic vectors'. *Geol. Soc. Am. Bull.*, **90**, 331.

Kim, W.-Y., Kulhanek, O. and Meyer, K., 1984. `Source processes of the 1981 Gulf of Corinth earthquake sequence from body-wave analysis'. *Bull. seism. Soc. Amer.*, **74**, 459.

King, G., Ouyang, Z., Papadimitriou, P., Deschamps, A., Gagnepain, J., Houseman, G., Jackson, J., Soufleris, C. and Virieux, J., 1985. `The evolution of the Gulf of Corinth (Greece): an aftershock study of the 1981 earthquakes'. *Geophys. J. Roy. astr. Soc.*, **80**, 677.

Kissel, C., Laj, C. and Muller, C., 1985. `Tertiary geodynamical evolution of northwestern Greece: paleomagnetic results'. *Earth planet. Sci. Lett.*, **72**, 190.

Kissel, C., Laj, C., Mazaud, A., 1986. `First paleomagnetic results from Neogene formations in Evia, Skyros and the Volos region and the deformation of central Aegea'. *Geophys. Res. Lett.*, **13**, 1446.

Lamb, S.H., 1987. `A model for tectonic rotations about a vertical axis'. *Earth planet. Sci. Lett.*, **84**, 75.

Luyendyk, B.P., Kamerling, M.J. and Terres, R., 1980. `Geometrical model for Neogene crustal rotations in southern California'. *Geol. Soc. Am. Bull.*, **91**, 911.

Lyon-Caen, H., Armijo, R., Drakopoulos, J., Baskoutass, J., Delibassis, N., Gaulon, R., Konskoura, V., Latoussakis, J., Makropoulos, K., Papadimitriou, P., Papanastassiou, D. and Pedati, G, 1988. `The 1986 Kalamata (south Pelopennesus) earthquake: detailed study of a normal fault and tectonic implications'. *J. geophys. Res.*, (in the press)

McKenzie D., 1986. `The geometry of propagating rifts'. *Earth planet. Sci. Lett.*, **77**, 176.

McKenzie D. and Jackson, J.A., 1983. `The relationship between strain rates, crustal thickening, paleomagnetism, finite strain and fault movements within a deforming zone'. *Earth. planet. Sci. Lett.*, **65**, 182, and correction to the above, 1984, *ibid.*, **70**, 444.

McKenzie, D. and Jackson, J.A., 1986. `A block model of distributed deformation by faulting'. *J. Geol. Soc. Lond.*, **143**, 349.

McKenzie, D. and Morgan, W.J., 1969. `The evolution of triple junctions'. *Nature*, **224**, 125.

Mercier, J.L., Delibasis, N., Gauthier, A., Jarrige, J.-J., Lemeille, F., Philip, H., Sébrier, M. and Sorel, D., 1979. La néotectonique de l'Arc Egéen'. *Rev. Geol. dyn. Geogr. phys.*, **21**, 61.

Nelson, M.R. and Jones, C.H., 1987. `Paleomagnetism and crustal rotations along a shear zone, Las Vegas Range, southern Nevada'. *Tectonics*, **6**, 13.

Ron, H., Freund, R., Garfunkel, Z. and Nur, A., 1984. `Block rotation by strike slip faulting: structural and paleomagnetic evidence'. *J. geophys. Res.*, **89**, 6256.

Scott, W.E., Pierce, K.L., and Halt, H.H., Jr., 1985. `Quaternary tectonic setting of the 1983 Borah Peak earthquake, central Idaho'. In Stein, R. and Buckham, R.C. (eds) *Proceedings of Workshop XXVIII on the Borah Peak, Idaho, Earthquake.* United States Geological Survey Open File Report, **85-290**, 1.

Sonder, L.J., England, P.C. and Houseman, G.A., 1986. `Continuum calculations of continental deformation in transcurrent environments'. *J. geophys. Res.*, **91**, 4797.

Soufleris, C., Jackson, J., King, G., Spencer, C. and Scholz, C., 1982. `The 1978 earthquake sequence near Thessaloniki (northern Greece)'. *Geophys. J. Roy. astr. Soc.*, **68**, 429.

Tritton, D.J., 1977. *Physical Fluid Dynamics*, New York: Van Nostrand Reinhold Co..

Walcott, R.I., 1984. `The kinematics of the plate boundary through New Zealand: a comparison of short and long term deformations'. *Geophys J. R. astr. Soc.*, **79**, 613.

RELATIONS BETWEEN SEISMICITY AND PALEOMAGNETIC ROTATIONS IN ZONES OF DISTRIBUTED CONTINENTAL DEFORMATION

J. JACKSON and D. McKENZIE
Department of Earth Sciences
Bullard Laboratories
Madingley Road
Cambridge CB3 0EZ
UK

ABSTRACT. The observed movement on faults within a wide zone of continental deformation does not in general determine the rotation rate of the rigid blocks bounded by the faults. Only if an *a priori* model for the deformation is assumed can the fault movements be related to observations of paleomagnetic rotations. Even when such a model is valid, the faulting determines the block rotations relative to the boundaries of the zone, whereas the paleomagnetic observations measure the rotation with respect to the Earth's rotational axis.

1. Introduction

The two most fundamental observations in regions of active continental deformation are (i) that the seismicity, and hence the deformation, is distributed over zones up to several hundred km in width (e.g. Fig. 1 and McKenzie, 1972; Molnar and Tapponnier 1975) and (ii) that the focal depths of earthquakes are generally concentrated in the upper half (usually the top 10-20 km) of the continental crust (e.g. Sibson 1985; Chen and Molnar 1983). Furthermore, except in rather narrow zones of strike slip deformation, such as the North Anatolian and San Andreas Fault Zones, the largest continental earthquakes usually involve motion on faults whose lengths are small compared with the dimensions of the deforming zone. A consequence of this feature of continental deformation is that distributed shortening (or extension) between large aseismic regions can be accommodated by either crustal thickening (or thinning) or by movement of material along the strike of the wide deforming zone that separates the rigid blocks (or, of course, by a combination of these mechanisms). Relative movement of the plates is thus insufficient to describe the motions within the wide deforming regions that commonly occur where plate boundaries cross continental lithosphere. For instance the N-S convergence between Africa and Eurasia cannot be used to predict N-S extension in the Aegean Sea. The same result applies at regional scales in, for instance SW Iran, where the NE-SW convergence across the Zagros mountains is accommodated by thrust faulting parallel to the strike of the belt. In contrast, the NE-SW convergence in NE Iran is achieved by a mixture of thrust and strike slip faulting (see Jackson and McKenzie, 1984, 1988). The kinematics of wide continental deforming zones is best described by a continuum approach, with no strain discontinuities (e.g. England and McKenzie, 1982; McKenzie and Jackson 1983). Such an approach is only valid at length scales larger than the dimensions of the faults within the deforming zones. This paper

33

C. Kissel and C. Laj (eds.), Paleomagnetic Rotations and Continental Deformation, 33–42.
© *1989 by Kluwer Academic Publishers.*

34

summarizes what may be discovered about the deformation at such length scales from observations of seismicity and faulting within the zone. This topic is covered in greater detail by McKenzie and Jackson (1983) and Jackson and McKenzie (1988).

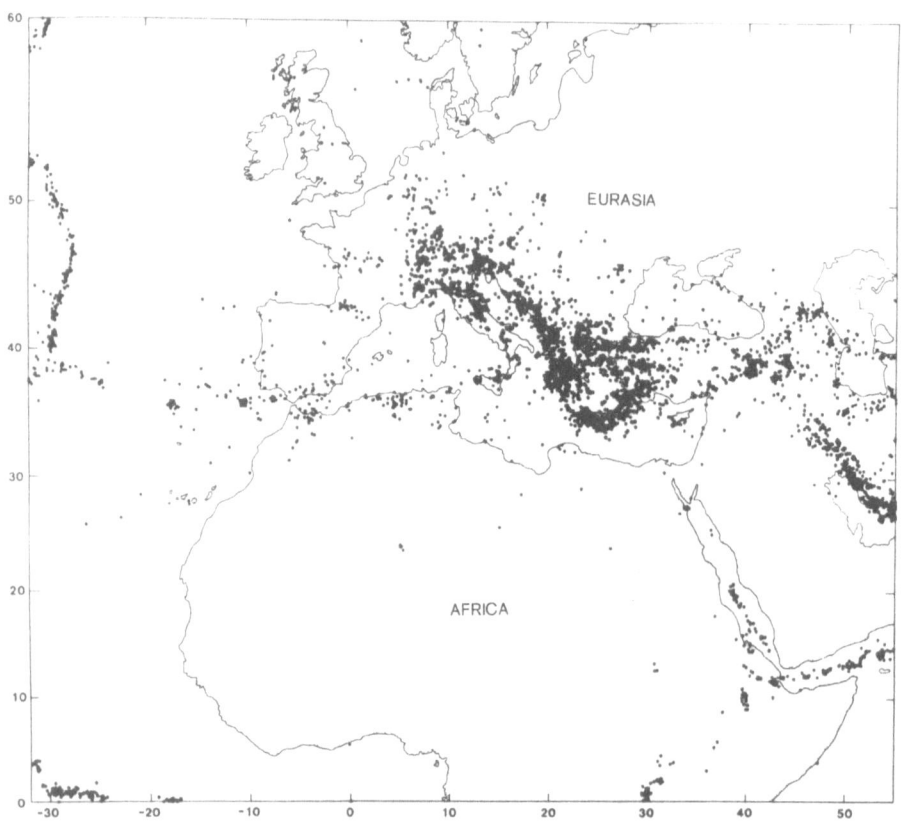

Fig. 1: Map of epicentres reported by USGS from 1961 to 1983.

2. A simple description of continuous deformation

Figure 2 is a cartoon of a deforming zone of dimensions l, a and t separating two rigid plates. The velocity of the front relative to the back of the zone (i.e. perpendicular to strike) is v, that of the top relative to the bottom is w, and that of one end relative to the other is u. In general, we would like to know the relationship between the vector y joining two points in the zone at time τ to that at time 0:

$$y(\tau) = F(\tau)\, y(0) \tag{1}$$

F is the deformation gradient tensor and includes the effect of rigid body rotations. If the velocity gradients are constant everywhere in time and space, McKenzie and Jackson (1983) showed that

$$\frac{d\,F}{d\tau} = L\,F \tag{2}$$

where L is the velocity gradient tensor

$$L = \begin{pmatrix} u_x/l & v_x/a & w_x/t \\ u_y/l & v_y/a & w_y/t \\ u_z/l & v_z/a & w_z/t \end{pmatrix} \tag{3}$$

L may be decomposed into its symmetric part S and its antisymmetric part A:

$$L = S + A \tag{4}$$

where

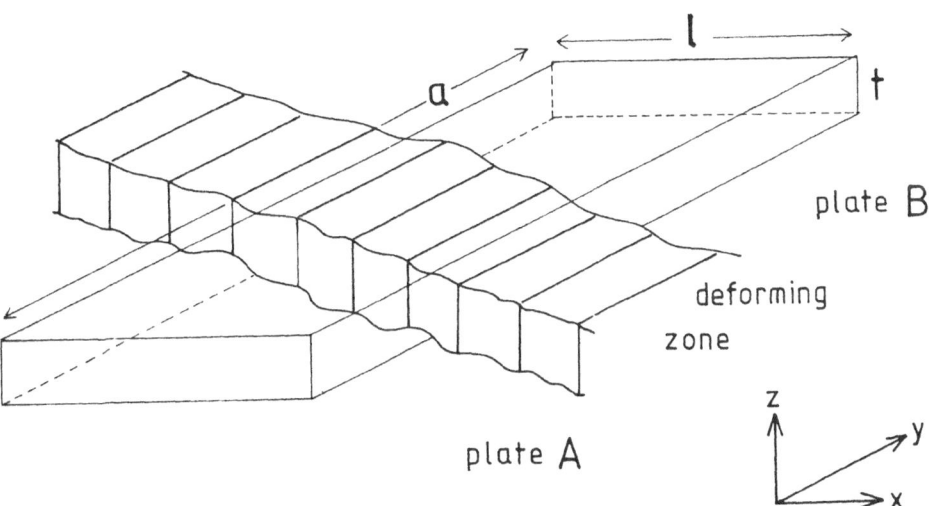

Fig. 2: Sketch of a deforming zone separating plates A and B. **w** is the average vector velocity of the top face of the volume with respect to the bottom. **v** is that of the x-z plane in B with respect to that in A. **u** is that of the right-hand face in the y-z plane with respect to the left-hand face.

$$S = \frac{1}{2} (L + L^T) \qquad (5),$$

$$A = \frac{1}{2} (L - L^T) \qquad (6)$$

and L^T is the transpose of L. S is, by definition, the strain rate tensor (see eg. Jackson and McKenzie 1988), and A corresponds to a rigid body rotation, with components about the x, y and z axes.

3. Seismicity and distributed deformation

What information about L can be obtained from observations of seismicity within the deforming zone? The basic information on the size and orientation of faulting during an earthquake is contained in the moment tensor M:

$$M = M_0 (u_i n_j + u_j n_i) \qquad (7)$$

where
$$M_0 = \mu As \qquad (8)$$

M_0 is the scalar moment, μ the shear modulus, A the fault area, and $\hat{u}$ and $\hat{n}$ are unit vectors in the direction of slip and of the normal to the fault surface, respectively (see Aki and Richards, 1980). $\hat{u}$ and $\hat{n}$ may be obtained from fault plane solutions, and, because M is a symmetric matrix, it is not necessary to distinguish the fault and auxiliary planes. $\hat{u}$ and $\hat{n}$ must be orthogonal if slip is to be in the fault plane, and hence M has a zero trace.

If all the deformation within a volume V occurs on faults that move in earthquakes, Kostrov (1974) showed that there is a relationship between the average strain $\bar{\varepsilon}_{ij}$ of the volume and the sum of the moment tensors of all the earthquakes within it:

$$\bar{\varepsilon}_{ij} = \frac{1}{2\mu V} \sum_{n=1}^{N} M_{ij}{}^n \qquad (9)$$

where M^n is the moment tensor of the nth earthquake. If the observations are made in a time period τ then:

$$\frac{\bar{\varepsilon}_{ij}}{\tau} = S \qquad (10)$$

S is the strain rate tensor and the symmetric part of L (equation 5). Thus S may be estimated directly from the seismicity providing the time period of observation, τ, is long enough to provide a representative sample of the deformation within the zone. In practice only the moment tensors from the larger earthquakes need to be summed as they account for most of the deformation within the zone (e.g. Molnar, 1979; Jackson and McKenzie, 1988).

S may be written as:

$$S = \frac{1}{2} \begin{pmatrix} 2u_x/l & v_x/a + u_y/l & w_x/t + u_z/l \\ v_x/a + u_y/l & 2v_y/a & w_y/t + v_z/a \\ w_x/t + u_z/l & w_y/t + v_z/a & 2w_z/t \end{pmatrix} \quad (11)$$

from which it may be seen that the shortening across the zone (v_y) is obtainable from S_{22}; the thickening of the seismological layer (w_z) from S_{33}, and the movement of material along strike of the zone (u_x) from S_{11}. The zero trace of the moment tensor requires that

$$S_{11} + S_{22} + S_{33} = 0 \quad (12)$$

because the material is incompressible. The off-diagonal elements of S are combinations of components of u, v and w and, in general, have no simple interpretation. However, the length of the deforming zone l is often very much greater than the width a across strike. If this is the case then $v_x/a \gg u_y/l$ and S_{12} is approximately $v_x/2a$ (see Jackson and McKenzie, 1988). Thus, although all nine components of u, v and w are needed to specify L in equation (3), in general only three of them (u_x, v_y, w_z) are obtainable directly from the seismicity. Note that only v_x and v_y are predicted from a knowledge of the plate or block motions across the zone, as plate tectonic descriptions specify only horizontal velocities. Of these, v_y , and under special circumstances v_x , may be compared with values of S_{22} and S_{12} respectively, obtained from summing seismic moment tensors.

Although S can, in principle, be recovered from the sum of seismic moment tensors, using Kostrov's result, it is only a part of the overall deformation within the zone, which is represented by L (see equations 5 and 6). As equation (4) shows, L can only be determined if both S and A are known. A represents a rigid body rotation and can be written as:

$$A = \begin{pmatrix} 0 & v_x/a - u_y/l & w_x/t - u_z/l \\ u_y/l - v_x/a & 0 & w_y/t - v_z/a \\ u_z/l - w_x/t & v_z/a - w_y/t & 0 \end{pmatrix} \quad (13)$$

The deformation within the zone in Fig. 2 will not be affected by the addition of a uniform rigid body rotation rate to the whole system. The tensor A can therefore not be obtained from observations of faulting within the zone. This is an important general result, discussed in greater detail by Jackson and McKenzie (1988). Note that, if $l \gg a$ then $A_{12} \approx v_x/2a$. McKenzie and Jackson (1983) show that a circular body embedded in the deforming fluid (or zone) will rotate with an angular velocity about the z axis given by

$$\frac{d\phi}{d\tau} = -\frac{v_x}{2a} \quad (14)$$

provided that $v_x/a \gg u_y/l$. Since v_x is recoverable from the seismicity under these circumstances, the rotation of a disc about the z axis *relative to the boundaries of the zone* may be obtained from the seismicity. Note, however, that knowledge of v_x does not determine whether it is the zone boundary or the rigid disc that rotates relative to an external reference frame. Moreover, Lamb (1987) showed that the rotation rate of an elliptical body relative to the boundaries of the zone depends on both its ellipticity and its initial orientation.

The same reasoning may be extended if the zone thickness t is very much smaller than the width a and length l. Then $w_x/t \gg u_z/l$, $w_y/t \gg v_z/a$ and $S_{13} \approx w_x/t$, $S_{23} \approx w_y/t$ (see equation 11). Then equation (13) shows that $A_{13} \approx w_x/t \approx S_{13}$ and $A_{23} \approx w_y/t \approx S_{23}$. Thus, under these special conditions, the rotation of an object with circular cross-section can be estimated *relative to the boundaries of the zone* about the x and y axis. Once again, knowledge of A_{13} or A_{23} cannot distinguish the rotation of the object from the rotation of the zone boundary relative to an external frame, though in the case of rotations about horizontal axes isostatic forces are likely to keep the boundaries of the zone approximately horizontal at large length scales.

Thus, in general, observations of faulting in deforming zones cannot be used to estimate rigid body rotations affecting the system as a whole. Under particular circumstances the summed seismic moment tensors may be used to estimate rotations of blocks within the zone relative to the zone boundaries, but not relative to an external reference frame unless a model is specified. This is illustrated in the next section.

4. An example of distributed deformation achieved by faulting

Figure 3 shows a simple block model of a deforming zone in which the motion between the two bounding plates is taken up by a number of parallel faults that rotate as they move. The model is discussed in greater detail by McKenzie and Jackson (1983, 1986). The velocity between the plates has a component $-2Ta$ normal to the zone and a component of right lateral shear Wa parallel to the zone boundary. It is possible to move the model either by keeping the orientation of the boundaries east-west, in which case the faults rotate clockwise relative to the north direction (Fig. 3(b)), or by keeping the faults with their original strike, in which case the boundaries of the zone rotate anticlockwise with respect to the north direction (Fig. 3(c)). In both cases, the strain rate tensor S is given by

$$S = \begin{pmatrix} 0 & W/2 \\ W/2 & -2T \end{pmatrix} \qquad (15)$$

and can be obtained from the addition of seismic moment tensors. However, in Fig. 3(b) L is given by:

$$L = \begin{pmatrix} 0 & W \\ 0 & -2T \end{pmatrix} \qquad (16)$$

whereas in Fig. 3(c)

$$L = \begin{pmatrix} 0 & 0 \\ W & -2T \end{pmatrix} \qquad (17)$$

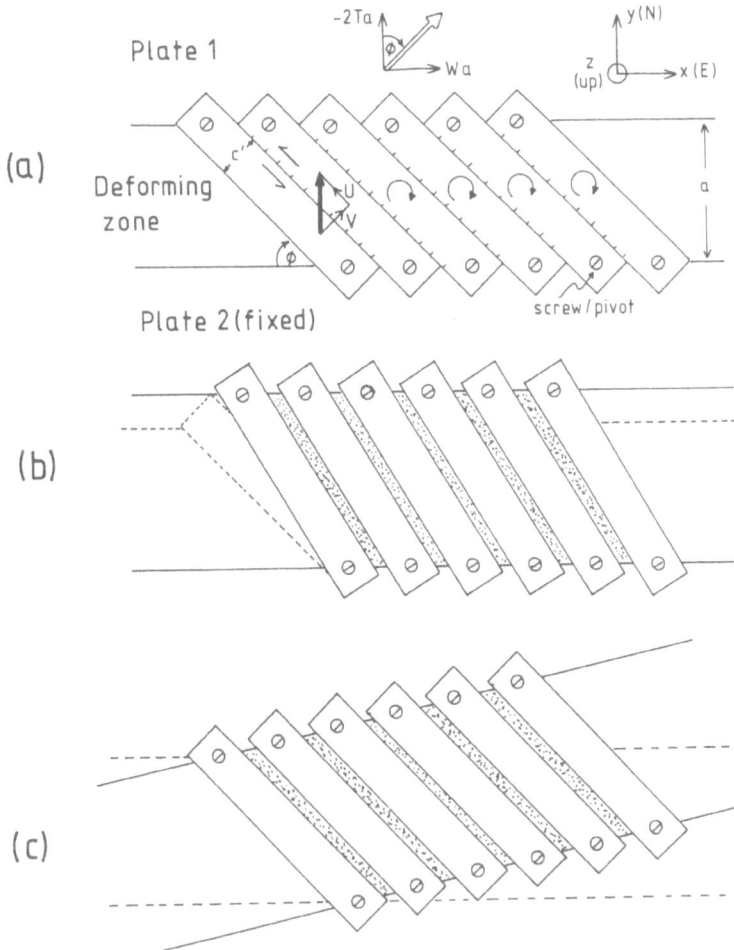

Fig. 3: (a) The pinned block model of McKenzie and Jackson (1986), which may be constructed from wooden slats screwed to two larger slats. The model represents a plan view of a deforming zone separating two plates. The motion of plate 1 relative to plate 2 is given by the large white arrow. The slip vector between adjacent blocks is shown by the large black arrow, and involves both normal (V) and left lateral strike-slip (U) motion. The blocks rotate in a clockwise direction. (b) The configuration of the model after an infinitesimal amount of motion. The area created by movement normal to the zone (-2Ta) is shown by stippling. The blocks have rotated from their initial position (dashed) but the zone still strikes E-W. This deformation is described by the velocity gradient tensor **L** in equation (16). (c) The configuration after infinitesimal motion in which the faults have not rotated relative to north but the zone boundaries have. The summed moment tensors in moving from (a) to (b) and (a) to (c) are identical.

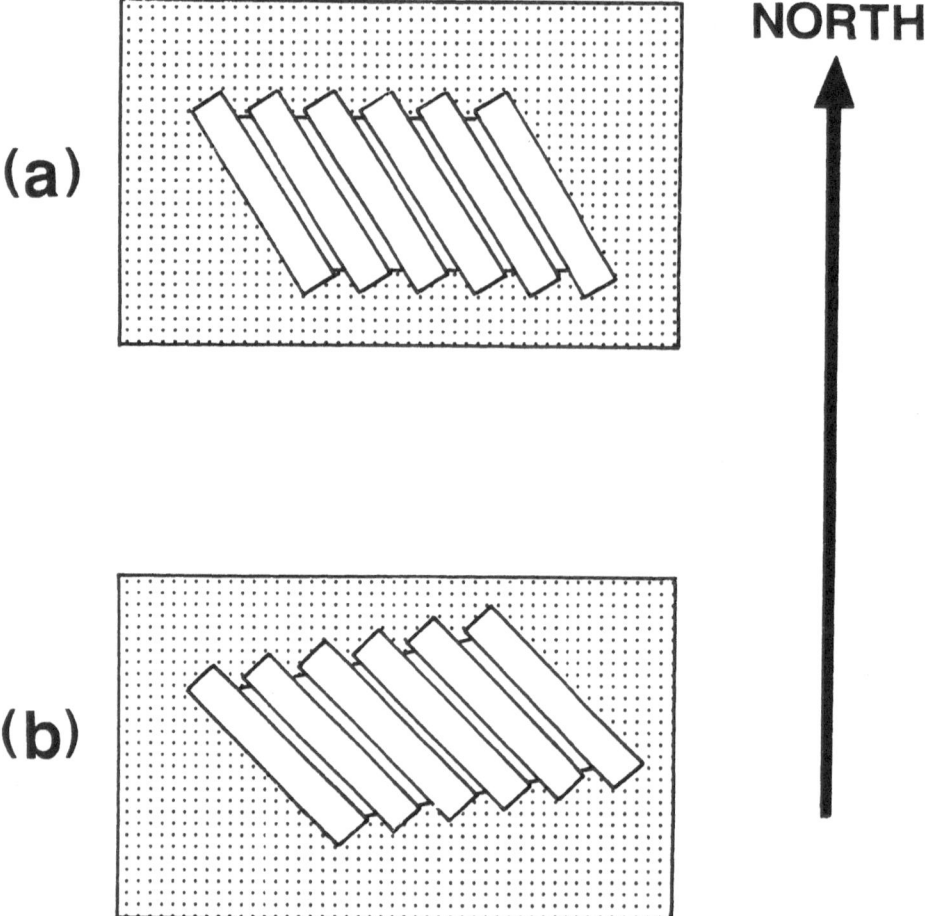

NORTH

(a)

(b)

Fig. 4: In both (a) and (b) a deforming zone, outlined by a rectangular box, contains within it a domain of faults of the type shown in Fig. 3(a). Fig. 4(a) shows the fault configuration after movement of the type in Fig. 3(b), in which the domain boundary does not rotate relative to north. Fig. 4(b) shows the fault configuration after movement of the type in Fig. 3(c) in which the domain boundary does rotate relative to north. The two cases require different deformation to be taken up in the rest of the deforming zone (stippled), and can be distinguished by paleomagnetic measurements in the fault bounded blocks, but not by observations of faulting within the domain alone.

(see Jackson and McKenzie, 1988). Thus the two cases Fig 3(b) and 3(c) cannot be distinguished by observations of faulting, but can be distinguished by paleomagnetic declinations within the blocks as they measure orientation relative to an external frame. In both cases the blocks rotate relative to the zone boundary at a rate given by

$$\frac{d\phi}{d\tau} = W \qquad (18)$$

and this rate may be determined from the fault movement (15). It is not simply the rotation relative to the zone boundaries that is of concern in continental tectonics, as becomes obvious if the deforming zone in Fig. 3 is embedded in a much larger deforming zone. This is the situation in central Greece, which the model in Fig. 3 was designed to illustrate (McKenzie and Jackson, 1983, 1986). The two cases shown in Fig. 3 require quite different motions to be taken up elsewhere in the wider deforming zone (Fig. 4). It is therefore of interest to know whether the configuration of Fig. 3(b), or 3(c), or a combination of both, occurs. This question can be resolved only by reference to an external frame (e.g. paleomagnetism), and not by observing the movement on the faults. Thus seismicity cannot be used to predict paleomagnetic rotations unless an *a priori* model of the deformation is adopted.

Acknowledgments

This work was supported by NERC, and is publication No. ES 1209 of the Dept. of Earth Sciences, Cambridge.

References

Aki, K., and Richards, P., 1980. *Quantitative Seismology; Theory and Methods*, (2 vols) W.H. Freeman, San Francisco.

Chen, W-P. and Molnar, P., 1983. `Focal depths of intracontinental and intraplate earthquakes and their implications for the thermal and mechanical properties of the lithosphere'. *J. geophys. Res.*, **88**, 4183-4214.

England, P.C. and McKenzie, D., 1982. `A thin viscous sheet model for continental deformation'. *Geophys. J. R. astr. Soc.*, **70**, 295-321, and correction to the above, *ibid.*, 1983, **73**, 523-532.

Jackson, J.A. and McKenzie, D., 1984. `Active tectonics of the Alpine Himalayan Belt between western Turkey and Pakistan'. *Geophys. J. R. astr. Soc.*, **77**, 185-264.

Jackson, J.A. and McKenzie, D., 1988. `The relationship between plate motions and seismic moment tensors, and the rates of active deformation in the Mediterranean and Middle East'. *Geophys. J.*, **93**, 45-73.

Kostrov, V.V., 1974. `Seismic moment and energy of earthquakes, and seismic flow of rock'. Izv. Acad. Sci. USSR *Phys. Solid Earth*, **1**, 23-44.

Lamb, S., 1987. `A model for tectonic rotations about a vertical axis'. *Earth Planet. Sci. Lett.*, **84**, 75-86.

McKenzie, D., 1972. `Active tectonics of the Mediterranean region'. *Geophys. J. R. astr. Soc.*, **30**, 109-185.

McKenzie, D. and Jackson, J.A., 1983. `The relationship between strain rates, crustal thickening, paleomagnetism, finite strain and fault movements within a deforming zone'. *Earth planet. Sci. Lett.*, **65**, 182-202, and correction to the above, *ibid.*, 1984, **70**, 444.

McKenzie, D. and Jackson, J.A., 1986. `A block model of distributed deformation by faulting'. *J. geol. Soc. London.*, **143**, 349-353.

Molnar, P. and Tapponnier, P., 1975. `Cenozoic tectonics of Asia: effects of a continental collision'. *Science*, **198**, 419-426.

Molnar, P., 1979. `Earthquake recurrence intervals and plate tectonics'. *Bull. seism. Soc. Am.*, **69**, 115-133.

Sibson, R., 1982. `Fault zone models, heat flow, and depth distribution of earthquakes in the continental crust of the United States'. *Bull. Seism. Soc. Am.*, **72**, 151-163.

THE DETECTION OF ROTATIONS BY SURVEYING TECHNIQUES

P. A. CROSS
Department of Surveying
University of Newcastle upon Tyne,
England

ABSTRACT. Classical and modern space geodetic observing techniques are reviewed and their potential for detecting crustal rotations is assessed. Special emphasis is placed on the Global Positioning System and its geodynamic applications. Methods for estimating rotations from geodetic observables are given and rigorous methods for measuring the quality of estimated rotaions are explained. A brief discussion of the associated computer aided design problem is also included.

1. Introduction

Over the last thirty years or so many authors have reported that large regions of continental crust, in many parts of the world, have undergone substantial rotations about vertical axes. The evidence for such conclusions has almost invariably been paleomagnetic, with measurements being made of the paleomagnetic declinations frozen into young rocks caught up in deformed continental belts. Such information relates to geological time scales and current rates of rotation are generally inferred by long term averaging over millions of years. Questions relating to these current rates can, in principle, be answered by surveying techniques.

Surveying (throughout this paper the adjective is used synonymously with geodetic) techniques are those that lead to the relative positions of points on the earth's surface. Rotations of the crust about vertical axes will lead to temporal changes in the horizontal components of such coordinates, and time separated sets of surveying measurements can lead to the estimation of rotation. Either high quality surveying measurements over short time intervals or lower quality measurements over longer periods can be used.

This paper reviews both the classical terrestrial surveying measurement techniques and modern space geodesy methods, with emphasis on the latter. The quality of the measurements is discussed and the mathematical treatment necessary to estimate coordinates, and rotations, from such observations is given. Emphasis is placed on the assesment of the quality of derived information such as rotations because without suitable quality measures for the surveying results the validity of any geological conclusions drawn remains open to question.

It is important to note at the outset that rotations will, in general, be estimated from coordinates via the computations of angles and azimuths. If the rotation of the whole region in which a survey has taken place is to be estimated then its shape will not have changed and it is essential that the survey contains observations that provide the overall azimuth of the region at the epoch of the measurements. On the other hand if rotations have taken place within the area then changes of shape will occur and the angles derived from the coordinates will change. In general the latter situation is most easy to deal with by surveying methods. This paper does not consider the

43

C. Kissel and C. Laj (eds.), Paleomagnetic Rotations and Continental Deformation, 43–67.
© *1989 by Kluwer Academic Publishers.*

statistical techniques for assessing the significance of derived rotations, i.e. for separating real rotations from those implied by errors in the surveying measurements. Clearly this is a most important topic in this context and Caspary (1987) is recommended as a starting point for those who wish to pursue it.

2. Terrestrial method

The classical surveying instrument (over the last two centuries or so) is the theodolite. Theodolites are used to measure horizontal and vertical angles between points on the earth's surface. They can also be used to measure timed horizontal angles between such points and stars and so lead to the measurement of azimuths (bearings from true north as defined by the earth's axis of rotation). Most of the developed parts of the world are covered by triangulation networks which are essentially networks of observed horizontal angles. In order to estimate coordinates a baseline would have been measured (by catenary taping). Such baselines were often of very high quality but typically would take several weeks with large teams of personel for only a few kilometres. Azimuth measurements were needed for orientation but were rarely measured due to the difficulties of transferring time before the days of radio time signals. Triangulation networks observed since the development of atomic clocks and radio time signals have their absolute orientation rather more rigorously controlled by regular azimuth measurements.

The accuracy with which an angle can be measured depends on a number of factors the most important of which are centering, circle graduations and refraction. For long lines (more than a few tens of kilometres) centering and circle graduation errors are usually insignificant and the critical factor is the horizontal bending of light (kown and lateral refraction) due primarily to the horizontal component of the temperature gradient. For most parts of the world this has proved almost impossible to model with the consequence that the quality of angle measurement has not significantly improved over the last hundred years or so (although theodolites have got lighter and more manageable the old instruments generally had graduation errors less than the refraction errors). Basically modern theodolites can yield a measuring resolution of about 0.1sec but the accuracy (one standard deviation) is more likely to be in the region of 0.7 - 1.0 sec. Azimuths can be meaured with a similar quality. The points are discussed in more detail in Ashkenazi et al (1972).

The coordinates that were derived in the past are, however, of rather poor quality due to a lack of azimuth and scale control (for reasons outlined earlier) and, most importantly, because the lack of computers did not allow the correct estimation procedures to be applied. This latter point is due to the fact that the estimation of coordinates from horizontal angles leads to sets of simultaneous equations of a size that is a function of the total number of unknown points. For instance even over a small area such as Great Britain the primary triangulation has over one thousand simultaneous equations and a recent computation in North America required the solution of about half a million equations. Errors due to this source approach 20 metres in Great Britain, see Ashkenazi et al (1972).

This means that old published coordinates most probably cannot be compared with new ones in order to estimate rotations (because the differences in the coordinates will be primarily due to the computational strategies adopted). However, if the original observations can be located, as described for instance in Walcott (1984), rotations may be usefully estimated (because the old angles are likely to be as good as modern ones).

For the last thirty years or so major advances have been made in the direct measurement of terrestrial distances with the development of the technique known as EDM (electromagnetic distance measurement). The process essentially involves an instrument generating, modulating and emitting an electromagnetic wave of known frequency (either microwave, visible or infra-red). The signal is transmitted to a reflector (or receiver/retransmitter for microwaves) and the return signal is received. The basic measured quantity is the difference in phase of the outgoing and incoming modulated signals. Measurements on a number of different modulation frequencies leads to an estimate of the distance between the instrument and reflector, for more details see Burnside (1982). The systems does of course need a line of sight between the instrument and reflector.

As with angle measurement the limiting factor on the achievable accuracy of EDM is a function of the accuracy with which the effects of the atmosphere can be modelled. In the case of EDM there is both a bending effect and, more seriously, an effect on the velocity (and hence wave length). The very highest quality instruments can achieve an accuracy (one standard deviation) of about 1ppm (part per million) of the distance up to a few tens of kilometres. EDM systems are, however extremely susceptible to scale errors and need to be calibrated regularly. There are many examples of survey networks that have incorrect scale resulting from the use of instruments with frequency errors due to incorrectly calibrated oscillators. This problem is, however, unlikely to be important in the estimation of rotations.

Another terrestrial system that could be useful in the detection of rotations is the gyrotheodolite. Gyrotheodolites use mechanical grroscopes to sense the spin of the earth and hence the north direction. They can then be rotated so that their graduated scales yield azimuths directly. Clearly they produce an observable that is of direct use in the detection of rotations. Unfortunately the accuracy of their derived azimuths is not as high as that of those obtained astronomically (although the observing process is significantly faster and the method has the advantage of not requiring cloudless skies). Currently, according to Bomford (1980), an accuracy (one standard deviation) of about 3 secs is achievable in mid-latitudes.

3. Satellite techniques

Satellite positioning is the term used to describe the determination of the absolute and relative coordinates of points on (or above) the earth's land or sea surface by processing measurements to, and/or from, artificial earth satellites. In this context absolute coordinates refer to the position of a point in a specified coordinate system, whereas relative coordinates refer the the position of one point with respect to another (again in a specified coordinate system). Relative positions are generally more useful in surveying and can usually be more accurately determined.

The first applications of the technique were made in the early 1960's but at that time the lengthy observing periods (weeks or months at a station) and rather low accuracy (standard deviatins of several metres) meant that it was only really useful for global geodesy. Important results were, however, obtained, especially in the connection of various national and continental terrestrial networks and in the determination of the overall position, scale and orientation of national coordinate systems. Nowadays relative positions can, in certain favourable circumstances, be determined from satellite measurements with standard errors of a few millimetres within a few minutes. Clearly this makes satellite positioning potentially a powerful tool for the detection of rotations.

Moreover satellite positioning has two very important advantages over its traditional terrestrial counterpart. Firstly the derived positions are genuinely three-dimensional. This is in direct contrast to traditional techniques where plan and height control have invariably been treated separately, both from a point of view of station siting (plan control points are normally on hill tops whereas height control points are usually located along roads and railway lines) and from a point of view of measurement and computation. Secondly, and perhaps most importantly, the traditional requirements of intervisibility between survey stations are not relevant. All that is required is that the stations should have a line of sight (in the appropriate part of the electromagnetic spectrum) to the satellite(s) being observed.

Here a general review of satellite positioning methods is given with emphasis on GPS as it is expected that this system will have special applications to the detection of rotations.

3.1 REVIEW OF SATELLITE POSITIONING OBSERVABLES.

Figure 1 shows an idealised geodetic satellite S_1 and ground station P. In principle the following quantities are observable.

(i) The direction, d, from P to S_1 can be observed either by the use of a special tracking telescope equipped with encoders for angle recording (the theodolite version is called a kine-theodolite), or, more commonly, by means of a camera. In the latter case the result is a photograph of the satellite against a background of stars and some rather straightforward photogrammetry leads to the computation of the direction cosines of the satellite from the known star directions. Computations with directions observed simultaneously from two ground stations lead to the direction between them, and a number of such interstation directions to 'satellite triangulation'.

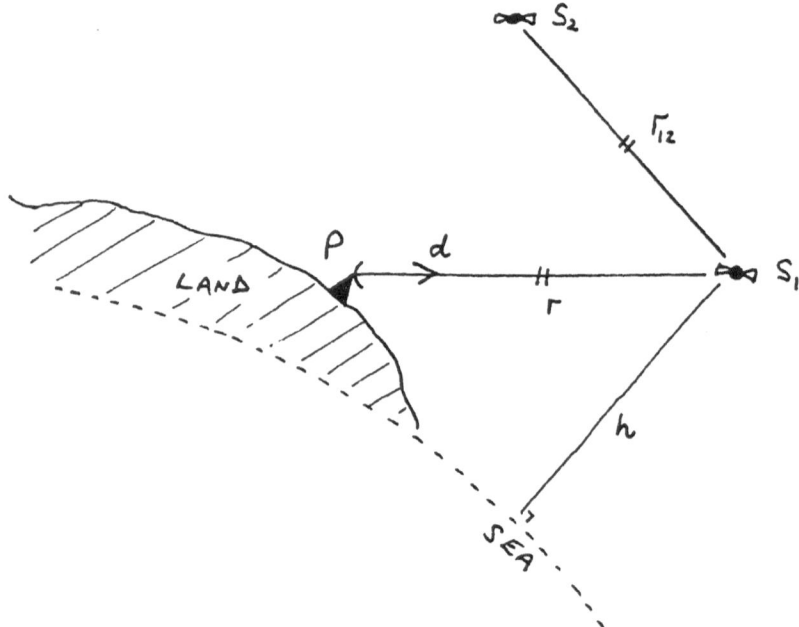

Fig. 1 Satellite geodesy observables

Obviously direction measurement can only take place at night and the satellite must either have a light on board or must be lit by the sun (in same way that the moon is). Bomford (1980) gives a full description of this technique but it is not considered further here because, although it does still have some rather specialist geodetic applications, it is completely obsolete as a positioning technique.

(ii) The distance or range, r, from P to S_1 can be observed in a number of ways (all based on the time of travel of an electromagnetic signal) and is probably the most useful observable in current satellite positioning. Three important distance measuring techniques are as follows.

(a) laser ranging: where a laser pulse is transmitted towards a satellite carrying a corner cube reflector and the return travel time measured. Laser ranging is currently extremely important in geodesy with a large number of geodynamic applications. Also it is the most accurate method for absolute positioning and for relative positioning over long distances.

(b) pseudo-ranging: where a satellite transmits a signal at a known time and the time at which it reaches a receiver at P is measured. GPS navigation is based on pseudo-ranging.

(c) phase measurement: where the receiver measures the phase of a signal received from satellite. This is not strictly a distance measurement but, rather like in an EDM instrument, when combined with a knowledge of the total number of complete cycles between the satellite and receiver, it can lead to a distance computation (although this is unlikely to be explicitly carried out). Currently precise relative positioning using GPS is based on measured phase.

(iii) The **radial velocity,** $\partial r/\partial t$, of the satellite with respect to the receiver can be observed by measuring the Doppler shift of a signal emitted by the satellite. The measurement process is actually identical to the foregoing phase measurement but the rate of change of phase is computed rather than using its instantaneous value. This is the main observable in positioning using the TRANSIT satellites.

(iv) The **height, h,** of the satellite above the earth's surface can be observed (by the satellite itself) by sending a radar pulse towards the ground and measuring the return travel time. The process is known as satellite altimetry and currently only operates effectively over very flat surfaces (sea, lake, ice etc). Its main geodetic application is in the determination of the earth's gravity field. It will not be further considered here.

(v) The **satellite to satellite distance,** r_{12}, or **radial velocity,** $\partial r_{12}/\partial t$ between S_1 and another satellite S_2 is an observable that is of great theoretical interest to satellite positioning. Such systems are currently not operational (missions are planned for the mid-1990's) but would be of great value in the determination of the earth's gravity field and in satellite orbit monitoring. These observables are not treated here but Taylor et al (1984) is recommended for further reading on this topic.

It is important to emphasise that Fig. 1 represents an idealised situation. No current satellite or ground observing system is capable of all of these measurements. The foregoing has been included simply as an overview of what is possible.

3.2 SATELLITE POSITIONING METHODS

Philosophically satellite positioning methods can be divided into three groups: **geometrical,** **dynamic** and **short-arc**. Geometrical methods solve the positioning problem by use of pure geometry as, for instance, in Fig. 2. Imagine a satellite S passing over four stations A, B, C and P, all of which can simultaneously observe, say, the distance to S. If the relative three-dimensional positions of A, B, and C are known, and P is unknown we can proceed as follows. Let all four stations simultaneously observe the distance to the satellite at points S_1, S_2 and S_3. Then distances

AS_i, BS_i and CS_i can be used with the known ground station coordinates to compute the satellite coordinates S_i (for i =1,3) so enabling distances S_1P, S_2P and S_3P to be used to compute the coordinates of P (from the now known satellite positions). The process, which used to be called trispheration, is extremely simple as no knowledge is required of the orbit (except to ensure the correct pointing of the distance measuring system) and it illustrates satellite positioning at its simplest level.

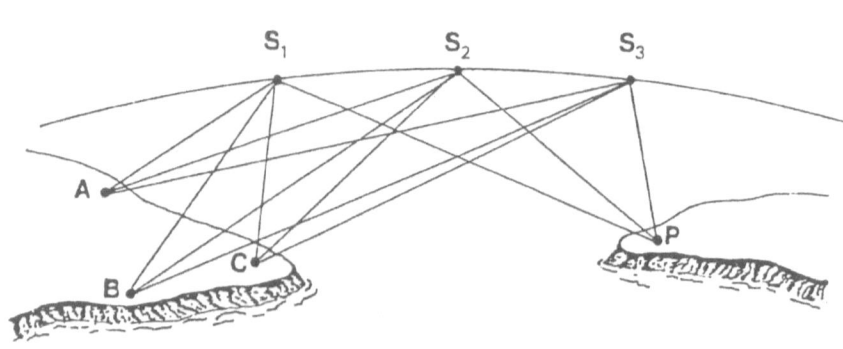

Fig. 2 Geometrical satellite positioning

To apply the purely geometrical approach in practice leads to very serious difficulties, e.g. four simultaneous observations are virtually impossible to arrange, four stations must be occupied to obtain the coordinates of one and (unless the stations are hundreds of kilometres apart) the geometry will be extremely poor. The opposite scenario is represented by the dynamic positioning case shown in Fig. 3. Here the orbit is modelled (by the use of orbit dynamics) and the satellite position at the time of observation predicted. Hence the position of a single ground station P can be found from, for example, distance observations PS_1, PS_2 and PS_3, to a satellite in three positions or to three satellites. The major problem with this method is that any error in the orbit prediction is transferred directly into the station position. It is worth noting here, however, that if two receivers are placed near to each other then dynamic positioning will lead to similar errors in each station position and the relative position between the two stations may well be extremely well determined. This procedure is sometimes known as **translocation**.

The third positioning strategy, short arc positioning, is a combination of the two foregoing methods. A number of receivers are deployed at a combination of known and unknown sites and observations made approximately simultaneously. The relative positions of the satellite along a short arc of the orbit are predicted (much simpler models are needed than those for dynamic positioning which must model the satellite over many revolutions) and a computation made for the relative positions of the receivers and perhaps some general parameters describing the short arc of the orbit.

Most "everyday" satellite positioning is nowadays carried out using the dynamic method. Short arc methods are, however, used when the highest accuracy is required and geometric methods reserved for very special applications where multi-station campaigns can be organised so that large numbers of simultaneous observations are made.

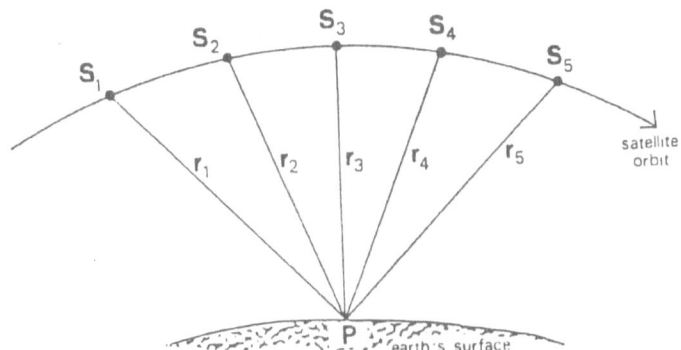

Fig. 3 Dynamic satellite positioning

3.3 THE GPS SYSTEM

GPS (sometimes referred to as NAVSTAR which is an acronym derived from NAVigation Satellite, Timing And Ranging) was conceived as a continuous multi-purpose positioning and navigation system. When fully operational it will consist of eighteen satellite (plus three in-orbit spares) arranged in six orbit planes with inclinations of 55°, and with 60° spacing in longitude. The satellites will have a height of about 20183 km and periods of twelve hours. The configuration is shown in Fig. 4 and is such that, for most points on the earth, there will be at least four (and sometimes up to nine) satellites in view at any time.

Each satellite will transmit two signals, which have been named L1 and L2 (because they are in the L-band of the electromagnetic spectrum) and have frequencies of 1575.42 MHz (154 x 10.23 MHz) and 1227.60 MHz (120 x 10.23 MHz) respectively. Both L1 and L2 carry a formatted data message (rather similar to the BE for TRANSIT) which contains the Keplerian elements and other information for satellite position computations. It also contains information to relate the GPS time system to Universal Time (previously GMT) and estimates of the state of the ionosphere to enable refraction corrections to be made. The satellites are tracked by five stations located at Kwajalein, Diego Garcia, Ascension, Hawaii and Falcon Air Force Station (AFS) in Colorado. who send their data to the master control station at AFS. There the data messages are computed for each satellite and sent to the most convenient "ground antenna" (located at Kwajalein, Diego Garcia and Ascension) for upload to the satellite. Normally uploads occur every eight hours but each satellite actually stores data for 14 days in case any problems arise with the tracking network. The orbital information currently has a standard error of about 20 metres which would, in fact, degrade to about 200 metres if not updated for 14 days.

Also modulated on L1 is a C/A (coarse acquisition, sometimes referred to as S, for standard) code and a P (precise) code, and modulated on L2 is just the P code. These codes are essentially pseudo-random binary sequences (i.e. sequences of zeros and ones) that are repeated every 1 millisec and 7 days respectively. They are in fact realised by a technique known as binary biphase modulation which results in 180° changes of phase of the carrier. The P code has a frequency of 10.23 Mhz (i.e. about 107 binary digits are transmitted every second with a physical spacing of about 30 metres) whereas the C/A code has a frequency ten times smaller with a spacing of about 300 metres between the changes of phase.

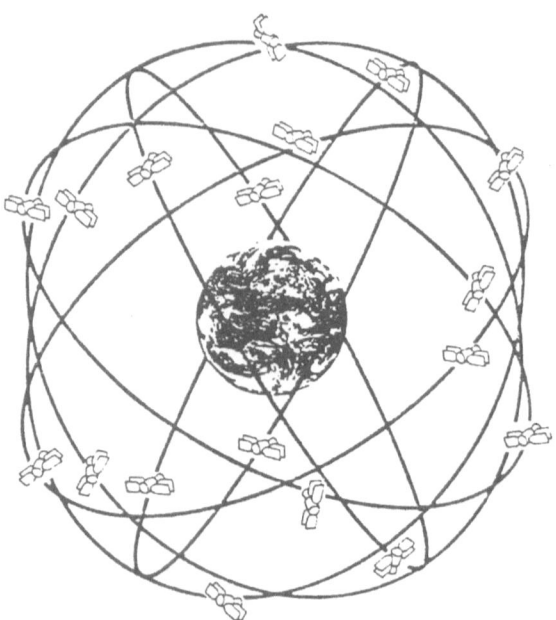

Fig. 4 The propsed GPS congifuration

3.4 PRINCIPLES OF GPS POSITIONING

The GPS signals can be used for positioning in a number of ways the two most important of which are **pseudo ranging** and **carrier phase measurement**.

Pseudo ranging is a technique whereby the range between a GPS satellite and a receiver is determined by use of the pseudo random timing codes. Essentially the receiver generates an identical code (called a **replica code**), at an identical time, to that produced by a particular GPS satellite. It is then compared, by a process known as cross-correlation, with the satellite signal. Cross-correlation involves moving the receiver-generated binary sequence in a **delay-lock loop** until it is exactly in phase with the one received. The amount by which it must be moved is then the time of travel of the satellite signal to the receiver (since they were generated at the same time). Multiplication by the velocity of light the yields the distance. The applications of pseudo range measurements are in navigation and error detection when carrying out phase measurements. They will not be considered further here.

Most geodynamics work with GPS is based on the so-called **continuous phase observable.** Essentially the phase observation process involves hetrodyning the incoming carrier signal (either L1 or L2, or both) with a signal generated in the receiver and making measurements with the resulting signal at specified times. The term continuous phase refers to the fact that as well as measuring the instantaneous value of the phase (in the range zero to 360°) the receiver also counts the number of complete cycles from some arbitrary starting time (i.e. it increments an integer counter every time the phase changes from 360° to zero). Of course in order to access the received carrier the receiver must remove the code (which is modulated by 180° changes of phase). This can be carried out either by a process known a "signal squaring" or by using a prior knowledge of the pseudo random binary sequence to reconstruct the original carrier (i.e. the carrier before the code was added). The former, **codeless** approach, has the advantage of not requiring any knowledge of the code and of leading to a more precise measurement (because the wavelength is halved) but the process destroys the data message and an external ephemeris (rather than a broadcast one) must be used. At present this creates considerable logistical dificulties so most receivers adopt the latter technique, but should the codes become unavailable at any time in the future the codeless approach may be more widely adopted.

Remondi (1985) has shown that the continuous phase observable, N[iaj], observed at station, a, and at epoch, i, from satellite, j, is given by

$$N[iaj] = ES[ij] + ER[ia] + I[aj] + f_j \ \Delta t \ [iaj] \tag{1}$$

where

ES[ij]   is a term describing the satellite clock error at epoch i,

ER[ia]   is a term describing the receiver clock error at epoch i,

I[aj]   is the **integer ambiguity**, which is the (arbitrary) integer value of the phase counter at the start of the observing period,

f_j is the frequency of the emitted signal, and

$\Delta t[iaj]$ is the time of travel of the signal from satellite, j, to receiver, a, at epoch, i.

The travel time, $\Delta t[iaj]$, is given by

$$t \ [iaj] = (1/c)\{(X_a - X_j) ^2 + (Y_a - Y_j) ^2 + (Z_a - Z_j\ ^2)\}^{1/2} \ + \Delta t^{tion} \ [iaj] + \Delta t^{trop} \ [iaj] \tag{2}$$

where $\Delta t^{tion} [iaj]$ and $\Delta t^{trop} [iaj]$ are the signal delays in the ionosphere and troposphere respectively. Substitution of (2) into (1) leads to (3) which is the general continuous phase observation equation for an observation at epoch, i; between satellite, j, and receiver, a.

$$N[iaj] - ES[ij] - ER[ia] - I[aj] - (1/c)\{(X_a - X_j)^2 + (Y_a - Y_j)^2 + (Z_a - Z_j\ ^2)\}^{1/2} - \Delta t^{tion} [iaj]$$
$$- \Delta t^{trop} [iaj] = 0 \tag{3}$$

3.5 RELATIVE POSITIONING BY GPS.

In a general relative positioning problem we may have m stations observing n satellites over p epochs as indicated in Fig. 5. This would lead to m x n x p observation equations of type (3), with 3m station coordinate (assuming no datum defect), n x p satellite clock error, m x p receiver clock error and m x n integer ambiguity parameters to be estimated. In a typical case of 3 stations, 4 satellites and 1000 epochs this would mean 12000 observation equations and 7021 parameters. The

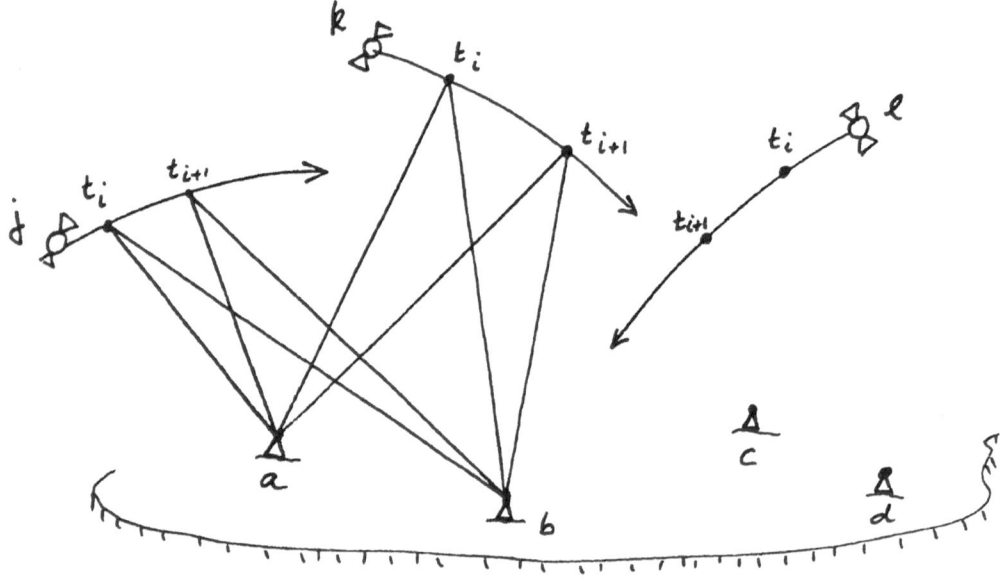

Fig. 5 GPS relative positioning

solution of such a system creates serious computational difficulties and, in practice, it is usual to reduce the number of parameters by differencing the observation equations (i.e. by subtracting them from each other). A number of strategies are possible based on combinations of differencing with respect to satellites, ground stations and time. Each removes a different group of parameters from the system as follows.

(i) Differencing with respect to satellites removes the receiver clock error term, $E_R[ia]$.
(ii) Differencing with respect to ground stations removes the satellite clock error term, $E_S[ia]$.
(iii) Differencing with respect to time removes the integer ambiguity $I[aj]$.

It is important to note that so long as the stochastic models are properly formed identical answers are possible from all strategies and the work saved in reducing the number of parameters is replaced by the work involved in dealing with non-diagonal weight matrices. Most commercially available software has, however, been written to run on portable microcomputers so that computations can be carried out in the field, and consequently it is usual for it to be based on the assumption that the differenced observation equations are uncorrelated. This point is explored in more detail in Ashkenazi and Yau (1986) and Beutler et al (1986).

The most commonly adopted strategy is to difference with respect to both satellites and ground stations to yield the so-called **double difference** observable. In the foregoing example this would lead to a system 6000 observation equations with 15 parameters (probably less in practice when a datum was introduced). The workload in least squares problems is, in general terms, related to the cube of the number of parameters so the double differencing process clearly leads to a considerable saving and usually means that the computations can be carried out in the field on a portable microcomputer. Note that the least squares procedure is usually adapted to force the integer ambiguity parameters to take integer values. Another commonly adopted strategy is to form triple difference observation equations, i.e. to difference with respect to time, satellites and ground stations. This process does not yield the highest quality results (one reason being that the integer ambiguities are no longer solved for and it is not possible to force them to integer values) but leads to an extremely simple system (with the only parameters being station coordinates) that is very useful for data validation.

3.6 STATUS OF GPS

GPS phase 1 (concept validation program) began in 1973 and, in 1979, phase 2 (system testing) started with the launch of a number of Block I satellites. Currently (1988) seven of these are operational. These are arranged in two planes with an inclination of 63° (the optimum inclination for the original plan of 24 satellites) and enable GPS positioning on most parts of the earth's surface for a few hours per day. The current plan is to begin the launch of the operational (Block II) satellites late in 1988 and to complete the 21 satellite constellation by 1992.

Currently both the C/A and P codes are generally available (i.e. the US Department of Defense have released details of the numerical processes employed to generate the pseudo random sequences). There is some doubt as to whether or not this will be the case for the Block II satellites with the latest information, Baker (1986), suggesting that only the C/A code will be generally available. If this is the case then there will be some (not insurmountable) difficulties in accessing the L2 signal as it only carries the P code. Also there is a possibility that the orbit included in the data message may be deliberately downgraded (by a factor of ten). These measures are aimed at denying, for military reasons, the use of high-accuracy, real-time GPS point positioning to "non-approved users". It is unlikely, however, that anything will be done to restrict surveyors in their determination of post-processed relative positions. They can always produce their own orbits and use the signal squaring technique to access the carrier for phase measurement.

3.7 MODELS FOR GPS RELATIVE POSITIONING

The quality of relative coordinates derived from GPS observations can be considerably enhanced by the data processing methods employed. The three major factors are the treatment of refraction, orbit modelling and data filtering and these will be treated briefly here.

Refraction occurs in two layers of the atmosphere: the ionosphere and the troposphere. Of the two the ionospheric effect is the largest. In order to account for this some GPS receivers measure on both of the two broadcast frequencies (note that to improve results from single frequency instruments, the systems broadcast estimates of the state of the ionosphere as part of their data message). A paeticular linear combinations of the measurements on these two frequencies are known to be free (to a first order - about 99%) of ionosheric refraction and it is these combinations which are used as the observables in the respective observation equations. For more details of this Clynch and Coco (1986) is recommended and Clynch and Renfew (1982) discusses techniques for dealing with the residual (about 1%) effects. Tropospheric refraction is usually treated by measuring the local meteorological conditions (temperature, humidity and pressure) and applying appropriate atmospheric models such as those due to Black or Hopfield, see for instance Hopfield (1980).

There are a number of different ways to treat the orbits of GPS satellites and the topic is rather too complex to be treated here in detail. Essentially the methods range from accepting the broadcast satellite positions as being perfect to completely ignoring them and carrying out a separate orbit integration based on a selected set of force models. A mid-way treatment is to assume that an individual broadcast orbit has errors that can be described by some form of mathematical transformation so that in general we can write

$$X_S^E = F_X (X_S^B, Y_S^B, Z_S^B, \text{transformation parameters})$$

$$Y_S^E = F_Y (X_S^B, Y_S^B, Z_S^B, \text{transformation parameters}) \qquad (4)$$

$$Z_S^E = F_Z (X_S^B, Y_S^B, Z_S^B, \text{transformation parameters})$$

where X_S^E, Y_S^E and Z_S^E are the required earth-fixed coordinates of the satellite and X_S^B, Y_S^B and Z_S^B are those that were computed from the information broadcast by the satellite. Substitution of (4) into the observation equations then introduces a new set of unknown parameters for each orbit and the subsequent solution involves both the station coordinates and the orbit transformation parameters. This procedure is known as **orbit relaxation** and is a simple example of the short arc positioning technique introduced earlier. For a more detailed discussion of the treatment of GPS orbits readers are referred to Beutler et at (1986).

Finally here the problem of data validation and data selection is mentioned. It is usual in geodetic and engineering work to filter the observed data extremely rigorously before accepting it for computation. A detailed discussion of this topic involves reference to a large number of statistical (and other) criteria and is clearly outside the scope of this work but two points are worth mentioning. Firstly much data selection is based on pure geometry. For instance observations below a specified altitude (typically 15°) are commonly rejected (due to refraction errors near to the horizon). Secondly continuous phase measuring techniques are subject to errors known as cycle slips which are caused by momentarily losing lock on a satellite. When the measurements resume the fractional part of the phase will be correct (i.e. equal to that which it would have been without a loss of lock) but the integer counting will be in error. The problem of detecting and repairing cycle slips is currently receiving much attention, see for instance Remondi (1985) and Cross and Ahmad (1988), but it is interesting to note here that if differences with respect to time are computed then the problem will be isolated to a single measurement. Hence cycle slips can be seen in individual residuals after a triple difference computation.

3.8 QUALITY OF GPS

The wave length of the L1 carrier is approximately 0.19 metres and observations with a resolution of 1° of phase (equivalent to 0.5 mm) are possible. Measuring resolution is hence not the limiting factor on accuracy with the result being a system whereby the standard errors of derived coordinate differences are proportional to the distances between the stations (with the major error sources being orbit prediction and atmospheric refraction). For very short distances (up to 1 km) standard errors of 1 mm have been achieved and the 1 part per million (ppm) standard error can be maintained up to several hundred kilometres so long as the correct equipment and processing models are used. Recently some authors, eg Mueller (1986), claimed that 0.1 ppm is achievable and that 0.01 ppm was a realistic goal for the future, indeed JPL are now claiming that 10mm is possible over distances up to 1000km, Blewitt (1988). Currently these accuracies need several hours of observing time but it is expected that only minutes of observation time will be required with Block II satellites, future equipment and observing techniques.

Of course the foregoing is something of an oversimplification as the accuracy achievable depends on a large number of factors including the type of equipment, whether on not dual frequency measurements are made, the prevailing state of the ionosphere, the slope of the line joining the two stations (steep slopes present greater tropospheric modelling problems), the data validation process, the length of observation period, the number of satellites observed, the orbit modelling process, etc. Also the ease with which the predicted future accuracy will be achieved (but probably not the accuracy itself) will depend on a number of political decisions that the US Department of Defence have yet to make. Nevertheless the quoted figures give an overall impression of what can be obtained today and what might be expected in the future.

3.9 SATELLITE LASER RANGING

SLR involves the transmission of a short laser pulse to a satellite carrying corner cube reflectors. The pulse is retrodirectively reflected by the satellite and the return travel time measured. If t is the return travel time and t the epoch of emission then the range, r, to the satellite at time $t + \Delta t/2$ is given by

$$r = c \, \Delta t / 2 + C + R \qquad (5)$$

where c is the velocity of light in a vacuum, C a calibration correction (usually derived by ranging to terrestrial targets) and R a tropospheric refraction correction (computed from measured meteorological observations).

Laser ranging systems are generally rather large, permanently mounted instruments such as the one at the Royal Greenwich Observatory described in detail in Sharman (1982). There are, however, a number of mobile systems that have been deployed (by USA, Holland and West Germany) for special positioning work associated with geodynamics, especially in the Eastern Mediterranean and California, see for instance Tapley et al (1985). At present (1988) it is estimated that about 20 permanent and 6 mobile systems are routinely operating.

Typically the lasers operate at about 10 Hz, i.e. ten pulses per second, with pulse lengths of about 150 psecs (equivalent to about 45 mm). An accurate pointing device is necessary to ensure that the telescope points in the correct direction and a computer controlled tracking system is needed so that it follows the satellite as it passes overhead. Note that since laser light is in the visible part of the spectrum laser rangers can only operate in cloudless skies. Ranging standard errors as high as 10 mm can now be achieved, although most systems are up to about five times worse than this.

There are currently about 12 satellites with corner cube reflectors of which three are regularly used for geodetic purposes: LAGEOS (USA), Starlette (France) and Ajisai (Japan). These three are all rather similar passive spherical objects with very high mass to surface area ratios to ease the problem of modelling drag and radiation presure effects.

SLR has a number of important geodetic applications of which positioning is the most relevant here. By collecting and processing, using methods such as in Sinclair and Appleby (1986), data from a variety of sites well distributed over the entire earth, absolute and relative positions with standard errors of only a few centimetres can be determined. Such positions have recently confirmed current rates of continental drift, eg Christodoulidis et al (1985), and now form the basis for the modern reference systems used in satellite positioning, see Mueller and Wilkins (1986).

3.10 VERY LONG BASELINE INTERFEROMETRY

VLBI is not a satellite system. It involves the measurement of the variation in the arrival times of radio signals from quasars at different observing sites (radio telescopes) around the world. Basically data is recorded and accurately time-tagged at each site and then cross-correlated, as explained for instance in Ashkenazi and McLintock (1982), with similar data at other sites. Extremely precise relative positions (typically 0.01 ppm) and earth rotation parameters can be determined and, as with SLR, the data contributes to the IERS and coordinate system definition.

Relative positions can also be used for the study of deformations of the earth due to continental drift and tidal phenomena. They are, however, of limited use on a global scale because there are very few radio telescopes around the world regularly devoting time to VLBI work. In USA, however, a number of mobile systems have been constructed and they are permanently deployed in California monitoring the San Andreas fault system.

4. The surveying estimation process

Rotations are most effectively determined via coordinates. The observations at each epoch being used first to estimate sets of independent coordinates. In order to estimate coordinates two models need to be specified: the functional model and the stochastic model.

The functional model is usually expressed as a set of linear equations

$$F(\bar{x}) = \bar{l} \tag{6}$$

where l is a vector containing the true values of the observed quantities and $\bar{x}$ is vector containing the true values of the required paremeters (coordinates, instrumental characteristics, coordinate system descriptors etc). As an example of a functional model for an observed horizontal distance in terms of plane (X,Y) coordinates we could write

$$\{(\bar{X}_j - \bar{X}_i)^2 + (\bar{Y}_j - \bar{Y}_i)^2\}^{1/2} = \bar{s}_{ij} \tag{7}$$

where $\bar{s}_{ij}$ is the observed distance from station i to j. For GPS relative positioning the functional model is given in (3), or a differenced version of it. Note that the real relationships between the parameters and the observed quantities may be extremely complex, they must for, instance, take account of the different planes in which angle measurements have been made (due to the varying direction of the vertical). A full study of this topic is well beyond this review and interested readers are referred to standard geodesy books such as Bomford (1980) and Vanicek and Krakiwsky (1986).

Functional models are linearised according to Taylor's theorem to give a set of equations

$$Ax = b + v \tag{8}$$

where $A = \partial F/\partial x$, x represents changes to parameter approximate values, $x°$, $b = l - f(x°)$ and v is a vector of *corrections* to the observations (sometimes called *residuals*).

The second model is the so-called stochastic model which essentially describes the statistics of the observational errors not included in the functional model. It expressed is in the form of a weight matrix W given by

$$W = C_1{}^{-1} \tag{9}$$

where C_1 is the covariance matrix of the observations, i.e. a matrix whose diagonal elements are the observation variances (standard deviations squared) and whose off-diagonal elements are the corresponding covariances. Methods for the assessment of stochastic models are discussed in Cross (1987).

The system (8) is overdetermined with an infinite number of solutions and the estimation strategy almost universally adopted is that of least squares. The least squares process is defined as the one that minimises a specified *objective* function, Φ, where

$$\Phi = v^T W v \tag{10}$$

Note that in the special case of uncorrelated measurements of unit variance , i.e. W=I, (10) reduces to

$$\Phi = v^T v = v_1{}^2 + v_2{}^2 + \ ... \ + v_n{}^2 \tag{11}$$

hence explaining the term *least squares*.

It has been shown in many places, for instance in Cross (1983) and Cooper (1987), that the least squares estimate of the parameters is given by the solution of the following *normal* equations

$$A^T W A \hat{x} = A^T W b \tag{12}$$

i.e.

$$\hat{x} = (A^T W A)^{-1} A^T W b \tag{13}$$

and that the corresponding least squares corrections, v, are given by

$$\hat{v} = A\hat{x} - b \ = \ \{A (A^T W A)^{-1} A^T W - I\} \, b \tag{14}$$

Associated with these least squares estimates are two covariance matrices.

$$C_{\hat{x}} = \hat{\sigma}_0{}^2 \, (A^T W A)^{-1} \tag{15}$$

and

$$C_{\hat{v}} = \hat{\sigma}_0{}^2 \, \{W^{-1} - A (A^T WA)^{-1} A^T\} \tag{16}$$

where $\hat{\sigma}_0{}^2$ is the so-called reference variance (sometimes referred to as u*nit variance, variance factor or variance of a measurement of unit weight)* and is given by

$$\hat{\sigma}_0{}^2 = (v^T W v) / (m-u) \tag{17}$$

Note in (17) m is the number of measurements and u the number of parameters. (m-u) is often denoted by r and is usually referred to as the *redundancy* or *degrees of freedom* of the particular problem in hand.

The foregoing are essentially the complete set of formulae for least squares estimation and quality analysis when dealing with problems that can be expressed as observation equations.. set-up and, as has already been implied, their derivation can be easily be found in a very large number of texts. It has been omitted here simply so save space.

4.1 THE SURVEYING DATUM PROBLEM

The foregoing estimation process assumes that the rank of the normal equation matrix is equal to m, the number of parameters, i.e. it assumes the the Cayley inverse, N^{-1}, of the matrix N, where

$$N = A^T WA \tag{18}$$

exists. For many surveying problems this is not the case because the parameters are not *estimable* from the measurements. For instance in a pure triangulation network (angles only) the overall orientation is not computable (and hence overall rotation could not be determined). In order to estimate orientation it is necessary to make a (in this example completely arbitrary) statement regarding the orientation *datum*. If this was not done the matrix N would be singular and the Cayley inverse could not be computed.

In general the number of datum quantities that must be defined will depend on the dimensions of the problem (one, two or three) and on the types of measurement available. For instance in a two-dimensional geodetic network with only angle measurements four quantities would be needed (usually the coordinates of two stations or the coordinates of one station plus a bearing and a distance). If the network also included distance measurements only three datum quantities would be required and so on. For three-dimensional problems up to seven datum quantities may be needed. These can be summarised as three for position, three for orientation and one for scale. If the measurements contain no datum information then the seven quantities may typically be two positions and one coordinate from a third point. An alternative combination would be one position, one distance, one direction (i.e. two quantities) and one "component" of another direction (say a bearing from north or a vertical angle). Of course a wide variety of combinations of measured data and arbitrarily specified quantities are possible when working in three-dimensions.

The number of required datum quantities is often referred to as the *defect* of N, where the defect is simply the order of N minus its *rank*.

In practice this problem is usually handled either by the inclusion of "ficticious" measurements or by removing selected parameters from the system of equations (and hence keeping them fixed at their provisional values). This can be described mathematically by adding a set of constraints

$$B\hat{x} = c \tag{19}$$

to the set of observation equations (19) which, following for instance Cooper (1987), leads to a set of normal equations that can be partitioned as follows.

$$\begin{bmatrix} A^T W A & B^T \\ B & 0 \end{bmatrix} \begin{bmatrix} \hat{x} \\ \hat{k} \end{bmatrix} = \begin{bmatrix} A^T W b \\ c \end{bmatrix} \tag{20}$$

Where k is a vector (with a number of elements equal to the defect) of Lagrangian multipliers often referred to, by surveyors, as correlatives. The solution of these normal equations can be written as

$$\begin{bmatrix} \hat{x} \\ \hat{k} \end{bmatrix} = \begin{bmatrix} M_{11} & M_{12} \\ M_{21} & M_{22} \end{bmatrix} \begin{bmatrix} A^T W b \\ c \end{bmatrix} \tag{21}$$

where

$$M_{11} = (N + B^T B)^{-1} N (N + B^T B)^{-1} \tag{22}$$

and

$$M_{12} = (N + B^T B)^{-1} B^T \qquad (= M_{21}{}^T) \tag{23}$$

Hence the least squares estimate for $\hat{x}$ can be written explicitly as

$$\hat{x} = M_{11} A^T W b + M_{12} c \tag{24}$$

This estimate is clearly not unbiased since taking expectations in (24) leads to

$$E\{\hat{x}\} = M_{11} N x + M_{12} c \tag{25}$$

which is not equal to x. Hence it follows that, in the presence of an arbitrarily defined datum the least squares esimates of the parameters are *biased*. In these circumstances it is natural to ask whether or not there are any functions of x that are unbiased. Let g be a vector of quantities derived from x according to the linear transformation

$$\hat{g} = G\hat{x} \tag{26}$$

then the expectation of $\hat{g}$ is given by

$$E\{\hat{g}\} = GE\{\hat{x}\} \tag{27}$$

and substitution of (25) yields

$$E\{\hat{g}\} = GM_{11} N \hat{x} + GM_{12} c \tag{28}$$

Hence it follows that for $\hat{g}$ to be unbiased (i.e. $E\{\hat{g}\}= g$) the following two conditions must be satisfied

$$G = GM_{11} N \quad \text{and} \quad GM_{12} = 0 \tag{29}$$

(29) can therefore be viewed as condition equations for unbiased functions of biased parameters. It can be shown, for instance in Caspary (1987), that both the corrections, $\hat{v}$, and the least squares

estimates of the measured quantities, $\hat{l}$, are functions that satisfy (29) so making them (and their covariance matrices) independent of any arbitrary choice of datum. This has important consequences for the statistical testing of the corrections when searching for model errors, and in the assessment of the quality of least squares estimates. It also is absolutely crucial, when comparing time dependent, to consider whether or not the two estimates are in the same datum and there are obviously great advantages in using estimable quantities whenever possible.

So far nothing has been said about the exact form of the constraint equations in (19). In a completely general case where the nature of the defect is completely unknown (i.e. the physical reasons for the datum defficiency are not understood), then it can be shown that the matrix B can be constructed such that its rows contain the eigenvectors corresponding to the negative eigenvalues of N (note that the number of negative eigenvalues of a matrix is equal to its defect). Actually constructing B in such circumstances is numerically rather difficult requiring the so-called *singular value decomposition* of N.

In surveying the physical nature of the defect is usually fully understood and predictable. In these circumstances the construction of B is rather straightforward. One of two strategies is usually applied: *minimum constraints datum* or *minimum trace datum*, the latter corresponding to the so-called free network case. Minimum constraints is the easiest strategy to apply and the one most commonly adopted in practice. It simply entails assigning fixed values to an appropriate subset of the parameters. For instance in a two dimensional network with two stations being fixed B would be given by

$$ B \;=\; \begin{bmatrix} 0 & 0 & .. & 1 & 0 & .. & 0 & .. & 0 & 0 & .. & 0 \\ 0 & 0 & .. & 0 & 1 & .. & 0 & .. & 0 & 0 & .. & 0 \\ 0 & 0 & .. & 0 & 0 & .. & 0 & .. & 1 & 0 & .. & 0 \\ 0 & 0 & .. & 0 & 0 & .. & 0 & .. & 0 & 1 & .. & 0 \end{bmatrix} \qquad (30) $$

And similarly for a three-dimensional network. To hold fixed a direction or distance then certain linear combinations of the appropriate parameters need to be specified.

It is important to note that it is not usual in practice to border the normal equations in order to achieve the minimum constraints solution. Far easier numerical strategies are available. The bordering technique was presented here simply because it enabled the foregoing general discussion of the datum problem.

5. Assessement of quality

Traditionally surveyors have divided their errors into three groups: random, gross and systematic errors. Hence to describe fully the quality of surveying information or measurements it is necessary to quantify the probability of there being errors of a specified size in all three of these classes. It is usual to use the words precision and reliability to describe the quality of a data set with respect to random and gross errors respectively but, surprisingly, there does not seem to be a widely accepted term to describe quality with respect to systematic errors. Here the term accuracy will be used for this purpose. Note that we will be assuming that the adopted functional and stochastic models are correct. This point is important because it will be seen that the usually adopted measures of quality depend only on these models.

5.1 MEASURES OF PRECISION

All commonly used measures of presion are computed from the covariance matrix of the parameters $C_{\hat{x}}$, given by (15). There are a number of useful measures of precision that can be computed. Two important points should, however, be made before describing these. Firstly some measures are local (i.e. they only describe the precision of a limited number of the parameters) whereas others are global (i.e. they describe the precision of the complete set of parameters with a single number). Obviously great care must be taken when interpreting global measures of precision as it is bound to be an over-simplification to attempt to describe the precision of a complete set of parameters in this way. They are, however, essential when comparing the overall quality of competing sets of measurements (as is necessary, for instance, when carrying out a design). Secondly care should be taken, as far as is possible, to use only estimable quantities when selecting quantities for precision assessment.

Cross (1987) contains a summary of some of the most commonly used measures of precision that can be computed from $C_{\hat{x}}$. In the detection of rotations the important measure would be the precision of a derived angle or azimuth, i.e. the standard deviation of an angle or azimuth derived from the coordinates estimated by the least squares process. It would be this angle or azimuth that would be compared with an earlier (or later) value to esimate a rotation rate.

In general if a quantity g is to be computed from $\hat{x}$ using the function

$$\hat{g} = f(\hat{x}) \tag{31}$$

then its variance is given by

$$\text{var}(\hat{g}) = f^T C_{\hat{x}} f \tag{32}$$

where

$$f = \left| \partial f / \partial x \right| \ x = \hat{x} \tag{33}$$

In the case of an azimuth from i to j computed in a plane coordinate model the appropriate function is

$$g = \tan^{-1}\{(\hat{X}_j - \hat{X}_i) / (\hat{Y}_j - \hat{Y}_i)\} \tag{34}$$

and for an angle at i between j and k

$$\hat{g} = \tan^{-1}\{(\hat{X}_j - \hat{X}_i) / (\hat{Y}_j - \hat{Y}_i)\} - \tan^{-1}\{(\hat{X}_k - \hat{X}_i) / (\hat{Y}_k - \hat{Y}_i)\} \tag{35}$$

5.2 MEASURES OF RELIABILITY

Essentially reliability is a measure of the ease with which outliers may be detected. The *internal reliability* of a particular measurement is the size of the marginally detectable gross error (clearly the smaller this is the more reliable is the measurement) and its *external reliability* is a measure of the effect of an undetected marginally detectable error on the parameters or on some information computed from them. It has been shown by, amongst others, Pelzer (1979) that high internal statistical testing procedure for outlier detection). Then it can be shown, e.g. in Cross (1983), that Δ_i^u is given by

$$\Delta_i^u = \delta^u / (e_i^T W C_V W_{e_i})^{1/2} \qquad (36)$$

which, in the case of uncorrelated measurements, reduces to

$$\Delta_i^u = \delta^u \, \sigma_i^2 / (\text{diag}_i \{C_V\})^{1/2} \qquad (37)$$

where σ_i^2 is the variance of the i-th measurement. δ^u is a function of the particular pdf and of α and β only (as given for instance in Cooper (1987) or Cross (1983)) and is a unitless quantity, for example for a normal distribution, when α is 0.05 and β is 0.10 then δ^u will be 3.85. Note that (37) is often written in the form

$$\Delta_i^u = \delta^u \, \tau_i \, \sigma_i \qquad (38)$$

where τ_i is often referred to as the tau factor for that particular measurement. It is simply the ratio of the standard error of an measurement to that of the corresponding correction.

For most practical problems it is external reliability that is crucial because we would like to know not only how large an undetected error might be but also its likely effect. Clearly the effect on the parameters of an error Δ_i^u in the i-th measurement is given by

$$\Delta X_i^u = \{C_X A^T W_{e_i}\} \Delta_i^u \qquad (39)$$

It then follows that should a quantity g be computed from the parameters as in (31) then the effect on g of a marginally detectable error in the i-th measurement is given by

$$\Delta g_i^u = f^T \Delta x_i^u \qquad (40)$$

Hence we have powerful and easily interpretable measures of external reliability which could, in the case of the detection of rotations be simply applied, i.e. we can compute directly the effect of an undetected gross error on the computed angles and azimuths used for rotations.

5.3 MEASURES OF ACCURACY

The assessment of quality with respect to systematic errors is much more difficult than that with respect to gross or random errors. It is usually only given a cursory treatment in the literature as it is assumed that all instruments have been properly calibrated so that systematic errors are removed before the measurements are presented to the estimation algorithm. This is, of course, usually the best practical procedure to adopt.

The approach suggested here for the assessment of accuracy is as follows. Firstly the functional model is extended to include any suspected systematic errors. Then the covariance matrix C_X is computed in the normal way and the variances of the parameters describing the systematic errors (often called bias parameters) examined in order to see how well they can be recovered. If they can be recovered with a small standard error then we can describe the parameters as being accurate with respect to these particular systematic errors. In some cases the modelling of a suspected systematic error will lead to singular set of normal equations indicating that the particular systematic error is not estimable from the measurements. The system would then be said to be completely inaccurate. Of course such a situation is usually easy to predict.

It is important to point out that bias parameters introduced for the purpose of assessing accuracy do not necessarily need to appear in the final estimation functional model (although of course they might). This is because the inclusion of parameters that describe biases that do not exist, often called over parameterisation, will tend to reduce the corrections and flatter the quality of the measurements and hence of the parameters (because the reference variance will be reduced). Here they are being introduced simply to see whether or not they are estimable and, if so, how well they can be estimated. It is obvious, for instance, that if a particular scale error cannot be estimated from the estimation process then any scale error would be undetectable and the resulting parameters would be completely inaccurate with respect to scale.

6. Optimal design methods

The term design is used in this context to describe the process of selecting surveying observation schemes to detect rotations in the most effecient manner. More generally the optimal design problem in surveying is as follows. Given the required quality (precision, reliability and accuracy) of a set of parameters that are to be estimated, find the set of measurements that will achieve this with the least cost.

Modern surveying techniques have greatly increased the importance of optimal design. For instance, in the past almost all surveying measurements required lines of sight between terrestrial points and usually the topography was such that there really was very little choice with respect to what could be measured. Also cost could rarely be properly modelled due to uncertainties with respect to the weather, travel times and communications. Developments such as satellite positioning have removed most of these problems and given tremendous scope to the designer of surveying measurement schemes.

It is important to note that all design procedures are essentially based on the fact that all quality measures dicussed in 5. are based on two covariance matrices C_x and C_v, given in (15) and (16). Strictly these cannot be computed until after the measurements have been made as they are dependent on the reference variance which in turn depends on the corrections. Their corresponding cofactor matrices, Q_x and Q_v, can, however, be computed from

$$Q_x = (A^T WA)^{-1} \qquad (41)$$

and

$$Q_v = W^{-1} - A (A^T WA)^{-1} A^T \qquad (42)$$

and it on these that all design procedures are based. After measurement has taken place it is necessary to check that the achieved reference variance is indeed insignificantly different from unity and hence that the design quality has been achieved.

There is extensive literature on optimal design. Readers are referred to Cross (1987) and Alberda (1980) for reviews of methods appropriate to terrestrial networks. Also Grafarend and Sanso (1985) is recommended as a general textbook on the topic. Here a brief summary of the two main classes of methods is given.

6.1 COMPUTER SIMULATION

Computer simulation, sometimes called preanalysis, is by far the most commonly employed technique for optimal design problems. The method essentially involves selecting a possible measurement scheme (set of measured quantities along with their expected covariance matrix) and computing $Q_{\hat{x}}$ and $Q_{\hat{v}}$ using (41) and (42) respectively. The selected quality criteria are then computed and compared with the requirements. The scheme is then iteratively altered until it just satifies the criteria. If more than one such scheme is possible the final choice is then made on a basis of cost.

The method has been used for around twenty years now and has two special advantages: it is completely general with respect to the allowable quality criteria and it is bound to lead to a sensible (realisable) scheme (because only such schemes would be simulated). On the debit side is the fact that it is not possible to be sure the the optimal scheme will be amongst those simulated and the fact that rather extensive human interference is required (some may consider this an advantage!). The method is particularly attractive when used in an interactive graphics environment, as for instance described in Mepham (1983). Then the design can be drawn, and interactively altered, (with a light pen, mouse or digitiser) leading to a continuous display of quality. In such circumstances quality measures are best represented graphically and great care is needed with regard to symbolisation, especially in the case of reliability and accuracy.

6.2 ANALYTICAL DESIGN

The term analytical design is used to describe a method that solves a particular design problem by a unique sequence of mathematical steps. In contrast to simulation methods it leads to a procedure that does not require human intervention (although it may involve iteration). To date no complete analytical solution to the optimal design problem has been found. Progress has, however, been made with certain su'bsets of the problem. In particular the so-called second order design problem has received a great deal of attention. In this problem the quantities to be measured are specified and a solution made for the necessary stochastic model. As far as precision is concerned we hence have to solve for the matrix W in

$$(A^T WA) = Q_{\hat{x}} \tag{43}$$

A large number of strategies for solving (43) have been suggested, most being based on either least squares or the methods of operations research (especially linear and quadratic programming). A major practical problem with (43) is the control of the solution with respect to realisation, it could be that measurements may be required with precisions that cannot be achieved in practice (sometimes measurements are even required with negative variances!). Also very few of the methods can also incorporate reliability and accuracy .

Most of the second order design approaches require that the precision criteria be expressed in the form of a criterion matrix although Cross (1986) descibes a linear programming procedure that allows the precision criteria to be in the form of variances of estimable derived quantities. Also problems of realisation and reliability can be handled (albeit in a limited manner).

The second order design problem is rather special and it may appear at a first glance that its solution is not of great significance in the light of the overall optimal design problem which also includes the much greater first order design problem (the selection of the measurements, i.e. the solution of (43) for the design matrix A). In practice, however, the second order design can help with the first order design when used in an appropriate iterative scheme. The procedure is to start with all feasible measurements, solve for their required standard errors and discard any whose required standard errors are large (because they will not then contribute to the overall precision and reliability). The process leads to an optimum subset of the original list of measurements.

In conclusion it must be emphasised that analytical methods are rarely used in pratice and most design is carried out by computer simulation. A great deal of research is, however, being carried out with analytical methods as future surveyors are bound to demand methods that can guarantee optimality with the minimum of human interference.

7. Conclusions

There are now a wide variety of terretrial and space methods that can be used for the determination of coordinates. Their accuracies can now approach a few millimetres over very short distances (up to a few km) and 1ppm and 0.1ppm respectively over longer lines. Space methods can achieve 0.01ppm over continental distances (over 1000km).

Methods for the detection of rotations from surveying observations have been presented. The estimabiltiy of the azimuths and angles from which such rotations might be computed has been discussed and quantitative measures of quality, suitable for describing their applicability to geodynamic problems, presented.

References

Alberda, 1980, A review of analysis techniques for engineering surrvey control schemes. *Proceedings of the Industrial and Engineering Survey Conference.* City University, London, p6.2/1-42.

Ashkenazi, V., Cross, P. A., Davies, M. J. K., and Proctor, D. W., 1972, The readjustment of the retriangulation of Great Britain and its relationship to the European terrestrial and satellite triangulation networks. OS Professional Paper, No 24, 51pp.

Ashkenazi, V. and McLintock, 1982, Very long baseline interferometry: an introduction and geodetic applications. *Survey Review*, vol 26, no 204, p279-288.

Ashkenazi, V. and Yau, J. 1986, Significance of discrepancies in the processing of GPS data with different algorithms. *Bulletin Geodesique*, vol 60, no 3, p229-239.

Baker, P. J. 1986, Global Positioning System policy. Proceedings of the Fourth Geodetic Symposium on Satellite Positioning, University of Texas, p51-64.

Beutler, G., Gurtner, W., Rothacher, M., Schildknecht, T., and Bauersima, I.,1986, Determination of GPS orbits using double difference carrier phase observations from regional networks. *Bulletin Geodesique*, vol 60, no 3, p205-220.

Beutler, G., Gurtner, W., and Bauersima, I., 1986, Efficient computation of the inverse of the covariance matrix of simultaneous GPS carrier phase difference observations. *Manuscripta Geodaetica*, vol 11, no 4, p249-255.

Blewit,T. J. 1988, JPL GPS activities. *Proceedings of the GPS Workshop*, Darmstadt (in press).

Bomford, G., 1980, Geodesy. 4th edition, Clarendon Press, Oxford, 855pp.

Burnside, C. D., 1982, Electromagnetic distance measurement. Granada Publishing, 204pp.

Caspary, W. F.,1987, Concepts of network and deformation analysis. Monograph 11, School of Surveying, University of New South Wales, 183pp.

Christodoulidis, D. C., Smith, D. E., Kolenkiewicz, R., Klosko, S. M., Torrence, S. M., and Dunn, P., 1985, Observing tectonic plate motions and deformations from satellite laser ranging. *Journal of Geophysical Research*, vol 90, no B11, p9249-9264.

Clynch, J. R. and Renfew, B. A. 1982, Evaluation of ionospheric residual range error model. Proceedings of the Third Geodetic Symposium on Satellite Doppler Positioning, University of Texas, p517-538.

Clynch, J. R. and Coco, D. S., 1986, Error characteristics of high quality geodetic GPS measurements: clocks, orbits and propagation effects. *Proceedings of the Fourth Geodetic Symposium on Satellite Positioning*, University of Texas, p539-556.

Cooper, M. A. R., 1987, Control surveys in civil engineering. Blackwell's Scientific Publications, Oxford, 381pp.

Cross, P. A.,1983, Advanced least squares applied to position fixing. North East London Polytechnic, Department of Land Surveying, Working Paper No 6, 205pp.

Cross, P. A., 1986, The design of surveying measurement processes. XVIII FIG Congress, paper 501P.3, 23pp.

Cross, P. A.,1987a, Computer aided design in surveying. *Chartered Land and Mineral Surveying*, Vol 5, No 9, p466-476.

Cross, P. A., 1987b, The assessment and control of the quality of industrial and engineering surveying information. *Proceedings II Industrial and Engineering Surveying Conference*, London, p66-83.

Cross, P. A. and Ahmad, N., 1988, Field validation of GPS phase data. Proceedings of the GPS Workshop, Darmstadt (in press).

Grafarend, E. W., and Sanso,F., 1985, Optimisation and design of geodetic networks. Springer Verlag, 606pp.

Hopfield, H. S. 1980, Improvements in the tropospheric refraction correction for range measurement. *Philosophical Transactions of the Royal Society of London*, A294, p341-352.

Mepham, M. P. A. 1983, Computer aided survey network design. MSc Thesis, University of Calgary, Division of Surveying Engineering, 98pp.

Mueller, I I (1986) From 100m to 100mm in (about) 25 years. Proceedings of the Fourth Geodetic Symposium on Satellite Positioning, University of Texas, p6-20.

Mueller, I. I., and Wilkins, G. A., 1986, On the rotation of the earth and the terrestrial reference system: joint summary report of the IAU/IAG Working Groups MERIT and COTES. *Bulletin Geodesique*, vol 60, no 1, p85-100.

Ordnance Survey, (1980) Report of investigations into the use of satellite Doppler positioning to provide coordinates on the European Datum 1950 in the area of the North Sea. *Ordnance Survey Professional Paper*, no 30, 32pp.

Pelzer, H., 1979, Criteria for the reliability of geodetic networks. In Optimisation of the Design and Computation of Control Networks, ed Halmos F, and Somogyi J, Akademiai Kaido, Budapest, p553-562.

Remondi, B. W.,1985, Global Positioning System carrier phase: description and use. *Bulletin Geodesique*, vol **59**, no 4, p361-377.

Sharman, P., 1982, The UK satellite laser ranging facility. *SLR Technical Note*, no 1, *Royal Greenwich Observatory*, 6pp.

Sinclair, A. T., and Appleby, G. M., 1986, SATAN - programs for the determination and analysis of satellite orbits from SLR data. *SLR Technical Note*, no **9**, *Royal Greenwich Observatory*, 14pp.

Tapley, B. D., Schultz, B. E., and Eanes, R. J., 1985, Station coordinates, baselines and earth rotation from LAGEOS laser ranging: 1976-1984. *Journal of Geophysical Research*, vol **90**, no B11, p9235-9248.

Taylor, P. T., Keating, T., Kahn, W. D., Langel, B. A., Smith, D. E., and Schnetzler, C. C., 1984, Observing the terrestrial gravity and magnetic fields in the 1990's. *EOS*, vol **64**, no 43, p609-611.

Vanicek, P and Krakiwsky, E., 1986, Geodesy: the concepts. 2nd edition, Elsevier, Amsterdam, 697pp.

Walcott, R. I., 1984, The kinematics of plate boundary through New Zealand: a comparison of short and long-term deformations. *Geophysical Journal of the Royal Astronomical Society*, vol **79**, p613-633.

GEODETIC MEASUREMENTS OF CONTINENTAL DEFORMATIONS: PROJECTS AND FIRST RESULTS

E. GEISS, CH. REIGBER, P. SCHWINTZER
German Geodetic Research Institute (DGFI)
Dept. 1: Theoretical Geodesy
Marstallplatz 8
D--8000 München 22
Fed. Rep. Germany

ABSTRACT. Space geodetic techniques such as Very Long Baseline Interferometry (VLBI), Satellite Laser Ranging (SLR) and microwave range and range--rate tracking techniques (GPS) are nowadays capable of observing the distance between points on the earth's surface with centimeter accuracy, even if these points are hundreds or thousands of kilometers apart. From repeated measurements over several years significant changes of intercontinental baselines were found. In this paper we present an overview of international projects for monitoring horizontal crustal movements and examine first results of distance changes as obtained from the analysis of SLR and VLBI data. They show in general a good agreement with the rates of motion according to geophysical plate models. There are, however, also some intraplate baselines with significant distance changes that give clear indications of continental deformations. In addition, results for some lines across plate boundaries show characteristic deviations from the rigid plate model. Finally, some possibilities for interpolating the observed point motions are discussed. This stresses especially the importance of a careful selection of the observation sites in connection with the expected motion.

1. Introduction

The direction and velocity of motion of the major lithospheric plates in the geologic past can be deduced from a variety of observations: investigation of the magnetization of crustal rocks (paleomagnetism, marine magnetic anomalies), seismological observations and results of geology, neotectonics and others. For the period covering the past 3 million years models of the "contemporary plate motion" have been deduced e.g. by Minster and Jordan (1978) and Chase (1978) using sea floor spreading data, orientation of transform faults, and slip vectors of earthquakes. These models, describing the motion of the plates by a set of relative plate rotation vectors, suppose that the plates are rigid in themselves. Results, coming mainly from paleomagnetic investigations have proved, however, that there has taken place a large amount of internal deformation and block rotation in many continental areas (McClelland et al., 1986). This implies that for these areas the concept of rigid plates has to be rejected or greatly modified.

Highly precise repeated geodetic distance measurements between and within the major lithospheric blocks as well as within local geodetic networks can provide in this connection a number of important information. Examples are: Are the station motions that were observed between points on different plates in accordance with the rates from geophysical models for the past

C. Kissel and C. Laj (eds.), Paleomagnetic Rotations and Continental Deformation, 69–81.

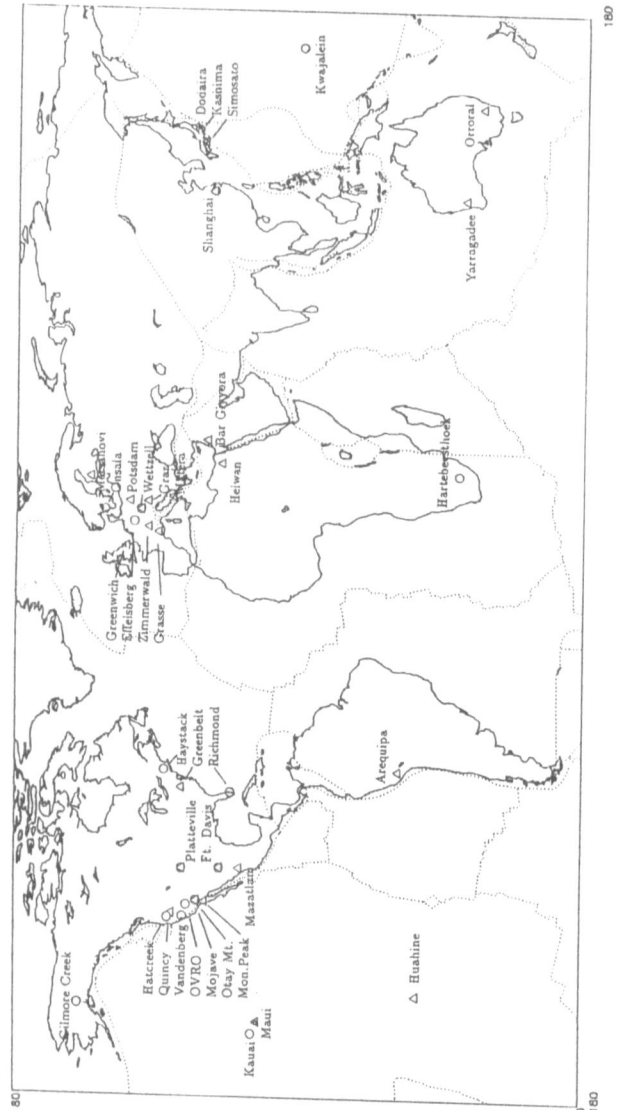

Figure 1: Global distribution of fixed SLR sites (triangles) and VLBI antennas (circles).

few million years? To what extent do the plates behave rigid? What motioncan be observed at well defined plate boundaries, what at more diffuse boundaries? Closely related to this is the question of the appropriate interpolation between the measurements that are carried out at discrete points.

2. Methods

2.1. SATELLITE LASER RANGING (SLR)

The principle of satellite laser ranging consists in the measurement of the travel time of a laser pulse from the ground station to a retroreflector equipped satellite (for example LAGEOS: Laser Geodynamics Satellite) and back. Analyzing ground--to--satellite range data from a network of stations in a dynamic measurement reduction process allows to derive the coordinates of the observing stations with high accuracy (few centimeters). Presently a global network of about 20 fixed stations is operating. Figure 1 depicts the present distribution of these stations around the globe.

In addition a few mobile Laser Ranging Systems are used for regional investigations. Especially for monitoring the complex situation in the Mediterranean area a Mediterranean Laser network has been installed (MEDLAS: Mediterranean Laser Campaign, Figure 2a.). This network, consisting of 14 different points in Turkey, Greece and Italy, has been observed for the first time in 1986/87 by 3 mobile systems. A preliminary evaluation of the 1986 observations revealed point accuracies of about 2 cm. A similar network for regional investigations has been established along the San Andreas Fault (SAFE: San Andreas Fault Experiment) within the framework of NASA's

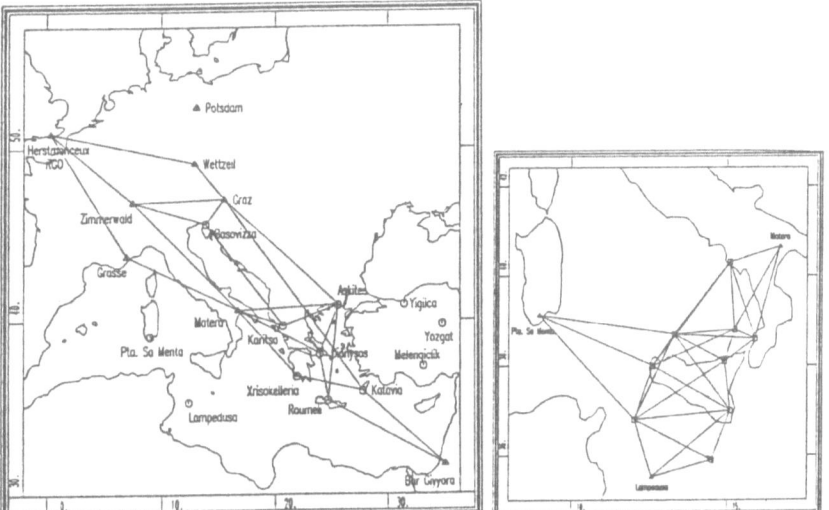

Figure 2. a.: Observation network for the Mediterranean Laser campaign (MEDLAS) in 1986. Sites occupied in 1987 are also shown. Triangles: fixed SLR sites; circles: mobile SLR sites.
b.: Observation network for a densification of the MEDLAS net in the Central Mediterranean using Microwave techniques (GPS). Triangles: SLR sites; squares: GPS stations.

Crustal Dynamics Project (CDP).

2.2. MICROWAVE METHODS USING SATELLITES

Microwaves are also used to carry out geodetic measurements (see e.g. Moritz and Mueller, 1987; Thornton et al., 1986). Using this technique either travel-time measurements (similar to SLR) are carried out, or phase shifts or the frequency shift of the microwaves that areemitted by the satellite is measured (Doppler effect). This method isespecially suitable for regional networks with interstation distances of less than 1000 km. Our institute is presently involved in two measuring projects using the Global Positioning System (GPS): one in the area of the Calabrian Arc (Southern Italy, Figure 2b.) and one in Venezuela (Drewes et al., 1987, 1988).

2.3. VERY LONG BASELINE INTERFEROMETRY (VLBI)

This technique is described in more detail in Herring (1986) and Moritz and Mueller (1987). In principleone measures the travel time differences of the signals from extragalacticradio sources (quasars) to various radio telescopes. As a result, the straight distances between the participating stations can be computed. Another goal of VLBI observations for geodetic purposes is the determination of earth rotation parameters in the framework of the IRIS project (IRIS: International Radio Interferometric Surveying). In this project stations in Northamerica, Europe and Japan are participating. The sites of fixed VLBI antennas for geodetic purposes are given in Figure 1. In the U.S.A. a number of points have also been observed by mobile VLBI systems.

3. Data

3.1. SLR

In this paper we mainly use the results of the DGFI solution SSC(DGFII)87L02, prepared by Reigber et al. (1988a,b). This data set is based on the analysis of 7 years of LAGEOS observations (1980--1986) and contains earth rotation parameters for this period and annual station coordinates for about 55 tracking sites. In this solution spherical distances and their annual rate of change are only given, if the line has been observed at least during 4 years with a minimum of 25 LAGEOS passes for each single year. This led to a set of 98 lines connecting 15 stations. The mean r.m.s. for the rates of change is about 2.9 cm. For further computations we used only those lines with an r.m.s. of less than 4 cm (70 lines). In some cases we made also use of the corresponding NASA SLR solution SL7 (Smith et al., 1987).

3.2. VLBI

As the data from the SLR solution included only few intraplate lines, wedecided to use also results from VLBI experiments. Therefore we examined the recent results of the NASA solution GLB122 for fixed VLBI stations and GLB171for mobile stations, published by Ryan and Ma (1987) and Ma and Ryan (1987). From the baselines and their annual rates of change given in these reports, we usedonly those, that were observed at least in 5 experiments over a time intervalof at least 2.5 years. This led to a set of 69 baselines between 24 stations.These baselines are all within one plate or connect stations on directly adjacent plates.

4. Comparison with a Plate Tectonic Model

4.1. GENERAL COMPARISON

In this section we compare the geodetically determined rates of distance changes between the different stations with the corresponding rates that result from a geophysical plate tectonic model. For this purpose we use in the following the model of Minster and Jordan (1978) as a reference. Test computations using the global plate model of Chase (1978) revealed only minor differences. Usually the stations could unequivocally be ascribed to specific tectonic plates, the exceptions being the SLR stations in Matera, Italy and Simosato, Japan and the VLBI station Kashima, Japan. We ascribed all three stations formally to the Eurasian plate.

First we compute the correlation coefficient r between results of the SLR data analysis and the corresponding rates of the Minster--Jordan model. This leads to a correlation coefficient of r = 0.8 with the slope of the regression line being about 0.9 (compare Figure 3a.). An analogous investigation of the VLBI derived distance changes yielded a correlation coefficient of r = 0.9 and a slope of the regression line of 0.7 (Figure 3b.).

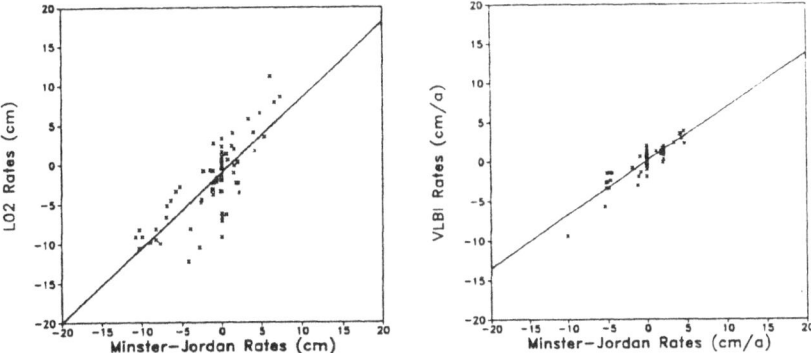

Figure 3. a.: Regression between SLR baseline rates and the corresponding rates according to the Minster-Jordan plate model. b.: Same as a. using VLBI rates.

From these results we can draw some preliminary conclusions. It can first be confirmed that it is possible to monitor horizontal crustal movements by high precision geodetic space techniques. The high correlation coefficients between the geodetic observations and the plate tectonic model support the principal applicability of such a model also for the present-day plate motion. The slope of the regession line through the data points is in all considered cases less than 1 (SLR--DGFI and SLR--NASA solution 0.9, VLBI--NASA solution 0.7). The accuracy of the data and their geographical distribution, however, do not allow at present the conclusion, that this is indicative for a generally lower rate of actual lithospheric motion.

4.2. DIFFERENCES BETWEEN THE RESULTS OF GEODETIC OBSERVATIONS AND THE RIGID PLATE MODEL

In this section we examine those baselines, which we suspect not to agree with the Minster-Jordan model. Thereby we assume that rates of change that agree within the two sigma error with the

Minster-Jordan model do not contradict this model (which does not necessarily mean that they support it). This criterion is certainly not very strict, it is shown, however, that those lines not fulfilling it, usually reveal a quite consistent behaviour. Nevertheless the results obtained can not be regarded as final but as first clues.

Out of the 167 baselines investigated (SLR and VLBI), 49 showed significant deviations in the sense defined above (Table 1). From these 24 are intraplate lines, and 25 are interplate lines. The intraplate lines belong to Eurasia (5), the Pacific plate (6) and the Northamerican plate (13). Interplate lines were only analysed for Eurasia - Northamerica and Pacific - Northamerica, as only in these cases a sufficient number of lines was available.

4.2.1. Eurasia. Within Europe only the line Wettzell (F.R.G.) - Matera (Italy) reveals a decrease in length that may be significant: -0.3 cm/a according to the SLR--DGFI solution, -0.5 cm/a according to the SLR-NASA solution. It has to be remarked, however, that this value is only statistically significant for the NASA solution (error: $\pm$ 0.2 cm/a (NASA), $\pm$ 0.7 cm/a (DGFI)). If this motion is real, it would be an indication for the northward movement of the Adriatic block with respect to Eurasia.

Striking distance changes can be observed between the Laser station at Simosato (Japan) and the European sites. For these interstation connections both SLR solutions show a very consistent decrease of the spherical distance of about -6...-7 cm/a ($\pm$ 2.5 cm/a). This is astonishing, as from the tectonic situation an increase of distance due to back arc spreading of the Japan Sea would be expected. It has to be stressed that these results depend from only one station and could be subject to undetected systematic errors. Comparable results from VLBI experiments that fulfill our data selection criterion are not available, nevertheless, Ryan and Ma (1987) report for two years of observation a rate of change for the line between the VLBI sites in Wettzell (F.R.G.) and Kashima (Japan) of -4.7 cm/a (recomputed for spherical distance). This would be in the range of the SLR derived solution.

4.2.2. Intraplate deformation of the Pacific Plate. Six baselines within this plate show obvious differences to the Minster-Jordan model (i.e. no change). Two of these (from SLR observations) cross large parts of the Pacific plate, while the others (VLBI lines) are confined to the continental part of this plate west of the San Andreas Fault (SAF). All six rates are positive (increase of baseline length). It is striking, that the annual relative length change (strain rate e) of those lines in the vincinity of the plate boundary exceeds the transpacific ones by about a factor of ten ($\varepsilon = 2...5.10^{-8}$ a^{-1} and $\varepsilon = 3...4 . 10^{-9} a^{-1}$, respectively).

The lines in California show as well a very specific behaviour (Figure 4.): The distance changes between stations that are situated in about the same distance from the SAF (JPL-Pinflats, Mon. Peak-Vandenberg) are distinctly smaller than in those cases where one station is close to the SAF and one further away. This is obviously due to internal deformation at the plate boundary. The general pattern shows, that the stations JPL, Pinflats and Mon. Peak, that lie - referred to the motion of the Pacific plate - in front of the bend of the SAF, move "too slow" compared to Vandenberg. This proves, that in the vincinity of plate boundaries a significant proportion of the plate motion is dissipated into a relatively wide area. This is also in accordance with the results of other authors (Minster and Jordan, 1987; Weldon and Humphreys, 1986).

4.2.3. Intraplate deformation of the Northamerican plate. According to our analysis baselines that span over almost the whole Northamerican continent from E to W show no or only small significant

Line		Distance (km)	Obs. Rate (cm/a)	M/J Rate (cm/a)	Solution
Table 1: Interstation distance changes with evident differences to the Minster–Jordan model.					
Within Europe:					
Wettzell	– Matera	990	-0.5 ± 0.2	0.0	SLR–NASA
Wettzell	– Simosato	9260	-6.2 ± 2.0	0.0	SLR–DGFI
Graz	– Simosato	9280	-6.6 ± 0.8	0.0	SLR–DGFI
Greenwich	– Simosato	9690	-9.1 ± 0.8	0.0	SLR–DGFI
Matera	– Simosato	9700	-7.1 ± 1.7	0.0	SLR–DGFI
Within Northamerica:					
Ft. Davis	– Mazatlan	850	-1.9 ± 0.3	0.0	SLR–DGFI
Gilm. Creek	– Mojave	3870	-1.9 ± 0.3	0.0	VLBI–NASA
Ft. Davis	– Hatcreek	1940	1.0 ± 0.2	0.0	VLBI–NASA
Platteville	– Hatcreek	1420	1.0 ± 0.4	0.0	VLBI–NASA
Ft. Davis	– Haystack	3170	-0.6 ± 0.1	0.0	VLBI–NASA
OVRO	– Haystack	3990	0.4 ± 0.1	0.0	VLBI–NASA
Ft.Davis	– Mojave	1310	0.5 ± 0.1	0.0	VLBI–NASA
Ft. Davis	– OVRO	1510	0.8 ± 0.1	0.0	VLBI–NASA
Ft. Davis	– Richmond	2370	-0.6 ± 0.2	0.0	VLBI–NASA
Mojave	– Yuma	360	0.6 ± 0.2	0.0	VLBI–NASA
Platteville	– OVRO	1220	1.6 ± 0.5	0.0	VLBI–NASA
Quincy	– OVRO	380	-0.5 ± 0.2	0.0	VLBI–NASA
Blackbutte	– Mojave	210	0.5 ± 0.2	0.0	VLBI–NASA
Within the Pacific:					
Mon. Peak	– Huahine	6610	1.6 ± 0.6	0.0	SLR–DGFI
Maui	– Huahine	4180	2.6 ± 0.5	0.0	SLR–NASA
JPL	– Pinflats	170	0.5 ± 0.2	0.0	VLBI–NASA
JPL	– Vandenberg	230	1.2 ± 0.2	0.0	VLBI–NASA
Mon. Peak	– Vandenberg	430	1.0 ± 0.3	0.0	VLBI–NASA
Pinflats	– Vandenberg	400	2.0 ± 0.3	0.0	VLBI–NASA
Eurasia – Northamerica:					
Onsala	– Ft. Davis	8570	1.0 ± 0.3	1.9	VLBI–NASA
Onsala	– Richmond	7780	-0.1 ± 1.0	2.0	VLBI–NASA
Wettzell	– Haystack	6240	1.2 ± 0.3	2.2	VLBI–NASA
Wettzell	– Ft. Davis	9190	0.3 ± 0.5	2.1	VLBI–NASA
Northamerica – Pacific:					
Quincy	– Mon. Peak	880	-2.8 ± 0.4	-5.3	SLR–DGFI
Quincy	– Huahine	7010	-0.8 ± 0.6	-2.3	SLR–DGFI
Greenbelt	– Huahine	9860	4.9 ± 0.5	1.3	SLR–NASA
Mazatlan	– Mon. Peak	1440	3.2 ± 0.9	5.4	SLR–NASA
Mazatlan	– Huahine	6570	4.0 ± 0.6	1.4	SLR–DGFI
Blackbutte	– Vandenberg	460	3.0 ± 0.5	4.3	VLBI–NASA
Hatcreek	– Mon. Peak	990	-2.7 ± 0.3	-5.3	VLBI–NASA
Hatcreek	– Vandenberg	700	-3.4 ± 0.5	-4.9	VLBI–NASA
Ft. Davis	– Mon. Peak	1200	3.5 ± 0.2	4.1	VLBI–NASA
Mojave	– JPL	170	0.6 ± 0.3	-0.9	VLBI–NASA
Mojave	– Mon. Peak	270	-2.4 ± 0.2	-4.7	VLBI–NASA
Mojave	– Pinflats	200	-1.4 ± 0.2	-4.8	VLBI–NASA
OVRO	– JPL	340	-1.5 ± 0.3	-4.5	VLBI–NASA
OVRO	– Mon. Peak	510	-2.6 ± 0.1	-5.1	VLBI–NASA
OVRO	– Pinflats	440	-1.5 ± 0.6	-5.2	VLBI–NASA
OVRO	– Vandenberg	360	-0.8 ± 0.2	-1.9	VLBI–NASA
Quincy	– Mon. Peak	880	-3.4 ± 0.4	-5.3	VLBI–NASA
Yuma	– Mon. Peak	210	2.4 ± 0.3	3.4	VLBI–NASA
Yuma	– Pinflats	220	2.3 ± 0.6	4.7	VLBI–NASA
Pacific – Australia:					
Mon. Peak	– Yarragadee	15110	-8.9 ± 0.5	-10.3	SLR–NASA
Eurasia – Pacific:					
Simosato	– Huahine	9530	-9.1 ± 0.7	-10.7	SLR–DGFI

76

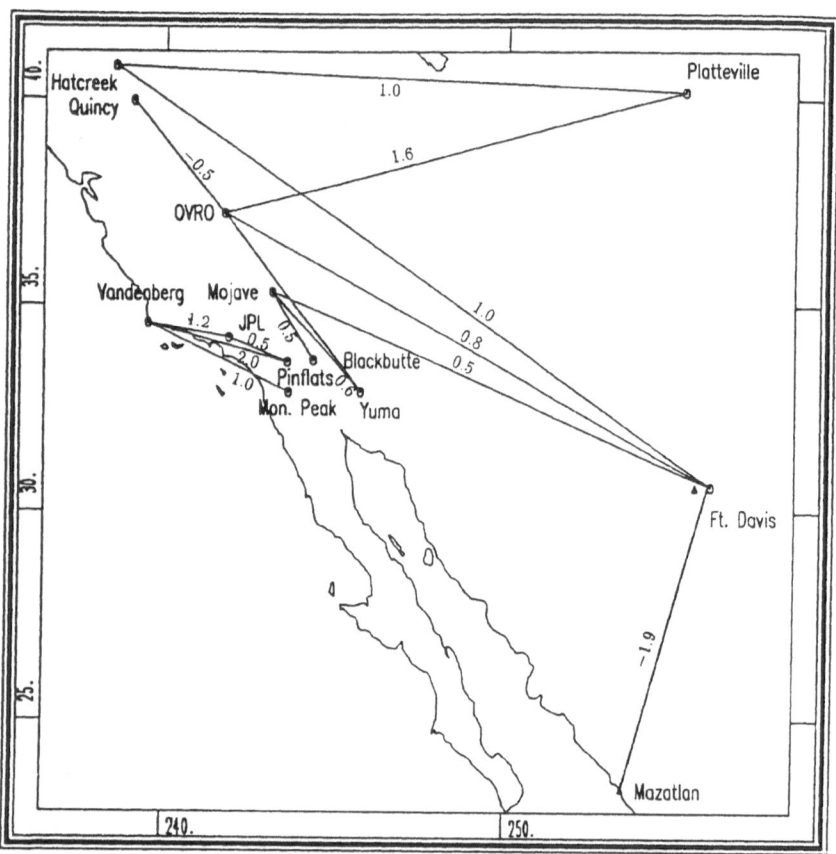

Figure 4. Distance changes for lines within the Pacific plate (west of the San Andreas Fault) and within the Northamerican plate (east of the SAF) in the western U.S.A.
Numbers: rates (cm/a); triangles: SLR stations; circles: VLBI sites.

changes (OVRO - Haystack 0.4 cm/a, Ft. Davis - Haystack -0.6 cm/a), the strain rate being usually smaller than $\varepsilon = 2.10^{-9}\, a^{-1}$. On the other hand there is a significant decrease in the distance between the VLBI sites at Gillmore Creek (Alaska) and Mojave (California) of -1.9 cm/a, for which we at present have no explanation. The SLR line Ft. Davis - Mazatlan also shows a decrease of -1.9 cm/a. This could be related to the vincinity of the station of Mazatlan to the border of the Northamerican plate. It has to be remarked, however, that the SLR station Ft. Davis had some severe problems with the equipment during the investigated time period (this does not affect the other baselines from Ft. Davis as these were observed with VLBI).

Looking at the intraplate baselines of the Northamerican plate, the most striking feature is, however, the consistent increase of length in the area of the American West (Figure 4). This is obviously connected to the extension of the Basin and Range Province. The amount of extension according to the geodetic results is about 1...1.5 cm/a ($\varepsilon \approx 10^{-8}\ a^{-1}$). This extension has been postulated already earlier on the basis of field investigations which revealed extensive normal faulting, the presence of many graben structures, tensile tectonic stresses, and high heat flow values (e.g. Zoback and Zoback, 1980).

Hard to explain are the observed distance changes in the vincinity of the SAF. In analogy to the observations within the Pacific plate a decrease of distance between stations close to the fault (Blackbutte, Yuma) and those further away (Mojave, OVRO) would be expected. Instead an increase, although small, is observed.

4.2.4. Interplate Baselines: Europe -- Northamerica. Due to the fact, that the SLR lines connecting Northamerica and Europe have great uncertainities, only VLBI results can be used for the compari-

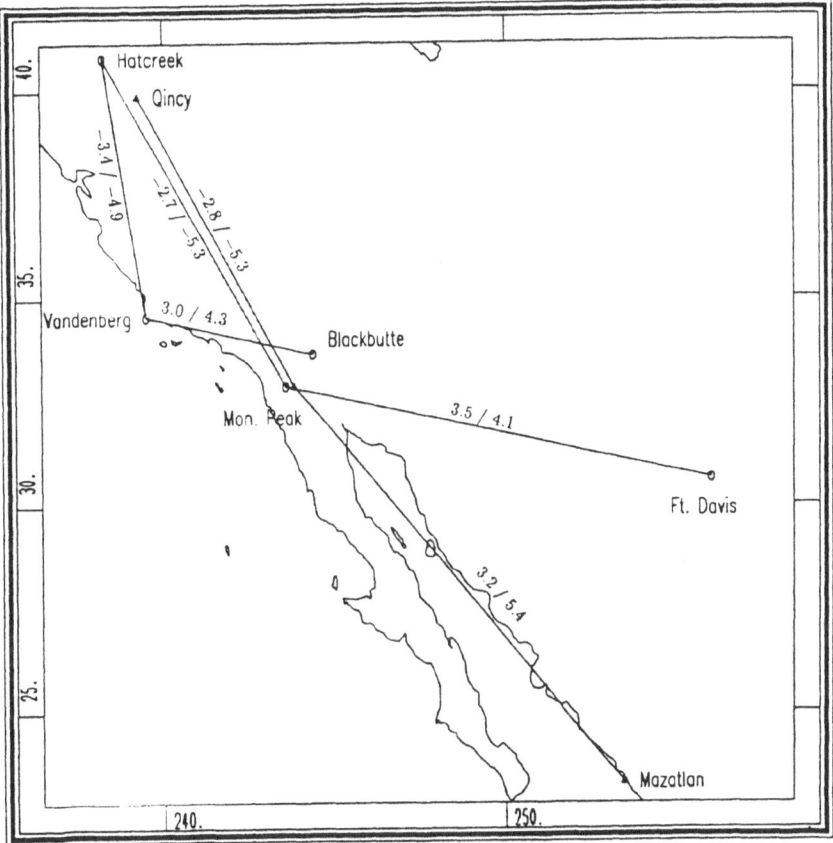

Figure 5. Comparison between some geodetically observed baseline rates (first number) and the prediction from the Minster- Jordan model (second number) in the vincinity of the San Andreas Fault.
Triangles: SLR sites; circles: VLBI sites.

-son between geodetic observations and the plate tectonic model. The VLBI lines reveal consistently, that the observed rates of distance change are lower than they should be according to the Minster-Jordan model. While the plate model predicts for the baselines considered here a mean opening rate for the Northatlantic Ocean of about 2 cm/a, the VLBI observations give a mean value of only about 1 cm/a. There is therefore good reason to suspect that the present day spreading rate of the Northatlantic is considerable lower than its average over the past 3 million years.

4.2.5. Interplate Baselines: Northamerica - Pacific. The differences between the actually observed distance changes and the rigid plate model in California suggest again that there is a considerable amount of continental deformation in this area (Figure 5). Stations, which are situated on either side of the SAF and approach each other (Mon. Peak - Quincy, Hatcreek- Vandenberg, Hatcreek - Mon. Peak) do this at a systematically slower rate than postulated by the plate model. The same holds true for stations that move away from each other (Mon. Peak - Mazatlan, Mon. Peak - Ft. Davis, Vandenberg - Blackbutte). This supports the assumption, that even at a very clearly defined plate boundary like the SAF with predominantly strike slip movement a considerable amount of motion is taken up in internal deformation or motion along higher order faults. Local geodetic networks and geological investigations in the close vincinity of the SAF point towards a motion along the fault that is about 40\% smaller than the plate model prediction (for a summary of these investigations see Minster and Jordan, 1984). This implies, that velocity data from plate boundaries provide only lower bounds for the determination of the relative motion between lithospheric plates.

5. Interpolation Techniques

Up to here we only considered the relative motion between two points. In general, however, one would like to infer from the motion of the observation sites the motion of not observed points in between. For this purpose suitable interpolation techniques have to be applied.

In this connection one has to distinguish between physical models which describe the real deformation process by a simplified physical approach (using physical parameters) and mathematical models, that are described by mathematical equations (compare Drewes, 1984). A purely mathematical approach could be given by some functional or stochastic interpolation, e.g. by splines, polynomials or least squares prediction. In this case the deformation is described by some mathematical function that approximates the observations (without *a priori* physical information). A simple physical model is the rigid plate model, in which every point motion is uniquely described by the rotation vector of the plate and the relative position of the point to it. Other physical models can be set up using e.g. continuum mechanical approaches in which the rheological parameters and the applied forces determine the deformation. Drewes and Geiss (1986, 1988) compared these three different approaches (mathematical prediction, rigid plates, continuum mechanics) using a simulated data set for the MEDLAS campaign in the Mediterranean area. The results reveal considerable differences regarding the overall deformation pattern. Most promising appears the finite element technique, including regional variable rheological parameters (Figure 6). Successful finite element models were - among others - presented by Kroger et al. (1987) for the VLBI observations along the San Andreas Fault.

These models point out, first that the geographical distibution of the observation sites has a serious effect on the quality of the model, second that a good knowledge of the rheological parameters is very important, even if in many cases the deformation pattern is more depending from the relative values of the parameters. A further problem is, that at present only kinematical approaches appear feasible. For dynamic formulations considering also the geodynamic forces our knowledge seems still insufficient.

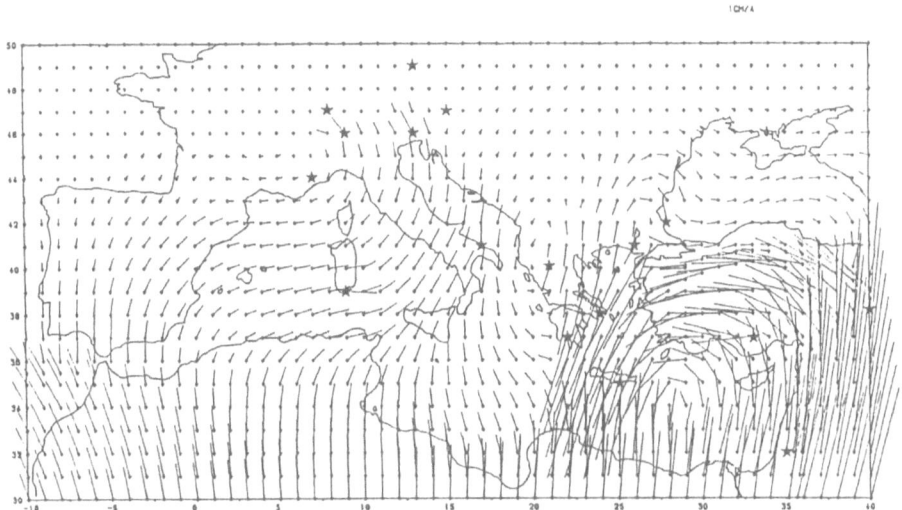

Figure 6. Deformation model for the Mediterranean area. A finite element technique has been applied to a simulated data set for the MEDLAS campaign using regionally variable material parameters (from Drewes and Geiss, 1986).

6. Conclusions

In the paper presented here, we could show, that it is possible by international geodetic measuring campaigns to determine the relative motion of points on the earth's surface due to geodynamic effects. As a - by no means self- evident - result, it could be demonstrated that distance changes observed over a time period of ≤ 10 years agree very well with the predictions derived from a plate tectonic model for the last 3 million years. This supports on the one hand the general applicability of this model for present - day plate motion, on the other hand it shows that plate motion is more or less continuous already over the timescale of a decade.

The quality of the geodetic observations (SLR, VLBI) allows us also to investigate (as 'second order effects') discrepancies with the plate model, even if it is still sometimes difficult to decide what is signal and what is noise. As main conclusions two points may be stressed: First, the VLBI observations between Europe and Northamerica reveal systematically smaller rates (by about 50%) than it would be expected from the plate tectonic model.

We think that this is an indication for a significant lower spreading rate of the mid--atlantic ridge in this area. Second, the very detailled measurements in the vincinity of the San Andreas Fault proof, that even at plate boundaries with predominantly strike slip motion there occurs considerable internal deformation of the plates involved. The amount of this deformation may vary from plate (or block) to plate, according to its overall rheological properties. The geodetic measurements in the west of the Northamerican plate also support the idea of an ongoing extension of the Basin and

Range province. It can be hoped, that the Mediterranean Laser Campaign and related densification networks will in the near future provide us with information on the deformation pattern in a geodynamically very complex area.

The experiences with existing observation networks, however, especially stresses the need for interdisciplinary cooperation. For instance paleomagnetic results can provide important constraints on the design of such a network. The findings of rheological investigations will be of increasing importance for the interpolation of the geodetic observations at discrete points.

Acknowledgements

The financial support of this research by the German Research Foundation (DFG, project no. Re 536/1-1) is gratefully acknowledged.

References

Chase C.G. 1978. Plate Kinematics: The Americas, East Africa, and the Rest of the World. *Earth Planet. Sci. Lett.*, **37**, 355-368.

Drewes H. 1984. Models for Monitoring Regional Geokinematics in the Alpine-Mediterranean Region. *Ann. Geophysicae* ,**2**, 235-238.

Drewes H., Geiss E. 1986. Simulation Study on the Use of MEDLAS Derived Point Motions for Geokinematic Models. *Adv. Space Res.*, **6**, 71-74.

Drewes H., Geiss E. 1988. Modellierung geodynamischer Deformationen im Mittelmeerraum. *In: Final Report of SFB 78* (in press).

Drewes H., Achilli V., Baldi P., Broccio F., Cagnetti V., De Marco R., Geiss E., Marsan P., Milana G., Reigber Ch., Tremel H., Zerbini S. 1987. The Calabrian Arc Project. *Proceedings of the 3rd WEGENER-MEDLAS Conference*, Bologna 1987.

Drewes H., Chourio O., Henneberg H., Hoyer M., Reigber Ch., Rekkedal S., Stuber K., Tremel H. 1988. The Venezuelan Andes GPS-Network. *Paper pres. at 14th Crustal Dynamics Project Principal Investigators Meeting*, JPL, Pasadena, U.S.A., March 22-24, 1988.

Herring T.A. 1986. Very Long Baseline Interferometry and its Contributions to Geodynamics. In: Anderson A.J., Cazenave A. (eds.): *Space Geodesy and Geodynamics*. Academic Press, London a.o., 169-196.

Kroger P.M., Lyzenga G.A., Wallace K.S., Davidson J.M. 1987. Tectonic Motion in the Western United States Inferred From Very Long Baseline Interferometry Measurements, 1980-1986. *J. Geophys. Res.*, **92**, 14151-14163.

Ma C., Ryan J.W. 1987. Crustal Dynamics Project Data Analysis - 1987. Vol. 2 - Mobile Site VLBI Geodetic Results 1979-86. *NASA Tech. Memorandum 100689*, 122pp.

McClelland E., Courtillot V., Tapponnier P. (eds., 1986), Magnetotectonics. *Tectonics*, **5**, 5 (Spec. Issue).

Minster J.B., Jordan T.H. 1978. Present-Day Plate Motions. *J. Geophys. Res.*, **83**, 5331-5354.

Minster J.B., Jordan T.H. 1984. Vector Constraints on Quaternary Deformation of the Western United States East and West of the San Andreas Fault. \In: Crouch J.K., Bachmann S.B. (eds.): *Tectonics and Sedimentation Along the California Margin: Pacific Section. S.E.P.M.*, **38**, 1-16.

Minster J.B., Jordan T.H. 1987. Vector Constraints on Western U.S. Deformation From Space Geodesy, Neotectonics and Plate Motions. *J. Geophys. Res.*, **92**, 4798-4804.

Moritz H., Mueller I.I. 1987. Earth Rotation - Theory and Observation. Ungar, New York, 617pp.

Reigber Ch., Schwintzer P., Müller H., Barth W., Massmann F.H. 1988a. The terrestrial Reference Frame Underlying the Pre-ERS-1 Earth Model Determination. *Manuscripta Geodaetica,* in press.

Reigber Ch., Schwintzer P., M\"uller H., Massmann F.H. 1988b. Globale relative Stationsbewegungen - abgeleitet aus Laserentfernungsmessungen zum Satelliten LAGEOS. *In: Final Report of SFB 78,* in press.

Ryan J.W., Ma C. 1987. Crustal Dynamics Project Data Analysis - 1987. Vol. 1 - Fixed Station VLBI Geodetic Results 1979-86. *NASA Techn. Memorandum 100682,* 163pp.

Smith D.E., Kolenkiewicz R., Dunn P.J., Torrence M.H., Klosko S.M., Pavlis E.C., Robbins J., Fricke S. 1987. The SL7 LAGEOS global solution (abstract). *Eos Trans. AGU,* **68,** 285.

Thornton C.L., Fanselow J.L., Renzetti N.A. 1986. GPS-Based Geodetic Measurement Systems. \In: Anderson A.J., Cazenave A. (eds.): *Space Geodesy and Geodynamics.* Academic Press, London a.o., 197-218.

Weldon R., Humphreys E. 1986. A Kinematic Model of Southern California. *Tectonics,* **5,** 33-48.

Zoback M.L., Zoback M. 1980. State of Stress in the Conterminous United States. *J. Geophys. Res.,* **85,** 6113-6156.

CONTINENTAL ROTATIONAL DEFORMATION: EXAMPLES FROM GREECE

S. PAVLIDES
Geological Department,
University of Thessaloniki
54006 -Thessaloniki,
Greece

ABSTRACT. Herewith are summarized some results from a current research refering to: 1st) the ~30° clockwise rotation of the Chalkidiki peninsula (N. Greece) which confirms the counter-clockwise rotation of the neotectonic stress field of the area. This recent and active stress field has been determined both graphically and by numerical methods taking into account data from meso- and mega-structures of the area, and 2nd) the latest paleomagnetic results from Melos volcanic island (South Aegean Active Volcanic Arc), which do not show any significant rotation and are close to that of Crete and Rodos. All these islands follow similar movements during Plio-Quaternary and compose a unique block, for the existence of which there is additional seismological and tectonic evidence. The above mentioned examples and the available paleomagnetic and tectonic data constitute a very complex rotational deformation for the Aegean area where some simple description of Plate Tectonics are tested and new questions arise.

1. Introduction

In the following three words "Φυσισ κρυπτεσθαι φιλει" (Nature likes to be hiden, or Nature hides its secrets) the Greek philosopher of the 4th century B.C. Heraklitus best describes the difficulty of understanding the nature and approaching its secrets. This desire has tormented human thought for centuries and it remains strong even though the human thought has reached higher spheres of intellect and we have been witnesses to an eruption of scientific activity.

Please, do not be alarmed! this is not an introduction to the philosophy of science. Such thoughts have rather been thrown away from our technocratic perception and from our recent scientific papers. It only points out epigramatically some aspects and problems of Geodynamics and puzzle us with current questions arising from this.

Geodynamics is the fundamental branch of Geosciences and, at the same time, an important tool for understanding earth's behavior, especially after the Plate Tectonics revolution. It is the heroic age for Geodynamics.

Over the years many theories have been advanced to explain crustal movements. One of them, the geosynclinal theory of orogeny, undoubtly important, dominated the thinking of geologists for about a century. The mobilistic theory of Global Tectonics is ingenious. The establishing of horizontal movements opened new roads to geoscientists, so one could say nowadays it is an eruption of geoscience activities. Now several new sets of geological and geophysical data are available to us, which makes it possible to discuss and to understand the past and present-day dynamics affecting the earth's crust. But time is ripe for a comprehensive review. The geoscientists have to look back in order to evaluate the degrees of successes achieved by the Plate Tectonics, as

83

C. Kissel and C. Laj (eds.), Paleomagnetic Rotations and Continental Deformation, 83–93.
© *1989 by Kluwer Academic Publishers.*

well as to point out some obscure points of unconforming rules of the theory, especially that of continental deformation. It is interesting to enrich our understanding with new ideas about contiental processes of deformation and driving mechanisms of crustal movements. The failure of some well-established rules in science has naturally led to the search for new ones. On the other hand, overwhelming success of a scientific theory does not necessarily imply absolute success in interpretation of all related phenomena.

In Geodynamics a large number of interesting partial models have been developed in recent years in an effort to interprete the behavior of the earth's crust. The majority of them partly describe the phenomena, while some contradict others. Some recent geodynamical models look like abstract constructions, especially to those who have never studied the subject. Anyway current and future efforts would produce significantly important results.

Concerning this relatively new subject of Geodynamics one could point out the problem of continental rotational deformation with certain selective questions, such as:

1) What is the degree of reliability and the limitation of paleomagnetic data in detecting rotations?

2) How are they related to block rotations ?

3) Is the contemporary rate of block motion constant or not ?

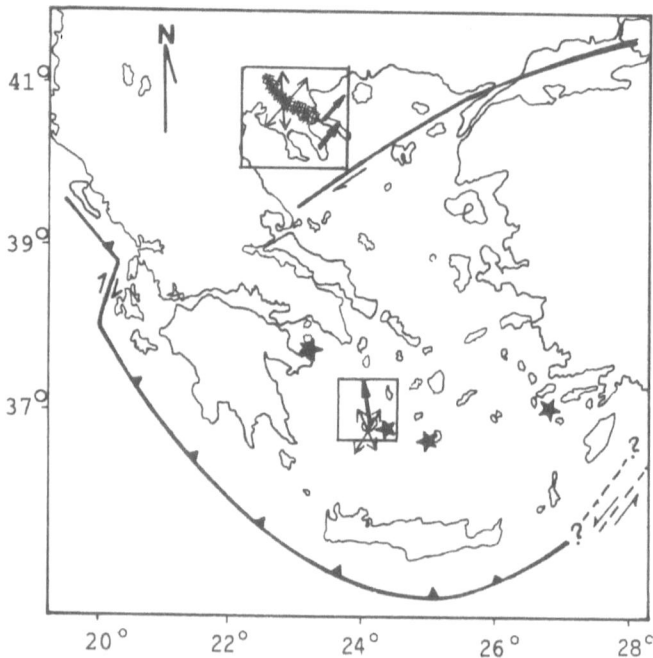

Fig. 1. General map of Greece. The squares indicate the studied areas. Dashed belt define the position of the most active zone in Chalkidiki peninsula. Large arrows represent mean paleomagnetic declinations for these areas. Thin arrows show the direction of extension. Stars indicate the Active Volcanic Arc of South Aegean. Plate boundaries are also drawn.

4) Are these blocks real rigid bodies, and if not what is the amount of their internal deformation ? How does it affect the block motion?

5) What is the nature of forces driving blocks and how are they related to block rotations?

6) How is this deformation related to earthquakes?

These are some important questions among others, that have to be answered throughout the current and future research.

In the framework of this workshop of "Paleomagnetic Rotations and Continental Deformation" are summarized some results from our recent research from the University of Thessaloniki, which constitute a small effort of understanding the rotational deformation of the Aegean.

This brief summary should serve to describe rotational phenomena both from Chalkidiki peninsula (Central part of Northern Greece) and from the western part of South Aegean Active Volcanic Arc, and to offer an explanation for them (Fig.1).

2. Neotectonic rotational deformation in Chalkidiki Peninsula (Northern Greece).

Structural studies concerning the brittle deformation on the central part of northern Greece (Chalkidiki) show many neotectonic and active faults. The major faults which dominate the area trend in a NW-SE direction. They are mainly normal faults with a distinct sinistral strike-slip component, while the recent faults affecting the Plio-Quaternary sediments are smaller as a rule, dip-slip structures trending almost E-W. The narrow active zone extends from NNW to SSE

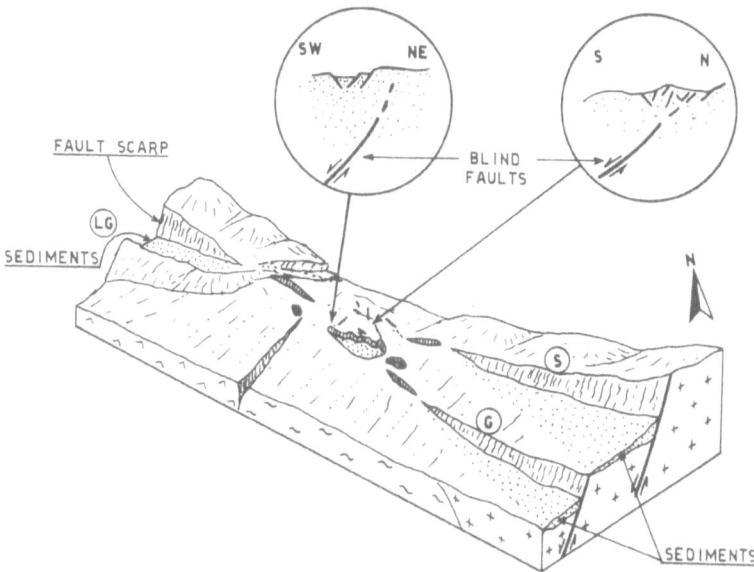

Fig. 2. Simplified block-diagram illustrating a possible existance of blind faults in between the main active structures in Chalkidiki (Langada graben (LG) and Stratoni (S)-Gomati (G) faults).

direction through a graben system (Langada lake) into a second semi-graben one (SE) (Fig. 2). This is obvious from structural data (see also Fig. 1 of Pavlides & Kilias, 1987, and Pavlides et al., 1988) and independent seismological ones, such as the seismogenic volumes and the rupture zones of 1902 (M=6.6), 1932 (M=6.9), and 1978 (M=6.5) earthquakes. In between them there are only small, second order faults affecting the basement granite-granodiorite rocks, as well as Plio-Quaternary sediments. It is suspected that this active zone continues in this area with blind normal faults (faults that do not reach the surface). The upward propagating normal fault planes are identified by precursor breccia, fresh slip small planes, etc,.. (Fig.2).

Microtectonic measurements obtained from fault plane directions, tension gashes, feather joints, comb fracture traces, steps, striae etc, have been made in order to establish the neotectonic stress pattern. All these elements supply evidence for two main extensional phases, one of Quaternary-Recent N-S to NNW-SSE direction and an older neotectonic one (Oligo-miocene-Pliocene?) trending NE-SW. These have been graphically determined in Pavlides & Kilias, 1987, while new data and quantitative stress (or strain) analysis clearly show the two extensional events afecting the SE part of Chalkidiki peninsula. For a better constraint of the stress tensor, the Carey's method (1979) has been used here and the numerical approach of Caputo & Caputo (under publication) applied to both the right-dihedrons method (Angelier & Mechler, 1977), and the P/T axes.

As it is arising from figure 3 (A and B), at least two main extensional directions have been detected in the same large fault ranging from approximately 30° to 40°. If one reconstructs the original system by rotating the fault plane counterclockwise at this angle, the extensional direction remains constant with respect to the structures. Additional evidence for this comes mainly from paleomagnetic data (Kondopoulou & Westphal, 1986 ; Pavlides et al., 1988) which have been collected from plutonic Oligocene formations of the same region. All these observations indicate an ~30° clockwise rotation of the area.

With regard to the timing of this rotation, one could take the following into account:

i) the action of the north Aegean Trough (possible extension of the N. Anatolian fault into North Aegean trough), which is strongly connected with the geodynamical behavior of the studied area, from middle-late Miocene to recent.

ii) the neotectonic movements studied in Plio-Quaternary sediments along the master faults of the area, and finally,

iii) the existence of geological evidence (opening and development of Neogene basins, Mio-Pliocene sedimentation, Psilovicos, 1977 ; Chatzidimitriadis & Kelepertsis, 1984). So, it is believed that the rotation of the area is of neotectonic age (Middle-Late Miocene to Recent).

The explanation of such a rotation -relatively high for this area- is not so clear. Although in strike-slip environment rotations about vertical axes are expected, in extensional environments - as this case - it is rather rare and difficult to be investigated (see also Kissel et al., 1987). However, no strike-slip fault cuts across this area. Instead the motion is taken up on number of major neotectonic and active faults, striking NW-SE and dipping both NE and SW. Although they are normal faults, they appear to have a distinct sinistral strike-slip component, while some NE-SW trending fault planes show a significant dextral strike-slip component and the very recent faults affecting the youngest sediments are mainly pure normal faults striking almost E-W (Fig. 4).

It seems that the deformation of the area is strongly connected with the major right-lateral strike-slip faulting of the North Aegean Trough (deforming zone). It is worth noticing that the horizontal projections of slip vectors, as deduced from both focal mechanisms and especially from recent striae, observed on the active or reactivated fault planes are normal to the boundaries of the North Aegean main deforming zone. This has been predicted by the McKenzie & Jackson's (1983,

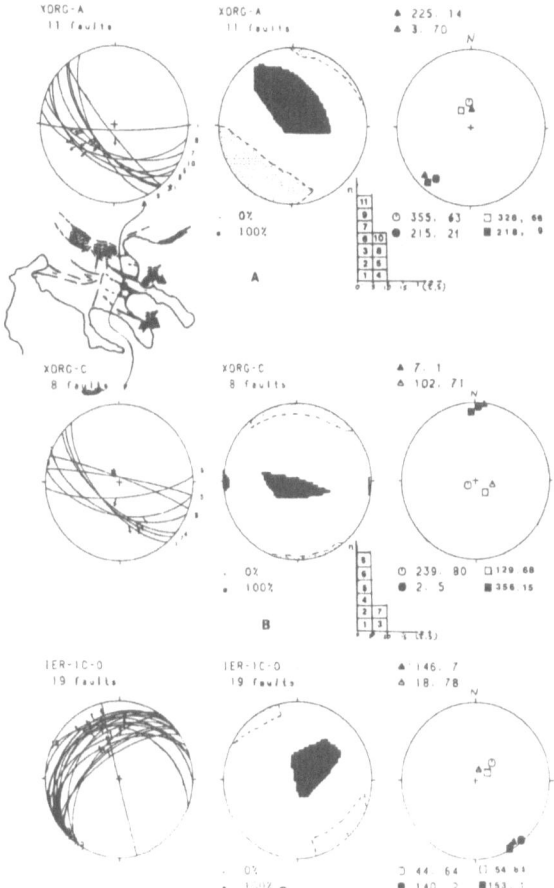

Fig. 3.Structural data from SE Chalkidiki. In the left colummn are shown the faults used to compute the regional stress field of the area. In the central column are shown the results of the right-diherdrons methode (Angelier & Mechler, 1977). In all three cases there are the 100% and 0% areas of probability to find the σ_1 and σ_3 main stress axes respectively. In the third column are shown the results of the Carey's (1979) method (squares) and that of the numerical methodology of Caputo & Caputo (1988) applied on the right dihedrons method (circles) and the P/T axes method (triangles). The open symbols represent the σ_1(compressional) axes, while the solid ones represent the σ_3 (tensional). Histograms give the deviation between the measured (s) and predicted (t) slip vectors. All are equal-area stereographic projections, lower hemisphere. Numbers outside the stereonet refer to corresponding number data inside the histogram. A: Gomati (G) master fault affecting basement (pre-Alpine) rocks. B: the same main fault which dominates the area, with the second family of striation (recent ones). C: small normal faults affecting the nearest Plio-Quaternary sediments of Ierissos (I).

1986) pinned block model. They have also applied their model to explain the observed normal faulting in Central Greece.

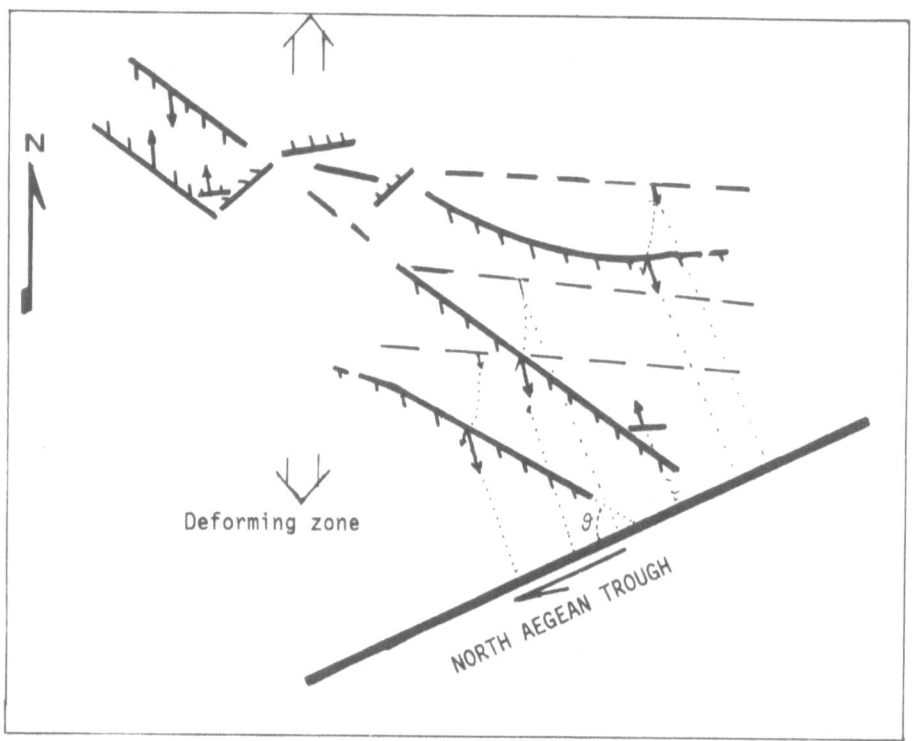

Fig. 4. A simple sketch showing the fault geometry of the Chalkidiki (Northern Greece) domain. The NW-SE roughly parallel major faults form an en-echelon pattern similar to that of McKenzie & Jackson's 1983 model, while the main boundary of the deforming zone is considered to be the North Aegean trough. Small black arrows indicate the horizontal projection of slip-vectors (striae) between adjacent blocks. The angular rotation of the blocks is dθ/dt. Large white arrows indicate the direction of extension (N-S). The dotted lines show the "original" position of blocks before the rotation, where the slip-vector '(older striae) remain approximately perpendicular to the boundareis of the deforming zone.

Furthermore, it is interesting to note that horizontal projection of the slip vectors defined from previous striae (older activation of the same faults) remain also perpendicular, to the boundary of the deforming zone, only if we reconstruct the structures by rotating them 25°-30° counterclockwise (Fig. 4).

So, the simple two-dimensional pinned block model of McKenzie & Jackson's 1983 could be applied well enough in this area by predicting clockwise rotations of the blocks. These blocks are bounded by the major normal-sinistral NW-SE faults of the region. So, this relatively small area seems to be highly complicated and needs further investigation.

3. Paleomagnetic and structural evidence for recent deformation in the South Aegean Active Arc (Melos Island).

Paleomagnetic results are already available of the South Aegean Sedimentary Arc (Laj et al., 1982). They indicate a clockwise rotation of about 10° to 25° for the western part of the arc and no rotation for the central and eastern ones. So we started to study the active Volcanic Arc of South Aegean from a paleomagnetic and structural point of view in order to investigate possible rotational deformations in the area. As a first step, the island of Melos has been chosen, which consists mainly of recent volcanic rocks (rhyolites, dacites, pyroclastic series, etc,..). Their age range between 3 Ma (Middle Pliocene) to 0.1 Ma (Late Pleistocene), according to Fytikas et al. (1986).

The paleomagnetic sampling has been undertaken on eight sites, located mainly in the western part of the island, which have been chosen carefully far away from major faults and geothermal fields. The ages of the collected rock-samples are approximately 2.5, 2.0, 1.8, 1.5, 0.9 Ma. The paleomagnetic results have been obtained by Kondopoulou et al. (1988) using standard paleomagnetic techniques. The preliminary results (mean declinations-inclinations) have been plotted in the stereographic projection of figure 5. They indicate an almost N-S mean declination with a small deviation towards the west and a value of inclination slightly different from the expected one for this region.

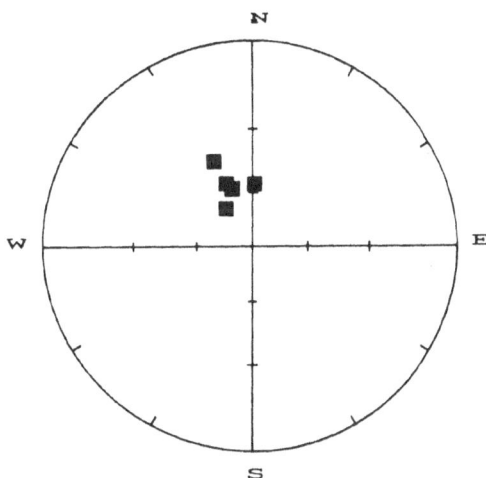

Fig. 5. Stereographic projection of the mean paleomagnetic direction from Melos Island (Kondopoulou et al., 1988).

On the other hand, it has been emphasized that the neotectonic evolution of the Hellenic Arc (back arc area) is clearly dominated by extensional processes related to large-scale normal faulting. Melos which lies in this extensional area has been studied from the neotectonic point of view by Jarrige (1978) (few outcrops), Angelier (1979) (some measurements), and Simeakis (1986) (in detail). They suggest two extensional neotectonic phases, one trending NE-SW and acting during Pliocene, and the second with a NW-SW and acting duriing mid-Pliocene-Recent. Recent observations and calculations (see also Pavlides & Simeakis, 1988) show a mean N-S direction of extension which averages from NE to NW. No strong evidence for two extensional phases has been detected, though in few cases there are two differently oriented sets of strieae. This average direction of extension is due either to local conditions (compatibility with nearby structures, anisotropies in the rocks, etc,..) or to the existence of two subphases. However, there is information which supports the idea that the neotectonic faults, affecting the young volcanics, reflect the basement fault geometry (resurrected or replacement structures according to Sengör et al., 1985). Tsokas (1985) using gravimetric and electrical resistivity data has mapped the deep faults which affect the ceiling of the metamorphic basement of the Island and the neighbouring islets. So it is very possible that the neotectonic regime follows pre-existing faults inherited from last Alpine deformation and shows a scatter direction of extension.

Another important geodynamic point is that Melos is believed to be part of the Cretan megablock (microplate according to Dewey & Sengör, 1979, or "scholle", according to Sengör et al., 1985). There is additional information from geophysical and structural points of view for such a notion:

The earthquakes of 1955 (April 13) M=6.0 ; 1957 (May 29) M=5.3, and 1966 (September 1) M=5.9 (from Ritsema 1974), lying western of Melos with dextral strike-slip component, and especially the 1956 (July 9) Ms=7.5 Amorgos earthquake, with a pure sinistral strike-slip movement (Papadopoulos, 19983) define a strike-slip zone in side the wider extensional area of

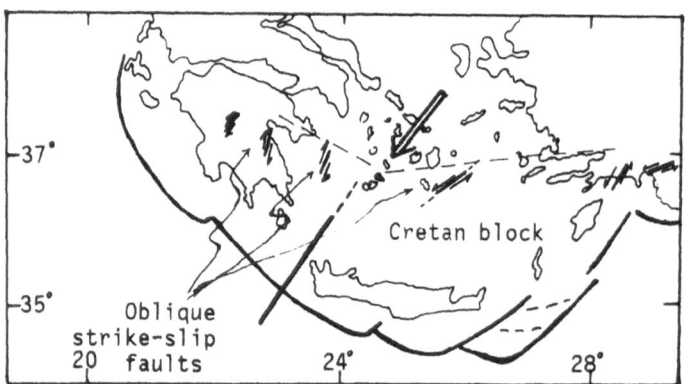

Fig. 6. South Aegean region with the corresponding blocks. Notice the strike-slip zone of deformation crossing the back-arc area. It could be defined by the focal mechanisms of the 1955 Apr. 3 M=6.0 ; 1956 Jul. 9 M=7.5 ; 1957 Apr. 25 M=7.2 ; 1957 Mai 29 M=5.3 ; 1966 Sept. 1 M=5.9, and 1967 June 1 M=5.4 earthquakes (Ritsema, 1974; Papazachos et al., 1984; Papadopoulous et al., 1986), if they are realy reliable fault plane solutions. M = Melos island.

Aegean (Papadopoulos et al., 1986). This zone separates the Attico-Cycladic block from those of Peloponnesus and Cretan ones. Furthermore Simeakis (personal communication) surveyed a large strike-slip fault in Kimolos island, NE of Melos (strike N 40° -50°, dip 70° -80° SE, pich 15° -20° SW, Dextral (see also Simeakis, 1985 and Pavlides & Simeakis, 1988). This geologically observed fault and the results arising from Amorgos earthquakes (Papadopoulos, 1983) support the idea of the existing blocks in South Aegean. On the other hand, seismological results (seismic activity, change of b parameter, Papazachos & Comninakis, 1982) and structural data (Lyberis et al., 1982) strongly support evidence for a block separation in the strait of Kythera between Peloponnesus and Crete (Figure 6).

The paleomagnetic data from Melos are close to those of Crete and Rhodos. They show no significant rotation or a possible slightly counterclockwise one, and they are different from that of peloponnessus with 10°-25° clockwise rotation. So it could be concluded that Melos, which, behaves like Crete and Rhodos, belongs to the same megablock as these islands.

4. Conclusion

From McKenzie's (1978), and Dewey & Sengör's (1979) models for the Aegean area, which separate this region into microplates, through the opinion of Mercier (1981) for a "continuum of deformation", that is a continuous rigid-plastic deformation, Le Pichon and Angelier (1979, 1981) reconstructed the pattern of motion over the eastern Mediterranean for the last 13 Ma. Although this last model explains adequately the majority of geological/geophysical data, and especially the main rotational deformation of the South Aegean area, there are still some important problems which need further investigation, especially those referring to compatibility with new paleomagnetic data and the complexity of blocks movements.

Furthermore in the north Aegean area the structural-rotational deformation is much more complex, with strong clockwise rotations in Chalkidiki (Kondopoulou & Westphal, 1986), in Evia and Skyros (Kissel et al., this volume) or complex pattern of rotations in Izmir-Foca region of W. Anatolia (Kissel et al., this volume), and northern part of Serbomacedonia (Pavlides et al., 1988). So, this plasitically deformed relatively small edge of Euro-Asiatic plate (Aegean and surroundings) with rigid-plastic block movements, high neotectonic and seismic activity and special geophysical and geological properties, show a very complex neotectonic and active deformation and makes up a real natural geodynamical laboratory, where some simple descriptions of plate tectonics are tested and new ideas on continental-back arc kinematics are developed.

Acknowledgements

Thanks are due to P. Choukroune, G. Papadopoulos and C. Sengör for their helpful comments during the workshop. My colleague D. Kondopoulou yielded eagerly all the available paleomagnetic data for Melos and surrounding areas, while the students (now geologists) P. Papaïoannou, C. Kampouridis and M. Tranos helped during field work. Thanks also are due to K. Peftitselis for correcting the English text.

References

Angelier, J., 1979. Néotectonique de l'Arc Egéen, *Thèse d'Etat*, Société Géologique du Nord Publ. 3, 418p.

Angelier, J. and Mechler, P., 1977. Sur une méthode graphique de recherche des contraintes principales également utilisable en tectonique et en sismologie: la méthode des dièdres droits, *Bull. Soc. Géol. Fr.*, **19**, 1309-1318.

Caputo, M. and Caputo, R., 1988. A method for the determination of the regional deformation in structural analysis, *Ann. Tectonicae*, **II** (under publication).

Carey, E., 1979. Recherche de directions principales de contraintes associées au jeu d'une population de failles, *Rev. Géol. Dyun. Géogr. Phys.*, **21**, 57-66.

Chatzdimitriadis, E., and Kelepertsis, A., 1984. Contribution to the knowledge of the geology of Greece. The internal Hellenides in the context of classical geology and plate tectonics, *Oruktos Ploutos*, **33**, 41-58.

Dewey, J.F., and Sengör, C.A.M., 1979. Aegean and surrounding regions: complex multiplate and continuum tectonics in a convergent zone, *Geol. Soc. Am. Bull.*, **90**, 84-92.

Fytikas, M., Innocenti, F., Kolios, N., Manetti, P., Mazzuoli R., Poli, G., Rita, F., and Villari, L., 1986. Volcanology and petrology of volcanic products from the Island of Milos and neighbouring islets, *J. Volcanol. Geotherm. Res.*, **28**, 297-317.

Jarrige, J., 1978. Etudes néotectoniques dans l'arc volcanique égéen, *Thèse 3e cycle*, Univ. Paris XI, pp. 235.

Kuhn, 1972. *The structure of scientific revolutions*, (Chicago Univ. Press).

Kissel, C., Laj, C., Sengör, C., and Poisson, A., 1987. Paleomagnetic evidence for rotation in opposite senses of adjacent blocks in Northeastern Aegea and Western Anatolia, *Geophys. Res. Lett.*, **14**, 907-910.

Kondopoulou, D., and Westphal, M., 1986. Paleomagnetism of the Tertiary intrusives from Chalkidiki (N. Greece), *J. Geophys.*, **59**, 62-66.

Kondopoulou, D., Pavlides, S., Papaioannou, P., and Kampouridis, C., 1988. Paleomagnetic evidence for recent deformation in the western Aegean Active Volcanic Arc,*Ann. Geophysicae, special issue of XIII Gen. Ass.* (abstract) 221.

Laj, C., Jamet, M., Sorel, D., and Valente, J., 1982. First paleomagnetic results from Mio-Pliocene series of the Hellenic sedimentary arc, *Tectonophysics*, **86**, 45-67.

Le Pichon, X., and Angelier, J., 1979. The Hellenic arc and trench system: a key to Neotectonic evolution of the easter Mediterranean, *Tectonophysics*, **60**, 42.

Le Pichon, X., and Angelier, J., 1981. The Aegean Sea, *Phil. Trans. Roy. Soc. London A*, **300**,357-372.

Lyberis, N., Angelier, J., Barrier, E., and Lammemant, S., 1982. Active deformation of a segment of arc: the strait of Kythira, Hellenic arc, Greece, *J. Struct. Geol.*, **4**, 299-311.

McKenzie, D., 1978. Active tectonics of the Alpide-Himalayan belt: the Aegean Sea and surrounding regions, *Geophys. J. Roy. Astron. Sco.*, **55**, 217-254.

McKenzie, D., and Jackson, J., 1983. The relationship between strain rates, crustal thicknening, palaeomagnetism, finite strain and fault movements within a deforming zone, *Earth Planet. Sci. Lett.*, **65**, 182-202.

McKenzie, D., and Jackson, J.,1986. A block model of distributed deformatioon by faulting, *J. Geol. Soc. Lond.*, **143**, 349-353.

Mercier, J-L., 1981. Extensional-compressional tectonics associated with the Aegean Arc: comparison with the Andean Cordillera of South Peru, North Bolivia, *Phil. Trans. Roy. Soc. Lond. A*, **300**, 337-355.

Papazachos, B., and Comninakis, P., 1982. Long-term earthquake prediction in the Hellenic Trench-Arc system, *Tectonophys.*, **86**, 3-16.

Papazachos, B., Kiratzi, A., Hatzidimitriou, P., and Rocca, A., 1984. Seismic faults in the Aegean area, *Tectonophysics*, **106**, 71-85.

Papadopoulos, G. A., 1983. Seismotectonic properties of Amorgos region (Cyclades, Greece), *Bull. Geol. Soc. Greece*, **18**, 17-30.

Papadopoulos, G.A., Kondopoulou, D.P., Leventakis, G-A., and Pavlides S.B., 1986. Seismotectonics of the Aegean region, *Tectonophysics*, **124**, 67-84.

Pavlides, S., and Kilias, A., 1987. Neotectonic and active faults along the Serbomacedonian zone (SE Chalkidiki, N. Greece), *Ann. Tectonicae*, I, 97-104.

Pavlides, S., Kondopoulou, D., Kilias, A., and Westphal, M., 1988. Complex rotational deformations in the Serbo-Macedonian massif (N. Greece): structural and paleomagnetic evidence, *Tectonophysics*, **145**, 329-335.

Pavlides, S., and Simeakis, C., 1988. Neotectonics and active tectonics in low seismicity areas of Greece: Vegoritis (NW Macedonia) and Melos Isl complex (Cyclades) - comparison, *Ann. Geol. Pays Hell.*, (in press).

Psilovikos, A.A., 1977. Paleogeographical evolution of the Mygdonia basin and lake (Langada-Volvi), *Dr Thesis*, Univ. Thessaloniki (Greece) pp 156 (in Greek with an English abstract).

Ritsema, A.R., 1974. The earthquakes mechanisms of the Balkan region, *R. Netherl. Meteorol. Inst.*, De Bilt., Sci. Rept, 74-4, 1-36.

Sengör, A.M.C., Görür, N., and Saroglu, F., 1985. Strike-slip faulting and related basin formation in zones of tectonic escape: Turkey as a case study, *Soc. Econ. Paleont. Mineral.*, Spec. Publ. 37, 227-264.

Simeakis, C., 1985. Neotectonic evolution of Melos Island complex, *IGME* unpublished report 50p.

Tsokas, G., 1985. A geophysical study of Milos and Kimolos Islands, *Dr Thesis*, Univ. Thessaloniki (Greece) **23**, No. 50, 200p.

CENOZOIC MAGMATISM, DEEP TECTONICS, AND CRUSTAL DEFORMATION IN THE AEGEAN SEA

G.A. PAPADOPOULOS
Section of Seismology and Earthquake Resistant Structures,
Ministry of Environment, Physical Planning, and Public Works
71 Louisis Riankour Street, 11523 Athens, Greece.
Mailing address: 98 Mavromichali Street,
11472 ATthens,
Greece

ABSTRACT. A large body of radiometric and analytical data have been elaborated to define the spatio-temporal distribution as well as the petrochemistry of the Cenozoic magmatism in the Aegean area-Four distinct magmatic phases (Eo-Oligocene, eartly- and late Miocene, Plio-Quaternary) and corresponding processes of lithospheric subduction have been suggested. The spatial distribution of a new petrochemical index, determined in a statistical way from five well-known petrochemical diagrams, has been utilized to estimate trends and dips of the early- and late Miocene palaeo-subduction zones. The gradual variation in trend from the early Miocene up to the present indicates either the counter-clockwise rotation of ~50° of the Hellenic consuming boundary or the clockwise rotation of the Aegean lithosphere. This simple method based solely on magmatic data verifies similar results obtained from other geophysical and geotectonic observations.

1. Introduction

Seismotectonic and geophysical data are extremely useful in studying active, deep tectonic conditions along presently consuming plate boundaries.On the other hand, observations on magmatic, sedimentary, and metamorphic processes are of primary importance for defining deep tectonic conditions which prevailed over past geological times. In the light of this experience I have utilized magmatic data to outline some basic properties of the Aegean deep tectonic during the last 50 Ma or so.

It is generally accepted that the South Aegean Plio-Quaternary volcanism is related to the active subduction of the Mediterranean slab beneath the Aegean lithosphere. However, there is only a limited knowledge about the relationship between the Central-North Aegean Tertiary-Quaternary magmatism and deep tectonics. Although some suggestions have been formulated by several authors (Boccaletti et al., 1974; Papazachos and Papadopoulos, 1977; Panagos et al., 1978; Innocenti et al., 1982; Papadopoulos and Andrinopoulos, 1984; Fytikas et al., 1985), there are three major unresolved problems:

(1) What is the spatio-temporal evolution of the Aegean Cenozoic magmatism?
(2) How we can evaluate its petrochemistry on the basis of a unique method?
(3) How to answer to (1) and (2) can provide information for the Aegean deep tectonics ?

95

C. Kissel and C. Laj (eds.), Paleomagnetic Rotations and Continental Deformation, 95–113.
© *1989 by Kluwer Academic Publishers.*

This study is an effort toward responding these questions and contributing to the knowledge of the geodynamic evolution of the Aegean area from the Eocene up to the present.

2. The data

To define the spatio-temporal distribution of the Aegean Cenozoic magmatism, published data concerning localization, main petrological types and ages of Cenozoic magmatic rocks of Greece and Western Turkey have been collected from more than 200 sources. Data for the volcanics of the Kozut region (SE Yougoslavia) are also included because it, along with the volcanic region of Almopia (NW Greece), represent two parts of a single volcanic area. All these data are listed in published, and in one case (volcanics of Methana, NE Peloponnesus, Greece) unpublished (Fytikas et al.,unpublished study), whole rock analyses collected from about 100 different sources. The available analyses, which total 1168, cover rocks from 89 of the 113 sites.

3. Space-time distribution of the magmatism

Figure 1 shows the frequency distribution of the radiometric age of the Cenozoic Aegean magmatism. As one may observe: i) the Aegean area was affected by an almost continuous magmatism during the last 50 Ma or so, that is from the early Miocene up to the present, ii) the magmatism occurred in probably four main phases corresponding to the Eo-Oligocene (53--26Ma), early Miocene (23-14Ma), late Miocene (14-7Ma), and late Pliocene-Quaternary (4-0Ma).

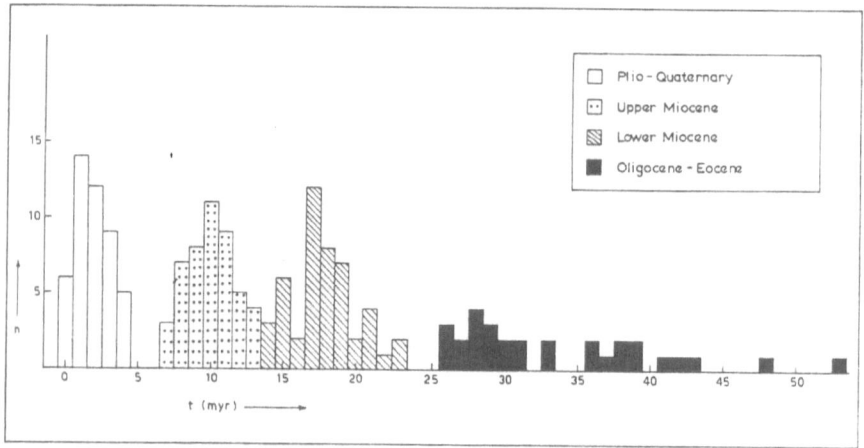

Fig. 1. Frequency distribution, n, of the radiometric age, t, of the Aegean Cenozoic magmatic rocks.

The spatial distribution of this magmatism is illustrated in Figure 2. Only sites having magmatic rocks, the age of which has been determined by radiometric methods, are represented, that is 56 out of 113 sites. Kos (=53) and Tuzla (=104) are shown by two different symbols because the magmatic activity took place there in two separate phases. From Figure 2 arises that: i) the Cenozoic magmatism occurred exclusively in the inner geological zone of the Hellenides, ii) the Eo-Oligocene phase took place in Northern Greece, especially in the Serbo-Macedonian and Rhodopia, crystalline massifs, iii) the early Miocene volcanism developed more to the south, that is in NW Turkey and Central Aegean, iv) the late Miocene magmatism took place even more to the south in the Central Aegean, v) the Plio-Quaternary volcanism occurred along the active island arc of the South Aegean as well as in some regions of the Central and NW Aegean, vi) the overall picture indicates that since the early Eocene the magmatism has gradually migrated with the main component of the migration directed from approximately north to south.

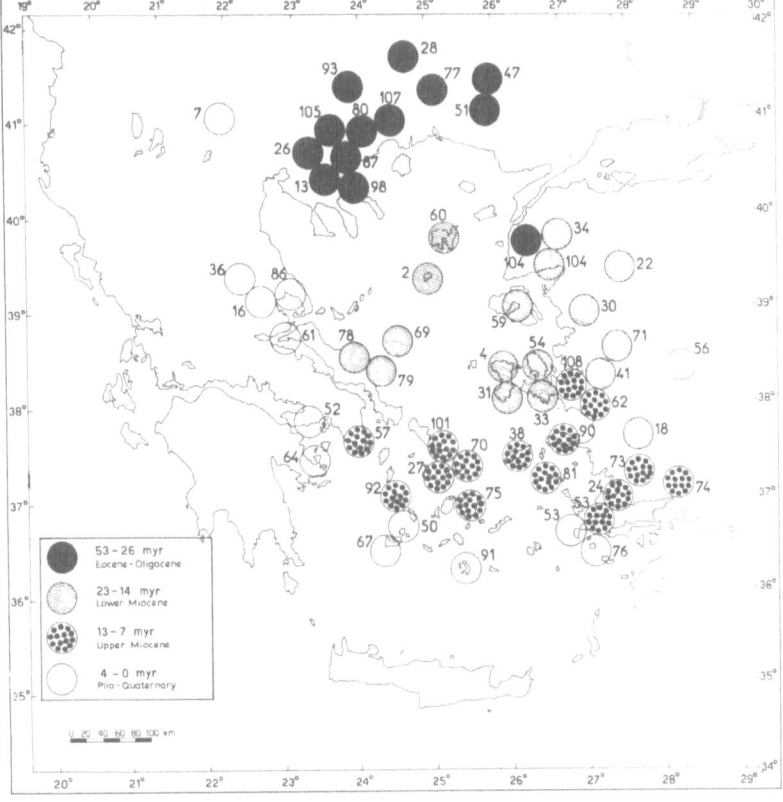

Fig. 2. **Spatial distribution of the Cenozoic magmatic rocks in the Aegean area. Figure near each symbol is the code number of the corresponding site in the Table presented by Papadopoulos (1982) (details in Section 2).**

4. Petrochemistry

The petrochemical classification of igneous rocks is not an easy taks. However, the distinction between alkalic and non-alkalic series has been shown or accepted many years now by a considerable number of authors including Bowen (1928), Rittmann (1957), Kuno (1959), and Miyashiro (1974). In convergent orogenic zones, island arcs, and compressional continental margins, there are mainly non-alkalic igneous rock associations represented by calc-alkalic igneous rock associations represented by calc-alkalic and tholeiitic series. As has been widely stated (Kuno, 1959, 1966; Hatherton and Dickinson, 1969; Jakes and White, 1972; Miyashiro, 1974), magmatic rocks tend to become more alkalic across an island arc or a continental margin toward the adjacent continent. The observation of increasing alkalinity gave support to the hypothesis of magma generation along the upper surface of the descending slab or at increasingly greater depth toward the continent within the upper mantle wedge overlying the descending slab (Kuno, 1966; Green and Ringwood, 1967; Dickinson, 1968, 1975; Ringwood, 1969).

A number of petrochemical diagrams is in use to separate non-alkalic series from alkalic ones on the basis of percentages of several oxides. However, most of them lead only to a qualitative estimation of the alkalinity (or non-alkalinity) of the rock. Moreover, two different diagrams possibly indicate opposite results for a single rock sample leading, thus, to confusion not only for the petrochemical nature of the rock but also for its tectonic setting.

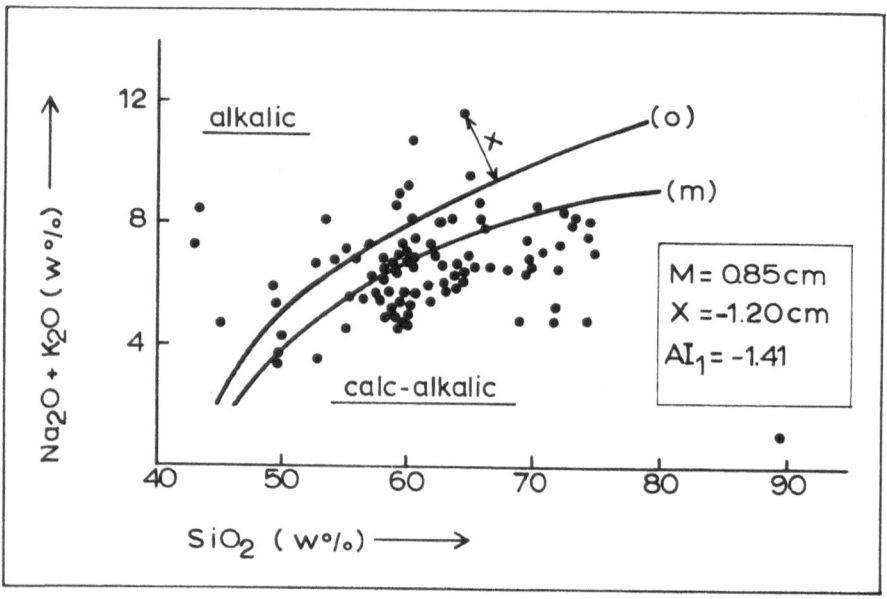

Fig.3. Application of the method for defining the alkalinity index, Al, in the Rittmann's (1957) diagram (see Appendix).

To get around these problems, I have applied a statistical method to define an alkalinity index, Al, that is a quantitative criterion of the alkalinity (or non-alkalinity) of a specific rock sample (see Appendix). Al has the advantage to incorporate in a statistical may the petrochemical characterization of the rock sample from a number of petrochemical diagrams. Five well-known diagrams, in which there is a clear distinction between the non-alkalic and alkalic series have been chosen as follows: (1)Na_2O+K_2O vs. SiO_2, wt percent (Rittmann, 1957), (2)MgO vs. $FeO+Fe_2O_3$,wt percent (Tilley and Muir, 1967), (3) Na_2O+K_2O vs. SiO_2, mole ratio (Gorshkov, 1969), (4) and (5): K_2O vs. SiO_2, and K_2O/Na_2O vs. SiO_2, wet percent (Barberi et al., 1974). Figure 3 shows graphically the application of the method to the first of the chosen diagrams. The Al values range from +4.77 to -4.49 (Table 1, Appendix 2). Positive or negative Al means non-alkalic or alkalic petrochemical character, respectively. The degree of alkalinity or non-alkalinity is proportional to the absolute value of Al. On the basis of the frequency distribution of the Al value, the Cenozoic magmatic rocks of the Aegean area have been classified into four distinct petrochemical groups as follows (Table 1): i) calc-alkalic group when Al≥+0.75; ii) weak calc-alkalic when +0.75>Al>0; iii) weak alkalic when 0 > Al ≥-0.75 ; iv) alkalic when -0.75>Al. This classification is compatible with the classical one of Rittmann (1957).

It is of interest to note that diagrams (2), (4), and (5) show that igneous rocks of tholeiitic compositions are rare in the Aegean area. Therefore, we can say that in the Aegean Cenozoic magmatism the non-alkalic series are mainly dominated by calc-alkalic melts.

5. Implications for the deep tectonics

5.1. ACTIVE PROCESSES

Many geophysical, seismological, geomorphological, volcanological and tectonic observations show that the hellenic arc has the basic properties of the active Circum-Pacific island arcs. It is widely accepted that the geophysical and other features of the South Aegean can be considered as a result of the active subduction of the Mediterranean (or African) lithosphere beneath the Aegean in SSW-NNE direction (Caputo et al., 1970; Papazachos and Comninakis, 1971; McKenzie, 1972, 1978; Le Pichon and Angelier, 1979; Papadopoulos et al., 1986). The South Aegean is characterized by its active volcanic arc. Lavas are predominantly rhyolithic, andesitic, and dacitic, while their orogenic calc-alkalic nature is well-known for many years (Paraskevopoulos, 1956; Pichler and Stengelin, 1968; Pe and Piper, 1972; Barberi et al., 1974). As far as the Christiana-Thera volcanic group is particularly concerned, a gradual increase of the ratio K_2O/Si_2O is likely to occur from the convex to the concave side of the arc (Puchelt et al., 1977). However, it is unknown if this increase constitutes a general feature of the whole volcanic arc.

Figure 4 shows the spatial distribution of the petrochemistry of the Aegean Plio-Quaternary volcanics on the basis of Al values. In Figures 4 and 5 volcanic rocks of Kalymnos and Tilos islands (Nos 5 and 13 in Table 1), located in the eastern part of the South Aegean volcanic arc, have not been taken into account because geologic and petrographic observations have shown that these rocks have very probably come from neighbouring volcanic cneters (Keller, 1969; Christodoulou and Tataris, 1972). It must be noted that not all of the rocks represented in Figure 4 are found in Figure 2, because the former includes all the rocks for which the Plio-Quaternary age is determined not only by radiometric but also by reliable geologic observations.

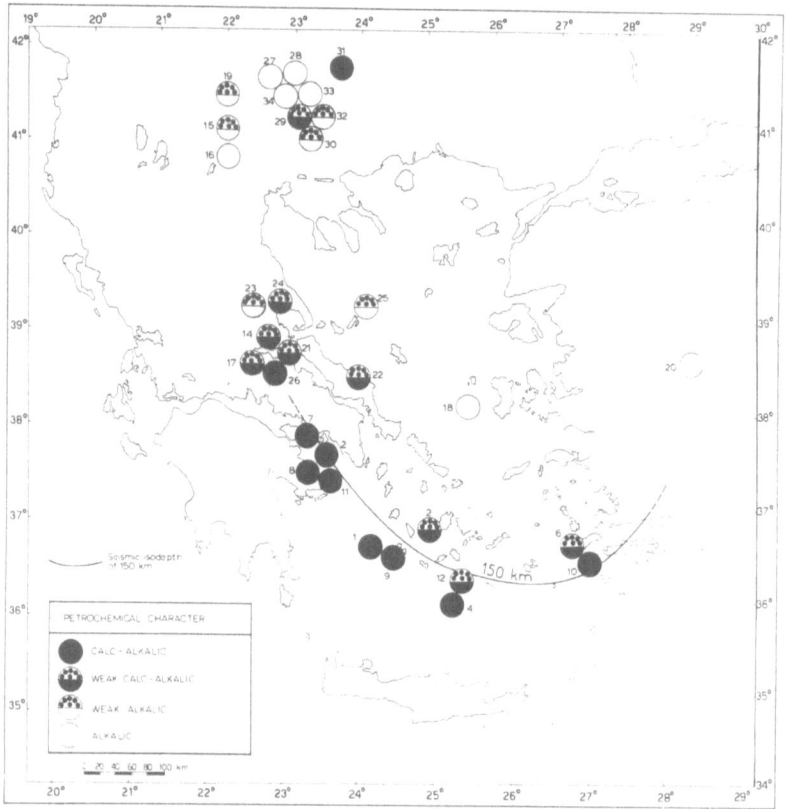

Fig. 4. Spatial distribution of the petrochemical character of the Aegean Plio-Quaternary volcanic rocks. Figures correspond to those in Table 1 (1-12=South Aegean volcanics; 14-34=volcanics of the Central and North Aegean).

As evident from Figure 4, lavas of the South Aegean volcanic arc are calc-alkalic and weak calc-alkalic along the outer (convex) and the inner (concave) part of the arc, respectively; that is the alkalinity index, Al, tends to decrease from SSW to NNE. This tendency is very clearly reflected in Figure 5 where the Al has been plotted against the distance of the volcanic centers from the 150 km seismic isodepth.

Depths of Benioff zones occurring in Indo-Pacific regions have plotted in the K_2O/SiO_2 diagram.(e.g. Ninkovich and Hays, 1972) in order to correlate the petrochemical nature of lavas with the depths of seismic foci. The K_2O/SiO_2 ratios of South Aegean volcanics have been plotted in this diagram (Fig. 6) and expected Benioff-zone depths, h, are estimated (Table 1). The mean depth value of h=147±15 km is in excellent consistency with the seismological data which indicate that the South Aegean volcanic arc almost coincides with the 150km seismic isodepth (Papazachos and Comninakis, 1971). The Al is, as expected, negatively correlated (r = -0.97) to h, that is the

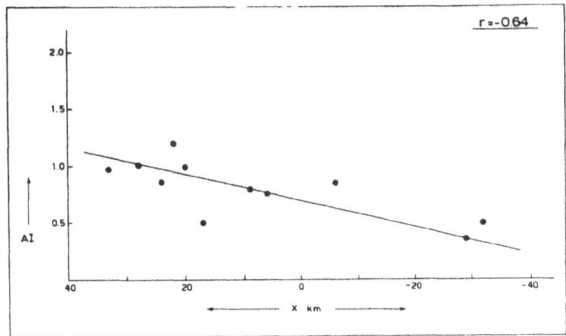

Fig. 5. Variation of the alkalinity index, Al, of the South Aegean volcanics in relation to their distance, x, from the 150km seismic isodepth.

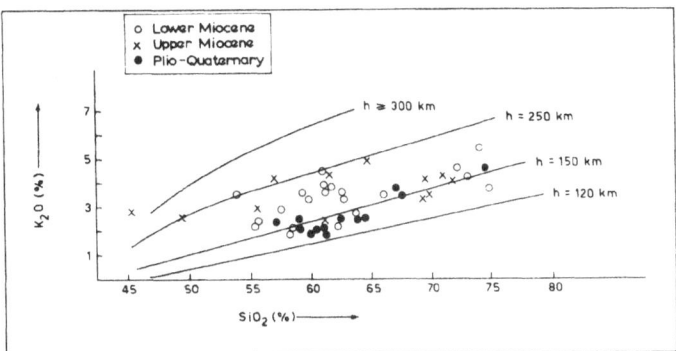

Fig. 6. K_2O and SiO_2 contents of magmatic rocks from the suggested Aegean subduction zones and curves of equal earthquake focal depths in the Indo-Pacific region as defined by Ninkovich and Hays (1972).

alkalinity of lavas gradually increases (or the calc-alkalinity decreases) as the Benioff-zone depth increases (Fig.7).

Many authors have found that the mean dip of the South Aegean Benioff zone is 30°-40°. From two independent samples of seismic data concerning the periods 1911-1963 and 1964-1978, Papadopoulos (1982) stated that this dip is about 38° (see also in Kondopoulou et al., 1985). Therefore, the length of the magmatically active part of the subducting Mediterranean slab is about 80km. Volcanism in the South Aegean began probably 4Ma ago and, hence, the average subduction rate is about 2.1 cm/yr. Since the maximum depth of the Benioff zone is about 180km, as deduced from both seismic and petrochemical observations, the total lenth of the subducting slab seems to be 290km or so. The subduction process started probably 14Ma ago, that is within the middle Miocene. Le Pichon and Angelier (1979) have proposed that the age of the slab is 13.5Ma and the and the linear underthrusting is about 2 cm/yr in the west and 4.5 cm/yr in the east of the Hellenic arc.

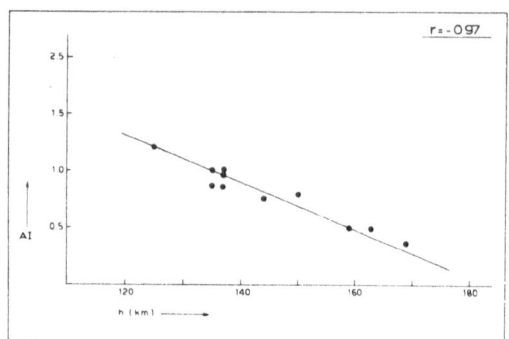

Fig.7. Variation of the alkalinity index, Al, of the South Aegean volcanics in relation to the expected depths, h, of the Benioff zone, as they determined by petrochemical data.

Fault plane solutions of shallow shocks indicate that the lithosphere of the arc convex side is dominated by an almost horizontal compression (Ritsema, 1974; Papazachos and Comninakis, 1977; Papadopoulos et al., 1986). Fault plane solutions have been used to determine with accuracy

Table 2. Fault plane solution of shallow earthquakes which occurred in the convex side of the Hellenic arc.

Date	Geogr. coord. Φ°N	Geogr. coord. λ°E	M	P Axis Trend	P Dip	T Axis Trend	T Dip	References
1952 12 17	34.4	24.5	6.8	51	16	290	61	1
1956 07 30	35.9	26.0	6.0	244	3	338	48	2
1958 06 30	36.4	27.3	6.0	46	23	296	38	1
1959 05 14	35.1	24.6	6.3	6	3	98	41	2
1961 10 02	36.6	21.9	5.7	200	35	20	55	1
1962 01 26	35.2	22.7	6.3	198	45	18	45	1
1966 05 09	34.4	26.4	5.8	203	6	23	84	1
1968 07 27	35.4	28.0	5.7	178	23	68	40	1
1969 06 12	34.4	25.0	6.1	195	18	346	70	3
1972 05 04	35.1	23.6	6.5	266	20	31	57	1
1973 04 06	34.4	25.3	5.4	210	18	30	72	1
1973 11 29	35.2	23.8	6.0	49	42	49	49	4 (improved)
1975 09 22	35.2	26.3	5.5	120	9	16	59	4
1976 06 12	37.5	20.6	5.8	202	37	32	53	new solution

References key: 1= Ritsema (1974); 2= Papazachos & Delibasis (1969); 3= McKenzie (1972); 4= McKenzie (1978).

the average trend of the lithospheric convergence. Table 2 summarizes the basic seismic parameters as well as information on the solutions of 14 shallow shocks which occurred in the convex side of the Hellenic arc between 1952 and 1976. Thirteen solutions have been published by several authors, while the fourteenth is a new one determined by P-wave first motion records listed in the Bulletin of the International Seismological Center. The trend of P axis was found to be equal to 206°±34°.

In the central and North Aegean, many volcanic centers produce compositionally variable volcanic rocks (rhyolites, dacites, rhyodacites, andesites, trachytes, and rarely latites) during the last 4Ma or so (code numbers 14-34, Figure 4). The frequency distribution of the alkalinity index, Al, shows that lavas of these areas as a whole, do not have a typical non-alkalic character. It appears that the Central-North Aegean volcanism is not closely correlated to the active subduction of the Mediterranean lithosphere along the Hellenic arc, as suggested for some volcanic centers located in the Central Aegean by Ninkovich and Hays (1972), Barberi et al. (1974), Fytikas et al. (1976), Pe and Panagos (1976), because this subduction is exclusively limited to South Aegea, as is obviously seen from the distribution of the epicenters of the intermediate focal depth earthquakes (Papazachos and Comninakis, 1971; Comninakis and Papazachos, 1980). This volcanic activity may be tectonically related to the N-S tensional field which prevails in both Central and North Aegean.

5.2. LATE MIOCENE

Fig. 8 shows that the petrochemical character of the late Miocene magmatism in the Central Aegean is predominantly calc-alkaline to weakly calc-alkaline. The index Al tends to decrease from about SW to NE. This mode of spatial distribution of the magmatism, as well as the Barrovian metamorphism of late Miocene age (Altherr et al., 1977), and the active, extensional stress-field (Ritsema, 1974; McKenzie, 1978; Papadopoulos et al., 1986) in the Central Aegean area lead to the suggestion of a possible paleo-Benioff zone underthrusting from SW to NE during the late Miocene. Expected depths, h, of this zone have been determined employing the K_2O/SiO_2 diagram (Fig. 6) (Table 1). The mean true dip, d, of the descending slab is given by the formula

$$\sin^2 d = \sin^2 d_1 + \sin^2 d_2 \tag{1}$$

where the E-W and N-S apparent dips, d_1 and d_2, are determined by least-squares fitting. Dip, d, is about 32°. The angle α between the horizontal projection, BB' (Fig. 8), of the slope direction and the E-W axis is given by the formula

$$\tan \alpha = \sin d_2 / \sin d_1 \tag{2}$$

and estimated to about 45°. Cross section (Fig. 9) shows that h gradually increases along the BB' direction (Fig. 8). The length of the magmatically active part of the subducted slab was approximately 230km. The late Miocene magmatic activity in the Central Aegean lasted for about 7Ma and, therefore, the average subduction rate is estimated as about 3.2cm/yr. The total length of the descended slab was about 490km and the subduction process lasted probably for about 15Ma, that is, it began during the early Miocene.

5.3. EARLY MIOCENE

In southern Aegea lavas are mainly of calc-alkaline type while in the central and northern parts they become mostly weak alkalic (Fig. 8). Thus, the alkalinity index, Al, tends gradually to decrease, or in other words the alkalinity increases from about SW to NE. This pattern is compatible with the existence of a lithospheric slab subducting from the Central Aegean to the NW Turkey during the early Miocene. This point of view is in agreement with the geophysical properties of the area which are similar to those of young, non-active marginal basins. For example, the high heat flow (Jongsma, 1974) and the shallow seismic activity associated with normal faulting (McKenzie, 1972, 1978; Ritsema, 1974; Papazachos and Comninakis, 1977; Papadopoulos et al., 1986) are well-known peculiarities of that area.

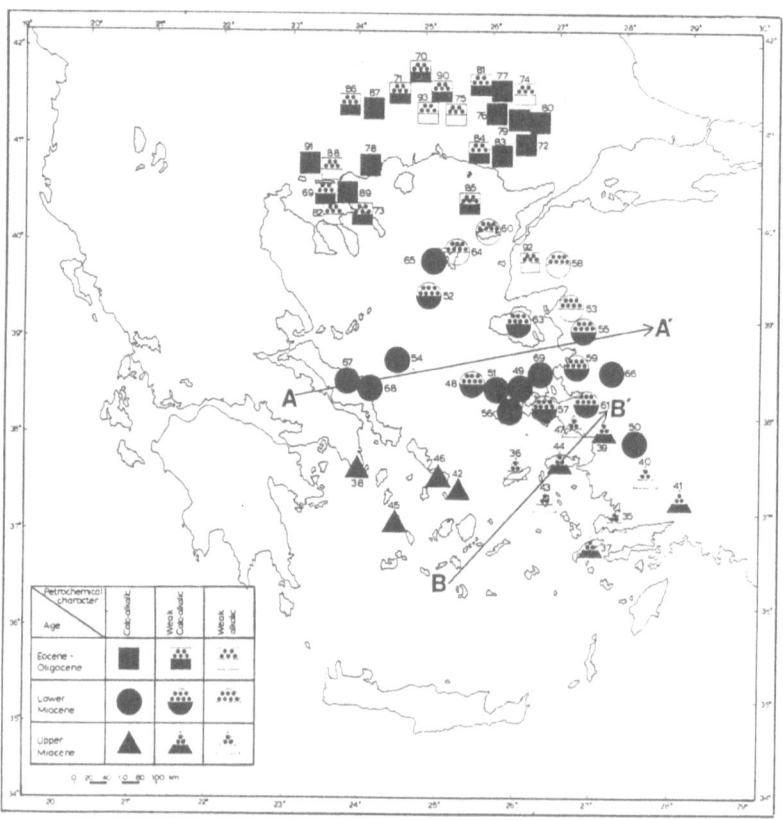

Fig. 8. Spatial distribution of the petrochemical character of the Tertiary Aegean magmatic rocks. Figures correspond to those in Table 1. AA' and BB' represent the horizontal projection of the slope direction of the suggested early and late Miocene paleo-Benioff zones, respectively.

According to Fytikas et al. (1985), the Oligo-Miocene volcanism started in the northernmost part of the North Aegean area with mostly calc-alkaline intermediate and acidic products. The volcanism shifted successively southwards becoming progressively enriched in potassium. This evolution is interpreted as being related to an increase in the dip of the Benioff zone under the Eurasian plate.

The expected depths, h, of the paleo-Benioff zone, based on the K_2O/SiO_2 diagram (Fig.6), are listed in Table 1. Dip, d, is equal to about 21°, and the angle, a, between the horizontal projection, AA'(Figure 8), of the slope direction and the E-W axis is about 12°. Cross-section (Figure 9) shows that h gradually increases along AA'. The magmatically active part of the subducted slab was about 350km in lenth. The early Miocene volcanism lasted for about 9Ma and, hence, the average subduction rate is 3.9cmm/yr. The total length of the descended slab was approximately 730km; the subduction process probably lasted for about 19Ma, that is it began during the middle Oligocene.

It seems that the average subduction rate in the Aegean gradually decreased from 3.9cm/yr during the early Miocene, to 3.2cm/yr during the late Miocene, and to 2.1cm/yr in the Plio-Quaternary. As the trend of the compressional axis in the active Hellenic arc is about 206°, as already mentioned, the angle a is being equal to about 64°. The angle a, therefore, has gradually increased from about 12° in the early Miocene to 45° during the late Miocene and to 64° at the present. On the other hand, the dip of subduction has also increased from 21° in the early Miocene to 32° in the late Miocene and to 38° at the present.

5.4. EO-OLIGOCENE

Geological (Boccaletti et al;, 1974), geophysical and metallogenic (Papazachos and Papadopoulos, 1977; Papadopoulos and Andrinopoulos, 1984) evidence indicate that a lithospheric subduction process took place in the North Aegean area during the Palaeogene. Currently, deep tectonic activity seems to be very weak because no large earthquakes with focal depths greater than about 100km have been observed. However, an almost conical Benioff zone, inclined from about north to south, has been roughly defined (Papazachos, 1976). According to Papazachos and Papadopoulos (1977), the deepest part of the Palaeogene North Aegean slab has been almost assimilated into the mantle, but there still may exist shallow remnants inducing upward transfer of hot material which is responsible for the extensional, thermal, and magnetic properties of North Aegean Trough.

The Eo-Oligocene magmatism in the North Aegean area has a predominantly calc-alkaline to weakly calc-alkaline petrochemical character, while weak-alkalic rocks are rare (Table 1, Fig. 8). This observation reinforces the point of view that a lithospheric slab was dipping beneath that area during the Palaeogene. Nevertheless, there is not an obvious spatial variation of the petrochemical nature, as expected, in such a case. It must be noted, however, that the magmatism of the North Aegean area is only a part of an extensive magmatism that occurred in this area as well as in Bulgaria and SE Yugoslavia during the Palaeogene. Therefore, the spatial distribution of the petrochemistry of this magmatism should be studied as a whole.

6. Discussion

Study of the Cenozooic magmatism in the Aegean indicates that a continious process of lithospheric subduction has probably taken place from at least the early Eocene up to the present. The gradual increase of the angle a from the early Miocene up to the present implies that this process may be due either to a counter-clockwise rotation of oceanic lithospheric remnants of the Tethyan ocean (or lithospheric segments from the front of the African plate) or to a clockwise rotation of the Aegean lithosphere, or to combination of both rotations. According to Le Pichon and Angelier (1979), a counter-clockwise rotation of about 30° of the Hellenic consuming boundary, in respect to the

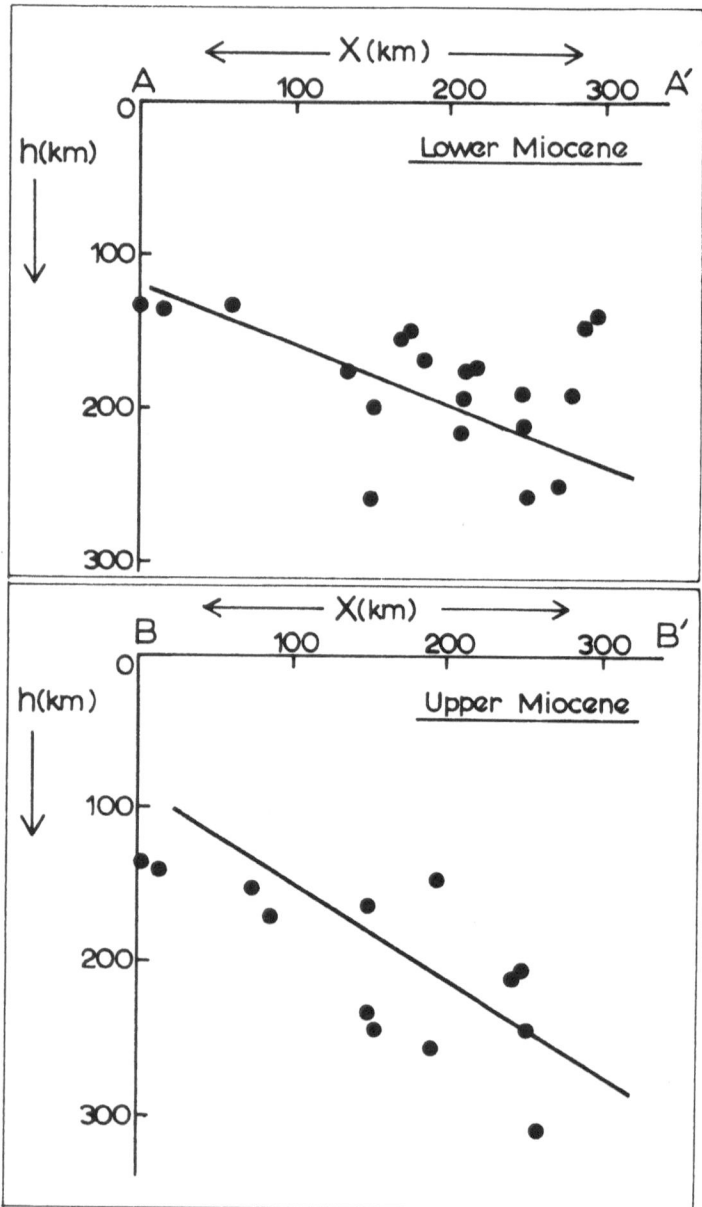

Fig. 9. Cross-sections showing the suggested Miocene paleo-Benioff zones determined from petrochemical data.

Eastern Mediterranean sea floor, took place during the last 13.5Ma. Moreover, paleomagnetic results are in agreement with the notion of the clockwise rotation of the Aegean area in Tertiary times (see references in Papadopoulos et al., 1986).

Active deep tectonic of the Aegean is very complicated, as deduced from its geophysical properties, and, therefore, it is not easy to interpret these properties with simplified tectonic models. Observations concerning fault plane solutions, lateral distribution of the intermediate earthquake foci, attenuation of body waves, residuals of travel times of body waves, volcanic activity, heat flow, as well as magnetic and gravimetric fields, lead to the following conclusions. The convex side of the Hellenic arc (outer geological zones of the Hellenides) is characterized by compressive stress-field, shallow seismic activity, thrust faulting, considerable lithospheric thickness, absence of magmatic activity, low heat flow, and negative gravity anomalies. In contrast, the concave side (Central and North Aegean area) has the fundamental propeties of the active back-arc marginal basins. It is dominated by tensional stress-field, shallow seismic activity associated with normal faulting, relatively thin lithosphere, Plio-Quaternary volcanism, high heat flow, and neotectonic rift valleys. Especially, the South Aegean region (Cretan Trough and its surroundings) seems to be a typical back-arc basin. It is characterized by a well-defined Benioff zone, shallow shocks caused by normal and strike-slip faulting, high heat flow, seismic waves amplitude damping, high intensity of the magnetic field, and positive gravity anomalies. The active calc-alkaline volcanic arc of the South Aegean constitutes the northern front of this region.

The active subduction zone in the South Aegean can very satisfactorily explain the features that have been observed there as well as in the outer part of the Aegean, that is in the Hellenic fore-arc zone.

Volcanic and other geophysical features of Central and North Aegean are not adequately explained by the active underthrusting of the Mediterranean lithosphere beneath the Aegean because, as already mentioned, the South Aegean subduction zone is limited in this area . It appears, therefore, that there should exist a magmatic convection and thermal transfer mechanism from the upper mantle to the crust. Such a mechanism is probably related to the following processes: 1) the actual existence of weak convection currents of asthenospheric material, which where induced by the Eo-Miocene subduction proceses, 2) the active drifting of small lithospheric blocks which may locally cause magma formation; 3) the contribution of convection currents which are induced by the active subduction of the Mediterranean lithosphere.

The suggested geodynamic evolution of the Aegean area in the last 50Ma or so is in accord with the notion that accreted terranes play a leading role in the Alpine-Himalaya chain. Included are the Paikon, Pelagonian, and Gavrovo platforms (Biju-Duval et al., 1977; Geise et al., 1979) as well as the Apullian, Moesian, and Rhodopian fragments (Burchfield, 1980) which may be responsible for the Carpathians and Balkan orogenic belts. According to Nur and Ben-Avraham (1983), "it is questionable whether a full collisioon between Africa and Europe, presumably responsible for the development of the European Alps, actually occurred... We suggested that much of the orogenic complexity in the Alpine Chain is due to the consumption of the separate fragments and not necessarily due to massive continent-continent collision".

Acknowledgement

This paper was prepared at the Geophysical Laboratory, University of Thessaloniki (GLUT), Greece (1981-1982), at the Department of Earth, Atmospheric, and Planetary Sciences, MIT, UASA (1984), and at the Hellenic Air Force Academy (1985-1987). The author remained at MIT as a Visiting Scientist supported by a scholarship from the Ministry of National Economy, Greece, and partly from the Fulbright Scholarships Program. I am grateful to Pr Kei Aki for everything he did to make my period at MIT as productive as possible. Thanks are due to Pr B. Papazachos (GLUT) for encouragement and useful suggestions during the beginning of this research, to Pr. T. Grove (MIT) for the critical reading of the manuscript and improving comments, and to Dr M. Fytikas (IGME, Greece) who kindly offered me unpublished whole rock chemical analyses of the Methana volcanics.

Appendix 1

The average values of the rocks oxides of each site have been plotted to the corresponding petrochemical diagram. The available whole rock analyses cover rocks from 89 sites (see section 2). However, the total number of the plotted points in each diagram is 93 because volcanic and plutonic rocks coexist in the sites of Minos, Kirki, Kos, and Petrota and, therefore, two distinct average values of the oxides have been determined for each one of these sites. In each of the five diagrams the mean curve (curve m), defined by the 93 points, as well as the curve separating the non-alkalic and alkalic fields (curve 0), have been drawn. Then, in each diagram, and index Al_n has been determined for each point as

$$Al_n = \frac{X}{M} \qquad n= 1,2,..., 5$$

where x is the distance of a point from the curve 0, and M is the mean deviation of the 93 points from the curve m. Positive or negative x is used when the point is plotted in the non-alkalic or alkalic field, respectively. The alkalinity index, Al, of each point is defined as

$$AI = \frac{\Sigma Al_n}{n} \qquad n= 1,2,..., 5$$

Appendix 2.

Table 1. Petrochemistry of the Cenozoic magmatic rocks of the Aegean and surrounding regions. Key: Al= alkalinity; index S.D. = standard deviation; h: expected depth of the Benioff zone (in km), C=calc-alkalinic, WC: weak calc-alkalic, WA= weak alkalic, A= alkalic.

N°	Site	Al	S.D.	Petrochemical type	h
Plio-Quaternary active volcanic South Aegean arc					
1	Antimilos isl.	+ 0.86	0.82	C	135
2	Antiparos isl.	+ 0.49	1.48	WC	159
3	Eyina isl.	+ 0.76	0.47	C	144
4	Hristina isl.	+ 0.96	0.45	C	137
5	Kalymnos isl.	+ 1.28	0.98	C	144
6	Kos isl.	+ 0.36	0.93	WC	169

Table I (continued)

N°	Site	Al	S.D.	Petrochemical type	h
7	Krommyonia	+ 0.79	1.03	C	150
8	Methana	+ 1.21	0.53	C	125
9	Milos isl. group	+ 1.02	0.72	C	137
10	Nissyros isl. group	+ 0.85	0.56	C	137
11	Poros isl.	+ 0.99	0.50	C	135
12	Santorini isl. group	+ 0.49	0.61	WC	163
13	Tilos isl.	+ 0.97	1.15	C	176

Plio-Quaternary volcanics of Central and NW Aegean

14	Ahillio	+ 0.11	1.00	WC	
15	Almopia North	- 0.65	1.27	WA	
16	Almopia South	- 0.95	1.51	A	
17	Ayios Ionnis, Lokrida	+ 0.67	0.91	WC	
18	Kaloyiri	- 0.93	1.65	A	
19	Kozut	- 0.38	1.48	WA	
20	Kula	- 1.23	1.33	A	
21	Lihades isl.	+ 0.38	0.74	WC	
22	Metohi, Evia isl.	+ 0.19	0.30	WC	
23	Micrithives	- 0.002	1.94	WA	
24	Portyrion	+ 0.49	2.00	WC	
25	Psathoura isl.	- 0.004	1.12	WA	
26	Vromolimni, Lokrida	+ 0.88	1.02	C	
27	Doirani	- 3.37	3.26	A	
28	Gavra	- 1.44	1.30	A	
29	Pontokerassia	+ 0.38	3.32	WC	
30	Rizana	- 0.12	3.66	WA	
31	Sitsi-Kamen	+ 4.77	4.51	C	
32	Strymoniko	- 0.29	5.19	WA	
33	Vathi	- 2.94	6.52	A	
34	Yierakari	- 4.49	12.92	A	

Early Miocene magmatic rocks

35	Bodrum	- 0.17	1.29	WA	258
36	Ikaria isl.	- 0.11	1.34	WA	166
37	Kos isl.	+ 0.04	1.00	WC	237
38	Lavrio	+ 0.89	1.21	C	141
39	Labedos	+ 0.22	1.91	WC	210
40	Milas	- 0.42	0.65	WA	310
41	Mugla	+ 0.25	0.85	WC	242
42	Mykonos isl.	+ 0.76	1.42	C	169
43	Patmos isl.	- 0.67	1.20	WA	243
44	Samos isl.	+ 0.23	0.20	WC	142
45	Seritos isl.	+ 10.9	1.22	C	138

Table I (continued)

N°	Site	Al	S.D.	Petrochemical type	h
46	Tinos isl.	+ 0.93	1.77	C	150
47	Urla	− 0.11	0.37	WA	209
	Late Miocene magmatic rocks				
48	Antipsara isl.	+ 0.40	1.44	WC	177
49	Antistrovilas, Chios isl.	+ 1.28	1.90	C	171
50	Aydin	+ 1.51	1.10	C	141
51	Ayii Pantes, Chios isl.	+ 0.89	1.41	C	150
52	Ayios Efstratios isl.	+ 0.54	1.03	WC	190
53	Ayvalik	− 0.02	1.66	WA	250
54	Bares, Skyros isl.	+ 1.12	0.72	C	135
55	Dikili-Bergama	+ 0.49	0.98	WC	190
56	Emborio, Chios isl.	+ 0.82	1.02	C	157
57	Ermegan Da-Koka Da	+ 0.46	0.80	WC	177
58	Ezine	− 0.99	0.97	WA	259
59	Foca	+ 0.48	0.71	WC	188
60	Imroz isl.	− 0.01	0.67	WA	217
61	Izmir	+ 0.26	0.98	WC	210
62	Karaburun	+ 0.77	0.91	C	179
63	Lesvos isl.	+ 0.34	0.94	WC	197
64	Limnos isl.	− 0.21	1.87	WA	261
65	Limnos isl.	+ 1.24	1.02	C	203
66	Manisa	+ 1.04	0.66	C	145
67	Orio, Evia isl.	+ 1.09	1.53	C	136
68	Oxylithos, Evia isl.	+ 1.26	1.00	C	140
	Eo-Oligocene magmatic rocks				
69	Asprolakkos	+ 0.18	0.53	WC	
70	Dipotama	+ 0.39	1.30	WC	
71	Emonio-Kotyli	+ 0.56	1.25	WC	
72	Ferres-Sappes	+ 0.77	0.94	C	
73	Ierissos	+ 0.56	1.41	WC	
74	Issmaros	− 0.002	2.06	WA	
75	Kalotyho	− 0.04	0.60	WA	
76	Kassitera	+ 1.31	1.35	C	
77	Kelembek	+ 1.56	2.24	C	
78	Kerdyllia	+ 1.11	1.16	C	
79	Kirki	+ 1.10	1.01	C	
80	Kirki	+ 0.93	0.50	C	
81	Kymi Komotinis	+ 0.49	2.22	WC	
82	Megali Panalyia, Chalkidiki	− 0.12	0.99	WA	
83	Petrota	+ 19.1	1.22	C	
84	Petrota	+ 0.37	0.83	WC	

Table I (continued)

N°	Site	Al	S.D.	Petrochemical type	h
85	Samothraki isl.	+ 0.12	1.00	WC	
86	Serres-Drama	+ 0.14	0.87	WC	
87	Sidironero	+ 0.77	0.95	C	
88	Stefanina	- 0.49	0.96	WA	
89	Stratoniki	+ 0.93	0.90	C	
90	Sunio Xanthis	+ 0.36	0.49	WC	
91	Triantafyllia	+ 0.84	0.71	C	
92	Tuzla	- 0.26	1.05	WA	
93	Xanthi	- 0.13	0.83	WA	

References

Altherr, R., Keller, J., Harre, W., Höhndorf, A., Kreuzer, H., Lenz, H., Rashkal, H., and Wendt, I., 1977. Geochronological data on granitic rocks of the Aegean Sea, Preliminary results, In On the Structural History of the Mediterranean Basins, Intern. Symp. Proc., Split, 1976, 318-318.

Barberi, F., Innocenti, F., Marinelli, G., and Mazzuoli, R., 1974. Vulcanismo e tettonica a placche: Esempi nell'area Mediterrana, Ist. di Miner. e Petr., Pisa, 327-356.

Bizu-Duval, B., Dercrout, J. and Le Pichon, X., 1977, From the Tethys Ocean to the Mediterranean Seas: A plate tectonic model of the evolution of the Western Alpine system, In On the Structural History of the Mediterranean Basins, Intern. Symps. Proc., Split, 1976, 143-164.

Boccaletti, M., Manetti, P., and Peccerillo, A., 1974. The Balkanids as an instance of a back-arc thrust belt: possible relations with the Hellenids, Geol. Soc. Amer. Bull., 85, 1077-1084.

Bowen, N.L., 1928. The evolution of the igneous rocks, (Princeton Univ. Press, Princeton), 334pp.

Burchfield, B.C., 1980. Eastern Europe Alpine system and the Carpathian orocline as an example of collision tectonics, Tectonophysics, 63, 31-61.

Caputo, M., Panza, G.F., and Postpischl, D., 1970. Deep structure of the Mediterranean basin, J. Geophys. Res., 75, 4919-4923.

Christodoulou, G., and Tataris, A., 1972. On the geological structure of the Tilos island, Bull. Geol. Soc. Greece, 9, 28-80 (in Greek).

Comninakis, P.E., and Papazachos, B.C., 1980. Space and time distribution of the intermediate focal depth earthquakes in the Hellenic arc, Tectonophysics, 70, T35-T47.

Dickinson, W.R., 1968. Circum-Pacific andesite type, J. Geophys. Res., 73, 2261-2269.

Dickinson, W.R.,1975. Widths of modern arc-trench gaps proportional to past duration of igneous activity in associated magmatic arcs, J. Geophys. Res., 78, 3376-3380.

Fytikas, M., Giuliani, O., Innocenti, F., Marinelli, G., and Mazzuoli, R., 1976. Geochronological data on recent magamtism of the Aegean Sea, Tectonophysics, 31, 29-34.

Fytikas, M., Innocenti, F., and Mazzuoli, R., Geology and petrology of the Methana Peninsula, Unpublished study.

Fytikas, M.,Innocenti, F., Manetti, P., Mazzuoli, R., Peccerillo, A., and Villari, L., 1985.Tertiary to Quaternary evolution of volcanism in the Aegean region, In: I.E. Dixon and A.H.F. Robertson (eds.), The Geological Evolution of the Eastern Mediterranean, *Spec. Publ. of the Geol. Soc.*, N° **17**, (Oxford, Blackwell Sci. Publ.) 687-699.

Geise, P., Gorler, K., Jakobshagen, V., and Reutter, K.J., 1979. Geodynamic evolution of the Apennines and Hellenides, In: *Mobil Earth, Closs Intern. Geodyn. Proj.*, Boppard, 71-87.

Gorshkov, G., 1969. Geophysics and petrochemistry of andesite volcanism of the Circum-Pacific belt, *Oregon Dept. Geol. Min. Bull.*, **65**, 91-98.

Green, D.H. and Ringwood, A.E., 1967. The genesis of basaltic magmas, *Contr. Miner. Petrol.*, **15**, 103-190.

Hatherton, T., and Dickinson, W. R., 1969. The relationship between andesitic volcanism and seismicity in Indonesia, the Lesser Antilles and other island arcs, *J. Geophys. Res.*, **74**, 5301-5310.

Jakes, P., and White, A.J.R., 1972. Major and trace element abundances in volcanic rocks of orogenic areas, *Geol. Soc. Amer. Bull.*, **83**, 29-40.

Jongsma, D., 1974. Heat flow in the Aegean Sea, *Geophys. J. Roy. Astr. Soc.*, **37**, 337-346.

Innocenti, F., Kolios, N., and Manetti, P., 1982. Acid and basic late Neogene volcanism in Central Aegean Sea: its nature and geotectonic significance, *Bull. Volcanol.*, **45**, 87-97.

Keller, J., 1969. Origin of rhyolites by anatectic melting of granitic crustal rocks, the example of rhyolitic pumice from the island of Kos (Aegean Sea), *Bull. Volcanol.*, **33**, 942-959.

Kondopoulous, D., Papadopoulos, G.A., and Pavlides, S.B., 1985. A study of the deep seismotectonics in the Hellenic arc, *Boll. Geof. Teor. Appl.*, **27**, 197-207.

Kuno, H., 1959. Origin of Cenozoic petrographic provinces of Japan and surrounding areas, *Bull. Volcanol.*, **20**, 37-76.

Kuno, H., 1966. Lateral variation of basaltic magma type across continental margins and island arcs, *Bull. Volcanol.*, **29**, 195-222.

Le Pichon, X., and Angelier, J., 1979. The Hellenic arc and trench system: a key to the neotectonic evolution of the Eastern Mediterranean area, *Tectonophysics*, **60**, 1-42.

McKenzie, D., 1972. Active tectonics of the Mediterranean region, *Geophys. J. Roy. Astr. Soc.*, **30**, 109-185.

McKenzie, D., 1978. Active tectonics of the Alpine-Himalayan belt: the Aegean Sea and surrounding regions, *Geophys. J. Roy. Astr. Soc.*, **55**, 2117-254.

Miyashiro, A., 1974. Volcanic rock series in island arcs and active continental margins, *Amer. J. Sci.*, **274**, 321-355.

Ninkovich, D., and Hays, J.D., 1972. Mediterranean island arcs and origin of high potash volcanoes, *Earth Planet. Sci. Lett.*, **16**, 331-345.

Nur, A., and Ben-Avraham, Z., 1983. Break-up and accretion tectonics, In: *Accretion tectonics in the Circum-Pacific Regions*, Hashimoto and Uyeda Eds., Terra Sci. Publ. Co., Tokyo, 3-13.

Panagos, A.G., PE, G.G., and Vanrnavas, S.P., 1978. The volcanic rocks of Strymonikon-Metamorphosis, Central Macedoine, Greece, *Chem. Erde*, **37**, 50-61.

Padadopoulos, G.A., 1982. Contribution to the study of the active deep tectonic of the Aegean and surrounding regions, *D. Sci. Thesis*, Univ. of Thessaloniki, 176pp (In Greek with English abstract).

Papadopoulos, G.A., and Andrinopoulos, A., 1984. Metallogenic evidence for palaeo-subduction zones in the Aegean area, *Geologica Balcanica*, **14**, 3-8.

Papadopoulos, G.A., Kondopoulou, D., Leventakis, G.A., and Pavlides, S.B., 1986. Seismotectonics of the Aegean region, *Tectonophysics*, **124**, 67-84.

Papazachos, B.C., 1976. Evidence of crustal shortening in the Northern Aegean region, *Boll. Geot. Teor. Appl.*, **13**, 66-71.

Papazachos, B.C., and Comninakis, P.E., 1971. Geophysical and tectonic features of the Aegean arc, *J. Geophys. Res.*, **76**, 8517-8533.

Papazachos, B.C., and Comninakis, P.E., 1977. Modes of lithospheric interaction in the Aegean area, In *On the Structural History of the Mediterranean Basins*, Intern. Symp. Proc., Split, 1976, 3119-331.

Papazachos, B.C., and Delibasis, N.D., 1969. Tectonic stress-field and seismic faulting in the area of Greece, *Tectonophysics*, **7**, 231-255.

Papazachos, B.C. and Papadopoulos, G.A., 1977. Deep tectonic and associated ore deposits in the Aegean area, *Proc. 6th Colloq. Geology Aegean Region*, **3**, 1071-1080.

Paraskevopoulos, G., 1956. Über den Chemismus und die Provizialen Verhältnisse der Tertieren und Quatären Ergussgesteine des Ägäischen Raumes und der Benachbarten Gebiete, *Tsch. Min. Petr. Mit.*, **3**, 13-72.

Pe, G., and Panagos, A., 1976. Comparative geochemistry of the Northern Euboecos lavas, *Bull. Geol. Soc. Greece*, **12**, 95-133, (in Greek with English abstract).

Pe, G., and Piper, D.J.W., 1972. Vulcanism at subduction zones: the Aegean area, *Bull. Geol. Soc. Greece*, **8**, 133-144.

Pichler, H., and Stengelin, R., 1968. Petrochemische und Nomenklatorische Revision der Vulkanite des Süd-"Ägäischen Raumes (Griechenland), *Geol. Rund.*, **57**, 795-810.

Puchelt, H., Murad, E., and Hubbertin, H.W., 1977. Geochemical and petrological studies of lavas, pyroclastic and associated xenoliths from the Christiana Island, Aegean Sea, *N. Jb. Min. Abh.*, **131**, 140-155.

Ringwood, A.E., 1969. Composition and evolution of the upper mantle, In: *The Earth's Crust and Upper Mantle*, Amer. Geophys. Union Geophys. Monogr., **13**, 1-17.

Ritsema, A.R., 1974. The earthquake mechanism of the Balkan region, *R. Netherl. Meteor. Inst. Rep.*, **74**, 1-36.

Rittmann, A., 1957. On the serial character of igneous rocks, *Egypt. J. Geol.*, **1**, 23-48.

Tilley, C.E., and Muir, I.D., 1967. Tholeiite and tholeiitic series, *Geol. Mag.*, **104**, 337-343.

A PATTERN OF BLOCK ROTATIONS IN CENTRAL AEGEA

C. KISSEL[1], C. LAJ[1], A. POISSON[2], K. SIMEAKIS[3] .
[1]Centre des Faibles Radioactivités,
Laboratoire mixte CNRS/CEA,
Avenue de la Terrasse,
91198 Gif-sur-Yvette Cedex, France.

[2] Laboratoire de Géologie Historique,
Université Paris XI,
91405 Orsay, France.

[3] Institue of Geological and Mining
Messoghion Street,
Athens, Greece.

ABSTRACT. Paleomagnetic results obtained from Neogene formations on both sides of the Aegean Sea yield of a complex pattern of block rotations in these regions. On the western side, in Evia and Skyros the observed clockwise rotations of various angles are interpreted within the framework of a previously proposed model of distributed deformation by faulting in a zone connecting the north-Anatolian trough to the outer arc. On the eastern side, significant clockwise, counterclockwise and null rotations of coherent blocks have been demonstrated in the island of Lesbos and in western Anatolia. These movements took place during the neotectonic extensional period and their mechanism is not yet fully understood.

1. Introduction

In the last few years, many tectonics, paleomagnetic and geophysical studies have led to the recognition of the extreme mobility of the Aegean lithosphere and of the very large deformations which have occurred in this area in the last few million years, during the neotectonic period (McKenzie, 1978; Le Pichon and Angelier, 1979; Mercier et al., 1979; Kissel and Laj, 1988). Two main domains have been individualized in Agea on the basis of their neotectonic evolution. While a compressive regime is active at the two terminations of the arc since at least the Middle Miocene (Mercier et al., 1979), most of the Aegean area has been dominated by an extensional regime only occasionally interrupted by compressive events in the Lower Pliocene and Lower Pleistocene periods (Mercier et al., 1979; Le Pichon and Angelier, 1979). This extension has resulted in large-scale normal faulting, grabens, and troughs.

The neotectonic period is also characterized in Anatolia and northern Aegea by a dextral strike-slip motion along the north-Anatolian fault, the western termination of which is located north of Evia. Recently, McKenzie and Jackson (1983, 1986) have proposed a model in which the right-handed NE-SW strike-slip motion along this fault is taken up north of Evia on a number of several normal faults striking ESE with a left-handed strike-slip component. This system forms a shear zone connecting the North-Aegean trough to the Hellenic arc. The rigid blocks bounded by the

115

C. Kissel and C. Laj (eds.), Paleomagnetic Rotations and Continental Deformation, 115–129.
© 1989 by Kluwer Academic Publishers.

normal faults can be shown to undergo surprisingly large clockwise rotations when geologically reasonable slip rates related to the SW expansion of Aegea are used in the model.

In this paper, we report paleomagnetic results obtained from Middle Miocene to Upper Pliocene sedimentary and volcanic formations in central Aegea, on both sides of the Aegean Sea. The data obtained in Evia region are discussed within the framework of the above model whereas the complexity of the rotational pattern observed in western Anatolia is not yet understood.

2. Geological setting.

Central Aegea is characterized by widespread calc-alkaline volcanic products of both Tertiary and Quaternary ages. Different authors have recognized the existence of two main distinct volcanic phases of Oligo-Miocene and Plio-Quaternary ages separated by a period of quiescence of several million years during the Upper Miocene period (Bellon et al., 1979; Innocenti et al., 1981). While the most recent products are found in the southern Aegean sector, older volcanics of Oligo-Miocene age are distributed all over the central and northen Aegean regions and in Western Anatolia (Borsi et al., 1972, Bellon et al., 1979, Fytikas et al., 1984). The older volcanic phase is considered to have stopped about 13 Ma ago (the youngest products are located in Evia), while the first Plio-Quaternary products were erupted around 3 Ma ago along the southern Aegean arc. Between the two calc-alkaline cycles, small volumes of lava with variable petrogenetic characters were emitted over all central Aegea and western Anatolia.

The Neogene sedimentary formations in central Aegea are essentially represented by continental (fluviolacustrine) and brackish limestones and marls which indicate the quasi-definitive emersion of these regions. These formations are not precisely dated. In western Anatolia, as a result of the main tectonic activity, widespread volcano-sedimentary deposits are also present.

Both sedimentary and volcanic formations have been sampled on both sides of the Aegean Sea.

2.1. WESTERN SIDE OF THE AEGEAN SEA.

Calcalkaline products have been sampled in the islands of Skyros (5 sites) and Evia (6 sites) (Figure 1). In Skyros, the andesites are dated at 15 Ma (Fytikas et al., 1979) and they have been intruded into the Mesozoic carbonatic series which appear more or less horizontal in this region. In Evia, the Oxylithos volcanic massif is dated at about 13 Ma (Fytikas et al., 1979). It lies on the lacustrine marly limestones, the age of which is between Aquitanian and Pontian (Lemeille, 1977). The bedding plane regularly tilting a few degrees towards the south and measured at different places on these sediments has been used to restore the volcanic formations to their paleohorizontal position.

Only three other sites were sampled in Evia in sedimentary formations in spite of careful search. Indeed the southern part of the island is mainly metamorphosed and the Neogene lacustrine basins in the northern part are often covered by thick vegetation or weathered. One of these sites (EU 232) is located in the blue-grey marls which underly the volcanic massif of Oxylithos. The two other sites (EU 233 - 234) are located in northern Evia, in lacustrine limestones, the age of which is Pliocene whithout any precision (Figure 1). Three sites were sampled in the Pliocene limestones of the island of Alonissos.

Finally, a very small volcanic massif of basaltic affinity and dated at about 3 Ma (Innocenti et al., 1979) has been sampled (4 sites) in mainland Greece, north of Evia, near the village of Glifa. We have assumed on the basis of the general aspect and the morphology of the flows that the massif has not been significantly tilted.

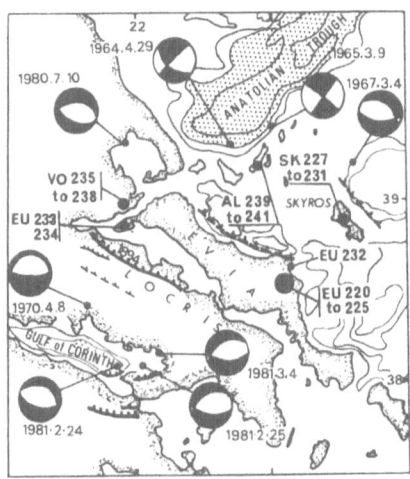

Fig. 1. Schematic map showing, according to McKenzie and Jackson (1983), the geometry of normal faulting zone on the western side of the Aegean Sea. Fault plane solutions are lower-hemisphere projections with compressional quadrants blacks. The location of the sampled area is indicated.

2.2 THE EASTERN SIDE OF THE AEGEAN SEA.

Since at least late Miocene, this motion is accompanied by large scale N-S extension in the Aegean and in Western Turkey with complementary E-W shortening (Tapponnier, 1977; Dewey and Sengör, 1979; Mercier et al., 1979; Le Pichon and Angelier, 1981; Angelier et al., 1981).

The most striking structural and morphological features of this region are the E-W and NNE trending grabens (McKenzie, 1978; Dewey and Sengör, 1979). Focal mechanism solutions obtained from some of these grabens, have shown that motion takes place on faults of listric geometry (Eyidogan and Jackson, 1985). The NNE faults are probably partially inherited while the E-W fault system corresponds to a neotectonic structural direction. The complex pattern of faulting includes numerous other neotectonic faults of various orientations and displacements (Kaya,1981), most of which are truncated by younger faults of different orientations. Slickensides indicating strike-slip movements in opposite senses and in different directions are also frequently observed in the field.

A total of 77 sites have been sampled mainly in volcanic formations on the eastern side of the Aegean Sea (Figure 2). In the island of Lesbos, the calc-alkaline volcanic event has been dated by K/Ar method between 15 and 18 Ma (Borsi et al., 1972) and we have sampled 26 sites in ignimbrites, lava flows and dykes. Only 4 sites were sampled in the Neogene lacustrine sedimentary formations of this island; three of them in limestones and one in marls which are ill dated but probably lower Pliocene.

118

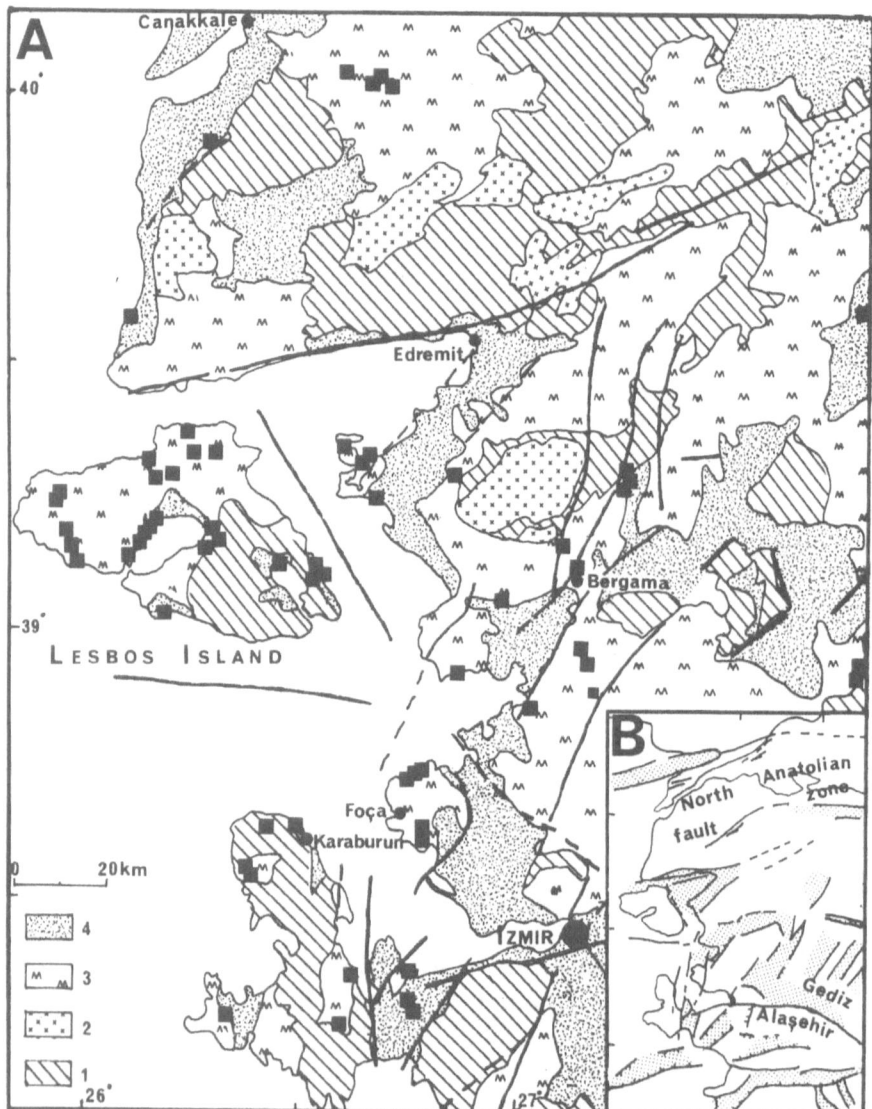

Fig. 2. Schematic map showing (A) the most proeminent structural characteristics of the eastern side of the Aegean sea (the E-W and NNE trending grabens) and (B) the location of the sampled sites . Symbols: 1. Neogene continental basins; 2. Neogene volcanic and volcano-sedimentary formations; 3. Neogene plutonic formations; 3. Neogene plutonic formations; 4. Paleozoic basement.

In western Anatolia, 13 sites are located in Neogene lacustrine sedimentary and volcano-sedimentary formations and 34 other sites have been sampled in calc-alkaline and basaltic products (Figure 2). The calc-alkaline event has been dated in this area between 21 and 16 Ma (Borsi et al., 1972). The basaltic rocks were erupted during a younger event related to the extensive tectonic regime and some of these products have been dated in this area at 11 Ma (Borsi et al., 1972).

The bedding plane attitude was measured in some cases on interbedded sediments. In some others, a local structural study gave a reasonable estimate of it, and in a very few cases where no precise correction for tilting was possible, we had to rely on a subjective choice of 'almost horizontal' lava flows based on the morphology of the flow itself.

All the cores were oriented using both sun and magnetic compass. Generally 10-12 cores were obtained from each sites, each core corresponding to 2 to 3 samples.

3. Paleomagnetic method

One sample per core i.e. 10-15 samples par site were thermally demagnetized with a minimum of 12 steps from room temperature to about 550°C and measured using either a Digico spinner or a LETI 3-axis cryogenic magnetometer depending on their intensity. The bulk susceptibility, measured at each step remained very stable up to the highest temperatures indicating that the magnetic minerals

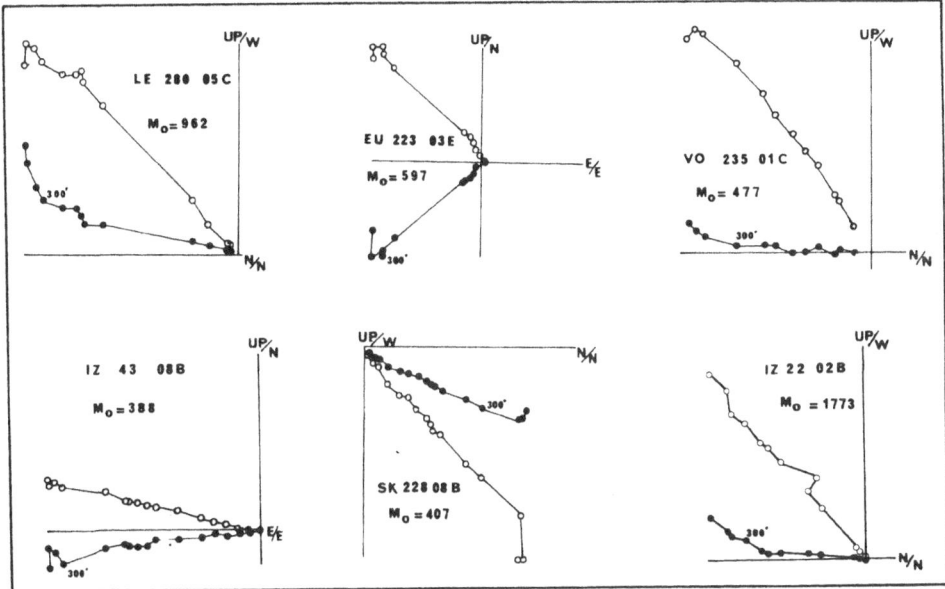

Fig. 3. Typical demagnetization diagrams. Full circles: horizontal projection; open circles, vertical projection.

were not seriously affected by the thermal treatment. The IRM acquisition curves in fields up to 1.5T also show that the main magnetic carrier in the studied samples is magnetite.

As shown in Figure 3, apart a small viscous component which is removed at about 300°C, only one single stable component has been isolated and its direction is easily determined within 2-3 degrees of accuracy.

4. Results and discussion.

A total of 22 out of 27 sites gave reliable results on the western side of the Aegean Sea. One site in Skyros is rejected because its mean direction is completely different from the ones obtained from the other sites. As this site is located on a slope, this difference might result from a small local landslide, unnoticed because of a rather thick vegetation. In Evia, the only Miocene sedimentary site did not give reliable result because of a too large within-site scatter (K = 3) in spite of perfectly rectilinear demagnetization diagram. No result could be obtained from the Alonnissos island because of a too weak intensity of magnetization.

On the eastern side of the Aegean Sea, the laboratory measurements have shown that some of the sampled lithologies are not suitable for a paleomagnetic study. In particular, different sites from the ignimbrites and the dykes sometimes gave conflicting results and usually an unreasonable within-site scatter. Ten sites in Lesbos and 15 sites in Turkey were rejected on this basis. Some sedimentary sites (limestones in Lesbos and tuff formation in western Anatolia) could not be measured because of their too weak intensity.

4.1. WESTERN SIDE OF THE AEGEAN SEA

The mean directions calculated using the Fisher's statistic for each reliable site are given in Table I as well as the regional mean direction.

In Skyros the within-site scatter is very small and the mean declination is of about 26°.

All the sites from the volcanic massif of Oxylithos in Evia have a reverse polarity and the mean regional declination calculated after bedding correction is about 228°. On the northern part of Evia, the lacustrine limestones are also of reverse polarities. Although it is unsignificant to calculate a statistical regional mean direction from only two sites, one can notice that the mean direction from each site are much better grouped after than before bedding correction. They are also quite consistent with the directions obtained from the Miocene volcanic sites.

The four sites studies in the region of Glifa yield tightly grouped results with reverse directions and declinations pointing perfectly south.

A recent examination of all the available paleomagnetic data from the Aegea leads to the conclusion that this entire domain has been linked to the African plate during most of its tertiary geodynamical evolution (Kissel and Laj, 1988). Using the Miocene African poles proposed by Westphal et al. (1986) and the geocentered dipole field for the 3 Ma formations, the expected paleomagnetic direction is D = 5.8°; I = 53.2° in Evia and Skyros, and D = 0° and I = 57° at Glifa respectively. The results thus indicate that the Glifa region has not undergone any significant rotation in the last 3 million years while Evia and Skyros have undergone clockwise rotation of about 43° and 20° respectively since at most the upper Miocene (Figure 4).

The age of this clockwise rotation is not precisely constrained by the data. However a paleomagnetic study of coeval formations in northwestern Greece has documented a 45° clockwise

Table I: Paleomagnetic data from the western side of the Aegean Sea

Sites	Age Ma	n	before B.C		after B.C		K	α_{95}
			D	I	D	I		
			region of Glifa (1)					
VO 235	3	.9	-----	-----	167	-61.5	266	2.8
VO 236	3	6	-----	-----	190	-54.0	26	11.3
VO 237	3	10	-----	-----	165	-67.0	120	4.0
VO 238	3	7	-----	-----	184	-54.7	29	9.8
			Skyros island (2)					
SK 227	13	10	-----	-----	31.8	40.0	189	3.2
SK 228	13	10	-----	-----	20.0	38.0	54	6.0
SK 229	13	10	-----	-----	29.7	49.0	77	5.0
SK 230	13	6	-----	-----	22.0	55.0	27	11.0
			Evia island (Miocene) (3)					
EU 220	15	7	221.8	-28.5	228.3	-37.0	949	1.7
EU 221	15	7	220.2	-42.3	230.8	-50.6	320	2.9
EU 222	15	6	223.0	-28.1	229.5	-36.2	63	7.2
EU 223	15	9	218.4	-32.4	225.6	-41.3	320	2.6
EU 224	15	10	235.7	-36.0	245.3	-41.5	45	6.5
EU 225	15	10	199.7	-41.8	206.4	-53.3	201	3.1
EU 232	15	10	223.0	-43.0	204.0	-65.0	3	24.0
			Evia island (Pliocene) (4)					
EU 233	Plioc?	10	254.0	-38.0	235.0	-58.5	115	4.0
EU 234	Plioc?	7	216.5	-22.0	226.5	-44.7	34	9.0

Mean regional directions after bedding correction:
(1)   N = 4   D = 178.0 I = -59.0 K = 85 α_{95} = 7.6
(2)   N = 4   D = 26.0 I = 45.5 K = 82 α_{95} = 7.7
(3)   N = 6   D = 228.3 I = -43.8 K = 53 α_{95} = 7.8
(3) + (4) N = 8   D = 228.7 I = -45.8 K = 56 α_{95} = 6.6
The K/Ar ages refer to the sampled units but have not been determined on single sites.

rotation occurring in two phases of comparable amplitude (Horner and Freeman, 1983; Kissel et al., 1985). The first rotational phase occurred during the Middle Miocene and the second one during the Pliocene and the Quaternary. The most recent phase has been precisely time bracketed: its onset coincides with the main compressive phase of upper Miocene - lower Pliocene age (5 Ma) and it has since proceeded at a constant rate of 5°/Ma (Laj et al., 1982). On the contrary the exact timing of the former rotation is rather completed at 12 Ma, because a period of at least 7 Ma, between 12 and 5 Ma, during which no significant rotation occurred has also been documented (Laj et al., 1982). This period is thus coeval with the epoch of quiescence separating the two phases of calc-alkaline volcanic activity, so that it seems unrealistic to us that major geodynamical movements might have occurred in central Aegea during a period of overall very reduced or null tectonic or volcanic activity. A much more reasonable hypothesis is that the large rotations documented in Evia and Skyros have occurred synchronously with one or the other of the two phases of rotation documented in northwestern Greece.

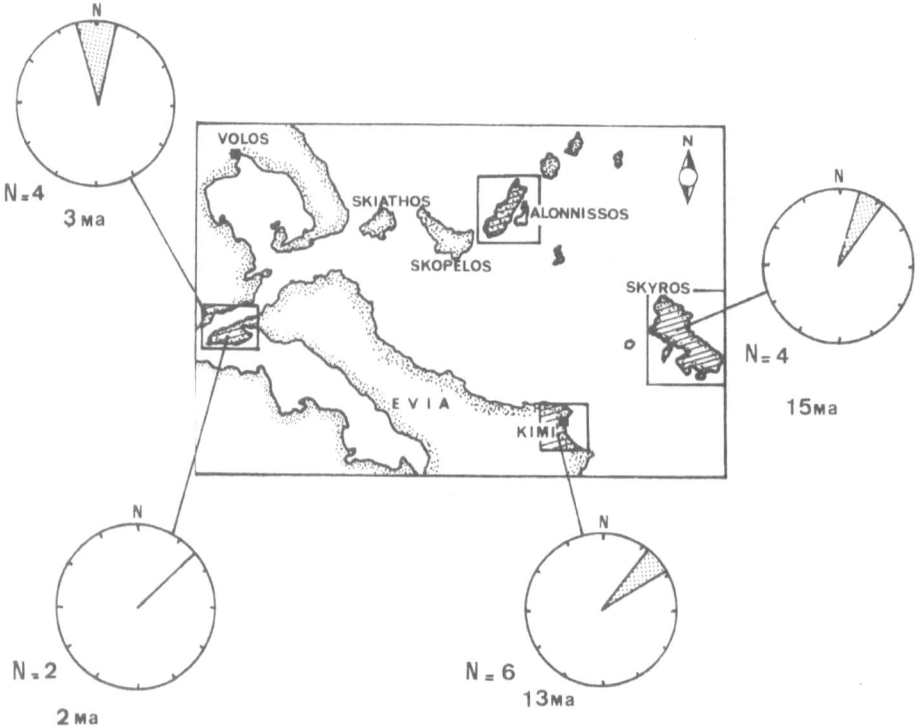

Fig. 4. Schematic map showing the paleomagnetic results obtained on the western side of the Aegean sea. The observed rotations are indicated in the circles with the number of sites (N) and their respective age. No results have been obtained from the island of Alonissos because the magnetization of the samples was too weak. Only two sites have been studied in orthern Evia, so that the dispersion parameters k and α_{95} cannot be calculated and only the vector mean is indicated.

The age of the volcanic formation in Skyros does not allow to attribute the rotation of this island to one or the oher of the two phases. The volcanic products in Evia on the contrary mark the end of the oligo-miocene phase (Fytikas et al., 1984) and the sedimentary formations are much younger. Although we are well aware that our data are scarce, we suggest that the rotations observed in Central Aegea are more likely related to the Plio-Pleistocene rotation of the external zone than to the middle Miocene one. If correct, this suggestion implies that, during the Pliocene, coeval rotational movements were of greater amplitude in central Aegea than in northwestern Greece. This situation can be explained by the damping of the right-handed strike-slip movement along the north Anatolian fault which has a western termination north of Evia and marks a main discontinuity in central Aegea since the Late Miocene.

The only model which specifically considers the damping of the movement along the north Anatolian fault is the one suggested by McKenzie and Jackson (1983) in which this region is considered as a shear zone formed by fault-limited rotating blocks, between the north Aegean trough and the Hellenic arc (Figure 1). Assuming a slip rate of 5 cm/yr accross a shear zone 50 km wide, these authors have calculated an instantaneous rotation rate of 40° to 20° par million years depending on whether the rotating blocks are pinned to the boundaries of the shear zone or are floating. Geological studies have not evidenced major changes in the tectonic pattern of central Aegea in the last few million years so that it is reasonable to assume that the shear zone has existed during this time. Taking a width of 90 km, which seems to us more realistic than 50 km, then the observed rotation of 48° implies a very reasonable mean slip rate of the order of 1.5 to 2.5 cm/yr over the last 5 Ma.

The Volos region, situated north of the shear zone, is supposed to be fixed in the model. This assumption is in agreement with the paleomagnetic results. Moreover, noting that the rotation of Skyros is closely one half that of Evia, this difference could be explained in term of this model, assuming that Skyros belongs to a block more loosely attached to the borders of the shear zone or entirely floating.

It must, however, be noted that if the deformation has to be approximated by the movement of perfectly rigid blocks separated by faults which move both normal and parallel to their strike, a definite angular relationship must exist between the strike of the faults, the trend of the zone and the velocity gradient. For a given trend and velocity only a particular orientation of the faults will fulfil this condition: this is the present day situation in central Aegea (McKenzie and Jackson, 1983). This angular relationship can however be satisfied only instantaneously, so that if the fault limited blocks have to be rotated through a finite angle either the trend of the zone or the velocity gradient must change, or else one has to consider that deformation of the blocks also occurs. These points are difficult to check because the submerged character of central Aegea does not allow to extend the paleomagnetic sampling. In the same way a denser sampling is needed in northern Evia to check whether the rather large rotations of the young sedimentary formation could be at least partially due to local rotation linked to deformation at the boundaries of the shear zone.

4.2. EASTERN SIDE OF THE AEGEAN SEA

All the reliable results are shown in Table II, together with 3 sites (Cesme 1, 2, 3) from the Karaburun peninsula reported by Lauer (1981). In each region the normal and reversed directions are antipodal, indicating a good magnetic stability and a satisfactory cleaning. After bedding correction the inclination values from both sedimentary and volcanic formations from the different regions are quite consistent, with mean values around 50°, except for the one calculated from the 3 sites south of Bergama which is shallower.

The mean declinations from the different regions are on the contrary quite divergent. In Lesbos and in the Canakkale region they are aligned with the N-S direction. The Izmir region is characterized by a 35° westerly declination, except for the 3 sites south of Bergama where a 20° easterly declination is observed. In the Karaburun peninsula the declinations are all deviated eastward (mean value 49°) with a large scatter. We have no explanation for this scatter which is not, in our opinion, related to erroneous declination values arising from the use of the usual tilt correction about the strike of the dip. Indeed all the measured dips in this region are of the order of 20°-30°, so that very limited effects are to be expected in this case (McDonald, 1980).

Using the same Miocene African pole proposed by Westphal et al. (1986), one can notice that the observed inclination values are in good agreement with the expected ones. An examination of the observed declinations then indicates quite a complex pattern of rotations of adjacent blocks. Indeed, while Lesbos and the Canakkale region have not undergone any significant rotation, the Karaburun Peninsula has been affected by a 44° clockwise rotation and the Izmir region has undergone a 37° anticlockwise rotation. The volcanic massif located south of Bergama has rotated clockwise with an angle of some 18°.

Table II: Paleomagnetic data from the eastern side of the Aegean Sea

Sites	Age Ma	n	before B.C		after B.C		K	α_{95}
			D	I	D	I		
			Canakkale region (1)					
IZ50	----	8	194.0	-65.0	194.0	-50.0	24	10.0
IZ51	----	8	183.2	-70.4	183.2	-55.4	252	3.1
IZ52	----	8	176.6	-69.6	176.6	-54.6	165	3.8
IZ54	----	9	8.3	60.5	8.3	45.5	205	3.2
			Lesbos island (2)					
LE249	----	10	8.0	37.0	11.2	27.0	280	2.6
LE252	----	8	0.2	49.0	4.8	40.2	153	4.0
LE253	----	9	13.6	59.5	17.2	49.7	196	4.0
LE255	----	11	----	-----	11.3	53.3	704	1.6
LE256	16.9	11	----	-----	340.5	55.0	229	2.8
LE257	16.9	10	----	-----	350.0	69.0	262	3.0
LE263	18.0	9	----	-----	177.8	-49.0	1333	1.3
LE269	18.0	8	----	-----	146.0	-63.6	177	3.7
LE271	18.0	11	----	-----	214.0	-20.0	15	10.0
LE272	18.0	10	----	-----	191.7	-44.7	181	3.3
LE273	15.5	10	31.4	58.5	19.1	56.0	257	2.7
LE274	15.5	10	358.6	44.3	353.5	38.7	134	3.8
LE275	15.5	9	----	-----	17.3	64.0	233	3.0
LE277	15.5	9	----	-----	349.5	45.5	170	3.5
LE278	15.5	9	----	-----	10.0	44.5	72	5.0
LE279	18.0	7	----	-----	199.5	-20.0	144	4.4
LE280	18.0	7	----	-----	191.0	-41.0	46	7.8
			Izmir-Bergama region (3)					
IZ06	11.3	12	----	-----	128.0	-47.0	158	4.2
IZ07	11.7	8	144.0	-34.0	141.0	-39.5	249	3.3
IZ09	7.0	12	169.0	-62.5	127.2	-41.0	93	4.2
IZ10	7.0	8	192.0	-55.0	144.0	-47.5	500	2.2
IZ24	----	7	----	-----	348.0	64.0	56	7.0
IZ28	----	8	346.0	54.0	349.6	57.0	57	4.0
IZ44	18.2	9	----	-----	342.0	71.4	828	1.6
IZ46	18.2	8	319.5	53.8	314.0	53.8	540	2.3
IZ57	----	9	328.0	36.6	325.0	14.7	35	8.0
IZ59	----	18	36.4	54.0	336.0	58.5	27	6.3
IZ60	----	10	50.6	58.2	331.0	67.5	76	5.0
IZ61	----	9	----	-----	333.0	31.7	47	6.7
IZ64	----	10	----	-----	165.2	-67.2	115	4.1

Table II (continued)

Sites	Age Ma	n	before B.C		after B.C		K	α95
			D	I	D	I		
			South of Bergama (4)					
IZ22	17.5	11	----	-----	190.0	-48.0	387	2.4
IZ47	17.5	9	----	-----	7.2	34.8	45	7.0
IZ48	17.5	8	----	-----	226.5	-28.8	12	14.0
			Karaburun Peninsula (5)					
IZ08	18.5	11	----	-----	267.0	-34.5	23	8.5
IZ14	----	8	181.0	-51.0	182.0	-36.0	264	3.2
IZ15	17.0	5	----	-----	226.0	-84.5	35	10.5
IZ16	21.3	6	----	-----	43.5	16.3	97	5.8
IZ43	18.5	8	----	-----	234.0	-53.5	450	2.6
Cesme1	18.5	7	----	-----	254.0	-51.9	189	4.4
Cesme2	18.5	7	----	-----	237.0	-51.5	52	8.4
Cesme3	18.5	5	----	-----	207.0	-59.5	132	6.7

Mean regional directions after bedding correction:
1) $N = 4/6$; $D = 6$; $I = 54$; $K = 254$; $α95 = 4.3$
2) $N = 17/27$; $D = 6$; $I = 49$; $K = 25$; $α95 = 6.8$
3) $N = 13/23$; $D = 327$; $I = 52$; $K = 20$; $α95 = 8.7$
4) $N = 3/4$; $D = 22$; $I = 39$; $K = 16$; $α95 = 20.2$
5) $N = 8/10$; $D = 49$; $I = 51$; $K = 9$; $α95 = 16$.
The K/Ar ages refer to the sampled units but have not been determined on single sites.

All the results are schematically summarized in Figure 5 which shows the amount of rotation and the inferred limits of the different blocks delineated by the paleomagnetic results. Two blocks, the Canakkale region and the small block south of Bergama, are not defined by a sufficient number of sites. However, if their geographical limits are doubtful, we believe that the differences of their paleomagnetic directions from those of the surrounding regions are real.

The ages of the rotations are constrained by the ages of the volcanic and sedimentary rocks that underwent rotation. Therefore only lower limits may be provided. Whatever the age of the studied rocks, in each region, the measured rotation is coherent indicating that the rotation post-dates the formation of the youngest studied rock. In the Izmir region, the youngest published datum is a K/Ar age of 11.3 Ma (Borsi et al., 1972). However, we have recently obtained a K/Ar age of 7 Ma for a volcanic formation sampled near Foça. Moreover, most of the neotectonic faults in the region were initiated during the late Miocene-Early Pliocene. The 7 Ma date for the Foça basalts indicates that the rotation of this region, quite coherent at all the sites, definitively belong to the neotectonic activity. The sites studied in the Karaburun peninsula are all situated in formations older than 16 Ma, so that the rotation cannot be directly time-bracketed. However, it seems unreasonable to us that the rotation of the Karaburun Peninsula has occurred during the pre-11 Ma activity and is thus not related to the other rotations of this area. The other sites exhibiting coherent rotations are all in rocks younger than 16 m.y..

These paleomagnetic data which demonstrate significant clockwise and counterclockwise rotations of coherent blocks that took place during the neotectonic extensional regime, are difficult to explain in the framework of the global geodynamical evolution of the Western Anatolian and Eastern

126

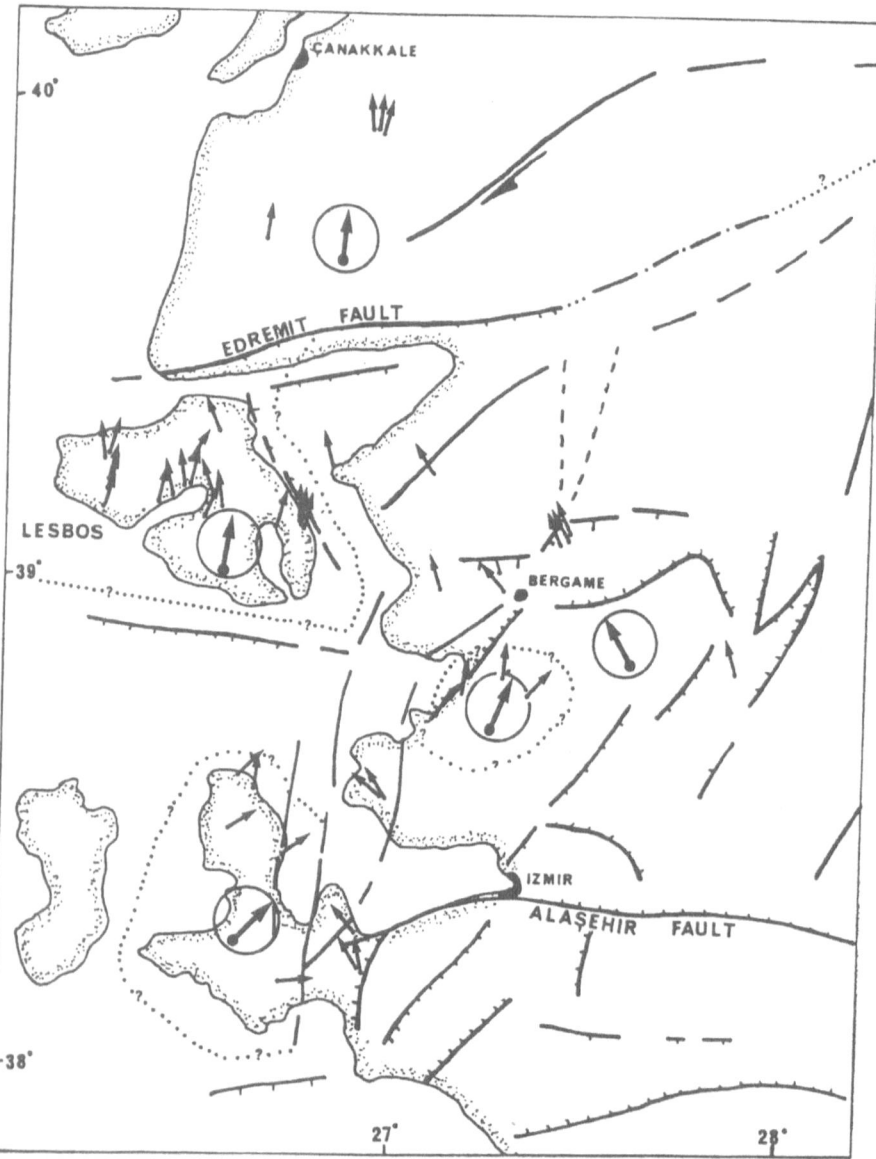

Fig. 5. Mean values of the declinations obtained on the eastern side of the Aegean Sea. The reverse directions have been inverted through the origin. For the different region, the amount and the sense of the rotation are shown on a stereogram. The limits of the different rotating units are only intended as a visual support to identify zones of coherent rotation. 1. Canakkale region; 2. Lesbos island; 3. Izmir-Bergama region; 4. south of Bergama; 5. Karaburun peninsula.

Aegean regions. They are most probably related to neotectonic regime that was governed by a main extension of about 40 - 50% since the beginning of the Tortonian (McKenzie, 1978; Le Pichon and Angelier, 1981). A model has been proposed independently from the Paleomagnetic data in order to explain rotations in opposite senses in extensional zones, taking the western Anatolian area as an example (Sengör et al., 1985, Fig. 18). An important feature of this model is the implication that large (>20°) rotations arise from successive episodes of faulting, older faults being cut and rotated passively by younger ones.

In our opinion, these results show that the brittle upper part of the lithosphere does not everywhere passively follow the motions of its lower ductile part or that the latter motions are more complex than hitherto assumed.

A more detailed analysis of the structural and the paleomagnetic data is certainly needed to develop a better constrained geometrical model, as it has already been done in purely strike-slip context of deformation (see this volume).

5. Conlusion

The clockwise rotations observed on the western side of the Aegean Sea, in Skyros and Evia islands, even of variable angles, can be rather simply explained in term of the model proposed by McKenzie and Jackson (1983), describing the damping of the dextral movement along the north-Anatolian fault.

On the eastern side, in Anatolia, the more complex rotational pattern suggests that the brittle carapace of the lithosphere may not everywhere follow the motions of the ductile lower part but that its movement may arise from adjustment to the extensional tectonic regime. However, the data are still uncomplete and detailed field structural studies are needed in oreder to determine the movement and the geometry of the fault bounding the different blocks. This example show the importance of well coordinated paleomagnetic and structural studies in such comlicated areas.

Acknowledments

The financial support was given by the CEA, the CNRS and the INSU-ATP: Sismogénèse, Plis et failles: Mécanique de la Lithosphère.
CFR Contribution N°973

References

Angelier, J., Dumont, J. F., Karamanderesi, H., Poisson, A., Simsek, S. et Uysal, S., 1981, Analysis of fault mechanisms and expansion of southwestern Anatolia since the late Miocene. *Tectonophysics*, **75**, T1-T9.

Bellon, H., Jarrige, J.J., Sorel, D., 1979, Les activités magmatiques égéennes de l'Oligocène à nos jours et leurs cadres géodynamiques. Données nouvelles et synthèse. *Rev. Géol. Dyn. Géogr. Phys.*, **21**, 41-55.

Borsi, S., Ferrara, G., Innocenti, F. et Mazzuoli, R., 1972, Geochronology and petrology of recent volcanics in Eastern Aegean Sea. *Bull. Volcan.*, **36**, 473-496.

128

Dewey, J. F. et Sengör, A. M. C. 1979. Aegean and surrounding regions: Complex multiplate and continuum tectonics in a convergent zone. *Geol. Soc. Am. Bull.* **90**, 84-92.

Eyidogan, H., et Jackson, J. 1985. A seismological study of normal faulting in the Dermici, Alasehir and Gediz earthquakes of 1969-70 in western Turkey: implications for the nature and geometry of deformation in the continental crust. *Geophys. J. R. Astron. Soc.*, **81**, 569-607.

Fytikas, M., Giuliani, O., Innocenti, F., Manetti, P., Mazzuoli, R., Peccerillo,A., et Villari, L., 1979, Neogene volcanism of the Northern and Central Aegean Region. *Ann. Géol. Pays Hellén.*, **30**, 106-129.

Fytikas, M., Innocenti, F., Manetti, P., Mazzuoli, R., Peccerillo, A. et Villari, L., 1984, Tertiary to Quaternary evolution of the volcanism in the Aegean region. Dans: J.E. Dixon et A.H.F. Robertson edts: The geological evolution of the Eastern Mediterranean. *Spec. Publ. Geol. Soc. London*, **17**, 687-699.

Horner, F. et Freeman, R. 1983. Palaeomagnetic evidence from pelagic limestones for clockwise rotation of the Ionian zone, western Greece. *Tectonophysics*, **98**, 11-27.

Innocenti, F., Manetti, P., Peccerillo, G., et Poli, G., 1979, Inner arc volcanism in NW Aegean Arc: geochemical and geochronological data. *N. Jb. Miner. Mh. Jg.* 1979, 145-158.

Innocenti, F., Manetti, P., Peccerillo, G. et Poli, G., 1981, South Aegean volcanic arc: geochemical variations and geotectonic implications. *Bull. Volcan.*, **44**, 377-391.

Kaya, O. 1981., Miocene reference section for the coastal parts of West-Anatolia. *Newsl. Stratigr.*, **10**, 164-191.

Kissel, C., Laj, C. and Muller, C. 1985, Tertiary geodynamical evolution of Northwestern Greece: palaeomagnetic results. *Earth and Planetary Sciences Letters.* **72**, 190-204.

Kissel, C. and Laj, C. 1988, The Tertiary geodynamical evolution of the Aegean Arc: a paleomagnetic reconstruction. *Tectonophysics*, **146**, 183-201.

Laj, C., Jamet, M., Sorel, D. et Valente, J.P. 1982. First paleomagnetic results from Mio-Pliocene series of the Hellenic sedimentary arc. In: Le Pichon,X. Augustithis,S.S. and Mascle,J. (eds.). Geodynamics of the Hellenic Arc and Trench. *Tectonophysics*, **86**, 45-67.

Lauer, J.P. 1981. Evolution géodynamique de la Turquie et de Chypre déduite de l'étude paléomagnétique. *Thèse*, Strasbourg, 295 pp.

Lemeille, F. 1977. Etudes néotectoniques en Grèce centrale nord-orientale (Eubée Centrale, Attique, Béotie, Locride) et dans les Sporades du nord (île de Skyros). *Thèse 3ème cycle*, Orsay, 173pp.

Le Pichon, X. et Angelier, J. 1979. The Hellenic Arc and Trench system: a key to the neotectonic evolution of the eastern mediterranean area. *Tectonophysics*, **60**, 1-42.

Le Pichon, X. et Angelier, J. 1981, The Aegean Sea. *Phil. Trans. R. Soc. London*, A, **300**, 357-372.

McDonald, W.D., 1980, Net tectonic rotation, apparent tectonic rotation, and the structural tilt correction in paleomagnetic studies, *J. Geophys. Res.*, **85**, 3659-3670.

McKenzie, D. 1978. Active tectonics of the Alpide-Himalayan belt: the Aegean Sea and surrounding regions (tectonics of the Aegean region). *Geophys. J. R. Astr. Soc.* **55**,217-254.

McKenzie, D. et Jackson, J. 1983. The relationship between strain rates, crustal thickening, paleomagnetism, finite strain and fault movements within a deforming zone. *Earth Planet. Sci. Letters*, **65**, 182-202.

McKenzie, D et Jackson, J. 1986. A block model of distributed deformation by faulting, *J. Geol. Soc. London*, **143**, 349-353.

Mercier, J. L., Delibassis, M., Gauthier, A., Jarrige, J. J., Lemeille, F., Philip, H., Sébrier, M. et Sorel, D. 1979. La néotectonique de l'Arc Egéen. *Rev. Geol. Dyn. Geogr. Phys.*, **21**, 61-72.

Sengör,A.M.C., Görür, N. et Saroglu, F. 1985, Strike-slip faulting and related basin formation in zones of tectonic escape: Turkey as a case study. In: K. T. Biddle et N. Christie-Blick (Edts), Strike-slip faulting and basin formation, *Soc. Econ. Paleont. Min. Spec. Pub.* **37**.

Tapponnier, P. 1977. Evolution tectonique du système alpin en Méditerranée: poinçonnement et écrasement rigido-plastique. *Bull. Soc. Geol. Fr.* **19**, 437-460.

Westphal, M., Bazhenov, M.L., Lauer, J.P., Pechersky, D.M. et Sibuet, J.C. 1986. Paleomagnetic implications on the evolution of the Tethys belt from the Atlantic ocean to Pamir since Trias. *Tectonophysics*, **123**, 37-82.

LATE CENOZOIC ROTATONS ALONG THE NORTH AEGEAN TROUGH FAULT ZONE (GREECE); STRUCTURAL CONSTRAINTS

C. SIMEAKIS[1], J.L. MERCIER[2], P. VERGELY[2] and C. KISSEL[3]
[1]*Institute for Geology and Mining Research (IGME),*
Messoghion Street 70,
Athens
Greece.

[2]*UA 730 Géogynamique et Géophysique Interne,*
Bât. 509, Université de Paris-Sud,
91405 Orsay,
France.

[3]*Centre des Faibles Radioacgivités,*
Laboratoire Mixte CNRS-CEA,
Domaine du CNRS
91198 - Gif-sur-Yvette Cedex
France.

ABSTRACT. The NE-SW to E-W striking North Aegean trough fault zone takes up the righe handed strike-slip motion of the North Anatolian Fault at its western termination. Paleomagnetic measurements suggest that its southern border has undergone a clockwise rotation in the range of 26 to 48° since at most the Upper Miocene. Analysis of the Late Cenozoic extensional faulting conducted on both sides of the North Aegean trough demonstrates that in this area the tensional directions trended (1) ESE during the Upper Miocene, (2) NE during the Pliocene-Lower Pleistocene and (3) N-S during the Middle Pleistocene - Present day period. From the orientations of these tensional directions with respect to the strike of the fault zone, it appears that the right-handed strike-slip motion started at the earliest during the Uppermost Miocene and has been clearly active since the Pliocene. A statistical analysis of the tensional directions in this area suggests a differential clockwise rotation of the southern border of the fault zone with respect to its northern border, with a probable value of ~ 25, this rotation being distributed on a large area (Limnos, Lesbos). On the other hand, a statistical analysis of fold axes of Early Cenozoic age has been conducted in regions located on both sides of the western termination of the North Aegean trough close to the Greek mainland. This cannot clearly demonstrate the 48° paleomagnetic clockwise rotation of Euboea. Yet, a differential rotation of Euboea with respect to the regions located north of the Sperchios Valley is possible in the range of the uncertainties on the mean fold directions, i.e. in the range of less than 30°. Therefore, if the 48° paleomagnetic rotation is regionaly significant, then a large part of this rotation has to be related to the well-known clockwise rotation of the Ionian branch of the Aegean arc.

1. Introduction

The NE-SW to E-W striking North Aegean trough fault zone takes up the dextral strike-slip motion on the North Anatolian Fault at its western termination. A clockwise rotation associated with this

131

C. Kissel and C. Laj (eds.), Paleomagnetic Rotations and Continental Deformation, 131–143.
© 1989 by Kluwer Academic Publishers.

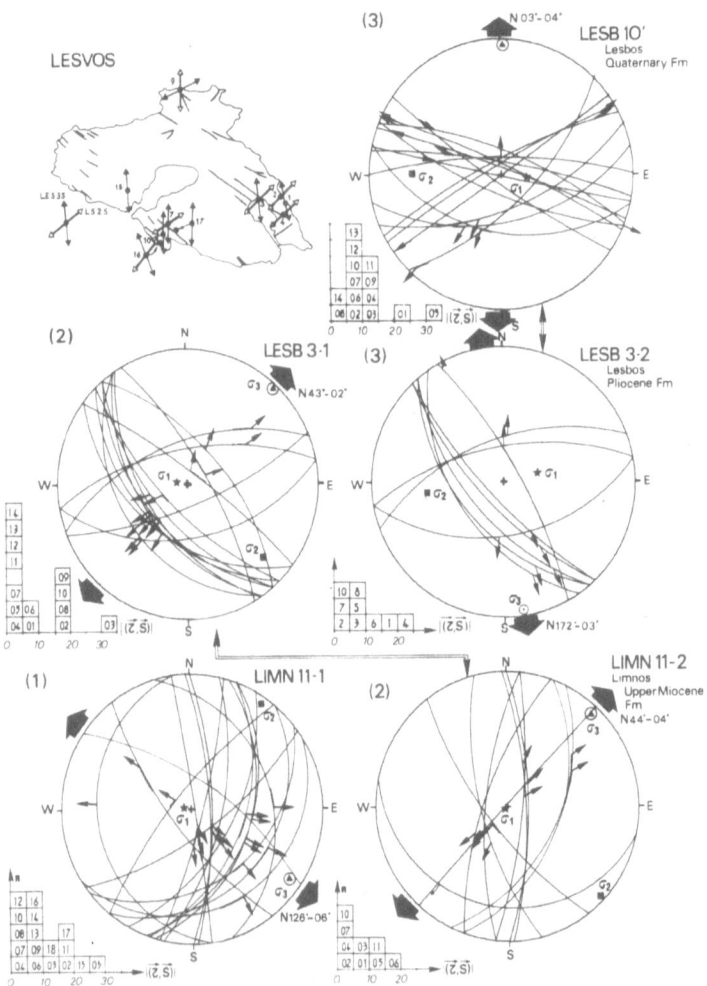

Figure 1. Normal minor fault data from Lesbos (LESB) and Limnos (LIMN) Is. Arrows attached to the fault traces correspond to the measured slip-vectors (Wulff stereonet, lower hemisphere). Histograms show the deviations between measured and predicted slip-vectors for each fault plane. Divergent black arrows give the azimuths of the computed minimum σ_3 principal stress directions. Three families of striations are observed: (3) in Quaternary formations, (3) and (2) in Pliocene formations, (2) and (1) in Upper Miocene formations. Map shows location of sites analyzed on Lesbos Is. (from Mercier et al., 1988).

dextral strike-slip motion has been predicted by McKenzie and Jackson (1983). Paleomagnetic measurements (Kissel et al., 1986a) suggest that the southern border of this strike-slip fault zone has undergone a clockwise rotation in the range of 26 to 48° since at most the Upper Miocene and, probably, during the last 5 M.y. period. In order to constrain these rotations from structural data, we have analyzed the Upper Miocene to Present-day extensional faulting in the Aegean basins located on both sides of the North Aegean trough (Mercier et al., 1987, 1988). A statistical analysis of the tensional directions deduced from the kinematics of the faults suggests a differential ~ 25° clockwise rotation of the southern border. Moreover, this shows that this differential rotation associated with the dextral motion started at the earliest during the Uppermost Miocene and has been clearly active since the Pliocene. We have also conducted a statistical analysis of fold axes of Early Cenozoic age in regions located north and south of the western termination of the North Aegean trough close to the Greek mainland. A clockwise rotation of the southern regions with respect to the northern regions is possible in the range of less than 30° coresponding to the uncertainties of the mean fold strikes. These results allow to discuss the timing and significance of these rotations.

2. Kinematics of the Late Cenozoic faults in the Aegean basins located on both sides of the North Aegean trough fault zone

2.1 METHODOLOGY

Kinematics of a fault population are defined using striations measured on the fault planes. Assuming that sliding occurs in the direction of the shear stress resolved on the fault plane, then a mean deviatoric stress tensor may be computed, within a factor k, from a set of striated faults (Carey and Brunier, 1974). Among the quantitative computer methods proposed to solve this problem by minimizing the deviations between the measured and the computed slip-vectors, we use the algorithm proposed by E. Carey (1979). Thus, several families of striations permit one to characterize several states of stress. However, to avoid misleading separation of successive states of stress, separation of families of striations must be based on field data demonstrating their constant relative chronology.

2.2 LATE CENOZOIC FAULTING IN THE NORTHERN AEGEAN BASINS

As an example we analyze two sites in the Lesbos Island which are affected by minor faults. These faults cut continental deposits of Pleistocene age and marls of Pliocene age (see detailed analysis in Mercier et al., 1988). Faults which affect the Pleistocene deposits result clearly from a N-S tension (Fig.1, LESB 10). On the other hand, faults affecting the Pliocene deposits show two superimposed family of striation. The older family results from a NE-SW trending tension (Fig.1, LESB 3-1). The younger one results from a N-S tension (Fig.1, LESB 3-2) as that observed in the faulted Pleistocene deposits.

Faults affecting sedimentary formations of Upper Miocene age sometimes exhibit an older family of striations (Mercier, 1981; Lyberis, 1984). In the Thassos island for instance, west of Limenaria (sites 1 and 8, Figure 2C), faults having affected soft sediments of Upper Miocene age result from a WNW-ESE trending tension. These synsedimentary or ante-lithification faults are anterior to faults activated by the NE-SW and N-S trending tensions demonstrated above. In the Limnos Is., highly fractured lavas of Lower Miocene age are affected by numerous fault planes showing two superimposed families of striations. The older one (LIMN 11-1, Figure 1) is

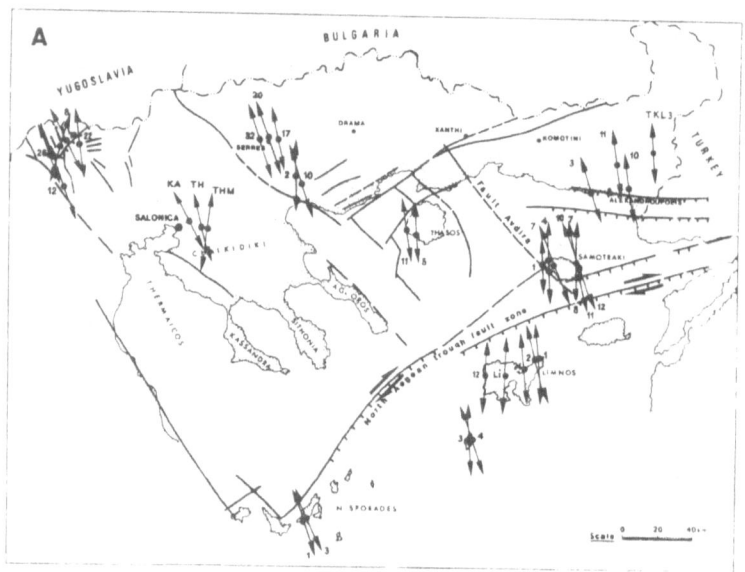

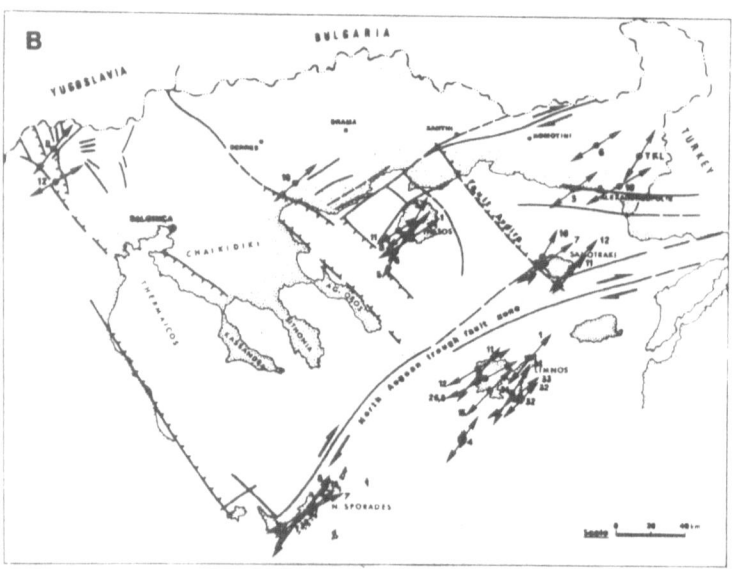

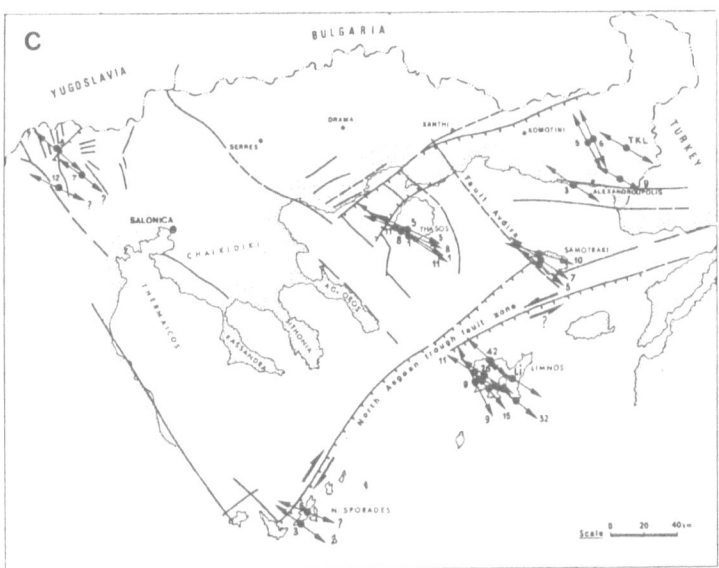

Figure 2. Stress patterns in the Northern Aegean domain deduced from structural analysis of faults of Upper Miocene to Quaternary age. Double arrows give the computed tensional σ_3 principal stress directions (from Mercier et al., 1988). A. Middle Pleistocene to Present-day period. B. Pliocene to Lower Pleistocene period. C. Upper Miocene period.

characterized by a WNW-ESE trending tension, the younger one (LIMN 11-2, Figure 1) by NE-SW trending tension already deduced from faults affecting Pliocene formations in Lesbos Is.

A statistical analysis of the tensional axes deduced from the numerous faulted sites in the Northern Aegea (Figure 2A, B, C) shows three tensional directions trending: (1) WNW, (2) NE and (3) N-S. These three mean tensional directions cannot result from large rotations of a single tensional direction "frozen" in the rocks in the form of brittle deformations because directions (2) and (3) have been observed in Lesbos Is. (Figure 3B) where paleomagnetic studies have shown no significant rotation since the Lower Miocene (Kissel et al., 1986c). Indeed, these three tensional directions characterize three different extensional tectonic regimes of respectively Upper Miocene, Pliocene-Lower Pleistocene and Middle Pleistocene to Present day ages.

2.3 THE RIGHT-HANDED STRIKE-SLIP MOTION ON THE NORTH AEGEAN TROUGH FAULT ZONE AND THE ASSOCIATED ROTATIONS.

The orientations of these tensional directions with respect to the strike of the North Aegean trough fault zone permit one to constrain the timing of the mergence of the North Anatolian fault into the North Aegean trough fault zone. The initiation of the North Anatolian fault zone took place between late Early Miocene (Dewey et al., 1986) and since that time its motion has been dextral, the Anatolian block moving westward (Mc Kenzie, 1972; Le Pichon and Angelier, 1979). During the

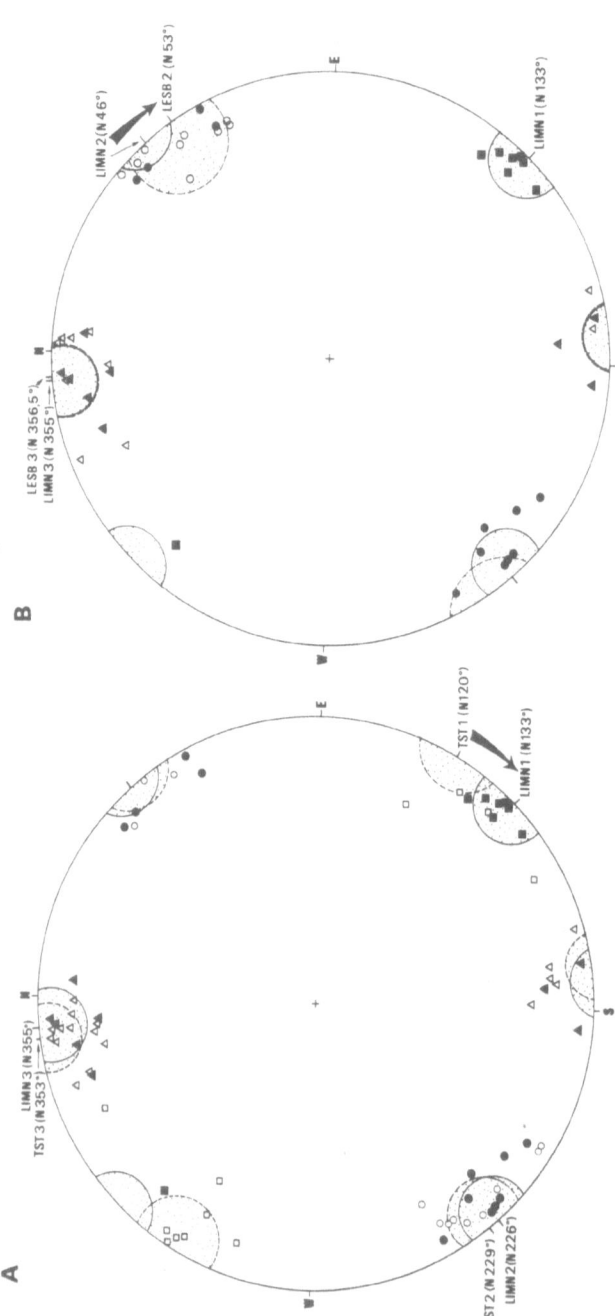

Figure 3. Rotations suggested by the mean tensional directions deduced from Late Cenozoic faulting in the Northern Aegean basins. Squares, circles and triangles gfve the computed tensional axes of respectively (1) Upper Miocene, (2) Pliocene-Lower Pleistocene and (3) Mid. Pleistocene-Present age (see numerical values in Mercier et al., 1988). Stippled areas give the mean tensional directions (1), (2) and (3) with a 95% confidence angle.

A. Filled and open symbols are respectively for the Limnos (LIMN) block and the Thraki-Samothraki-Thassos (TST) block. A clockwise rotation of the LIMN 1 tensional axis with respect to the TST 1 one is possible with a probable value of 15°.

B. Filled and open symbols are respectively for the Limnos (LIMN) and Lesbos (LESV) blocks. A clockwise rotation of the LESB 2 tensional axis with respect to the LIMN 2 one is possible with a probable value of 10°.

Upper Miocene (Figure 2C), the North Aegean trough fault zone had a normal motion with a small sinistral strike-slip component. This implies that at that time the western end of the North Anatolian dextral fault had not yet merged into the North Aegean trough fault zone. On the other hand, during the Pliocene-Pleistocene (Figure 2B and 2A), the North Aegean trough fault zone had a normal motion with a dextral strike-slip component indicating that the dexral motion of the North Anatolian fault was transferred into the Northern Aegea. Thus, the mergence of the western end of the North Anatolian fault into the Northern Aegea took place at the earliest during the Uppermost Miocene (Pontian) and was clearly effective during the Pliocene.

In order to constrain in the clocwkise rotations expected from these dextral motions, a statistical analysis (Fisher, 1952) of the computed tensional directions (Figure 3A and B) from both sides of the North Aegean trough has been realized. Between the Traki-Samothraki-Thassos block (TST on Figure 3A) on one hand and the Limnos block (LIMN onf Figure 3A) on the other hand, there are no significant rotations of the tensional directions (2) and (3). But, a clockwise rotation of the tensional direction (1) of Limnos (LIMN 1, Figure 3A) with respect to that of these northern side (TST 1, Figure 3A) of the North Aegean trough is possible in the range of 0-30° with a probable value of -15. The tensional direction (2) of the Lesbos block (LESB 2, Figure 3B) also shows a possible clockwise rotation with respect to that of Limnos (LIMN 2, Fig 3B) with a probable value of about 10°. Unfortunately, we have not obtained sufficient data to calculate the tensional direction (1) of Lesbos which could have better constrained the result.

Thus, the analysis of the Late Cenozoic faulting in the Northern Aegea suggests a clockwise rotation of the southern border of the North Aegean trough with respect to its northern border. This rotation is probably in the order of 25° but distributed on a large area, from Limnos to Lesbos. If correct, this result poses some interesting problems. The paleomagnetic declinations of the Oligocene lavas of Thraki is about 7° (Kissel et al., 1986a) and those of the Lower Miocene lavas of Lesbos is about 6° (Kissel et al., 1986c). They are not significantly different from the declination value D=5.8° which is expected in the Aegean from the African poles during the Lower-Upper Miocene (Kissel et al., 1987a). This indicates that Lesbos has not rotated with respect to Thraki since the Lower Miocene. But it appears in contradiction with the ~ 25° clockwise rotation of Lesbos with respect to Thraki suggested by the structural data. Supposing Thraki to be fixed, this may be explained if Lesbos has undergone an anticlockwise rotation of about 25° previous to the clockwise rotation of about the same amplitude suggested by the structural data. This older rotation might be related to the well-known anti-clockwise rotation of the Lycian branch of the Aegean Arc (Kissel and Poisson, 1987b) during the Middle-Upper Miocene.

3. Directions of the Early Cenozoic folds on both sides of the western termination of the North Aegean trough

3.1 METHODOLOGY

Fold axes have been studied as markers to investigate the possibility of a 48° clockwise rotation of Euboea suggested by paleomagnetic data (Kissel et al., 1986b). These folds have been analyzed on both sides of the western termination of the Sporadhes basin of the North Aegean trough (Brooks and Ferentinos, 1980) close to the Greek mainland, i.e. north and south of the WSW-ENE striking structure marked by the Maliakos gulf and the Sperchios valley (Figure 4). In these regions several periods of folding have been evidenced (Mercier and Vergely, 1972; Vergely, 1979, 1984). The first one is of Upper Jurassic-Lower Cretaceous age. During this period two phases of folding

138

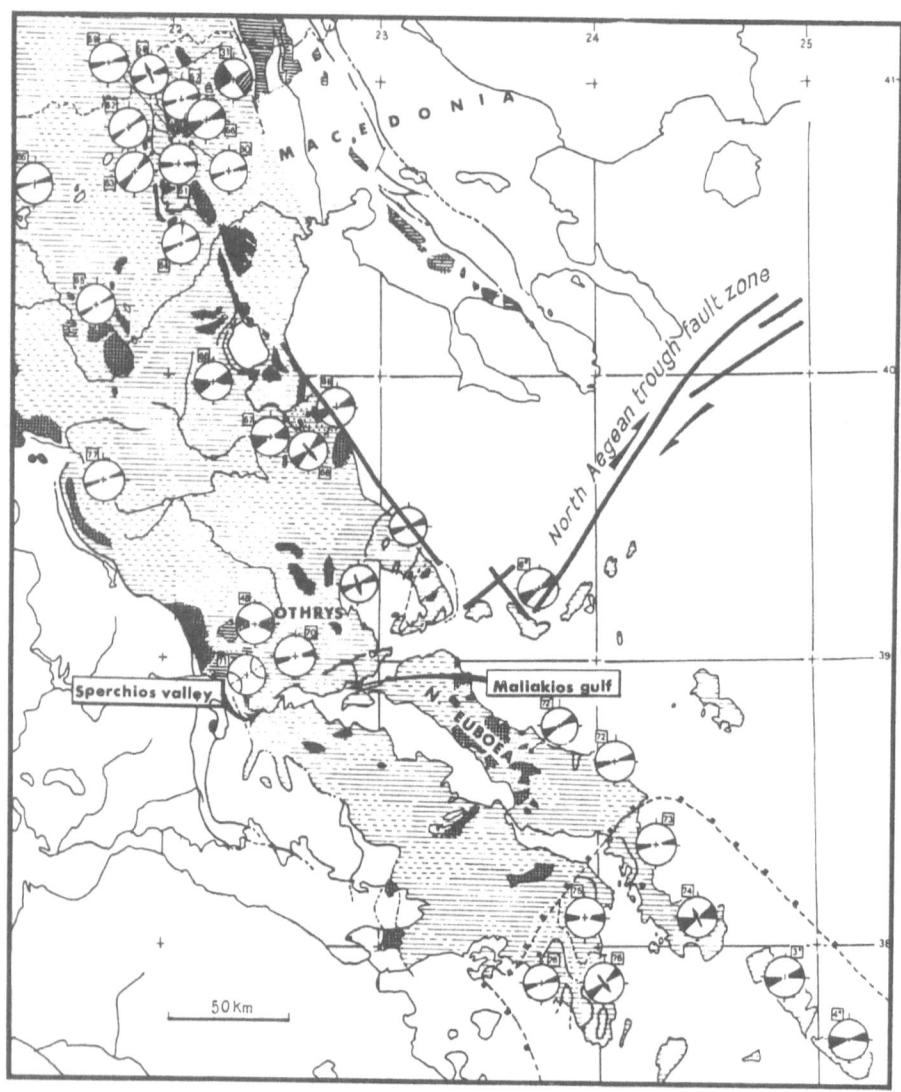

Figure 4. Mean fold directions CT 2 of Early Cenozoic age in the internal Hellenic zones. Black opposite triangles within the small circles given the mean direction of the fold axes at each site with the uncertainties. Small arrows attached to these opposite triangles indicate the fold vergence when determinated (from Vergely, 1984).

affected the Triassic to Lower Cretaceous formations which are unconformably overlain by transgressive Upper Cretaceous formations. During the second period of Lower-Middle Eocene age, two phases of folding (called CT1 and CT2) affected the Upper Cretaceous-Lower Eocene formations; in Central Macedonia they are unconformably overlain by late Middle Eocene transgressive molasses. CT1 and CT2 occurred prior to the Uppermost Eocene-Lower Oligocene CT3 folding which affects the Pindos zone. The CT1 and CT2 folds are chosen as markers of the rotational deformations because of two raisons. Firstly, they are easily separated from the folds of Upper Jurassic-Lower Cretaceous formations. Eventually, in Euboea, metamorphic minerals are associated with these folds; they have been dated by radiometric measurements (Bavay et al., 1980). Secondly, in these regions the CT1 and CT2 folds has been systematically established. Thus, there have not been systematic errors in attributing a given fold axis direction to a given phase of folding.

3.2 EARLY CENOZOIC FOLD DIRECTIONS AND ROTATION OF EUBOEA

At each site, a mean fold direction has been determinated by a statistic analysis of minor fold axes. In most cases, the CT2 mean fold axes are not tilted. On the other hand, the CT1 folds being refolded by the CT2 folds, their original orientations have been obtained by graphical methods usually used in structural analysis. As a consequence, the uncertainties on the CT1 fold directions are higher that on the CT2 fold ones (Figure 5). Data and graphical analyzes are described in details in Vergely (1984). Figure 4 shows the mean fold directions obtained at different sites for the CT2 phase of folding. The histogramms of the CT1 and CT2 fold directions (Figure 5) are drawn separating the regions located north and south of the Maliakos gulf and the Sperchios valley. These histogramms cannot demonstrate a clear rotation of the southern block with respect to the northern one. However, this remains possible in the range of the uncertainties i.e. in the range of less than 30°. Therefore, if the 48° paleomagnetic clockwise rotation measured in Euboea is significant of a regional rotation, it cannot be related only to the right-handed strike-slip motion of the North Aegean trough fault zone and might partly result from the Pliocene and possibly Upper Miocene clockwise rotation of the Ionian branch of the Aegean Arc (Laj et al., 1982).

4. Conclusions

Structural markers have been used to investigate the relative rotation between the regions located north and south of the North Aegean trough fault zone. These markers are of different origins. Some are the statistical mean directions of tension of Upper Miocene, Pliocene-Lower Pleistocene and Middle Pleistocene-Present day age deduced from the kinematics of the faults affecting the North Aegean basin. Others are statistical mean directions of folds of Early Cenozoic age. The dextral strike-slip motion of the North Aegean trough fault zone began at the earliest during the Uppermost Miocene (Pontian) and has been clearly active since the Pliocene. This dextral motion corresponds to the mergence of the dextral strike-slip North Anatolian fault into the North Aegean trough fault system. The analyzed markers are in agreement with the predicted clockwise rotation of the southern border of the North Aegean basin trough with respect to its northern border (Mc Kenzie and Jackson, 1983), in the range of ~ 25° but distributed on a large area from Limnos to Lesbos.

If the 48° paleomagnetic clockwise rotation measured in Euboea is significant of a regional rotation, this value is too high to be explained by the only destral strike-slip motion on the North Aegean trough fault zone. Thus, the well-known Late Cenozoic clockwise rotation of the Ionian

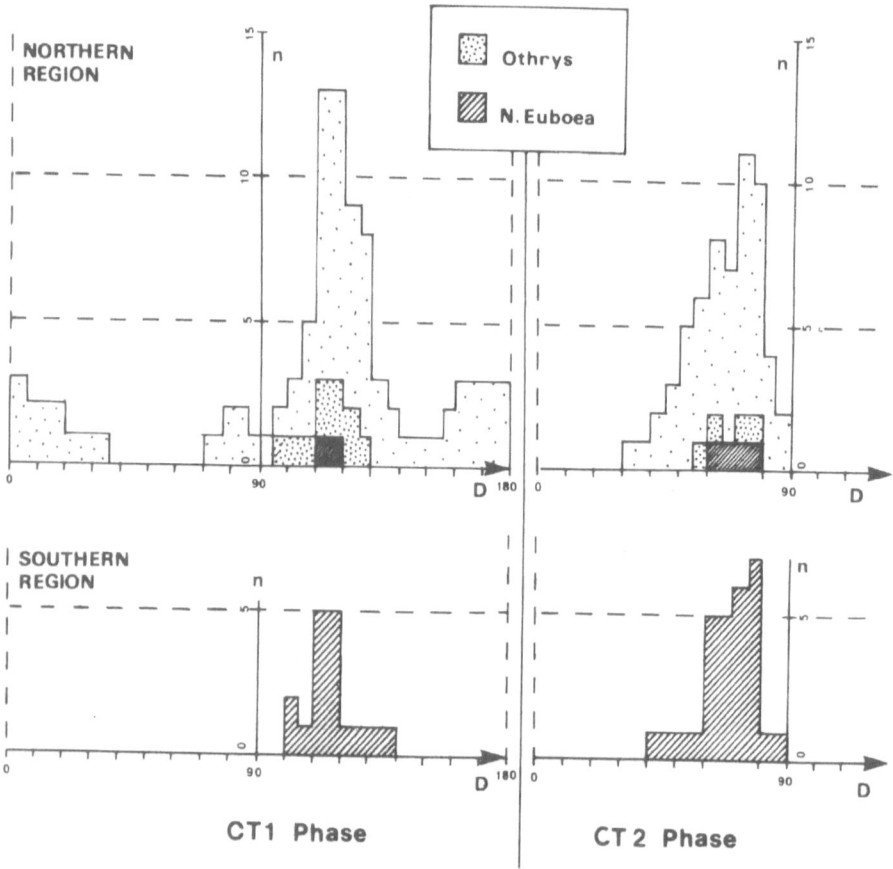

Figure 5. Histogramms of the CT 1 and CT 2 fold directions in the regions located north and south of the WSW-ENS striking structure marked by the Maliakos gulf and the Sperchios valley. Data on the top Figure correspond to the Othrys and N. Euboea regions located near the prolongation of the dextral strike-slip North Aegean trough fault zone into the Greek mainland. n is the number of sites, and D the mean direction of the fold axes.

branch of the Aegean Arc (Laj et al., 1982) probably add to the clockwise rotation related to the dextral strike-slip motion (Figure 6).

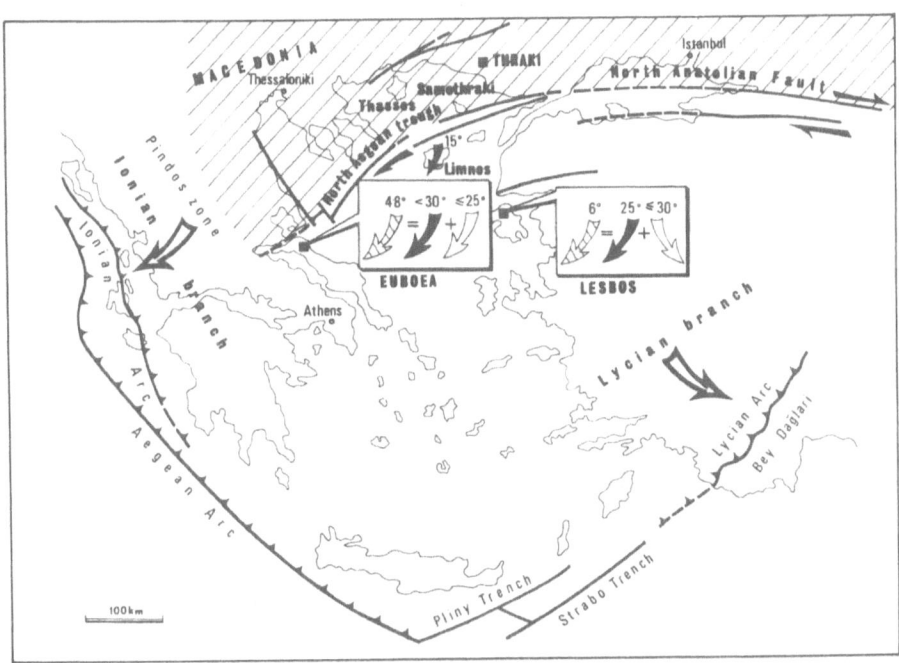

Figure 6. Sketch showing clockwise rotations (black arrows) of the southern border of the North Aegean trough fault zone with respect to its northern border (hachured area) deduced from structural data. Paleomagnetic rotations (hachured arrows) measured on Euboea and Lesbos Is. might result from superimposition of these clockwise rotations due to the dextral strike-slip movement of N. Aegean trough fault zone and of respectively clockwise rotation (clockwise open arrows) of the Ionian branch and anti-clockwise rotation (anti-clockwise open arrows) related to the one of the Lycian branch of the Aegean Arc.

On the other hand, the 6° and 7° declination values from respectively the Lower Miocene lavas in Lesbos (Kissel et al., 1986c) and from the Oligocene lava in Thraki (Kissel et al., 1986b) appear in contradiction with the ~ 25° clockwise rotation of Lesbos with respect to Thraki estimated from structural data. Supposing the Thraki-Samothraki-Thassos block to be fixed, this may be explained if the ~ 25° clockwise rotation due to the dextral strike-slip motion on the North Aegean trough fault zone must add to a roughly equivalent previous anti-clockwise rotation of Lesbos (Figure 6).

Thus, structural analyses suggest that paleomagnetic rotation data may result from superimposition of rotations of different ages and of different origins. Finally, combination of detailed paleomagnetic and structural data would surely permit one to analyze more precisely the timing and the significance of rotational deformations in the Aegea.

References

Bavay, D., Bavay, P., Maluski, H., Vergely P. and Katsikatsos G., 1980. Datation par la méthode 40 Ar/ 39 Ar de minéraux de métamorphisme de haute presion en Eubée du Sud (Grèce). Corrélations avec les évènements tectono-métamorphiques des Hellénides internes. *C. R. Acad. Sci., Paris* **290**, 1051-1054.

Brooks, M. and Ferentinos, G., 1980. Structure and evolution of the Sproradhes basin of the North Aegean trough, Northern Aegean Sea. *Tectonophysics*, **68**, 15-30.

Carey, E., 1979. Recherche des directions principales de contraintes associées au jeu d'une population de failles. *Rev. Géol. Dyn. Geogr. Phys.*, **21**, 57-65.

Carey, E. et Brunier B., 1974. Analyse théorique et numérique d'un modèle mécanique élémentaire appliqué à l'étude d'une population de failles. *C. R. Ac. Sci. Paris*, **279 D**, 891-894.

Dewey, J.F., Hempton, M.R., Kidd W.S.F., Sroglu F., Sengor A.M.C., 1986. Shortening of continental lithosphere: the neotectonics of Eastern Anatolia- a young collision zone. *In* Collision Tectonics " Cowared, M.P. & Ries A.C. Eds, *Geol. Soc. Spec. Publ.*, **19**, 3-36.

Fischer, R.A., 1953. Dispersion on a sphere, *Proc. Roy. Soc. London, Sér. A*, **217**, 295-305.

Kissel, C., Laj, C. and Mazaud A., 1986a. First paleomagnetic results from Neogene formations in Evvia, Skyros, and the Volos region and the deformation of Central Aegea, *Geophys. Res. Lett.*, **13**, 1446_1449.

Kissel, C., Kondopoulou, D., Laj, C. and Papadopoulos P., 1986b. New paleomagnetic data from Oligocene foramtions of Northern Aegea,.*Geophys. Res., Lett.*, **13**, 1039_1042.

Kissel, C., Laj, C., Poisson, A., Savascin, Y., Simeakis, K. and Mercier, J.L., 1986c. Paleomagnetic evidence for Neogene rotational deformations in the Aegean domain. *Tectonics*, **5**, 783-795.

Kissel, C., Laj, C., Sengor, A.M.C. and Poisson, A., 1987a. Paleomagentic evidence for rotation in opposite senses of adjacent blocks in Northeastern Aegea and Western Anatolia. Geophys. Res. Lett., 14, 907-910.

Kissel, C. and Poisson, A., 1987b. Etude paléomagnétique des formations cénozoiques des Bay Daglari (Taurides occidentales, Turquie). *C. R. Acad. Sci.*, **304 II**, 343-348.

Kondopoulou, D. and Lauer, J.P., 1984. Paleomagnetic data from Tertiary units of the North-Aegean zone. *In* "The Geological Evolution of the Eastern Mediterranean", (Eds Dixon J.E. & Robertson A.H.), *Geol. Soc. London, Spec. Publ.*, n°17, 681-686.

Laj C., Jamet, M., Sorel, D. and Valente, J.P., 1982. First paleomagnetic results from Mio-Pliocene series of the Hellenic sedimentary Arc. In "Geodynamics of the Hellenic Arc and Trench" (eds Augustichis S.S. & Mascle J.) *Tectonophysics*, **86**, 45_67.

Le Pichon, X. and Angelier, J., 1979. The Hellenic arc and trench system: a key to the neotectonic evolution of the Eastern Mediterranean area, *Tectonophysics*, **60**, 1-42.

Lyberis, N., 1984. Géodynamique du domaine égéen depuis le Miocène supérieur. *Thèse d'Etat*, Paris VI, 367 p.

Mc Kenzie, D.P., 1972. Active tectonics of the Mediterranean region, *Geophys. J. R. Astr. Soc.*, **30**, 109-185.

McKenzie, D.P. and Jackson, J., 1983. The relation between strain rates, crustal thickening, paleomagnetism, finite strain and faults mouvements within a deforming zone, *Earth Planet. Sci. Lett.*, **45**, 182-202.

Mercier, J.L. and Vergely, P., 1972. Les mélanges ophiolitiques de Macédoine (Gèce): décrochements d'âge ante-Crétacé supérieur. *Z. Deutsch. Geol. Gest.*, **123**, 469-489.

Mercier, J.L., 1981. Extensional-compressional tectonics associated with the Aegean Arc: comparison with the Andean Cordillera of South Peru-North Bolivia. *Phil. Trans. R. Soc. London*, A **300**, 337-355.

*Mercier, J.L., Sorel, D., Simeakis, K., 1987. Changes in the state of stress in the overriding plate of a subduction zone: the Aegean Arc from the Pliocene to the Present, *Ann. Tectonicae*, **1**, 20-39.

*Mercier, J.L., Simeakis, K., Sorel, D. and Vergely, P., 1988. Extensional tectonic regimes in the Aegean basins during the Cenozoic. *In* "The structural and sedimentary evolution of the Neotectoonic Aegean Basins, Geological Society of London, April 1988, Abstract.. *Basin Research*, Submitted.

Vergely, P., 1979. Ophiolites et phases tectoniques superposées dans les Hellénides. *VIth Coll. Aegean Region, Athens 1977*, III, 1293-1302.

Vergely, P., 1984. Tectonique des ophiolites dans les Hellénides internes (deformations, métamorphismes et phénomènes sédimentaires). Conséquences sur l'évolution des régions téthysiennes occidentales. Thèse d'Etat, Paris-Sud (Orsay), 250 p.

For more detailed references see papers marked by an asterisk (*)

SOME EXPERIMENTS ON BLOCK ROTATION IN THE BRITTLE UPPER CRUST

P.R. COBBOLD, J.P. BRUN, P. DAVY, G. FIQUET, C. BASILE and D. GAPAIS
Centre Armoricain d'Etudes Structurales des Socles (CNRS),
Universite de Rennes I,
35042 Rennes Cedex
France.

ABSTRACT. We present three series of experiments showing rotations of fault blocks about vertical axes. Experimental materials have yield criteria or flow laws similar to those estimated for the continental lithosphere. Thus we use a Coulomb material (dry sand) for the upper crust and viscous fluid layers for the lower crust, and most of the mantle. From nature to experiment, strengths are scaled down in proportion to lengths and densities: this ensures that the ratio of gravitational forces and surface forces is preserved.
The first series of experiments concentrates upon brittle deformation in the upper crust. Sandpacks are subjected to horizontal plane-strain via lateral boundaries. The resulting pattern of strike-slip faults and block rotations depends upon the deformation history. Coaxial stretching produces conjugate faults and small block rotations. Coaxial stretching, with some simple shearing, results in conjugate domino domains. Coaxial stretching and simple shearing in nearly equal proportions result in a single domino domain synthetic with the imposed shearing. Rotations are always clockwise in left-lateral dominos, counterclockwise in right-lateral dominos.
The second series of experiments deals with brittle deformation above a viscous detachment. Double layers of sand upon silicone putty are subjected to right-lateral basal wrenching via two sliding baseplates. A jog in the plate boundary produces an area of pull-apart deformation in the overlying layers. Sigmoidal fault traces in the sand show strike-slip motions at their ends and dip-slip motions in their centers. Intervening fault blocks rotate clockwise, separate to produce rift valleys, and show right-lateral offsets.
Finally, the third series of experiments deals with continental indentation. Sand and silicone layers (representing the lithosphere) rest upon glucose syrup (the asthenosphere). Northwards indentation by a rigid piston results in a complex pattern of faults and block rotations. If the lateral boundaries of a continent are rigidly confined, indentation produces crustal thickening and arcuate thrust faults. Rotations of 20° or more occur near the ends of the thrusts. If there is less confinement at the eastern margin, indentation produces a major left-lateral shear zone (with internal counterclockwise block rotations) as well as clockwise rotation of a large escaping eastern block. Finally, if confinement is even smaller, indentation produces a major left-lateral shear zone with a system of wrench-rifts that rotate by a domino mechanism.

1. Introduction

For some years now at Rennes we have been doing experiments, to invesigate mechanisms of tectonic deformation at various scales. We use materials with yield envelopes or creep laws similar to those measured or estimated for real rocks, but many times weaker. The experiments are properly scaled to account for gravitational forces (see Vendeville et al, 1987; Davy and Cobbold, in press): the strengths of materials are scaled down in proportion with linear dimensions and densities. Materials representing the brittle upper crust deform dominantly by faulting, whereas materials representing the lower crust and most of the mantle deform in ductile fashion.

C. Kissel and C. Laj (eds.), Paleomagnetic Rotations and Continental Deformation, 145–155.
© *1989 by Kluwer Academic Publishers.*

Many of our experiments show relative rotations of faults and intervening fault blocks during progressive deformation. In some instances, these rotations build up to values of more than 20⁻. Depending upon tectonic context, the rotations occur about horizontal or about vertical axes. In this paper, we concentrate on rotations about vertical axes. They mostly occur by domino mechanisms in localized zones of wrenching. We select examples where rotations exceed 20°. in the hope that they will be of interest to paleomagnetists either as models to test, or as indications of where to expect large rotations in nature.

For practical reasons, we have done three series of experiments at three different scales. A first series of experiments investigates faulting in brittle material, applicable to the upper part of the continental crust. The scale of these experiments allows us to describe the fault pattern and associated block rotations in some detail, but we have to use artificial boundaries and make assumptions about the boundary conditions. A second series of experiments investigates faulting in a brittle upper layer, as a result of viscous drag from below. At this scale, the basal boundary conditions are less artificial, but details of the fault pattern are not so clear. Finally, the third series of experiments investigates continental indentation. At this scale, various kinds of block rotations appear, depending mainly upon the conditions at lateral boundaries of indented continents.

2. Domino domains in Coulomb materials.

A Coulomb material is resistant to deformation unless the shear stress τ on a given plane exceeds the critical value $\tau_c = c + \mu.(\sigma_n - p)$, whereupon the material fails (yields). In this yield criterion, σ_n is the normal stress acting across the plane, p is the pressure of interstitial fluid, c is the coefficient of internal friction. The Coulomb criterion is mathematically simple. It applies to materials which are granular or fragmented, where the grains or fragments are rigid but in frictional contact. The Coulomb criterion is a good fit to data obtained in laboratory tests on rocks, under conditions of temperature and pressure representative of the upper crust (Byerlee, 1978). Hence we have used a Coulomb material in our experiments: dry sand. Our choice is guided by scaling requirements. Where gravity is active, the cohesion, which has dimensions of stress, must scale down by the same factor as the product of length by density; whereas, in general, the coefficient of internal friction, being dimensionless, must be identical in nature and experiment. Dry quartz sand fulfills these requirements (Horsfield, 1977; McClay and Ellis, 1987; Vendeville et al, 1987). Notice however that by using dry sand, we implicitly assume a negligible fluid pressure, p.

The Coulomb criterion is a condition upon the stress components, alone. To understand the strain history during yielding, we need a relationship between stress and strain. This subject does not seem to be very clear. Real materials such as brittle rock or sand almost always develop localized faults or shear zones. These tend to initiate after an initial stage of more uniform strain (typically less than 10%).

On theoretical grounds, the localization of deformation into narrow zones is attributable to a feedback mechanism of progressive softening (see review by Poirier, 1980). Experiments on real rocks and granular materials show that dilation occurs in the fault zones (Mandl et al, 1977). This reduces the amount of frictional contact between rock fragments or grains and presumably causes mechanical weakening. Small stress drops are usually recorded immediately after fault initiation (Jaeger and Cook, 1981; Mandl et al, 1977).

The Coulomb criterion does predict the orientation of initiating faults. The acute angle θ between the fault plane and the principal compressive stress σ_1, is predicted by $\theta = 45° \pm \phi/2$, where $\phi = \arctan \mu$ is the angle of internal friction. Because of the symmetry of the stress tensor, there are two conjugate orientations, $\pm\ \theta$, of predicted faults. In some experiments, only one set of faults develops finite amounts of slip, the other set being undetectable. If the faults are evenly spaced throughout a given domain and the amounts of slip are all equal, the domain is statistically homogeneous in terms of strain. The deformation accumulates as a progressive simple shear parallel to the single fault set, plus a possible bulk rotation with respect to a fixed external reference frame. We refer to such a domain as a domino domain. The domino or bookshelf mechanism has been discussed from kinematic and mechanical points of view by Mandl (1987). An important feature of a domino domain is that the principal direction of bulk infinitesimal strain is at 45° to the faults, whereas the principal stress is at $45° \pm \theta/2$ at initiation: hence stress and strain tensors are non-coaxial. This does not imply that the material is anisotropic, but it does show that the behaviour of frictional materials is more complex than that of simple fluids, or even of non-frictional plastic materials such as metals.

Domino domains form spontaneously in Coulomb materials, given certain conditions. One condition appears to be kinematic: the bulk strain must be nearly homogeneous. For example, domino domains form in subhorizontal layers of sand with free upper surfaces, provided the lower surface is subjected to uniform extension (Vendeville et al.,1987; McClay and Ellis, 1987; Mandl 1987). The faults are of dip-slip normal type. Progressive stretching produces tilting of faults and fault blocks about horizontal axes. If a sand layer is stretched in an inclined position, all the faults dip downslope, forming a single domino domain. This result we attribute to the non-coaxiality of bulk strain and stress (Vendeville et al, 1987; Mandl 1987). If a long sand layer is stretched in a horizontal position, the bulk strain is coaxial with the stress and is accommodated by a multidomainal arrangement of dominos (Vendeville et al., 1987). This arrangement avoids the interference that would otherwise occur at intersecting conjugate faults (Cobbold and Gapais, 1987). In general, dominos tilt by up to 30°, then they lock and are crosscut by new normal faults,as predicted theoretically by Nur and Ron 1987, or observed locally in the Basin and Range Province, USA (Profett, 1977).

Recently we have investigated the development of strike-slip faults in uniform rectangular sandpacks. These experiments are described in detail elsewhere (Cobbold et al, work in progress). Here we give only a brief summary of the experimental procedure and some results showing block rotations (Figure 1).

We chose the Fontainebleau sand described by Vendeville et al. (1987) and poured it into an initially rectangular space, 20 x 20 x 20 cm. Throughout each experiment, the upper surface of the sand remained free and the lower surface remained in well-lubricated contact with a rigid horizontal baseplate. A system of computer-driven screw jacks, plates and rubber strips imparted linear velocity gradients to the lateral boundaries, in such a way as to maintain a bulk horizontal plane strain. Ten experiments have been done so far, with different bulk kinematic vorticity numbers, ranging from O (coaxial stretching) to 1 (simple shearing). Somewhat similar experiments were done by Oertel (1965), Hoeppener et al (1969) and Freund (1970); but using clay or other materials in frictional contact with a homogeneously straining baseplate, so that loads were transmitted from below, not from the sides. In our experiments, loads were transmitted from the sides, not from below. Hence, in our experiments, homogeneity of strain was imposed in bulk, but not point by point. Boundary conditions similar to ours have been used for coaxial deformation of wet sand (Freund, 1970) or plasticine (Freund, 1974); but the cohesion of these materials is large in proportion to gravitational forces. Other experiments have been done with cohesive materials in progressive simple shear (Cloos, 1928; Riedel, 1929; Tchalenko and Ambraseys, 1970; Freund, 1974).

We illustrate finite stages of deformation for 4 sandpack experiments with different deformation histories (Figure 1). Passive grids on the upper and lower surfaces of the model show displacement patterns identical to within a few millimeters. Also, vertical motion was negligible at the upper surface. We infer that deformation was probably a plane strain at all points. Whatever the bulk deformation history, faults appeared after about 10 % uniform strain. Fault traces were identical at the upper and lower surfaces, where fault motions were horizontal. A single test, using a sandpack with variously coloured horizontal marker layers, revealed vertical faults with negligible vertical offsets. Hence we conclude that all faults were almost purely strike-slip. We attribute this rather remarkable result to the vertical gravitational loading, and the boundary conditions of bulk horizontal plane strain.

The four experiments illustrated show differing fault patterns, due to the differing deformation histories. They also show differing patterns of block rotation. Some rotations are clockwise, others counterclockwise, with respect to a fixed external frame. In general, the largest rotations occur within the best formed domino domains.

Experiment A (coaxial stretching, up to a stretch $\alpha = 1.45$) shows conjugate fault sets, with approximately equal development of each set. Dominos are not well formed and block rotations are small. Conjugate faults tend to intersect at points lying on the boundaries of the sandpack and this feature seems to govern the fault spacing to some extent. The angles between conjugate faults remain acute (about 60°, as predicted by Coulomb behaviour).

Experiment B (stretching combined with a small amount of left-lateral simple shearing) also shows approximately equal development of two conjugate fault sets, but notice that left-lateral faults are longer. Most right-lateral faults are confined to a central domino domain, bounded by two left-lateral dominos. Left-lateral dominos show clockwise block rotations, right-lateral dominos show counterclockwise rotations (in the external reference frame). As a result, the average angle between conjugate faults has increased from about 60° to about 90°.

Experiment C (stretching and simple shearing in roughly equal proportions) shows a well-developed left-lateral domino, with clockwise block rotation of up to 30°.

Finally, experiment D (simple shearing alone) shows a well-developed right-lateral domino, with counterclockwise rotations of up to 30°.

In general, counterclockwise rotations are associated with right-lateral dominos; clockwise rotations, wiht left-lateral dominos. The relative proportions of these two conjugate dominos (and other features of the finite fault pattern) depend on the deformation history. Details of this dependence will be discussed elsewhere (Cobbold et al., work in progress). For the purposes of this paper, we notice that faults initiate at angles controlled by the state of stress. Subsequently, a fault set survives as long as there are little or no extensions parallel to the faults. For example, in the nearly perfect single domino domains of experiments C or D, the grid lines show a history of little or no extension parallel to the faults. The same result can be obtained theoritically assuming homogeneous strain, throughout the sand-pack. In contrast, lines initially in orientations conjugate to the visible faults have a history of accumulating extension. Hence faults in such a conjugate orientation could not have survived. By extrapolating such arguments backwards in time, we suggest that they also govern the selection of one fault set over its conjugate, at the very stage of initiation.

In experiment B, individual domino domains also show histories of little or no extension parallel to the faults, but there are several domains in the sandpack. If a single domain neither

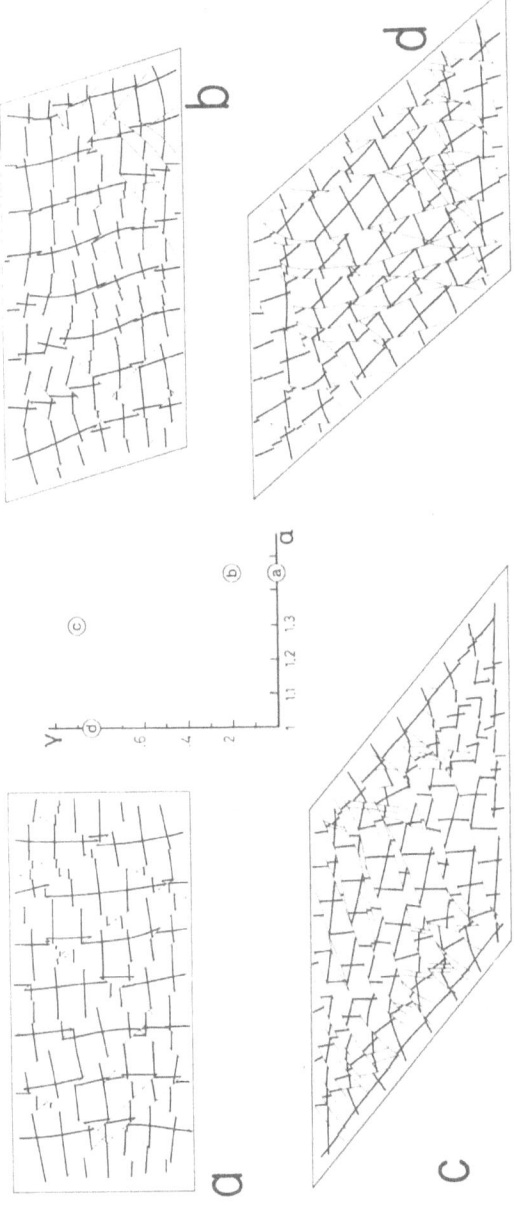

Fig. 1. Strike-slip faults and associated block rotations in 4 sandpacks with differing bulk deformation histories. Outer boundaries of sandpacks (A, B, C, D) are now parallelograms, but initially were squares. Thick lines are passive markers originally forming square grids parallel to outer boundaries. Thin lines are fault traces. Plane-strain bulk deformation was achieved by combining coaxial stretching along bases of parallelograms with left-lateral simple shearing in same direction. Total amount of basal stretch (α and left-lateral shear (γ) are plotted in central diagram for each sandpack. Line drawings (A, B, C, D) were traced from photographs.

formed, nor survived, this is because both potential conjugate directions of initiating faults suffered histories of non-negligible extension.

In conclusion, the deformation history has a strong control on the finite fault pattern, the distribution of domino domains and the senses and amounts of block rotation. Conversely, knowledge of the rotation pattern in nature (obtained, for example, by paleomagnetism) should help constrain the bulk deformation history.

3. Pull-apart domino

The previous experiments investigated rotations of purely strike-slip dominos in laterally-loaded brittle (Coulomb) material. Here we investigate the possiblility of basal loading by viscous drag from below. At the scale of the entire crust, the more brittle upper crust may be subject to viscous drag by the more ductile lower crust, itself in contact (at the Moho) with more resistant mantle material. At a smaller scale, evaporitic or other ductile layers may allow the detachment of blocks of more brittle overlying material. Basal loading, if it occurs, is likely to be more efficient in areas of crustal thinning, where the uppermost brittle layers are at their weakest.

We illustrate an experiment (Figure 2) where a brittle layer (dry quartz sand, initially 2 cm thick) overlies a ductile layer (Newtonian silicone putty, with viscosity about 10^4 Pa.s, initially 1 cm thick). This composite layer adheres to a system of two superposed rigid plates, one fixed, the other mobile (Figure 2a). The mobile plate has a dog-leg boundary. Displacement of the mobile plate induces velocity discontinuities along this boundary. The discontinuities are of wrenching type or of divergent type, according to position along the boundary. At the divergent boundary jog there is a progressive increase in horizontal area (dark ornament, Figure 2a). These basal boundary conditions, applied to a layer of sand alone, result in a strike-slip fault, with a single pull-apart basin at the jog (Faugere et al, 1985). Applied to a composite layer of sand upon siliscone, the basal boundary conditions result in more complex structures above the jog (Basile, 1987). We illustrate the results for a slow relative plate motion of 1 cm.h^{-1} (Figure 2). The main faults in the pull-apart area have sigmoidal traces. At their ends, the faults become nearly vertical, coincide with the wrenching velocity discontinuity, and show dominantly strike-slip displacements; whereas in the middle, the faults are less steep and show components of dip-slip (normal fault) displacement. The fault separate elongate tilted blocks from narrow half-grabens. These half-grabens wre progressively filled with sand during deformation, to simulate sedimentation. Marker lines, initially forming an orthogonal grid, are deformed, segmented and rotated in the final state.

Block rotations are everywhere clockwise. The amount of rotation is variable in space, reaching a maximum of about 20° above the basal jog (Figure 2c). A central traverse across the area (strip M, Figure 2b) shows large rotations and small strike-slip offsets; whereas lateral traverses (strips L and R) show less rotation and larger strike-slip offsets.

This pattern of pull-apart dominos is comparable in some ways with the geometrical pinned-block model of McKenzie and Jackson (1983, 1986), but there is one important difference: in our experiment, strike-slip faults are synthetic to the imposed sense of wrenching.

4. Continental indentation.

Continental indentation, such as that of Asia by India, appears to produce a variety of tectonic structures. The overall deformation can be very variable in space (see review by Cobbold and Davy,

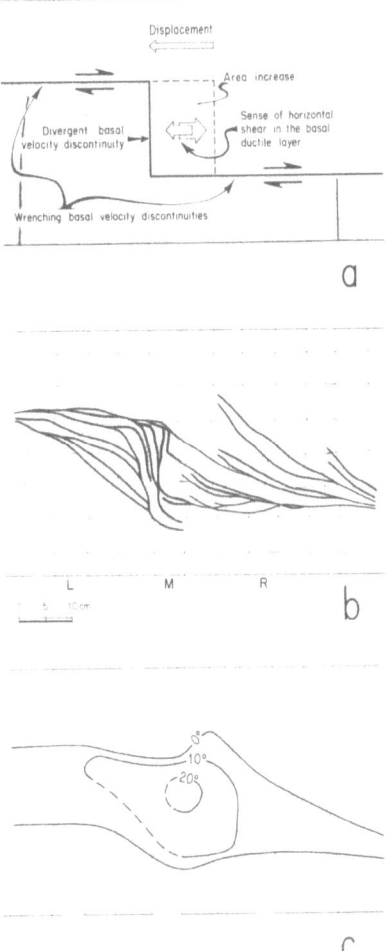

Fig. 2. Pull-apart dominos and associated block rotations. Experimental apparatus (a) consists of two superposed plates, one fixed (shaded), the other mobile (unshaded). Plate motion results in wrenching velocity discontinuity, or divergent velocity discontinuity, at dogleg plate boundary. Plates are overlain by ductile silicone layer (1 cm thick, no shown) and this in turn by sand layer (1 cm thick) with free upper surface (b). Plate motion results in faults (thick traces, b) that offset passive markers (thin lines, b) and enclose rift valleys (dark shading, b). Magnitude of clockwise block rotations is variable in space (contours in degrees, c). Three marker strips (L, M, N, R, in b) show differing patterns of fault offsets and block rotations. Line drawing (b) was traced from photograph.

in press). Experiments scaled for gravity confirm this observation. The experimental patterns of deformation are strongly dependent on both the rheological structure of the experimental lithosphere and the lateral boundary conditions upon the indented continent (Davy and Cobbold, in press). We here reproduce results of 3 experiments (Figure 3) where the lithosphere is a superposition of brittle layers (dry sand) and ductile layers (silicone putty of various densities and viscosities). It rests upon an asthenosphere (glucose syrup of smaller viscosity). Nothwards indentation by a rigid piston produced patterns of faults (thrusts, wrenches or rifts) in the brittle upper crust, with block rotations about vertical axes.

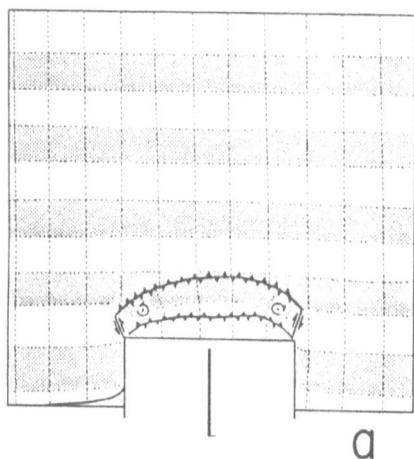

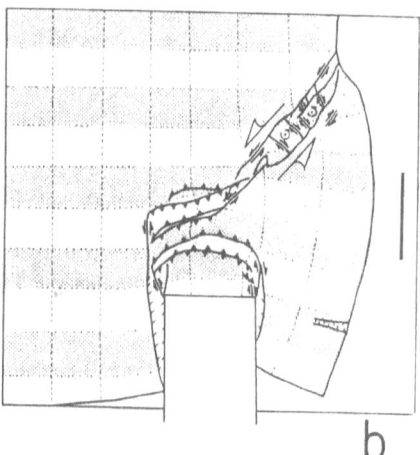

Fig. 3. Three experiments on continental indentation (after Davy and Cobbold, in press). Line drawings (A, B, C) show rigid rectangular indenters (bottom) that have moved northwards (up) into indented continents. At upper surfaces, initally orthogonal grids of marker lines (dots and shaded strips) are now deformed, rotated and offset at fault traces (thick lines). Faults have been identified as dominantly strike-slip (double arrows), dominantly inverse dip-slip (triangles pointing in slip direction of footwalls), or dominantly normal dip-slip (ticks pointing in slip direction of hangingwalls). Deformation pattern is strongly dependent on degree of confinement at continental margins (limits of areas with marker grids, right, B and C).

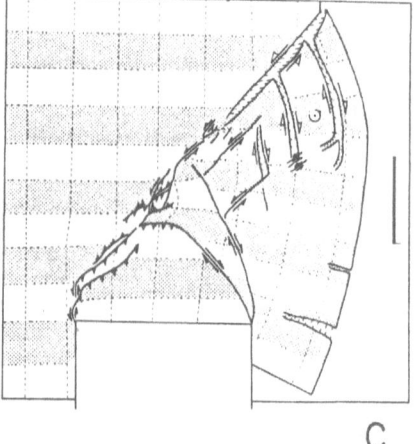

In experiment A (Figure 3A), the indented continent has lateral confinement. The rheological layering is described by Davy and Cobbold (in press). Northwards indentation produces crustal thickening in a plateau region in front of the indenter. The eastern and western margins of the plateau have histories of wrenching in NS directions. Thus arcuate thrusts develop. A detailed analysis of surface grids shows that slip vectors on arcuate thrusts are not simply directed NS, but rather have a radial disposition. Slip in these directions produces arc-parallel extension in the hanging walls of forethrusts, arc-parallel contraction in the hangingwalls of backthrusts. Thus, intervening pop-up wedges have gradients, from arc-parallel extension at the outer arc, to arc-parallel contraction at the inner arc. This implies a bending, with rotations of up to 20° about vertical axes, clockwise in the east, counterclockwise in the west. No other significant rotations are visible in this experiment.

In the two other experiments (Figure 3B,C), the continent has a weak eastern margin. Thus, part of the continent tends to escape eastwards along a major left-lateral wrench zone, that propagates from the western corner of the indenter towards the NE.
In experiment B (Figure 3B), the eastern margin of the continent has viscous confinement. The major left-lateral wrench zone is complex. In the SW, it is strongly overtaken by the accreting plateau and becomes convergent. In the NE, however the deformation is close to plane strain. Two well-spaced antithetic wrench faults form; then the slice between then is cut by series of closely-spaced antithetic strike-slip faults that rotate counterclockwise by a domino mechanism. This motion is similar to the one reported for Southwestern California by Nicholson et al (1986). At a larger scale in experiment B (Figure 3B), notice that the escaping eastern triangle undergoes some bending in the North near the major wrench zone, but otherwise rotates clockwise, in nearly rigid fashion, by about 20°.

In experiment C (Figure 3C), the eastern continental margin is even less confined. Indentation is mostly taken up by lateral escape. The main northeastward left-lateral wrench zone is convergent in the SW, with en-echelon thrusts and segmented wrench faults; whereas it is strongly divergent in the NE, with pronounced rift valleys. The wrench zone widened progressively with time, consuming the eastern triangle by formation of spaced synthetic and antithetic rift-wrenches, the latter are nearly parallel to the continental margin. The antithetic rift-wrench system rotated counterclockwise by a domino mechanism. On a bigger scale, as in experiment B, the remainder of the escaping triangle rotated clockwise by about 20°.

In conclusion, these experiments on continental indentation show block rotations at various scales. Small blocks are associated with wrench zones. Large blocks rotate as they escape towards an unconstrained margin.

5. Conclusions

Our experiments show finite rotations of blocks at various scales. All the rotations are associated with progressive displacements on faults. Wrench zones especially are sites of localized block rotation and domino mechanisms are especially efficient in producing them. Depending upon their tectonic context, so wrench zones can be pure, convergent or divergent. Domino rotations can therefore involve pure strike-slip faults, or oblique-slip faults with normal or reverse components of dip-slip motion.

Our experiments will be representative of nature, only if rheologies and boundary conditions have been correctly chosen. Of particular importance is the rheology of the upper crust, for which we have assumed Coulomb behaviour. Another most important factor is the relative strength of the upper brittle crust, with respect to the lower crust and underlying mantle. Even if our rheological choices are in error, we believe that our experiments reinforce the current view that block rotations are important. They also suggest that domino mechanisms may be ubiquitous in wrench zones, but that no single kinematic model will be sufficient to describe them. Finally our experiments suggest that block rotations may indeed occur at continental scale, as a result of indentation in the presence of unconstrained continental margins.

References

Basile, C., 1987. Modelisation de marges transformantes. Mem. D.E.A., Univ. Paris 6 and 7, 55p.

Byerlee, J.D, 1978. Friction of rocks. *Pure Applied Geophysics*, **116**, 615-626.

Cloos, H., 1928. Experimenten zur Innereen Tektonik. Centralbl f. mineralogie 1928. Abt. B., 609-621.

Cobbold, P.R. and Davy, P. Indentation tectonics in nature and experiment. 2. Central Asia. *Bull. Geol. Inst. Univ. Uppsala*, in press

Cobbold, P.R. and Gapais D, 1987. Slip-system domains. I. Plane-strain kinematics of arrays of coherent bands with twinned fibre orientations. *Tectonophysics*, **131**, 113-132.

Davy, P. and Cobbold, P.R. Indentation tectonics in nature and experiment. 1. Experiments scaled for gravity. *Bull. Geol. Inst. Univ. Uppsala*, in press

Faugere, E., Brun, J.P. and Van den Driessche, J., 1986. Bassins assymetriques en extension pure et en decrochement: modeles experimentaux. *Bull. Centre Rech. Explor.- Prod. Elf Aquitaine*, **10**, 2, 13-21.

Freund, R, 1970. Rotation of strike slip faults in Sistan, Southeast Iran. J. Geol., 78, 188-200.

Freund, R, 1974. Kinematics of transform and transcurrent faults. *Tectonophysics*, **21**, 93-134.

Hoeppener, R., Kalthoff, E. and Schrader, P., 1969. Zur physikalischen Tektonik: Bruchbildung bei verschiedenen affinen Deformationen im Experiment. *Geol. Rundsch.*, **59**, 179-193.

Horsfield, W.T., 1977. An experimental approach to basement-controlled faulting. In: Fault tectonics in N.W. Europe (eds R.T.C. Frost and A.J. Dikkers), *Geologie en Mijnbouw*, 56, 363-370.

Jaeger, J.C. and Cook, N.G.W., 1981. Fundamentals of rock mechanics. 2nd ed., Chapman and Hall, London, U.K., 515 p.

MacClay, K.R. and Ellis, P.G., 1987. Geometries of extensional fault systems developed in model experiments. *Geology*, 15, 341-344.

MacKenzie, D.P. and Jackson, J.A. , 1983. The relationship between strain rates, crustal thickening, paleomagnetism, finite strain and fault movements within a deforming zone. *Earth Planet. Sci. Lett.*, **65**, 182-202.

MacKenzie, D.P. and Jackson, J.A. , 1986. A block model of distributed deformation by faulting. J. Geol. Soc. London, 143, 349-353.

Mandl, G., 1987. Tectonic deformation by rotating parallel faults: the "bookshelf" mechanism. *Tectonophysics*, **141**, 277-316.

Mandl, G., De Jong, L.N.J. and Maltha, A., 1977. Shear zones in granular material. *Rock Mech. Wien.*, **9**, 95-144.

Nicholson, C., Seeber, L., Williams, P. and Sykes, L.R., 1986. Seismic evidence for conjugate slip and block rotation within the San Andreas fault system, Southern California. *Tectonics*, 5, 629-648.

Oertel, G., 1965. The mechanism of faulting in clay experiments. *Tectonophysics*, **2**, 343-393.

Poirier, J.P., 1980. Shear localization and shear instability in materials in the ductile field. J. Stuct. Geol., **2**, 135-142.

Profett, J.M., 1977. Cenozoic geology of Yerington district, Nevada, and implication for the nature of Basin and Range faulting. *Geol. Soc. Am. Bull.*, **88**, 247-266.

Riedel, W., 1929. Zur mechanik geologischer bruschersheinungen. *Zntalbl. F. Mineral. Geol. Pal.*, 354-328.

Tchalenko, J.S. and Ambraseys, N.N., 1970. Structural analysis of the Dasht-e Bayaz (Iran) earthquake fractures. *Geol. Soc. Am. Bull.*, **81**, 41-60.

Vendeville, B., Cobbold, P.R., Davy, P., Brun, J.P. and Choukroune, P., 1987. Physical models of extensional tectonics at various scales. *Geol. Soc. London Special Publ.*, **28**, 95-107.

LARGE RATES OF ROTATION IN CONTINENTAL LITHOSPHERE. UNDERGOING DISTRIBUTED DEFORMATION.

P. ENGLAND
Department of Earth Sciences,
Oxford University,
Parks Road Oxford, OX1 3PR
U.K.

ABSTRACT. The width of a region of continental deformation is governed by the length of the boundary over which relative motion occurs, and by the rheological properties of the continental lithosphere. If the continental lithosphere is regarded as a power-law fluid, the widths of zones of deformation may be predicted from simple relationships. In particular, strike-slip zones should be long and narrow, so that the shear applied by the velocity boundary condition is concentrated near the boundary, with consequently high rates of rotation about vertical axes. Rates of rotation and finite rotations in the Southern Alps of New Zealand and southern California are consistent with the deformation of a thin sheet of fluid obeying a power-law rheology such as that exhibited in the laboratory by olivine. In contrast, compressional or extensional boundaries probably do not produce much rotation about vertical axes unless they, themselves, rotate.

1. Introduction

It has been clear since the earliest days of Plate Tectonics that regions of continental deformation are different in character from plate boundaries in oceanic lithosphere. Although deformation in oceanic plates is restricted to narrow bands around their edges, regions of continental deformation are hundreds to thousands of kilometres wide - approaching in size the continents themselves.

One approach to describing the continents, which is the subject of this paper, is to abandon completely the hypothesis of rigid plates, and to treat the continental lithosphere as a continuous fluid. Undoubtedly there are, in any deforming zone, rigid blocks bounded by faults. It seems reasonable to suppose that these blocks are confined to the portion of the continental lithosphere in which the large earthquakes occur (generally the upper 10 to 20-km - e.g. Chen and Molnar, 1983). By treating continental lithosphere as a fluid we are assuming that these blocks - rather than controlling the deformation as in the oceans - follow passively the deformation of the underlying, ductile portions of the lithosphere.

In the simplest case, when the blocks are circular and do not interact, the rate of rotation, about a vertical axis, of the blocks riding on top of the fluid will equal half the vertical component of the vorticity of the fluid (Jackson and McKenzie, 1983; McKenzie, this volume, equation 4, this paper). Lamb (1986) points out the departures from this simple case that arise for elliptical blocks; the cases in which interaction between the blocks influences their rotation appreciably, and in which the presence of the blocks influences the deformation of the underlying fluid have not been investigated.

C. Kissel and C. Laj (eds.), Paleomagnetic Rotations and Continental Deformation, 157–164.
© *1989 by Kluwer Academic Publishers.*

If the continents are presumed to behave like the oceanic plates, declination changes may only be attributed to motion about a distant pole, whereas if the deformation of the continents is better described as the motion of a viscous fluid, changes in declination arising from local deformation must be considered - indeed measurement of declination changes can provide valuable information on the deformation.

2. Deformation of the Continental Lithosphere

Most laboratory studies of the creep properties of crustal and upper mantle rocks indicate that they probably deform, at conditions of strain rate, confining pressure and temperature that apply below the upper crust, by the mechanisms of crystal plasticity. It is likely that strain rates and deviatoric stresses below the brittle layer in the lithosphere can be approximated by an expression of the form:

$$\tau_{ij} \propto \dot{E}^{1/n-1} \dot{\varepsilon}_{ij} \tag{1}$$

τ_{ij} and $\dot{\varepsilon}_{ij}$ are elements, respectively, of the deviatoric stress tensor and the strain rate tensor; E is the square root of the sum of the squares of the elements of the strain rate tensor and may be thought of as a mean strain rate. Laboratory tests indicate that the value of n is often about 3 for the conditions that we consider.

A convenient approximation to the behaviour of the continental lithosphere is to treat it as a fluid layer whose lateral extent is much greater than its thickness, in which case its behaviour may be described entirely in terms of vertical averages of the stresses and strain rates within it (Bird and Piper, 1980; England and McKenzie, 1983). The vertically-averaged rheology of a lithosphere whose lower portion creeps according to equation (1) and whose upper portion fails on faults obeying a frictional law has been investigated by Sonder and England (1986); they show that a power law of the form of equation (1) is a good approximation to the behaviour of such a system, but with a value for n that is higher than that for the creeping portion alone.

The deformation of the continental lithosphere is influenced by the forces applied to its edges, by the relative motion of the plates, and by buoyancy forces generated in its interior by changes in crustal thickness (e.g. McKenzie, 1972; Molnar and Tapponnier, 1976; England and McKenzie, 1983; Houseman and England, 1986, Vilotte et al., 1986). It seems that, under most circumstances, the vertically-averaged pressure is higher under mountain belts than under their surrounding lowlands, consequently the deformation is transferred further away from a compressional boundary than would be the case if those forces did not act. Equally, the vertically-averaged pressure in a region of thinned crust may be lower than in its surroundings, so that extensional strain associated with a divergent boundary is also spread out by the buoyancy forces. However, though buoyancy forces alter the size of an extensional or compressional region, they have a relatively small influence on the vertical component of vorticity (Houseman and England, 1986; Sonder and England, 1988). As strike-slip boundary conditions produce relatively small changes in thickness of the lithosphere, the deformation in strike-slip regions is not much affected by buoyancy forces. For these reasons, only deformation resulting from the boundary conditions is considered in what follows.

England et al. (1985) calculated approximate velocity fields for the deformation of a thin sheet of material, whose rheology is described by equation (1), when it is subjected to different boundary conditions. The principal feature of these solutions is that the velocities in the interior of the sheet are nearly parallel to the velocities applied to the boundaries, and their magnitudes die out exponentially with distance from the boundaries (Figures 1 and 2).

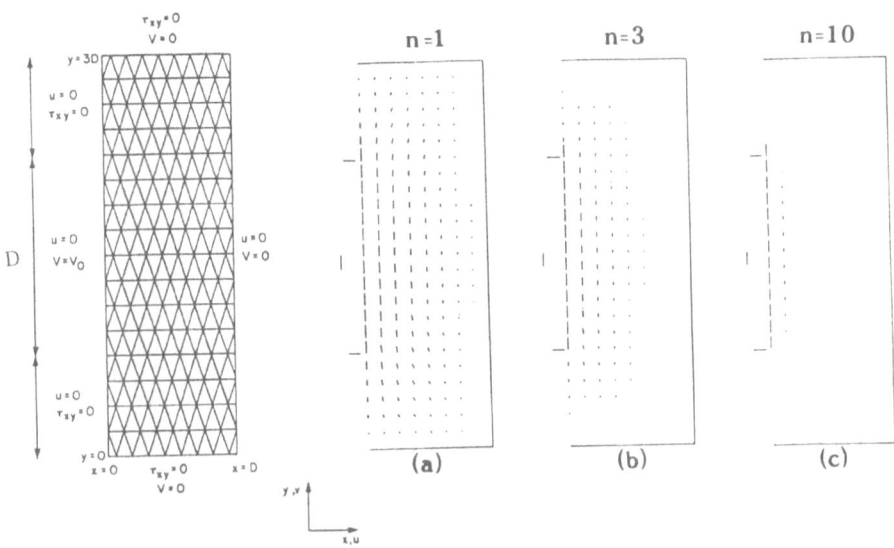

Fig. 1. Map view of calculated velocities in thin sheets of viscous materials having different values for the constant n in equation (1) and subjected to a strike-slip boundary condition (from Sonder et al., 1986). Left hand figure shows the finite element mesh on which the calculations are carried out; a constant velocity, V_0, up the page is applied over the length D of the left hand boundary and velocities in the interior are calculated for different values of n. These velocities are depicted by the small lines in (a) ($n=1$), (b) ($n=3$) and (c) ($n=10$). Note the decrease in across-strike length scale of the deformation as the value of n increases (see equation(2)).

For example, if a strike-slip boundary condition of characteristic length D is applied to the edge of a piece of lithosphere whose behaviour is described by equation (1), the resulting deformation dies out approximately exponentially with distance from that boundary with a scale length that depends on the value of D, and of the exponent, n, in the rheology. If the magnitude of the relative velocity parallel to the boundary is V_0, the velocity in the interior at a distance x from the boundary will be given approximately by:

$$v(x) \simeq V_0 \exp{-(2\Pi x \sqrt{n}/D)} \qquad (2)$$

(England et al., 1985, equation 32; Sonder et al., 1986, Appendix A; Figure 1, this paper). The velocity is small compared with the boundary velocity at distances greater than about $D/\Pi\sqrt{n}$ from the boundary, which gives the result that the width of a deforming strike-slip zone should be approximately equal to one fifth of its length if n is equal to 3, or about one tenth of its length if n

is equal to 10. (A power law exponent of 10 has no significance in terms of a single deformation mechanism operating within the lithosphere, but could represent the vertical average of creep at depth and sliding on faults near the surface -- see above, and Sonder and England, 1986.)

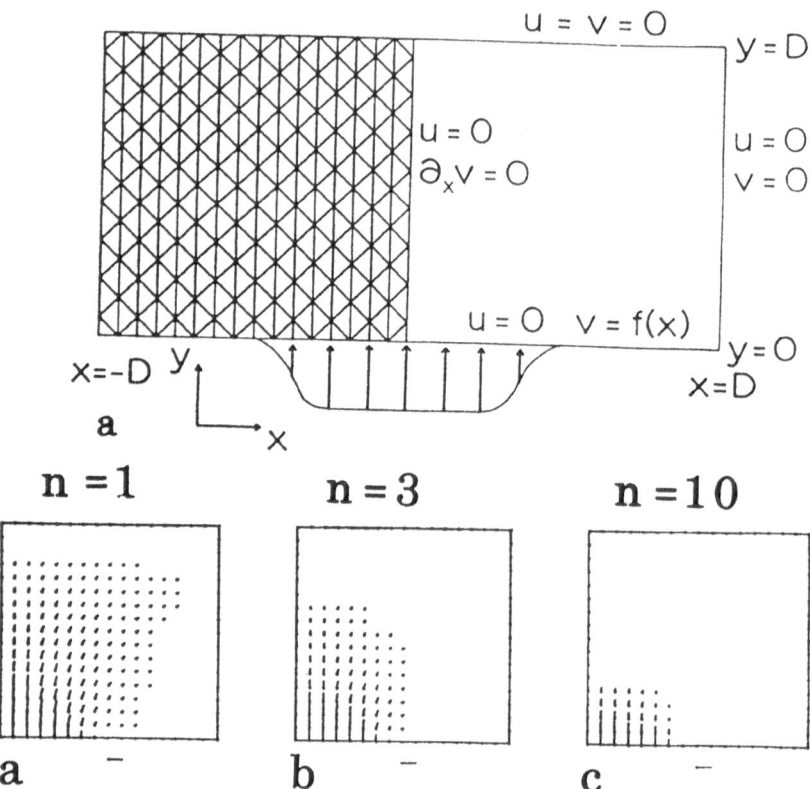

Fig. 2. Map view of calculated velocities in thin sheets of viscous materials having different values for the constant n in equation (1), and subjected to a convergent boundary condition (from Houseman and England, 1986). Upper figure shows the finite element mesh on which the calculations are carried out; a constant velocity up the page is applied over the portion of the lower boundary indicated by arrows. The problem is symmetrical about the line $x = 0$, and only the right-hand portion of the solutions are shown in (a), (b) and (c). The velocities in the interior are depicted by the small lines in (a) ($n = 1$), (b) ($n = 3$) and (c) ($n = 10$). Note that the only appreciable gradients of velocity that would lead to rotations about a vertical axis are found near the edge of the moving boundary (see also Figure 3).

A similar form of solution is obtained for a convergent or extensional boundary, but the width of the deforming region is four times greater than in the strike-slip case. If the magnitude of the relative velocity perpendicular to the boundary is V_0, the velocity in the interior at a distance y from the boundary will be given approximately by:

$$v(y) \simeq V_0 \exp -(\Pi y \sqrt{n}/2D) \qquad (3)$$

(England et al., 1985, equation 29; Figure 2, this paper).

If -- as in equations (2) and (3) -- the symbols u and v are used for the components of velocity in the x- and y- directions, the rate of rotation, $\dot{\theta}$ of the fluid about a vertical axis may be written:

$$\dot{\theta} = \frac{1}{2} \left(\frac{\partial u}{\partial y} - \frac{\partial v}{\partial x} \right) \qquad (4)$$

In each of the approximate solutions given above, the component of velocity perpendicular to the applied boundary condition is neglected; this is a reasonable approximation, as the full solutions illustrated in Figures 1 and 2 show. The rate of rotation about a vertical axis is then obtained by the appropriate differentiation of equation (2) or equation (3). For the strike-slip condition

$$\dot{\theta} \simeq - \frac{1}{2} \frac{\partial v}{\partial x} \simeq V_0 \Pi \sqrt{n}/ D \exp -(2\Pi x \sqrt{n}/D) \qquad (5)$$

For the convergent or extensional case:

$$\dot{\theta} \simeq - \frac{1}{2} \frac{\partial v}{\partial x} \simeq 0 \qquad (6)$$

Thus, while rotations about a vertical axis are to be expected along the whole length of strike-slip zones, in compressional or extensional environments appreciable rotations are expected only where there are variations along strike in the magnitude of the boundary velocity.

Sonder et al. (1986) applied these ideas to the deformation of Southern California and showed that the width of the zone of late Tertiary deformation in the region, and the clockwise rotations of over 90° recorded by Miocene rocks of the region may be explained if material of the North American plate were to have behaved as a fluid whose rheology is described by equation (1).

The South Island of New Zealand provides another example of the rates of rotation we may expect in a zone of distributed continental deformation. Oblique convergence bewteen the Australian and Pacific plates is taking place in this region at about 45 mm.yr[-1]; the component of velocity parallel to the boundary is about 37 mm.yr[-1]. The length of the plate boundary in this region is about 800 km, so the width of the region of deformation associated with the plate boundary should be, according to equation (2), 150 km if n is 3, or about 80 km if n is 10. The corresponding rates of rotation about a vertical axis are 1.7×10^{-7} yr[-1] and 3.1×10^{-7} yr[-1].

The width of the deforming zone in the South Island is most easily estimated from the present patterns of topography and uplift rates: as the component of velocity perpendicular to the plate boundary is relatively small, we should expect compressional deformation to be confined to the narrow region of low effective viscosity generated by the predominantly transcurrent boundary condition. The width of the Southern Alps is about 150 km through most of the South Island of New Zealand, broadening to about 200 km in the southern portion, where the plate boundary becomes predominantly convergent. The rates of shear strain rate ($\partial u/\partial y + \partial v/\partial x$) determined geodetically in the South Island (Walcott, 1984) show peak values of around 4×10^{-7} yr[-1] -- again dying out over a distance of about 100 km from the plate boundary. If these simple shear strain rates are expressed as rates of rotation about a vertical axis, they give a maximum rate of about 2 x 10^{-7} yr^{-1}.

The rotations associated with convergent and extensional boundary conditions take place mainly near the ends of the boundaries -- where the variations *along strike* of the velocity component perpendicular to strike gives rise to vorticity and to rotations of the boundary during deformation. Rotations like these are, clearly, sensitive to details of the boundary conditions, and it is hard to make useful general statements about them.

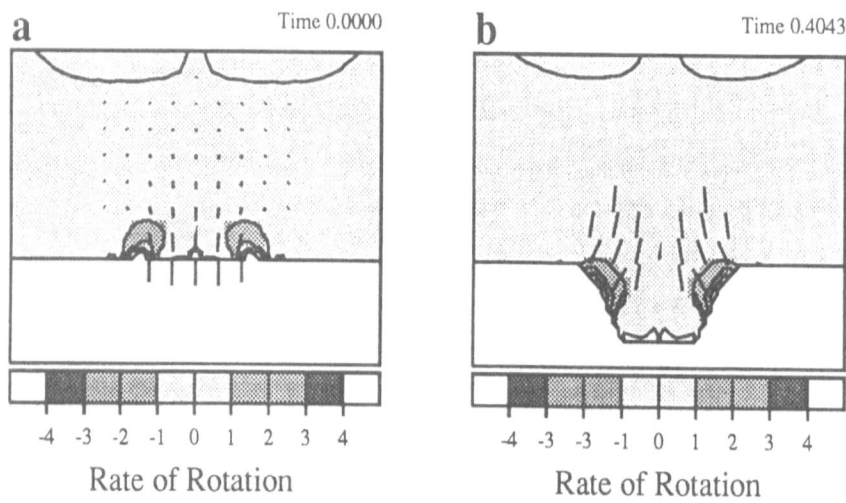

Fig. 3. Map view of calculated rates of rotation and of finite rotation about a vertical axis within a fluid with $n = 3$ in equation (1) (from Sonder and England, 1988). The region of solution is indicated by shades of grey, and is subjected to an extensional boundary condition, as indicated by the velocity vectors in (a), that leads to a downward motion of the lower boundary. (a) Velocity vectors at the beginning of deformation are superimposed on contours of the rate of rotation about a vertical axis. Calculations are dimensionless, so units of rate of rotation are arbitrary. (b) Finite rotations after a dimensionless time of 0.4 are superimposed on rates of rotation at that time. The lines in the interior indicated rotations that a small rigid block attached to the fluid would have undergone during deformation. Zero rotation direction is up the page, and rotations of less than 5° are not plotted.

An example is given in Figure 3 which illustrates rotations owing to motion of the boundary during extensional deformation (Sonder and England, 1988). Figure 3a shows contours of the vorticity at the start of deformation; the highest rates of rotation are at the edges of the boundary and are clockwise at its left end, counterclockwise at its right. In the course of the deformation (see Figure 3b) a portion of the boundary moves in the direction of its normal, while the ends of the boundary rotate, in order to maintain continuity. The finite rotations illustrated in Figure 3b are mainly the result of the rotation of the ends of the boundary, although coupling of the fluid to the non-rotating central portion of the boundary acts to inhibit rotations near the bottom of Figures 3a and b.

Calculations of rotation about vertical axes for the case of a convergent boundary condition are given by Houseman and England (1986). In this case, too, rotations are localised around the ends of the boundaries and are dominated by the rotations of those boundaries

3. Discussion

It appears that the rheology of the continental lithosphere determines that the width of zones of strike slip deformation should be small compared with their length and that, therefore, blocks within these deforming zones may experience high rates of rotation about vertical axes. Rocks in Southern California have rotated by more than 90° in the last 30 Myr; the maximum rates of simple shear in the Southern Alps would, could lead to rotation by 90° in 4 Myr. The calculations of the previous section, if applied to Western New Guinea, suggest that the 130 mm.yr^{-1} of oblique motion over the 1200 km of boundary of the region could result in rates of rotation as high as 7 x 10^{-7} yr^{-1}, or rotations of 90° in a couple of million years.

Jackson and McKenzie (1983) discuss rotations during distributed continental deformation and treat the continental lithosphere as a zone of weakness bounded by two rigid blocks; in this case, the deformation is two-dimensional, and rates of rotation are constant across the deforming zone. If, in contrast, the width of regions of strike-slip deformation is governed by the rheology of the deforming continental lithosphere itself, as discussed in the previous section, rates of rotation (and finite rotations) should die out with distance from the boundary, in the manner suggested by equation (4). Careful sampling of declination changes in regions of distributed strike-slip faulting can help to determine whether lateral heterogeneities in strength determine the width of a deforming zone.

References

Chen, W.-P., and P. Molnar, 1983. Focal depths of intracontinental and intraplate earthquakes and their implications for the thermal and mechanical prooperties of the lithosphere. *J. Geophys. Res.*, **88**, 4183-4214.

England, P.C., and D.P. McKenzie, 1983. Correction to: A thin viscous sheet model for continental deformation. *Geophys. J. R. Astron. Soc.*, 73, 523-532.

England, P.C., G.A. Houseman, and Sonder L.J., 1985. Lengthscales for continental deformation in convergent, divergent and strike-slip environments: Analytical and approximate solutions for a thin viscous sheet model. *J. Geophys. Res.*, **90**, 3551-3557.

Houseman, G.A., and P.C. England, 1986. Finite strain calculations of continental deformation 1. Method and general results for convergent zones. *J. Geophys. Res.*, **91**, 3651-3663.

Jackson, J.A., and D.P. McKenzie, 1983. The geometrical evolution of normal fault systems. *J. struct. Geol.*, **5**, 471-482.

Lamb, S.H., 1972. A model for tectonic rotations about a vertical axis, *Earth Planet. Sci. Lett.*, **84**, 75.

McKenzie, D.P., 1972. Active tectonics of the Mediterranean region, *Geophys. J. R. Astron. Soc.*, **30**, 109-185.

Sonder, L.J., and P.C. England, 1986. Vertical averages of rheology of the continental, lithosphere: Relation to thin sheet parameters. *Earth Planet. Sci. Lett.*, **77,** 81-90.

Sonder, L.J., and P.C. England, 1988. Effects of a temperature-dependent rheology on large scale continental extension. Submitted to *J. Geophys. Res.*

Sonder, L.J., P.C. England, and G.A. Houseman, 1986. Continuum calculations of continental deformation in transcurrent environments. *J. Geophys. Res.*, **91,** 4797-4810.

Vilotte, J.P., R. Madariaga, M. and Daignieres, Zienkiewicz O.C., 1986. Numerical study of continental collision: Influence of buoyancy forces and a stiff inclusion. *Geophys. J. Roy. Astron. Soc.*, **84,** 279-310.

STRAIN AND DISPLACEMENT IN THE BRITTLE FIELD

P. CHOUKROUNE
Laboratoire de Tectonique
Campus de Beaulieu
35042 Rennes Cedex, France

ABSTRACT. In the brittle field, deformed areas are difficult to restore mainly because key markers are often missing : the exact amount of displacement along discontinuties and the component of rigid rotation can rarely be measured using purely structural techniques.
However, the main targets of structural field work in deformed areas are often more easily reached in brittle domains than in ductile ones.
1. The orientation of small significant discontinuities created during a single deformation episode, and/or, the senses of motion along randomly distributed pre-existing discontinuities, allow us to measure the local orientation of strain axes.
2. The distribution and association of significant small discontinuities allow us to estimate the strain ellipsoid type.
3. The general pattern of discontinuities examined at the scale of the deformed system, the order of appearance of these discontinuities (progressive deformation) and the change in orientation of principal strain axes with time, are key data for assessing the strain regime.
4. The spatial integration of local strain data allow us to construct strain trajectories, which are relevant to i) the possible rigid rotation of blocks and ii) the general question of displacement at the boundaries of deformed systems.
Different field exemples and results illustrate the above approach. Some of them are taken in orogenic belt forelands (compressive zones),whilst others concern stretched crustal material (extensional zones).

1. Introduction

Within deformed domains, where discontinuous deformation is dominant, two questions concerning displacements are generaly posed: one concerns the amount of displacement between rigid blocks bounded by discontinuities. Using structural analysis and field arguments, this can sometimes be estimated, but never directly measured. Another question is related to displacement at the boundary of domains affected by discontinuities. Looking at the discontinuities pattern alone, the answer obtained is generally equivoqual and remains hypothetical because the first question usually has no satisfying answer. On the other hand, during the last ten years, many publications have been devoted to so-called palaeostress analysis which has led to the local determination of the state of stress responsible for the observed brittle deformation.

In fact, for a given discontinuity, the basic data is the measurement of the direction and sense of motion along the considered plane. For a given fault pattern, the integration of these local data then leads to an estimate of the "strain" suffered by the considered discontinuous area (Fig.1).

The scope of this paper is to recall the different steps of strain analysis, to review definitions of local strain characteristics using discontinuities in rocks and to show how to use these data to set up a strain field in the brittle domain. It might appear unusual or incorrect to apply strain concepts to discontinuous deformation. Considering that this is essentially a scale problem, we infer that no

165

C. Kissel and C. Laj (eds.), Paleomagnetic Rotations and Continental Deformation, 165–180.
© *1989 by Kluwer Academic Publishers.*

166

serious mistake is made if the geometrical aspects of strain analysis of homogeneous, ductile media are adapted to brittle ones. On the other hand, we would prefer to think in terms of strain than in terms of stress : indeed, the analysis of faults and discontinuities leads rigourously to displacement versus strain relationships, not to stress or palaeostress (see Fig.1).

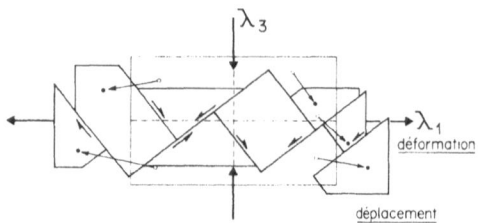

Fig.1. Schematic representation of relationship between relative displacement between rigid blocks (half arrows) absolute displacement of blocks (arrows) and orientation of λ_1 and λ_3 axes.

Generally speaking, strain characteristics are locally defined when the following data are available :
1) The local orientation of principal strain axes.
2) The successive orientations of superimposed finite strain, if any.
3) The type of each finite strain ellipsoid.
4) The strain history (strain regime) of each episode of deformation.
5) The axial ratio of each strain ellipsoid.
In the brittle field, all these data can be obtained except the last mentioned.
Moreover, all these local data can generally be integrated in order to define a strain pattern which shows the spatial variation of the above strain characteristics.

2.The local characteristics of strain in the brittle field.

2.1. ORIENTATION OF PRINCIPAL STRAIN AXES.

Two kinds of situation must be distinguished: the simplest one concerns areas where the "brittle strain" is low, single, and monophase. In that case (Fig.2) it is generally easy to observe discontinuities along which the displacement can be directly correlated to strain : gashes which open normally to λ_3 (stretching axis) (Ramsay 1980) and parallel to λ_1 (shortening axis),as well as stylolites whose peaks reveal directly λ_3 (Dunnington,1954; Arthaud and Mattauer, 1969; Choukroune, 1969; Mattauer, 1973; Buchner, 1981) are very well-known and useful discontinuities. Beside these, shear zones (en echelon tension gashes, Hancock,1972; Beach,1975)or shear discontinuities are organized into conjugate sets whose geometry can be correlated with strain axis orientation : if the displacements along these potential or observed discontinuities are not directly related to the orientation of strain axes, the overall geometry of the sets and the sense of shearing along them can lead us to find these orientations (Figure 2).
One precaution must be taken in this simple case : the local orientation of strain axes so determined must be representative of the "regional" strain. In other words, one must check that the obtained result is not due to local perturbations introduced,for instance,by some relay zone between first order discontinuities (Figure 3).

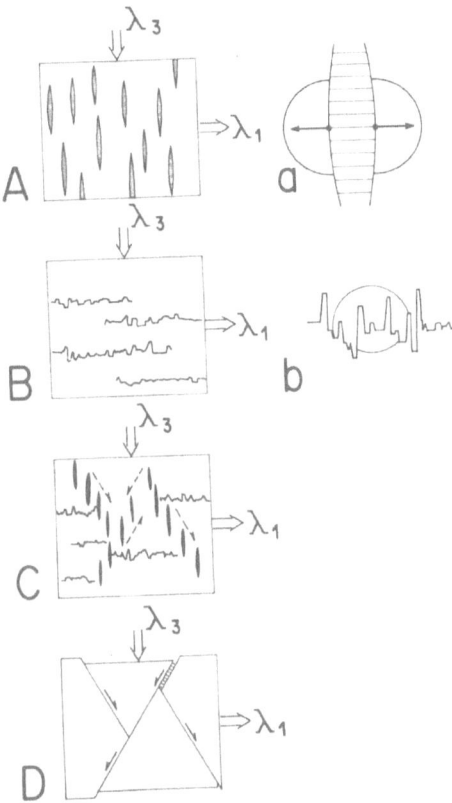

Fig.2. Orientation of elementary discontinuities and orientation of λ_1, λ_3, principal strain axes.
A: gashes; a) shows the direct relationship between displacement (opening) and stretching direction.
B: stylotitic joints; b) shows the direction relationship between displacement (closing) and shortening direction.
C: association of en echelon gashes and stylolites.
D: shear discontinuities organized in two conjugate sets.

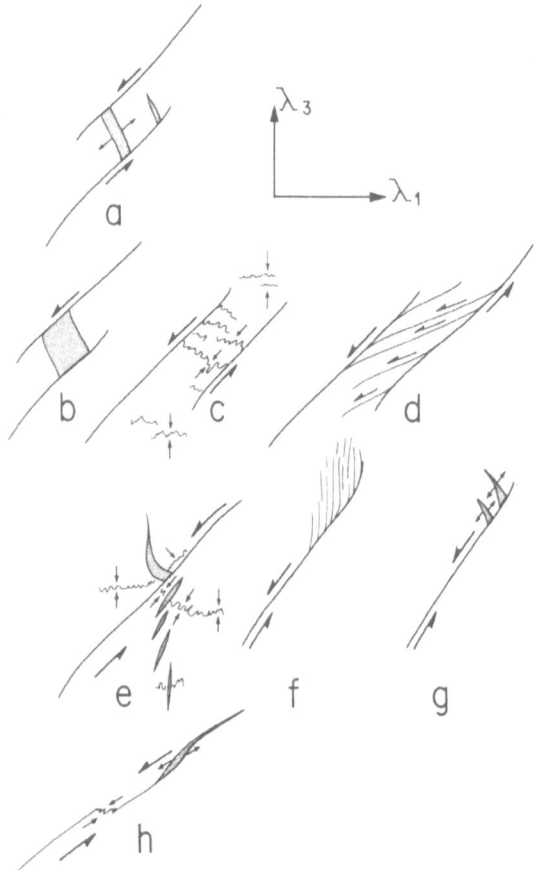

Fig.3. Special examples of open discontinuities, stylolites and discontinuous shear zones where the directions of opening, shortening or shearing are not directly related to the strain axes. These ocurrences are linked to relay or adsorption zones of displacement.

The second case is more complex and needs statistical analysis : the medium affected by brittle deformation is never originally continuous. For this reason, one must consider the behaviour of a population of randomly distributed discontinuities during a tectonic episode. This question has been treated successfully by Arthaud (1969) and his method is the basis of other studies (Carey, 1979; Etchecopar et al., 1981; Angelier et al., 1982; Rispoli and Vasseur, 1983; Angelier, 1983) which introduced so-called "palaeostress analysis". Two hypotheses are necessary and must be verified: the first assumes that the displacement along a given discontinuity depends on the relative orientation of the discontinuity with respect to the strain axes. The second assumes that the direction of striae or slickensides on a fault plane is the projection, on this plane, of one of the strain axes.

These two assumptions can be verified,if, starting from a random distribution of striated faults (Figure 4), the M planes, normal to each fault plane and parallel to striae on the plane, have three common intersections: these intersections are the principal strain axes. Then, the sense of displacement along a few fault planes leads us to determine which intersection corresponds to what

169

strain axis (for a review on sense of displacement on fault planes, see Petit, 1987; Tjia,1967; Petit et al. 1983).

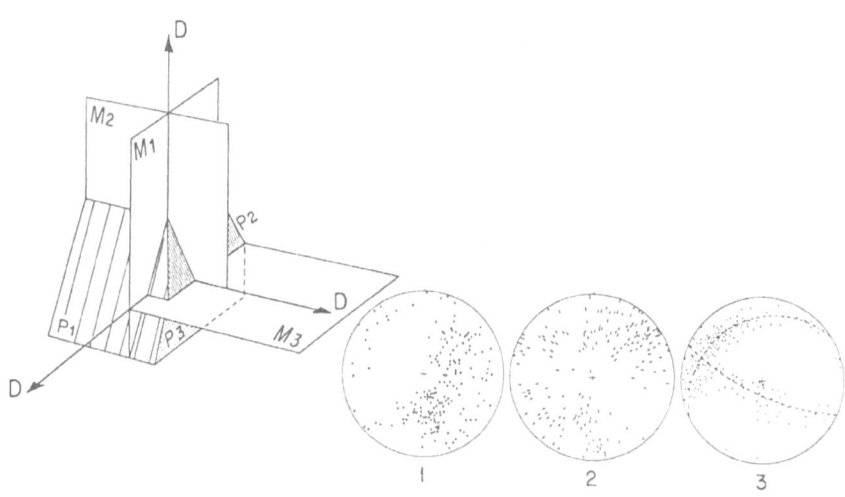

Fig.4. Representation of a simple case where M planes (normal to shear discontinuities and parallel to striae on the shear planes) have three intersections (D) representing the principal strain directions. The stereonet measurements are taken from Arthaud (1969). 1: all discontinuities, 2: all striaes, 3: the poles of M-plane intersect along one principal strain direction (Languedoc area, France).

2.2. THE DETERMINATION OF SUPERIMPOSED FINITE STRAIN IN THE BRITTLE FIELD.

Very often, the observed field of discontinuities does not result from a single event, but from different successive episodes each of which creates new fractures and re-activates the older structures.

Generally, successive episodes create superimposed small discontinuities whose order of appearance is easily determined from field arguments. Each strain orientation can also be established looking at the orientation of each set of representative discontinuities. For superimposed faulting, the case is more complex because the population of faults must be separated into a number of families, each of them compatible with a simple strain ellipsoid (Angelier and Manoussis, 1980; Etchecopar et al.,1981). This has been successfully achieved in many cases (Burg and Etchecopar,1980; Mercier 1981; Armijo and al. 1982; Bergerat, 1985; Faugere et al. 1985) (Figure 5) and verified by the general structural context where measurements are available.

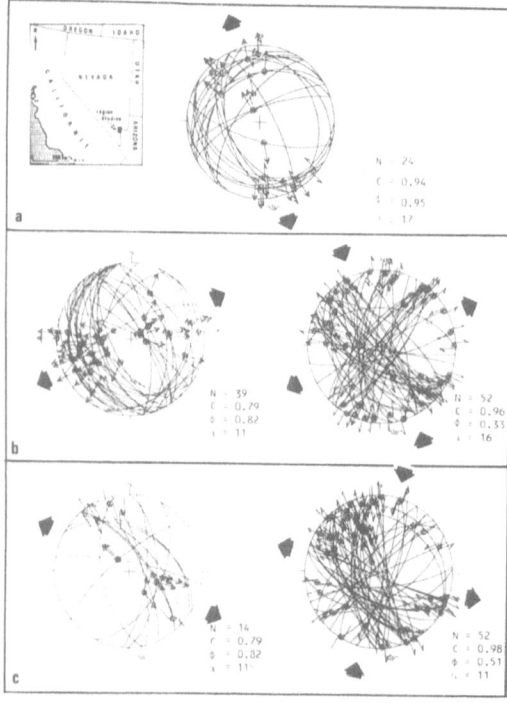

Fig.5. Successive strain orientations in the River Mountains (Nevada U.S.A.) (Schmid projection, lower hemisphere). This history is related to the extensional deformation which has affected the Basin and Range Province during the Miocene (from Faugere et al. 1985).

2.3 THE STRAIN ELLIPSOID TYPE IN THE BRITTLE FIELD.

The strain ellipsoid type depends on what has occurred (stretching, shortening or no deformation) along the λ_2 axes (Flinn, 1965).

Using gashes and stylolitic joints, flattening (two directions of stretching), constriction (two directions of shortening) or plane strain (only one direction of stretching and shortening)the strain ellipsoid type can easily be determined locally within a discontinuously deformed material (Figure 6). In the same manner, one can imagine two families having two sets of shearing planes with a spatial distribution which implies flattening or constriction (Figs.6 and 7). The only problem concerning this occurrence is that these two families cannot be formed in the same place at the same time : indeed the intersection of each set of the two families must be a direction of no deformation at the moment of formation. This incompatibility generally appears in the field where the non-simultaneity of families of shearing planes can be clearly demonstrated (Choukroune 1969, Choukroune and Delair, 1976) in a given place.

This problem can be also solved when the two families can be observed in different domains of the same area.

2.4. STRAIN REGIMES IN THE BRITTLE FIELD.

A knowledge of the strain regime (or strain history) is, of course, a key element in characterizing the strain, especially when one attemps to evaluate the rigid rotation of blocks within faulted zones.

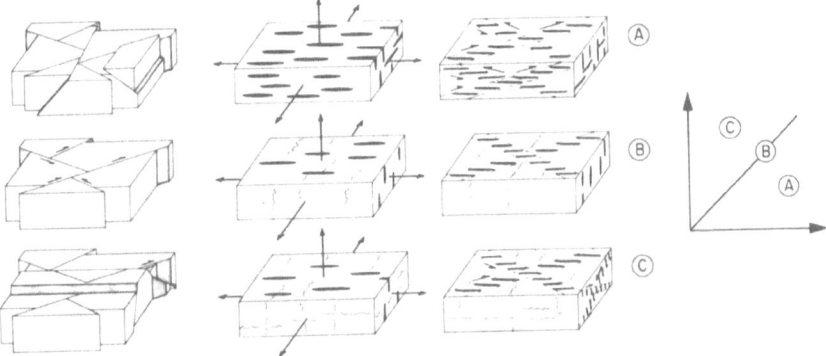

Fig.6. Theoretical block diagrams showing the elementary discontinuity pattern in the case of flattening (A) plane strain (B) and constriction (C).

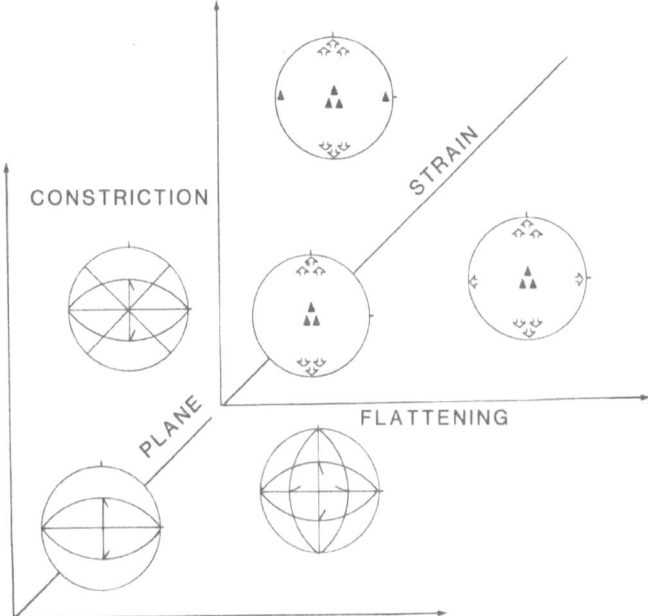

Fig.7. Theoretical geometrical relationship beetween stylolitic peaks (triangles), poles of gashes (white arrows) and shear discontinuities in different parts of the Flinn diagram.

172

Indeed, because simple shear is the sum of pure shear and rigid rotation, one can expect some rotation to occur if the deformation is non-coaxial. On the other hand, if the discontinuous deformation is coaxial, it is impossible to exclude some degree of rotation.

The question of strain regime can firstly be discussed in the light of field structures on the scale of the faulted system. It is well known that the symmetry or asymmetry of the fault pattern can indicate the coaxiality or non coaxiality of the deformation (Figure 8) : since Riedel (1929) we have known that shearing processes operating at the boundaries of a system create an asymmetric fault pattern which characterize the system. Otherwise, the strain history can be recorded by superimposed discontinuities which appear in a particular order and have a geometrical coherence; they enhance the progressive character of the apparent polyphase deformation sequence.

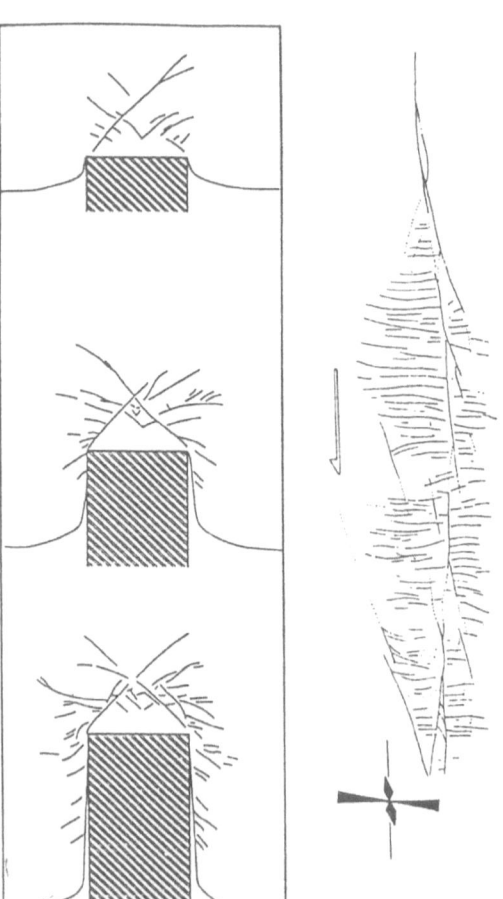

Fig.8. Two experiments showing two different discontinuity patterns.
Left : coaxial deformation creates a symmetric pattern of shear discontinuities at every stage of the evolution of the faulted system (after Tapponnier et al., 1982).
Right : classical Riedel experiment done by Tchalenko (1970) showing a contrasting asymmetric fault pattern.

Three examples illustrate this possibility :

i) The first one is simple, and concerns the development of gashes : if the deformation is coaxial, any gash must appear and develop parallel to λ_3. On the contrary, during simple shearing any gash must appear parallel to the incremental λ_3 direction which remains constant during deformation but must rotate afterwards : the result is the observation of superimposed gashes, the oldest being the most rotated ones and the youngest being parallel to the incremental λ_3 direction. Even if superimposed gashes are not observed at the same locality, the measurement of many gashes may yield an area of dispersion which lets us to estimate the amount of rigid rotation suffered by the material affected by shearing deformation (Figures 9 and 10a) (Choukroune et al., 1987).

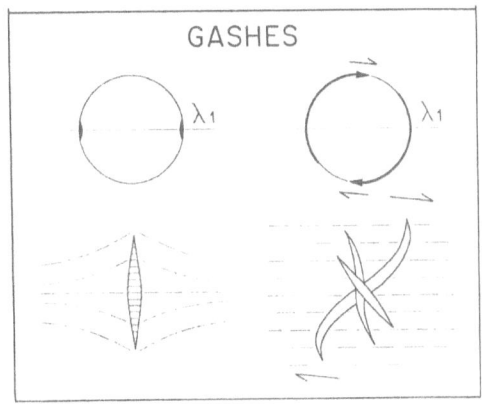

Fig.9. Preferred orientation of poles of gashes during simple shearing : the asymmetric dispersion indicates non coaxiality for the deformation (see natural example in Fig.10).

ii) The same reasoning can be applied to the appearance and development of any discontinuity or family of discontinuities. In the case presented on Figure 10, for instance, spatially dispersed shear zones can be separated and ordered, using field arguments, into three successive families : the oldest are compatible with an almost E-W λ_3 direction and the two others show a progressive dextral rotation of this principal strain axis. We can see that the bulk observed strain is the sum of three finite strain components consistent with the dextral shearing known in the studied area (Choukroune et al. 1987). One should note that this case is quite different from the first one, showing the rigid rotation of discontinuities during the shearing process: indeed, only a progressive rotation of λ_3 is recorded even if rigid rotation is also suspected and locally argued (Figure 10).

iii) The third example illustrates the case of the appearance and development of discontinuities during the progressive rotation of a part of the brittle material i.e. during folding. On one horizontal flank of the folded structure (Figure 11), two single conjugate sets of discontinuities can be measured. In contrast, a very complex fault pattern appears on the vertical flank and is, in fact, the sum of the same conjugate set of shears which appeared with these vertical intersections affecting the considered layer at different stages of its rotation (Figure 11) (Choukroune and Delair, 1976).

In this last case, the rigid rotation is not a part of the simple shear deformation : it is due to the folding which comes with coaxial discontinuous deformation. In other words, the recorded history shows a rotational history but not a non coaxial one. The amount of rigid rotation is measured by the amount of apparent rotation of the strain axes.

174

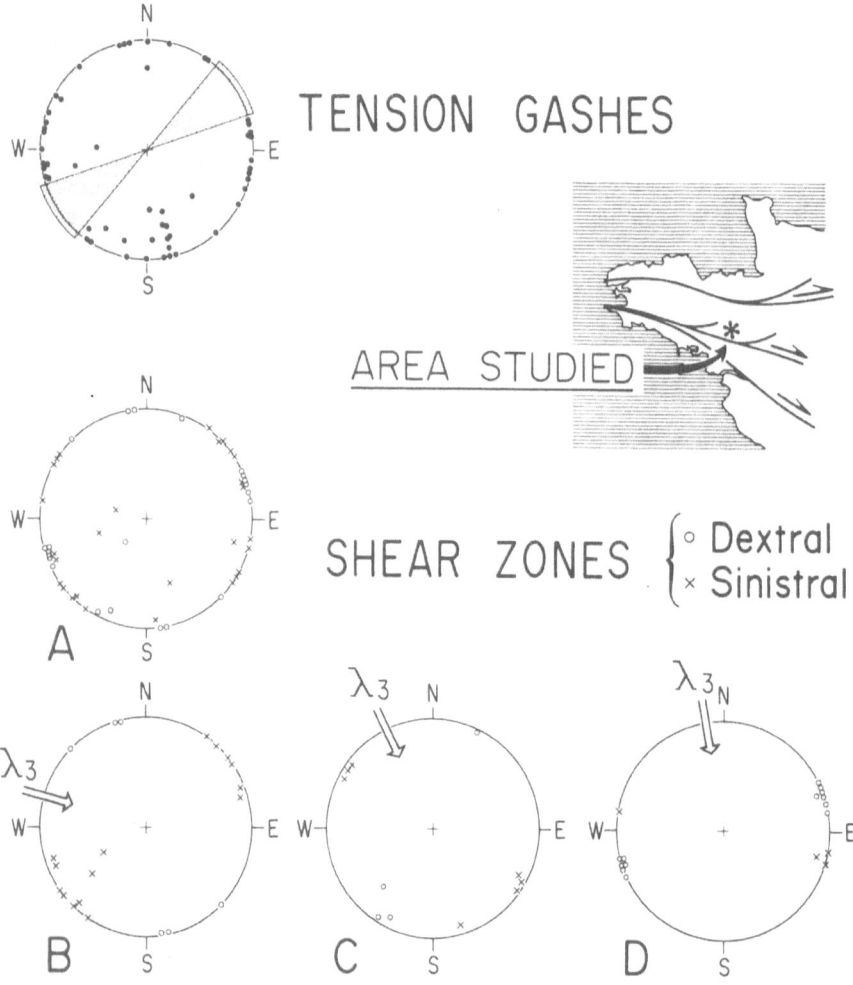

Fig.10. Distribution of incompatible superimposed structures (A) ; they can be split into three successive shapes leading to infer the dextral rotation of λ_3 (B-C-D). Example from the Hercynian shear zone of South Brittany (from Choukroune et al. 1987).

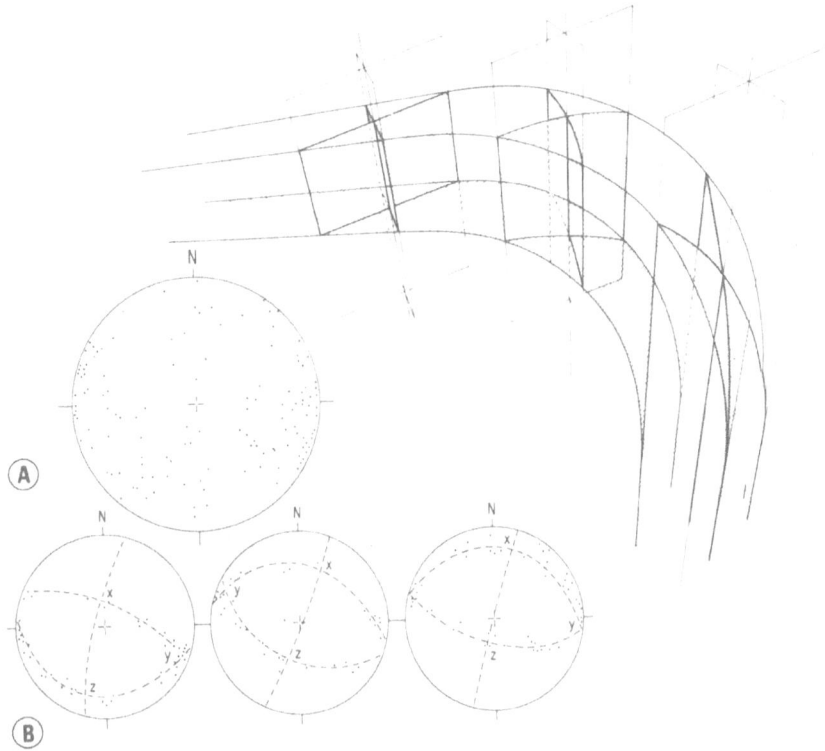

Fig.11. Continuous fracturation during progressive concentric folding of a single layer (Mas d'Azil area, French Pyrenees). The persistence of a fracture geometry (micro-wrench faults with vertical intersection) that affects the layer is represented at different rotation stages (upper). Moreover, the earlier fractures are progressively rotated: all the data (lower) show the difficulty of demonstrating the process without any separation of the fracturation stages (A). When this separation is performed (B), an apparent rotation of the principal strain axes appears. The projection plane is the stratification surface (after Choukroune and Delair, 1976). X is λ_3, Y is λ_2, Z is λ_1.

3. The spatial integration of local strain data: the use of strain trajectories.

The representation and integration of local strain data (orientation of strain ellipsoid, type of strain ellipsoid, strain values) in a given area is a representation of the strain field or strain pattern. Particularly in areas of discontinuous deformation, principal directions of strain are the most easily obtained values. Using these local directions, the strain trajectories can be drawn through large areas where their consistency can be tested and interpreted (Figure 12).

176

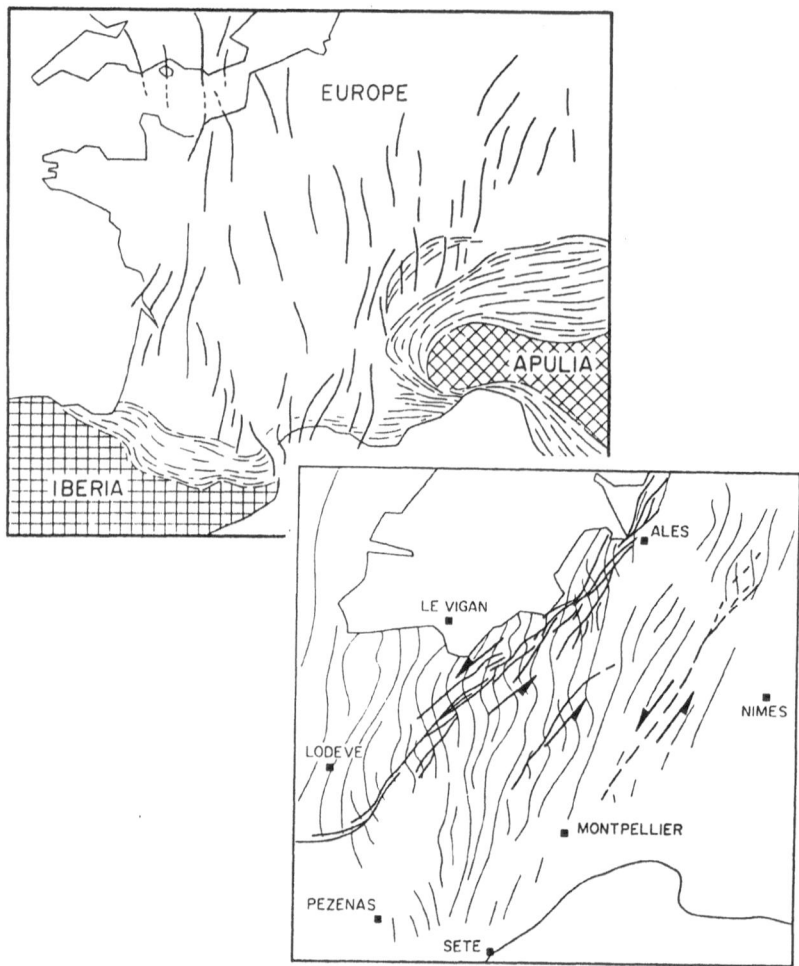

Fig. 12. λ_3 trajectories in the Alpine foreland at 50 Ma. (upper). Data from Arthaud and Choukroune (1972); Burg and Etchecopar (1980); Bergerat (1985).

Below : detailed zone in the Languedoc region. A strong perturbation of λ_3 direction trajectories appears along the Cevennes fault system. For discussion see text. Data from Arthaud et al. (1981).

Whatever the scale of the considered system, strain trajectories are probably the most pertinent data at our disposal and have been widely used (Arthaud and Choukroune, 1972; de Charpal et al., 1974; Burg and Etchecopar, 1980; Arthaud et al., 1981; Mercier, 1981; Lepvrier and Mougenot, 1984; Bergerat, 1985; Le Pichon et al., in press). Generally, the strain trajectories show a regional tendency which is locally disturbed in anomalous areas.

Starting from this, two questions must be asked : i) how can we explain the anomalies in the trajectories and ii) how can we explain the regional tendency over large areas?

These two problems are illustrated by the strain trajectories drawn for Eocene deformation in the French Alpine foreland (Figure 13). It appears that the λ_3 direction is essentially N-S and horizontal for this period. In more detail, however, this direction is locally disturbed; for instance, in the vicinity of the border of the Massif Central (Xiao Han, 1983), it crosses an old active margin which is known to have been re-activated as a sinistral strike slip zone during Eocene times (Arthaud and Mattauer, 1969, Arthaud and Seguret, 1981). To explain this anomalous zone, one can say : i) these trajectories reflect an original variation in the discontinuous shear zone, ii) these trajectories have been progressively rotated during the Pyrenean tectonic event (Eocene).

Fig. 13. Successive λ_3 direction trajectories (except 35 Ma episode) in the Alpine foreland from 50 Ma to present.

λ_1- direction trajectory is given for 35 Ma episode (see text for explanation). Data summarized by Le Pichon et al. (in press) and Philip (1983).

The first explanation implies that the spatial variation of the λ_3 direction is related to another parameter which concerns strain intensity (Ramsay and Graham, 1970). Field arguments do not support this hypothesis and support the progressive rotation of λ_3 which can be demonstrated locally (Choukroune, 1969).

In this case, a rigid rotation is suggested by strain trajectories which are associated with shearing deformation that can be argued from field criteria.

If we wish to explain the quite constant N-S direction of λ_3 on a larger scale, we propose that this direction reflects the direction of convergence between the European and Iberian plates at that time.

Even if such a hypothesis cannot be tested, it is clear that these data, consistent on the scale of the considered plates, must be interpreted also on the same scale. In other words, strain trajectories in forelands are probably the only structural data which can be directly related to plate kinematics. Moreover, it is interesting to visualize how strain trajectories change with time.

For instance, strain data for post-Eocene deformation have been recently summarized in the Alpine foreland (Bergerat, 1985). The result is given in Figure 13 ; this shows that the trajectories of the λ_3 and λ_1 directions are consistent, apart from an episode 25 Ma ago when the λ_3 directions are perpendicular to those observed during other periods.

This episode has been interpreted as a critical period for the movement of Africa relative to Europe (Le Pichon et al., in press). Strain trajectories indicate a relative displacement which has been generally N-S during the past 80 Ma, apart from a period at about 30 Ma when it was temporarily E-W.

This data is a confirmation that not only relative displacement for plate boundaries can be recorded at a given period by brittle deformation, but also that any change in kinematics probably has a structural response in the foreland.

4. Conclusion.

When strain concepts are applied to rocks affected by discontinuous deformation, the obtained data, resulting from simple field observations and measurements, are generally consistant on all scales from the outcrop to the larger lithospheric plates.

More particularly, the locally determined strain history is necessary to infer rotational and/or non-coaxial deformation : rotational and/or non-coaxial deformation can be demonstrated when a superimposed set of discontinuities appears within a progressively rotating material. Non-coaxial deformation can be argued when the progressive rotation of the λ_1 and λ_3 axes is recorded.

For a given period, the spatial integration of strain data, especially concerning strain orientations, are pertinent to discussions of possible rotations of blocks within which anomalous directions have been established.

Finally, we consider that strain trajectories in forelands are the only structural data capable of yielding a correlation between deformation and displacement at the boundaries of deformed lithospheric plates.

References

Angelier J. 1983.- Analyses qualitative et quantitative des populations de jeux de failles. *Bull. Soc. Géol. Fr.*, **(7), 26, 5**, p.661-672.

Angelier J. et Manoussis S. 198O. Classification automatique et distinction des phases superposées en tectonique de failles. *C.R.Acad.Sc. Paris*, **290**,(D), p.651-654.

Angelier J., Tarantola A., Valette B. and Manoussis S. 1982. Inversion of field data in fault tectonics to obtain the regional stress. I. Single phase fault populations : a new method of computing the stress tensor. *Geophys. J.R. astr. Soc.*,**69**, p.607-621.

Arthaud F. 1969. Méthode de détermination graphique des directions de raccourcissement, d'allongement et intermédiaire d'une population de failles. *Bull. Soc. Géol. Fr.*, (7), **11**, p.729-737.

Arthaud F. et Choukroune P. 1972. Méthode d'analyse de la tectonique cassante à l'aide des microstructures dans les zones peu déformées. Exemple de la plate-forme Nord-Aquitaine. *Revue I.F.P.*, **27**, **5**,p.715-732.

Arthaud F. et Mattauer M. 1969a. Exemples de stylolithes d'origine tectonique dans le Languedoc, leur relation avec la tectonique cassante. *Bull. Soc. Géol. Fr.*, (7),**11**, p.738-744.

Arthaud F. et Mattauer M. 1969b. Sur les décrochements NE-SW senestres contemporains des plis pyrénéens du Languedoc. *C.R. somm. Soc. Géol. Fr.*, p. 290-291.

Arthaud F. et Seguret M. 1981. Les structures pyrénéennes du Languedoc et du Golfe du Lion (Sud de la France). *Bull. Soc. Géol. Fr.*, (7), **23**, 1, p.51-63.

Arthaud F., Ogier M., Seguret M. 1981- Géologie et Géophysique du Golfe du Lion et de sa bordurenord. *Bull. B.R.G.M.*, **3**, 175-193.

Armijo R.,Carey E. and Cisternas A. 1982. The Inverse problem in microtectonics and the separation of tectonic phases. *Tectonophysics*, **82**, 145-160.

Beach A. 1975. The geometry of en-echelon vein arrays. *Tectonophysics*, p.245-263.

Bergerat F. 1985. Déformations cassantes et champs de contrainte tertiaires dans la plate-forme européenne. Thèse de Doctorat d'Etat, Université de Paris VI.

Buchner F. 1981. Rhinegraben : horizontal stylolites indicating stress regimes of earlier stages of rifting. *Tectonophysics*, **73**, p.113-118.

Burg P. et Etchecopar A. 1980. Détermination des systèmes de contraintes liés à la tectonique cassante au coeur du Massif Central français : la région de Brioude (Haut-Allier). *C.R. Acad. Sc., Paris*, **290**, (D), p.397-400.

Carey E. 1979. Recherche des directions principales des contraintes associées au jeu d'une population de failles. *Rev. Géogr. Phys. Géol. Dyn.*, **21**,1, p.57-66.

Charpal de O., Tremolieres P., JeanF. et Masse P. 1974. Un exemple de tectonique de plate-forme: les causes majeures (Sud du Massif Central, France). *Rev. I.F.P.*, **29**, **5**, p.641-659.

Choukroune P. 1969. Un exemple d'analyse microtectonique d'une série calcaire affectée de plis isopaques ("concentriques"). *Tectonophysics*, 7, p.57-78.

Choukroune P. et Delair J. 1976. Un modèle cinématique de la fracturation liée au plissement concentrique : l'exemple des petites Pyrénées. *Bull. Soc. Géol. Fr.*, (7), p.1591-1597.

Choukroune P., Gapais D., Merle, O. 1987.- Shear criteria and structural symmetry. *Journ. of Structural Geology*, Vol.9, n°5/6, pp.525-530.

Dunnington H.V. 1954. Stylolite development post-dates rock induration. *J. Sedim. Petr.*, **24**, 1, p.27-49.

Etchecopar A., Vasseur G. and Daignieres M. 1981. - An inverse problem in microtectonics for the determination of stress tensors from fault striation analysis. *J. Struct. Geol.*, **3**, 1, p.51-65.

Faugere E., Angelier A., Choukroune P. 1985. - Extension, décrochements et états de contraintes. *Bull. Soc. Géol. Fr.*, **6**, 871-877.

Flinn D., 1965. On the symmetry principle in the deformation ellipsoid. *Geol. Mag.*, **102**, 36-45.

Hancock P.L. 1972. - The analysis of en-echelon veins. *Geol. Mag.*, **109**, p. 169-176.

Le Pichon X.,Bergerat F.,Roulet M.J. Plate kinematics and tectonics leading to the Alpine belt formation.*GSA special book in honnor of J.Rodgers*, in press.

Lepvrier C. et Mougenot 1984. Déformations cassantes et champs de contrainte post-hercyniens dans l'Ouest de l'Ibérie Portugal. *Rev. Géol. dyn. Géogr. Phys.*, **25, 4,** p.291-305.

Mattauer M. 1973. - Les déformations des matériaux de l'écorce terrestre. Hermann Ed., Paris, 493 p.

Mattauer M. et Mercier J.L. 1980. Microtectonique et grande tectonique. Livre Jubilaire du Cent Cinquantenaire 1830-1980, *Mém. H.S. Soc. Géol. Fr.*, **10,** p. 141-161.

Mercier J.L., 1981. Extensional compressional tectonics associated with the Aegean Arc : comparison with Andean cordillera of South Peru - North Bolivia. *Phil. Trans. R. Soc. Lond.* **300,** 337-355.

Petit J.P., Proust F. et Tapponnier P. 1983. - Critères de sens de mouvements sur les miroirs de failles en roches non calcaires. *Bull. Soc. Géol. Fr.*, (7), **25,4,** p.589-608.

Petit J.P. 1987. Criteria for the sense of movement on fault surfaces in brittle rocks. *J. Struct. Geol.*, **9,** 597-608.

Philip H. 1983. La tectonique actuelle et récente dans le domaine méditerranéen et ses bordures, ses relations avec la sismicité. Thèse es Sciences, Montpellier, 240 p.

Ramsay J.G. and Graham R.M. 1970. Strain variations in shear belts. *Can. J. Earth Sc.*, **7,** p. 786-813.

Ramsay J.G., 1980. The crack seal mechanism of rock deformation. *Nature*, **284,** 135-139.

Riedel W. 1929. Zur Mechanik geologischer Brucherscheinungen. Zentralblatt fÅr Mineralogie, *Geologie un PalÑontologie*, **(B), 8,** p.354-368.

Rispoli R. and Vasseur G. 1983. Variation with depth of the stress tensor anisotropy inferred from microfault analysis. *Tectonophysics*, **93,** p.169-184.

Tapponnier, P., Peltzer G., Le Dain, A.Y., Armijo, R. and Cobbold, P.R. 1982. Propagating extrusion tectonics in Asia: next insights from simple experiments with Plasticine. *Geology* , 10, 611-616.

Tchalenko J.S. 1970.- Similarities between shear zones of different magnitudes. *Geol. Soc. Amer. Bull.*, **81,** p.1625-1640.

Tjia, H.D. 1967. Sense of fault displacements. *Geologie Mijnbow* , **46,** 392-396.

Xiao Han L. 1983. Perturbations de contraintes liées aux structures cassantes dans les calcaires du Languedoc. Thèse 3ème cycle, Montpellier, 152 p.

REGIONAL DEFORMATION BY BLOCK TRANSLATION AND ROTATION

Z. GARFUNKEL
Department of Geology
Hebrew University of Jerusalem
Israel

ABSTRACT. Deformation in the brittle part of the crust is often distributed on many faults which tend to form domains of conjugate fractures. Within each domain the fault blocks translate and rotate simultaneously, their motion being strongly governed by kinematic constraints. Blocks rotate about vertical axes where fault slip has a strike-parallel component (pure strike slip, oblique normal faults) and on horizontal axes where dip slip occurs. Relative to domain boundaries, the sense of rotation is opposite to the sense of the fault slip and away from the principal axis of shortening. In multi-domain areas the deformation is complicated in detail. The kinematic constraints may lead to geometric incompatibility between domains and may require formation of new structures in order to satisfy the externally imposed boundary conditions. However, as local shears and rotations in different senses and amounts can occur, the overall regional deformation may be quite simple. Mechanical constraints on fault motion control the amount of fault rotation and the formation of new structures, but these effects are difficult to evaluate, because the kinematic constraints can modify the stress field within faulted areas. The factors which allow fault slip and rotation to continue are also incompletely understood. Block systems that rotated on vertical and/or horizontal axes are decoupled to greater or lesser extents from the underlying mid-crustal ductile levels, but localized shear can probably lead to differential rotations of rock masses even below the decoupling zones. While block rotations related to faulting is probably more common than hitherto recognized, other mechanisms, e.g. motions of micro-plates or exotic terranes, may also cause rock masses to rotate about verical axes. Therefore paleomagnetic data on rotation must be combined with structural data in order to identify the contribution of faulting to block rotation.

1. Introduction

Faulting is an important deformation process in the brittle part of the continental crust. Geologic mapping demonstrates the abundance of faults in all tectonic settings, and seismic activity proves their important role for present day deformation. One approach to the study of faults, initiated by Anderson (1951), focuses on their relations with the causative stresses. This approach, however, has the severe limitation that it deals only with the initiation of faults, whereas to understand regional deformation produced by faulting the rules which govern continuing fault slip and buildup of finite fault offsets must be understood.

An important concept regarding finite deformation by faulting is that when fault blocks slip past each other they also rotate like books on a shelf or dominoes that topple sideways. The faults themselves rotate as well, as they are merely boundaries of blocks. This concept was first applied to blocks between normal faults that are tilted about horizontal axes (Fig. 1A; Ransome et al., 1910; Thompson, 1960). Later it was also applied to systems of sub-parallel strike slip faults and led to the prediction that in such cases fault blocks rotate about vertical axes (Fig. 1B; Freund, 1970, 1974; Garfunkel, 1974), which was confirmed by paleomagnetic measurements (e.g. Luyendyk et al., 1985; Hornafius et al., 1986; Burke et al., 1982; Calderone and Butler, 1984; Ron et al., 1984,

C. Kissel and C. Laj (eds.), Paleomagnetic Rotations and Continental Deformation, 181–208.
© *1989 by Kluwer Academic Publishers.*

182

1986). In these structural settings the relative translation of blocks and their rotation are just two different, but simultaneous, expressions of the same deformation process; generally one cannot occur without the other. Paleomagnetic data revealed rotations about vertical axes also in many cases where a direct relation with faulting is not obvious (e.g. Beck, 1980; Kissel and Laj, 1987; Kissel et al., 1987). Block rotations can, of course, result in several ways, not only by the mechanism shown in Figure 1, e.g. during displacement of microplates or exotic terranes.

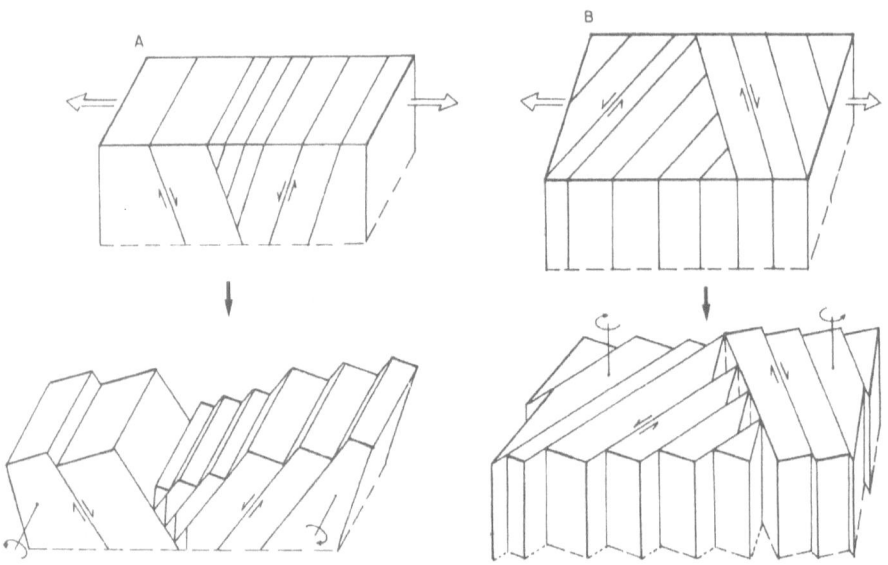

Fig. 1. Deformation by block rotation and translation. Conjugate fault systems rotate in opposite senses occur in separate domains. Note that the sense of rotation is the opposite of the sense of fault slip and that the rotations are away from the direction of shortening. A - normal faults; B - strike slip faults.

From another point of view, regional deformation produced by fault activity is a special case of discontinuous deformation in which the relative displacements of particles are achieved primarily by slip on discrete surfaces, whereas the intervening material units and the slip surfaces themselves are hardly deformed. Such deformation can occur on all scales, down to that of single crystals which are strained by slip on discrete lattice planes. However, on the microspcopic scale of crystals the slip systems can migrate through the lattice, whereas on the macroscopic scale faults always separate the same material units. Moreover, since in rocks a great number of crystals deform practically simultaneously, the overall result appears as continuous penetrative deformation on the scale of hand specimens. In contrast, faulted areas contain only one or a small number of fault sets. Therefore, on the mactoscopic scale the discontinuos nature of the deformation is strongly expressed and needs to be considered. This is also necessary for proper interpretation of field data, because such data deal directly with individual faults and fault blocks.

The purposes of this work are to discuss the properties of regional deformation that is produced by simultaneous translation and rotation of fault blocks, stressing its discontinuous nature, and to evaluate some related problems. In the first two parts the kinematics and geometric features of this deformation style, the development of faulted areas and the influence of the initial fault geometry are examined by studying simple models, expanding the treatment of Garfunkel and Ron (1985). Then problems relating to fault properties and to the downward continuation underlying lower crust are discussed.

2. Deformation by block displacement and rotation: kinematics.

2.1. BASIC FEATURES OF DEFORMATION OF FAULTED AREAS

Field observations deal directly with the elementary structural building stones - fault blocks and single faults. However, to understand the regional deformation of faulted areas it is necessary to study several scales of successively higher degrees of structural organization. On the smaller scale it is found that faults often appear as sets of more or less parallel fractures which move contemporaneously on long enough time scales, slip in the same sense and have similar offsets. Such fault sets define domains, which are the simplest structural assemblages in faulted areas. Many faulted areas, however, have a higher degree of structural organization and comprise several domains of conjugate faults, i.e. slipping in opposite senses (Fig. 1). The domains may be separated or delimited by major faults that are considerably longer and have greater slips than the faults within the domains.

The present work deals with those cases in which the deformation on the scale of single domains is as envisaged in Figure 1: As deformation progresses the blocks within each domain slip one past another and also rotate in the bookshelf manner relative to the domain boundary on which the faults terminate. This allows the domains to remain in contact with each other and with nearby areas, despite the changes of their external shapes. Without such rotations misfits between the fault domains and their surroundings would arise. Relative to the domain's boundary or to adjacent unfaulted areas block rotation is always in the opposite sense to the fault slip within the domain. Therefore, blocks in domains of conjugate fault sets rotate in opposite senses. The blocks also rotate away from the direction of maximum shortening. Fig. 1 demonstrates these features for simple cases, but they are clearly true for other fault geometries as well. The deformation of multi-domain areas is the sum of the deformations, rotations and displacements of the individual domains. In such areas some domains can also rotate as a whole as they remain in contact with nearby domains whose shapes change.

The behavior of faulted areas is well approximated by the above definition of discontinuous deformation because of the combination of two effects, to be treated more fully later. First, slip on existing faults, even after they have rotated, is preferred over formation of new fractures and over continuous deformation, because faults are weaker than the material between them. Therefore, once fault systems are activated they provide the main means for stress release and then kinematic constraints depending on fault geometry control the regional deformation paths. Only when faults rotate to very unfavorable orientations relative to the stress field they become locked and new fault sets are formed (Proffett, 1977; Angelier and Colletta, 1981; Nur et al., 1986). Second, simultaneous block rotation and translation provides an efficient deformation mechanism. Therefore, the results of faulting often dominate the regional deformation, even when the fault blocks are internally strained to some extent, i.e. the non-rigidity of the fault blocks can often be neglected.

2.2. DEFORMATIONS PATHS OF SINGLE DOMAINS

In fault domains deforming as envisaged above definite relations exist between fault rotation, offset, spacing and orientation relative to domain boundaries or to adjacent areas. This enables to interpret field data in terms of definite structural models of block rotation and to test these interpretations. The relations between the various parameters can be found directly from the domain geometry, but since fault spacing is often small compared with the size of fault domains, the regional deformation can also be treated as approximately continuous. Each treatment has its advantages and both will be used in what follows.

The continuous approximation is the limit obtained when the number of faults increases indefinitely and their spacing tends to zero, while the same overall deformation is distributed evenly on the more numerous fractures. In the continuous deformation so obtained surfaces parallel to the faults are not strained and always consist of the same material particles, similar to real faults. It is this property that characterizes approximations of discontinuous deformation (Garfunkel and Ron, 1985). Continuous deformation in which surfaces with such properties do not exist, e.g. pure shear with principal strain axes in fixed directions, cannot describe well single fault domains (though it can approximate multi-domain areas, see below).

The relations between the various parameters can be easily obtained for domains of planar faults which end on a planar boundary (Fig. 2). Though very simple, this geometry approximates well many natural cases. The deformation is studied in suitably chosen planes which cross the faults and

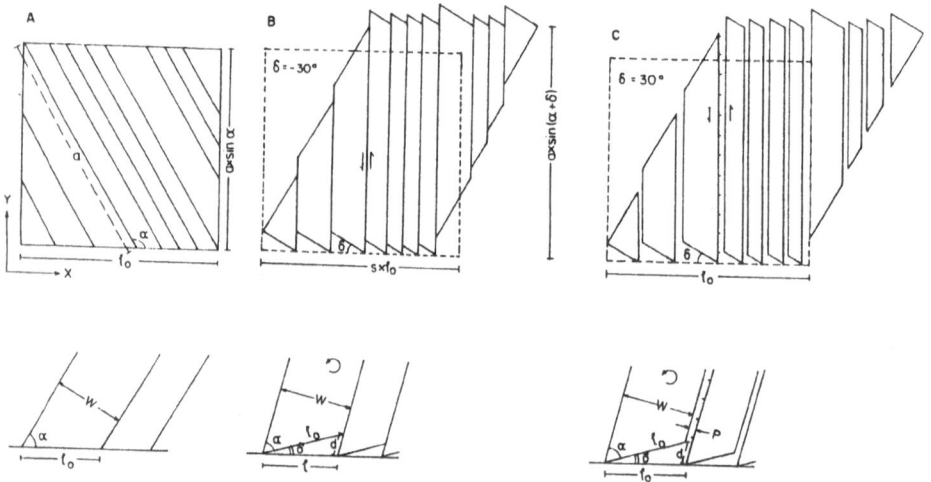

Fig. 2. Deformation of single fault domains. A - initial situation. B - The relations when area is preserved; C - when the domain's length along its boundary is preserved. See text for discussion.

are not perpendicular to fault slip, e.g. horizonal planes which cut sub-vertical strike slip faults or oblique normal faults. Let these be the X-Y planes, with the X axis along the trace of the domain boundary on which the faults end. Denote the original angle between the fault traces on the X-Y plane and the X axis by α and the block rotation by δ, positive angles being measured counterclockwise. Previous works dealt with deformation that preserves area in the X-Y plane (Freund, 1974: Garfunkel, 1974; Garfunkel and Ron, 1985) or preserves the length of the domain parallel to the X axis (McKenzie and Jackson, 1983, 1986). Here a more general case, in which the length of the fault domain along the X axis changes by any factor s, is treated. The value of δ and generally also of s, will change as deformation progresses. The changes in the X-Y plane can be approximated by the continuous deformation which rotates material lines originally at an angle α to the X axis through an angle δ without changing their lengths. This brings particles originally at (X, Y) to (x,y), given by

$$(x, y) = F (X, Y), \qquad (1)$$

where F is the deformation gradient tensor (Malvern, 1969; Mase, 1970) which, according to Figure 2, is

$$F = \begin{pmatrix} s & \dfrac{\cos(\alpha+\delta) - s\cos\alpha}{\sin\alpha} \\ 0 & \dfrac{\sin(\alpha+\delta)}{\sin\alpha} \end{pmatrix}. \qquad (2)$$

This shows that the initial fault orientation, given by α, determines the deformation path of the domain. Deformation is linear, as F does not depend on the coordinates.

The domain area in the X-Y plane changes by a factor A, given by

$$A = s\,\dfrac{\sin(\alpha+\delta)}{\sin\alpha} \qquad (3)$$

When $A \neq 1$, deformation is three dimensional and volume changes may occur. In such cases fault slip will generally be oblique to the X-Y plane, so that the deformation can be a combination of those shown in Figures 1A and 1B, i.e. rotation on both vertical and horizontal axes takes place.

The velocity gradient tensor L (satisfying $dF / dt = LF$, Malvern, 1969) is

$$
L = \begin{pmatrix} \dfrac{\partial u}{\partial x} & \dfrac{\partial u}{\partial y} \\[2ex] \dfrac{\partial v}{\partial x} & \dfrac{\partial v}{\partial y} \end{pmatrix} = \begin{pmatrix} \dfrac{\dot{s}}{s} & -\dot{\delta}-\dfrac{\dot{s}}{s\tan(\alpha+\delta)} \\[2ex] 0 & \dfrac{\dot{\delta}}{\tan(\alpha+\delta)} \end{pmatrix} \tag{4}
$$

where u and v are the velocities in the X and Y directions and the dot signifies differentiation with respect to time. The three non-zero terms of L cannot all be simultaneously constant (time independent). This is a consequence of the structure of F which must satisfy the constraint that the lengths of the fault traces on the X-Y plane cannot change. Hence, deformation in which L is constant cannot describe fault domains as envisaged here.

It should be noted that equations (1)-(4) cannot be used as in continuum mechanics. This is expressed in two ways. First, though eq. (2) defines a strain at any point, this is not the local strain of the material within the domain, because any small volume not containing a fault is not strained at all. The strain defined by (2) approximates the overall changes of shape of large enough areas or of the entire domain. Second, while by definition any material unit within the fault domain rotates at a rate $\dot{\delta}$, in a really continuous deformation with a velocity field given by (4) the local rotation rate is half the vorticity, i.e. $(\partial u/\partial y - \partial v/\partial x)/2$, which differs from $\dot{\delta}$; e.g. when s=1 this yields $\dot{\delta}/2$ rather than $\dot{\delta}$. Half the vorticity derived from equation (4) is, however, the rotation rate of the pricnipal axes of the increment of the overall strain of large enough areas or of the entire domain defined by the time variation of equation (2).

When deformation begins both s and A are equal to 1. The subsequent deformation history is defined when the changes of either s or A with time are specified. Important particular cases, that were mentioned above, are when A=1 or when s=1 and they will now be examined in some detail. Figure 2 shows that in these cases the fault domains are sheared parallel to the X axis and their length in the Y direction changes. In the first case (A=1) the domain length in the X direction also changes, whereas in the second case this length does not change (as s=1) and the faults behave as if anchored to fixed points on the X axis. In both cases the fault traces on the X-Y plane rotate because their slip has a component parallel to these traces; the rotation is necessary to keep the blocks in contact with the domain boundary along the X axis.

The first case, with A=1 (Fig. 2B), describes systems of purely normal faults viewed in vertical cross section or systems of vertical strike slip faults in plan view (e.g. as in southern California). The deformation is two dimensional and the fault slip is parallel to the X-Y plane, being along the fault traces on this plane. Within the fault domain only simple shear parallel to the faults takes place. The shear gradient K=d/w, where d is fault displacement, positive when right lateral, and w is fault spacing (Fig. 2), is given by

$$
K = \cot\alpha - \cot(\alpha+\delta). \tag{5}
$$

The angle $\alpha+\delta$ is the final angle between the faults and the domain boundary and can be found from field data. If K can also be determined in the field, then equation (5) allows to calculate α, hence also δ In areas of normal faulting δ may be recorded by dips within the tilted blocks, whereas in areas of strike slip faults it can be found from paleomagnetic data. Other expressions (Garfunkel and Ron, 1985) allow to calculate any one of α, δ or K in terms of the other two. This allows to cross check the compatibility of the values of K, α and δ hen they can be determined independently, e.g. from field and paleomagnetic data, and thus to test quantitative structural models.

Let l_0 and l be the domain's original and final lengths parallel to the X axis, i.e. to the boundary on which the faults terminate; then

$$s \; = \; \frac{1}{l_0} \; = \; \frac{\sin \alpha}{\sin (\alpha + \delta)} \tag{6}$$

Usually s ≠ 1 and therefore the domain boundary along the X axis becomes a discontinuity of velocity and displacement. As A=1, the domain's length parallel to the Y axis changes by a factor 1/s according to equation (3). It should be noted that both K and s do not depend on the fault spacing or on the number of blocks, but only on α, the initial angle between the faults and the domain boundary and on δ, the amount of block rotation.

The continuous approximation of this deformation is obtained by substituting equation (6) in equation (2); then the upper right term of F is [sin δ cos (2α+δ)] / [sin α sin (α+δ)] or K cos (2α+δ). At any moment the principal strain axes bisect the angle between the faults and the lines whose lengths are the same as their initial lengths. However, unlike the faults, these lines consist of successively different sets of particles as deformation progresses (Garfunkel and Ron, 1985). Let the angle between these sets of lines in the initial and final states be γ_0 and γ, respectively, then

$$\tan \gamma_0 = -2 / K, \; \tan \; \gamma = 2 / K, \tag{7}$$

so that γ =180°- γ_0. Thus, initially the strain axes form angles of α+ γ_0/2 and α+ γ_0 / 2-90° with the X axis. During deformation they rotate by 90° + δ - γ. The principal elongations λ_1 and λ_2 are given by

$$\lambda_1^2 , \; \lambda_2^2 = [2 + K^2 \pm \sqrt{K^2+4}] / 2 \tag{8}$$

It is easitly seen that these elongations can be quite large even when K and δ have moderate values.

The model with s=1 (Fig. 2C) was proposed by McKenzie and Jackson (1983, 1986) to describe areas of oblique normal faulting (e.g. in the Aegean region and in the Basin Range province, western USA), with the X-Y plane horizontal. Since s=1, the X axis need not become a velocity and displacement discontinuity, which differs from the previous case. According to equation (3) the domain's area on the X-Y plane changes by a factor sin (α+δ) / sin α. Thus, the spacing of fault traces on the X-Y plane changes and fault offsets have a component normal to this plane, i.e. they are oblique slip faults relative to it. Denote the components of fault offset in the X-Y plane parallel and perpendicular to the fault traces by d and p, respectively, and the original fault spacing by w (Fig. 2); then

$$K = \frac{d}{w} = \frac{\cos \alpha}{\sin (\alpha+\delta)} - \cot (\alpha+\delta) \tag{9}$$

$$P = \frac{p}{w} = 1 - \frac{\sin \alpha}{\sin (\alpha+\delta)} \tag{10}$$

These equations can be used to interpret field relations, similar to equation (5).

The continuous approximation of the deformation is obtained by substituting $s=1$ in equation (2). The principal strain axes in the X-Y plane always bisect the angles between the X axis and the fault traces, as the lengths of lines in these directions do not change. Thus, initially they form angles of $\alpha/2$ and $\alpha/2-90°$ with the X axis and during deformation they rotate by $\delta/2$. The principal extensions in the X-Y plane, λ_1 and λ_2 are

$$\lambda_1 = \frac{\sin [(\alpha+\delta)/2]}{\sin (\delta/2)} \quad ; \quad \lambda_2 = \frac{\cos [(\alpha+\delta)/2]}{\cos (\delta/2)} \tag{11}$$

With $s=1$, equation (4) gives the velocity components at any point (x,y):

$$u = -\dot{\delta}, \quad v = \dot{\delta}/\tan (\alpha+\delta). \tag{12}$$

They are time dependent, because δ changes as deformation progresses. Equation (12) also shows that at any moment the velocity component in the X-Y plane is perpendicular to the fault direction, as already shown by McKenzie and Jackson (1983). They also showed that the relative block motion at any moment is in the Y direction, which can also be seen from equations (9) and (10) when δ tends to zero. Thus, in active areas the Y direction can be found from field or seimsic data on fault slip.

As the domain area in the X-Y plane changes when $s=1$, some adjustment must occur in the third dimension. For example, when area increases grabens may form, or the fault blocks can be tilted on horizontal axes as well (producing a combination of what is shown in Fig. 1A and 1B). This changes the vertical dimension of the faulted region and shears it parallel to the X-Y plane. Whatever happens, lengths of vertical lines cannot change unless they are tilted, because otherwise the area of inclined fault planes will change, which is not possible when the fault blocks are little deformed.

Comparison of the two deformation types considered above reveals some common features: overall shearing in the X-Y plane takes place and all the material of the fault domain rotates through the same angle. The rotation about the Z axis results from the presense of a component of slip parallel to the fault traces on the X-Y plane. The deformation can follow only definite paths: from any initial state only particular successions of fault geometries, given by equations (5), (6) and (9) and (10), can be reached. Other configurations can be attained only if the character of fault slip changes, new faults are activated, or the fault blocks are internally deformed. Particle velocities change as deformation progresses. Equations (7), (8) and (11), (12) show that block rotation provides an efficient means of regional deformation: rotations of a few tens of degrees can produce regional strains reaching tens of percent. This justifies a-posteriori the use of simple models of rigid blocks and allows to neglect small misfits between rotating blocks and internal deformation of blocks, when this is not large. Though derived for very special histories of the parameters s and A, it is easy to see that these properties characterize also more general deformation histories.

The main difference between the two models considered above is that displacement and velocity discontinuities along domain boundaries arise when A=1, but not when s=1. Indeed, domains of strike slip faults are often delimited by large strike slip faults which express the strain incompatibility along their boundaries. The resulting misfits with the surroundings must be accommodated in nearby domains. In contrast, in the second model the domain's length along boundaries on which the faults terminate does not change. This model can describe a deformed zone which is delimited, without any strain incompatibility, by two rigid blocks moving one relative to the other. However, since particle velocities in the faulted domain change, this is possible only if the relative velocity of these blocks varies in the same way. Another difference between the models is that the sense of overall shearing of the domain is always in the sense of block rotation when s=1 but not when A=1 (Fig. 3). In this case the sense of overall shearing can, in fact, be reversed as block rotation progresses (Freund, 1974; Garfunkel and Ron, fig. 2). In fact, it follows from eq. (4) that when δ varies monotonously the vorticity can change sign when A=1, but not when s=1.

The simple models treated above can be modified to describe more complicated situations. Thus, the presence of rhomb-shaped pull apart basins along strike slip faults can be taken into account (Garfunkel, 1974, fig. 8). As this increases the area of the faulted region equation (3) cannot be used, but equations (5) and (6) are still valid. Also, the faults and domain boundaries may be curved, etc. In such cases the above relations between the deformation parameters must be modified, though sometimes they are still useful approximations. However, in all cases the same general principles govern the regional deformation and it will have the same the basic features as the simple models cosnidered here.

2.3. MULTI-DOMAIN AREAS

As noted above, faulted areas often comprise several domains which are assembled in various ways. Sometimes domains of faults with similar lengths and offsets are present, but domains of different ranks can also occur: in such cases some domains are delimited by large faults which themselves form sets and thus define domains of a higher rank (Fig. 4A). Domains of conjugate left- and right-lateral strike slip faults are known, for example, in eastern Iran (Freund, 1970), southern California (Garfunkel, 1974; Luyendyk et al., 1985; Nicholson et al., 1986), Japan (Research Group for Active Faults of Japan, 1980) and NW USA (Wells and Coe, 1985). Domains of different ranks occur in the Najd fault system, Arabia (Moore, 1979) and probably also in NW USA (Wells and Coe, 1985). Domains of normal faults that are titled in opposite directions are also common e.g. in the Basin and Range province (Stewart, 1980).

The formation of fault domains probably occurs early in the history of faulted areas. Any fracture criterion, such as the Coulomb criterion (see below), predicts that in homogeneous regions fractures of conjugate sets should be equally numerous and evenly distributed, as is observed in meso-structural studies (Lockwood and Moore, 1979; Segall and Pollard, 1983; Hancock, 1985). However, in such a situation fractures of the different sets can move only if they disrupt each other and if the displaced segments grow into intact material as they continue to slip. This will produce pervasively broken rock volumes, which generally are not observed. The interference between faults is avoided when the area is divided into domains in which faults of only one set remain active, while

190

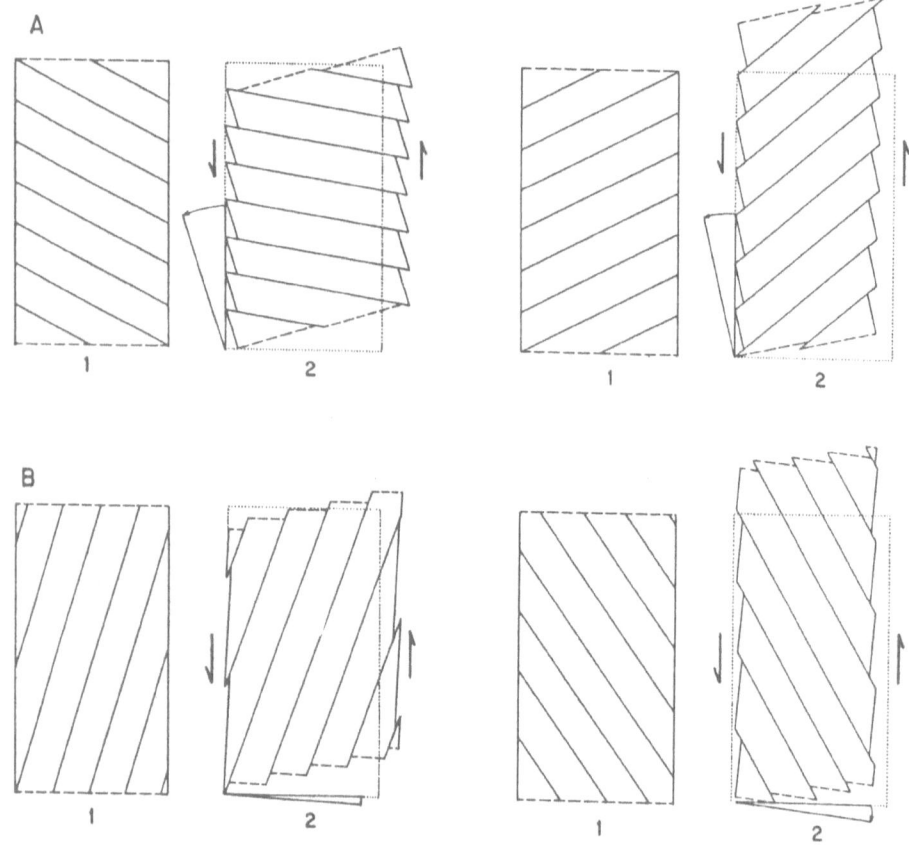

Fig. 3. The relations between block rotations and overall shearing of domains. A - Block rotations and overall shearing are in the same sense. B - Block rotations and overall shearing are in opposite senses. 1 - Initial state, 2 - final state. (After Garfunkel 1970, unpublished Thesis).

motion of other faults is inhibited. Such a sequence of events was actually observed in clay-model experiments (e.g. Hoeppener et al., 1969; Oertel, 1965; Freund, 1974). Once domains are formed, slip on the remaining faults is no longer hindered and can increase substantially. Domain formation is, thus, an essential stage in the development of faulted areas that must occur before faulting can produce significant deformation. If rotated faults become locked, then this process can be repeated when new faults are formed.

Domain formation is also a selection process that determines which fracture set will remain active in any one place. Two factors probably control the selection: local heterogeneities of the fractured area and the externally imposed boundary conditions. Of these the latter may be more important, but this needs further study.

The overall deformation of multi-domain areas is the sum of the deformations, rotations and displacements of the individual domains. The behabior of the lowest rank domains can be treated as above, while the behavior of an assemblage of domains depends also on their arrangement, shapes, etc. Since a variety of domain and fault geometries are possible, each region will behave in its own way. Therefore, simple generalizations cannot be made, but some important features deserve attention.

In domains of conjugate faults the blocks rotate in opposite senses relative to the domain boundaries. However, as these boundaries may themselves rotate, being embedded in a deforming region, block rotations by different amounts and not necessarily always in opposite senses are expected. Indeed, paleomagnetic data revealed that rotations about vertical axes by different amounts and in opposite senses took place in many areas of strike slip faulting: in the Galilee, Israel (Ron et al., 1984); in southern California and in nearby parts of Arizona (Luyendyk et al., 1985; Hornafius et al., 1986; Calderone and Butler, 1984); in Dixie Valley area, Nevada (Hudson and Geissman, 1985); in NW USA (Wells and Coe, 1985) and in western Anatolia (Kissel et al., 1987), though in this case the relation with lateral fault slip is not clear. In areas of normal faulting tilting in opposite directions and by different amounts is also well known. The occurrence of such variable rotations, and especially in opposite senses, strongly supports the interpetation of conjugate fault domains in terms of simultaneous block translation and rotation. Such rotation patterns are not predicted by other deformation mechanisms.

The occurrence of different rotations in domains of conjugate fault systems has some important consequences. One is that the angle between different fault sets changes as deformation progresses. As blocks tend to rotate away from the axis of maximal shortening, and hence also compression, the angle which encloses this axis increases. In initial faulting stages this angle is usually close to 60°. Therefore, when it is much larger, the possibility of block rotation is indicated. Another important consequence is that when neighboring domains of conjugate faults are sheared in opposite senses their combined deformation can resemble pure shear (Fig. 4B). Rotations and shears in opposite senses can also compensate for each other on a regional scale, smoothing out to some extent the irregularities that are apparent on the scale of single domains. Therefore the overall deformation of multi-domain areas can be rather simple and the behavior of single domains cannot be taken to represent the area as a whole.

The most complicated problem regarding multi-domain areas is the compatibility of the domains. As shown above, each domain can follow only a particular deformation path which is determined by its shape and fault geometry. Therefore misfits can arise between domains as faults slip continues (Garfunkel and Ron, 1985). For example, Figure 4B shows how an originally straight boundary may become bent. It is not at all clear that the initial fault geometry assures that such misfits will not develop. Below it will also be shown that the kinematic constraints may not allow the regional deformation to satisfy the externally imposed boundary conditions. It is expected, therefore, that during the progress of deformation new faults will form, blocks will be internally deformed and the stress fields may be modified and reoriented, so that misfits are minimized and the externally imposed constraints are satisfied. Further discussion of such effects will be possible, however, only when detailed structural analyses of multi-domain areas become available.

192

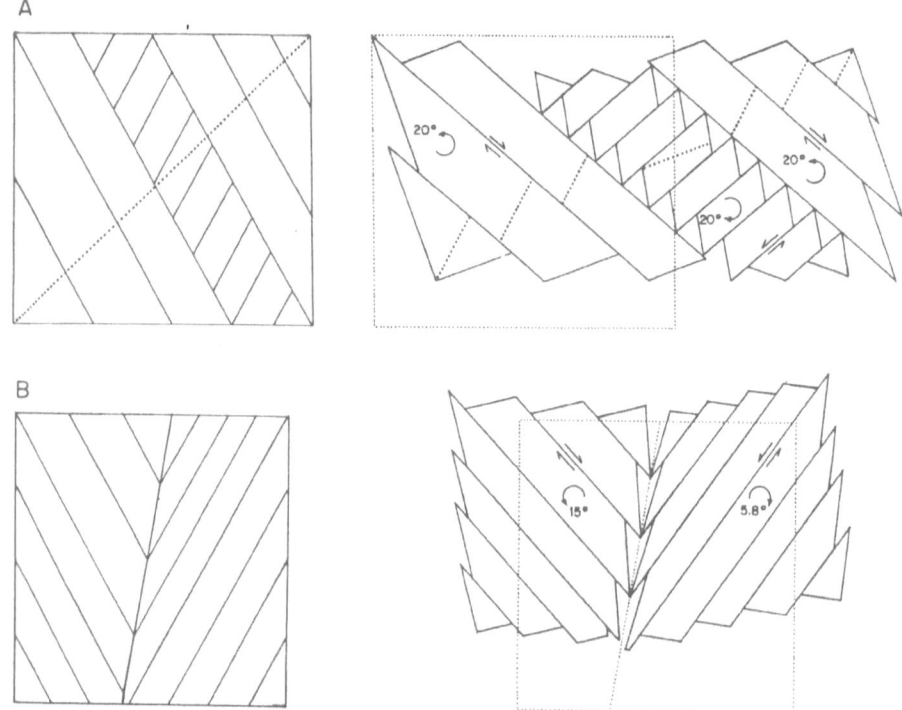

Fig. 4. The simultaneous deformation of adjacent domains of conjugate faults. A - Domains of different ranks; note how the rotation of the higher ranking domain is superimposed on that of lower rank. B - Deformation of two adjacent domains of conjugate faults; note that to fit best the rotations must be different when the boundary between the domains does not bisect the angle between faults; also note that the overall deformation resembles pure shear, but domain boundaries do not align along a straingt line after deformation.

An important problem which remains open is the possible amount of block rotation. Kinematic considerations merely show that blocks cannot rotate across domain boundaries. Since in multi-domain areas the domain boundaries themsleves can rotate, this is a rather weak constraint. Fault mechanics provide additional constraints. Since in each domain the blocks rotate away from the principal direction of shortening, they also rotate away from the maximum compressive stress axis, which agrees with the sense of shear stress acting on the faults. As active faults cannot rotate across stress axes, it follows that as long as the stress axes do not change, fault rotation cannot exceed 45° - 60°, (Garfunkel and Ron, 1985). Proffett (1977), Angelier and Colletta (1981) and Nur et al. (1986) showed that faults become locked even before they become parallel to the minimum stress axis and then new fault sets can form. The motion of such successive fault sets can produce large block rotations, while the older faults are disrupted and rotate as passive markers. However, the stress fields can be modified during deformation. For example, if an externally imposed shearing continues to ratate blocks in the model of Fig. 2C after $\alpha + \delta$ reaches $90°$, then oblique reverse slip will begin on what used to be oblique normal faults, which implies a reorentation of the stress axes. In multi-domain areas the complexities which, as noted above (see also next chapter), may arise as deformation porgresses can also modify the stress field. Thus, though the mechanical constraints are stronger than the kinematic ones, they can be applied only if the history of the stress field is known. Moreover, as stress histories are often difficult to decipher, the formation of successive fault sets can help, in fact, the stress analysis.

These considerations show that large block rotations, e.g. by some $90°$ as reported from California (Luyendyk et al., 1985; Hornafius et al., 1986), require a special explanation. One possibility is that they resulted from the activity of several fault generations. However, this produces large regional strains. For example, normal faults originally dipping at $65°$ which are tilted about horizontal axes by $30° - 40°$ extend the faulted area by a factor of 1.6-2.1 (equation (6)); successive activity of two such fault systems, producing tilts of $60° - 80°$, will extend the area by a factor of 2.5-4.6. Areas in which two generations of vertical strike slip faults were active will be horizontally extended and shortened by similar factors. Rotations of oblique slip faults on vertical axes produce considerably lesser, but still appreciable, areal strains. The feasibility of deformations of such magnitudes can be used to test whether large rotations (e.g. as determined paleomagnetically) are a direct result of offsets on several fault generations. Even then, the stress axes (as in Fig. 2C) can be reoriented during deformation. On the other hand, if stress orientations vary during fault slip, then large rotations can occur without formation of new faults. Kinematically this is possible, but more study is required to test whether this indeed occurs in nature. Large rotations can also arise, of course, by superposition of micro-plate and plate rotations on the rotations resulting from deformation by faults.

2.4. THE USE OF PALEOMAGNETIC DATA

Paleomagnetic data provide the best, often the only, means for measuring rotations on vertical axes. They are, therefore, essential for proper structural interpretation of faulted areas, but they should be used with great care. The important point is that paleomagnetic data measure rotations relative to the paleomagnetic north, which can be the sums of a variety of contributions. On the other hand, the rotations discussed above and which are quantitatively related to fault slip are defined relative to domain boundaries, which is quite a different thing. For structural analysis such rotations must be separated from those produced by other causes, e.g. plate and micro-plate motions. Also, rotations of different ages should be distinguished.

To overcome this problem it is desirable to compare paleomagnetic data from adjacent domains and from nearby unfaulted areas. Moreover, these data should not be used alone, but they should be combined with structural data. Above it was seen that within single domains definite relations exist between block rotations relative to domain boundaries and other parameters of the deformation (equations (5), (9) and (10)). Such relations allow to combine paleomagnetic data with field data and also provide the means to cross check the different data. The structural analysis is also necessary because of the various complications that can arise in faulted areas. For example, rotations of domain boundaries, superimposed rotations of domains of different ranks (Fig. 4A) or the successive rotations in areas deformed by multiple fault sets that replaced each other, must be separated. Only analysis of all the data enables to relate block rotations to definite structural contexts and deformation patterns and histories and to identify the contribution of other processes.

3. Pre- and post-faulting deformation

As the structural evolution of faulted areas is strongly controlled by kinematic constraints, it depends on the initial fault geometry as long as the same fault sets remain active. On the other hand, it should also depend on externally imposed boundary conditions. Fault formation is not expected to change these conditions, but it changes profoundly the gross rheological behavior of the faulted area: it becomes anisotropic and discontinuous and its subsequent deformation is strongly influenced by kinematic constraints that depend on fault geometry. This raises the question whether the fault controlled deformation can be the same as the precursory pre-faulting deformation, e.g. whether the boundaries of the faulted area continue to move in the same directions. Because of the complexity of multi-domain areas, discussion will be limited to the simple case of a single domain. It will be shown that then the pre- and post-fautling deformations are generally different.

For this purpose consider a deformable zone, with boundaries parallel to the X axis, which remains attached to flanking rigid blocks. Assume that these blocks move relative to each other, the motion being in the X-Y plane and at an angle γ to the X axis (Fig. 5). Initially the deformation of this zone is continuous and uniform, so that material particles originally at (X, Y, Z) move to (x,y,z), given by

$$x = X + Y k \cos \gamma \, , \, y = Y (1 + k \sin \gamma), z = Z / (1 + k \sin \gamma) \qquad (13)$$

k being the relative displacement of the rigid blocks. The deformation consists of simple shear parallel to the X axis combined with streching or shortening parallel to the Y and Z axes, so that volume is preserved. One principal strain axis is in the Z direction while the other two are in the X-Y plane and at angles Θ with the X axis which are given by

$$\tan 2\Theta = \frac{-2\cos \gamma}{2\sin \gamma + k} \qquad (14)$$

Thus Θ can be written as $\Theta = \pm 45° + \gamma^* / 2$, such that $\gamma^* > \gamma$, and $\gamma^* \approx \gamma$ when k is small. The elongations along these two principal axes, λ_H and λ_h , are given by

$$\lambda_H^2 , \, \lambda_h^2 = 1 + .5h [B \pm \sqrt{(B^2 + 4 \cos^2 \gamma)}] \qquad (15)$$

where B=2 sin γ + k. The change of length along the vertical axis, λ_v is

$$\lambda_v = 1 / (1 + k \sin \gamma). \qquad (16)$$

After some time conditions of brittle failure will be reached. Pre-existing fractures may be reactivated, but often new faults form, even in previously faulted areas. In the Aegean area, for example, the active faults cut across a Tertiary nappe pile. The orientation of new faults is determined by some failure criterion, such as the commonly used Coulomb criterion which describes fairly well laboratory results and numerous field data (Anderson, 1951; Handin, 1969; Paterson, 1978: Jaeger and Cook, 1979). It predicts formation of two conjugate sets of fractures,

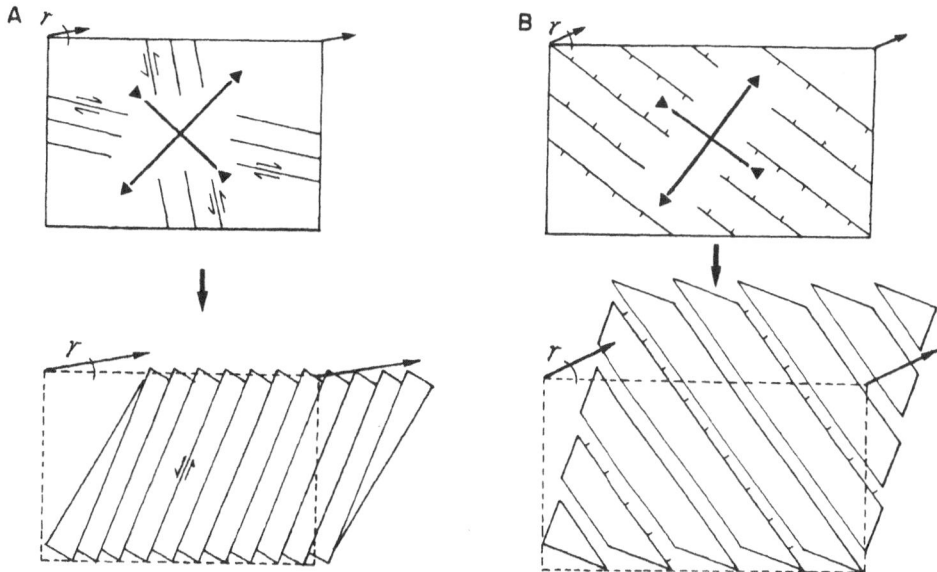

Fig. 5. The difference between geometry of pre- and post-faulting deformation of a single fault domain. A - area is preserved (cf. Fig. 2B); B - length along domain boundary is preserved (cf. Fig. 2C). Top - initial faulting; bottom - after finite deformation. Note that the faulting controlled deformation causes the domain boundary to move in a direction which differs from the pre-faulting direction.

but in experiments with three dimensional strain four sets of fractures are formed, probably because slip on just two fault systems can produce only very special types of strain (Oertel, 1965; Reches and
Dietrich, 1983). Though field examples of four contemporaneous fault systems were not recorded, this shows that the Coulomb criterion needs revision. However, for the lack of a better alternative, it will be used in the following discussion as a first approximation.

Thus, the faulting is taken to depend only on the orientation of the principal stresses. As the pre-faulting deformation is continuous, then in an isotropic medium the principal stress and strain axes will be parallel to each other and the principal stresses and strains will have the same relative magnitudes. Therefore equations (14)-(16), giving the principal strains, can be used to infer the conditions of faulting.

When $\gamma = 0$ the deformation is simple shear in the X-Y plane. Then $\gamma_v = 1$ and $\gamma_H > \gamma_v > \gamma_h$, i.e. the intermediate principal stress is vertical, so that two conjugate strike slip fault sets, called Riedel shears, will form (Fig. 5). This is often observed in experiments and in nature (e.g. Wilcox et al., 1973). When $\gamma = 90°$ extension parallel to Y occurs and then $\lambda_H = 1+k$, $\lambda_h = 1$, $\lambda_v = 1/(1+k)$ and $\lambda_H > \lambda_h > \lambda_v$, i.e. the maximum compressive stress is vertical so that normal faults striking parallel to λ_h (which in this case is parallel to the X axis) should form. Thus, there are two distinct fault regimes depending on γ. From equations (15) and (16) it can be found that the transition between the two faulting regimes occurs when γ is in the narrow range of 19°-19.5°, depending on k.

The post-faulting deformation in the two regimes is shown in Fig. 5. In the strike-slip regime the faults are vertical, so the blocks are envisaged to remain in mutual contact and area in the X-Y plane is preserved. The fault system which allows the externally imposed shear to continue is selected as the one which remains active. It is seen that the resulting overall deformation is no longer the continuation of the pre-faulting deformation: the domain boundary does not move any longer in the same direction as it did before faulting. For example, when $\gamma=0$ the deformation is no longer simple shear. In the normal faulting regime the faults are envisaged to remain anchored on the X axis, so the length of the domain parallel to the X axis does not change. Then the pre- and post-faulting deformation are even more different. Thus, in both regimes the faulting does not allow the deformation of single fault domains to be just an increase of the externally imposed pre-faulting deformation. This will generally also be the case if old fractures are reactivated. Even if at some stage a fault system forms such that it allows the precursory deformation to grow, it will not do so after some time. The reason for this is that the velocity of the edges of the deformed zone is time dependent, as shown above, but there is no reason why the motion of the bounding rigid blocks should change and vary in exactly the same way.

These examples demonstrate an important problem concerning faulted areas: the kinematic constraints that govern their structural evolution may not allow them to accommodate an externally imposed deformation. This may have various consequenses such as modification of the stresses within the faulted area and formation of new faults or other structures. Together with problems of geometric compatibility of domains these features complicate the structure and the deformation history of faulted areas. These effects need further study.

4. Faults as surfaces of weakness

The fact that in intraplate settings faults have offsets of the order of 0.1 km to tens of kilometers indicates that continuing slip on existing faults is generally favored over formation of new fractures, i.e. that faults are usually weaker than the intervening material. This is the reason why kinematic constraints are so important in discontinuous regional deformation: being weak, the faults usually provide the main means for stress release in the brittle part of the crust and thus they accommodate most of the deformation. Their weakness also allows faults to remain active even after they rotate relative to the stress field and they are no longer favored failure planes. Only when the rotation is large the faults become locked and then formation of new faults is favored over slip on existing ones (Proffett, 1977; Angelier and Colletta, 1981; Nur et al., 1986). The amount of rotation before this happens is, in a way, a measure of fault weakness. Thus, the factors that make faults weaker than their surroundings must be known in order to understand the mechanics of block translation and rotation.

A preliminary evaluations of the important factors is possible on the basis of the often used criterion for slip on existing planes of weakness as well as for fracturing of brittle rocks

$$|\tau| = S + \mu\,(\sigma\text{-}P), \qquad (17)$$

where τ and σ are the shear and normal stresses, μ is the coefficient of (static or internal) friction, S is the shear strength and P is the pore pressure, σ-P being the effective normal stress. For intact rock this is the Coulomb criterion which, as noted above, is only an approximation. In this case μ is the coefficient of internal friction. Equation (17) is also the criterion for frictional sliding on existing fractures, μ being in this case the coefficient of static friction (Jaeger and Cook, 1979; Byerlee, 1978). It was widely used to estimate state of stress in the crust (e.g. Brace and Kohlstedt, 1980; Meissner and Strehlau, 1982). However, its application is not simple, because initiation of fault slip and friction are complex phenomena (Dietrich, 1986; Carter and Tsenn, 1987), and stresses probably repeatedly rise and fall and also vary laterally along active faults, especially if they slip seismically. Because of these complications equation (17) can be used merely as a framework for a preliminary discussion.

Continued fault slip requires that the criterion for slip on exiting fractures is satisfied, but the criterion for formation of new faults is not. In terms of equation (17) this means that the value of one or several of S, μ and P on faults differs from the corresponding value in intact rock. As noted above, the combined effect of these differences can be measured by the maximum possible fault rotation in a constant stress field. This, in turn, can be analyzed by treating faulted areas as mechanically anisotropic, i.e. as being crossed by a set of planes of weakness (Jaeger and Cook, 1979; Nur et al., 1986). The influence of the various parameters is conveniently described by Mohr diagrams (Fig. 6A) in which the subscripts i and f denote parameters in intact rock and on faults, respectively, and τ_m and σ_m are the maximum shear stress and the mean effective stress ($=[\sigma_1+\sigma_3]$ / 2 - P). Denoting the maximum fault rotation by δ and defining ϕ such that $\tan \phi = \mu$ and the angle β as in Fig 6A, it is seen that

$$\delta = (90° - \beta + \phi_i - \phi_f) / 2, \tag{18}$$

$$\sin\beta = \frac{\sigma_{mf} \sin \phi_f + S_f \cos \phi_f}{\tau_{mf}} \tag{19}$$

$$\tau_{mf} = \sigma_m \sin \phi_i + S_i \cos \phi_i \tag{20}$$

Combining (19) and (20) yields

$$\sin \beta = \frac{\cos \phi_f [\mu_f + S_f/ \sigma_{mf}]}{\cos \phi_i [\mu_i + S_i/ \sigma_{mf}]} \tag{21}$$

To apply these equations, the values of S, μ and P on faults and in intact rock need to be known. Nur et al. (1986) examined the role of changes of S, but the other prameters should also be considered.

For intact rock the Coulomb criterion predicts formation of faults at an angle of 45°- ϕ / 2 to σ_1, the

198

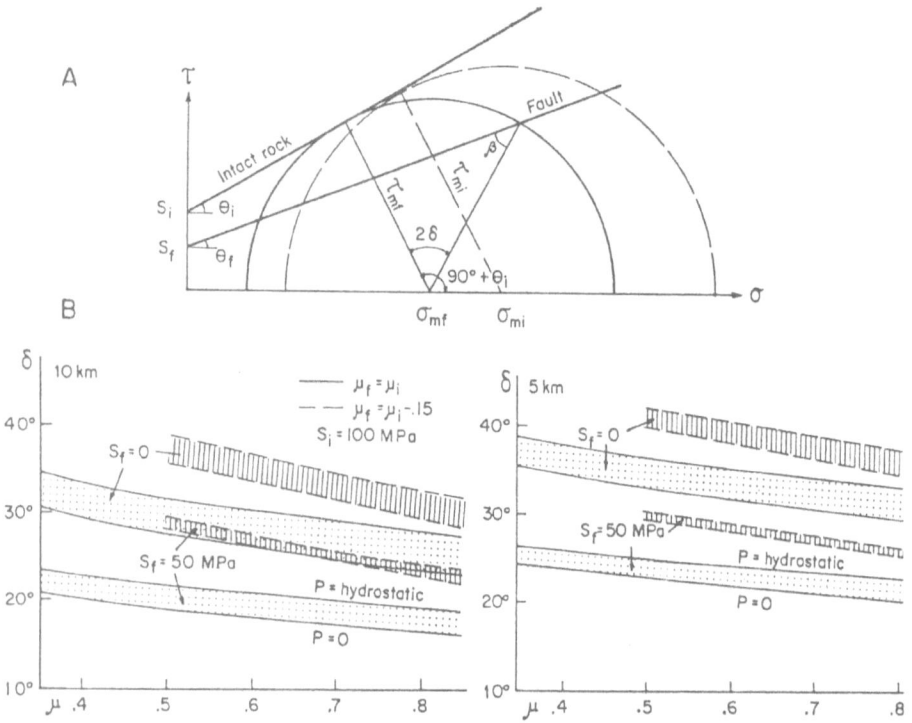

Fig. 6. The maximum possible fault rotation. A - The Mohr diagram showing the relevant parameters. B - the maximum possible rotation for different sets of parameters and at different depths. See text for discussion.

maximum stress axis, where tan $\phi = \mu$. In experiments ϕ is 25°-35°, showing that $\mu = .5-.7$, while S is usually 100-500 MPa (Jaeger and Cook, 1979; Paterson, 1978). Field data on fault orientations interpreted in this way yield the same values of μ, but there is no data on the in-situ value of S. If it were as high as found in experiments, i.e. several hundred MPa, then equation (17) shows that to produce new faults τ must be that high when $P \approx \sigma$, and much higher when P is hydrostatic or less. Such high shear stresses by far exceed the values that were measured in the upper few km of the crust and which are expected at depths of 5-10 km, in the lower part of the seismogenic zone (Bracc and Kohlstedt, 1980; McGarr, 1980). It is difficult to see that such high stresses are temporarily achieved in large rock volumes. More likely, faults form by growth and coalescence of minor flaws, joints and fractures (Segall and Pollard, 1983; Granier, 1985), a process in which Si is lower than the value found in the laboratory during breaking of sound rock samples. Therefore in-situ values of Si of about 100 MPa are more probable.

Experimens on frictional sliding (Byerlee, 1978) showed that up to effective normal stresses of 200 MPa (the lithostatic pressure at about 7.5 km) S = 0 and μ = .85, while at higher normal stresses S = 50 MPa and μ = .6, the results being independent of rock type and temperature. The scatter of the data is large, however. Overall they show only that $S \approx$ 0-50 MPa and $\mu = .6-.8$. Thus S_f is definitely lower than S_i of intact rocks and there is no clear difference between the internal friction of intact rock and the (static) friction of rocks or of products of rock wear (cataclasite, gouge) which develop along faults, but a small change of μ, by 10%-20%, cannot be excluded. The possibility that μ on faults is low is raised by the finding that in clay rich gouge μ may be as low as .35 (Morrow et al., 1982).

Using these values in equations (18) and (21), the values of δ, the maximum possible fault rotation, is shown in Fig. 6B. The difference between S_i and S_f produces a significant effect, which does not depend much on P, the pore pressure, as long this is less than hydrostatic; however, when P approaches the lithostatic pressure (i.e. $\sigma_{mf} \approx 0$), the δ is close to 45° if $S_f \approx 0$ and $\mu_i = \mu_f$. Fig. 6B also shows that δ decreases with depth if S_i and S_f do not vary but σ_{mf} increases downwards, which is the case when P is a constant fraction of the lithostatic pressure. Fig. 6B shows that differences between μ_i and μ_f can also influence δ; e.g. when $\mu_i - \mu_f = 0.15$, δ is larger by 6°-8° than when $\mu_i = \mu_f$.

It is instructive to compare Fig. 6B with the field data which suggest that δ can reach 35°-40° or even more (Nur et al., 1986). It is seen that such rotations are possible at depths of 5 km or less, but not at a depth of 10 km, if reduction of S is the only factor, μ is in the range 0.6-0.85 and P is not more than hydrostatic. Yet the important faults usually extend to depths of 5-10 km and more, where the most large earthquakes nucleate. As it is unlikely that the shallow parts of faults will continue to rotate when their deeper parts become locked, the neccessary conditions allowing large rotations apparently exist at these depths as well. Values of S_i - S_f at these depths considerably higher than assumed here would allow large rotations (cf. Nur et al., 1986), but, as noted above, this is not considered likely. Perhaps P is very high at such depths, being close to lithostatic, and/or τ is lower that assumed here.

Though pore pressure P can have an important role, its behavior is not well known. It is possible that P is modified after faulting. The products of rock wear and the strongly jointed zones which develop along faults are more permeable than intact rock and allow flow of water into fault zones, which can change P after fault formation. This will change σ_m, but the magnitude of the effect cannot be evaluated. This effect is illustrated in Fig. 6A where the values of σ_{mi} and σ_{mf} are shown to be different.

This discussion shows that the fault properties and the magnitude of pore pressure, and their variation with depth, which determine the amount of possible fault rotation in a constant stress field, are still incompletely known. In fact, the amount of fault rotation can provide an important constraint on models of fault properties at depth and should be incorporated into discussions of this topic. The uncertainty about fault properties and the possibility that stress fields change during deformation make difficult the study of the mechanical constrainsts on the deformation of faulted areas.

5. Relation between shallow and deep crustal deformation

The downward continuation of fault structures is incompletely understood (Sibson, 1983). In fact, it is not even clear whether the brittle and the directly underlying ductile levels always undergo the same overall deformations, or whether they are sometimes decoupled on a regional scale. Yet these relations must be known to understand the geometry of the structure, and how and to what extent the deep deformation drives the deformation of the shallow levels. In this respect two points are important. First, the shallow brittle and seismogenic deformation changes downward penetrative, ductile and a-seismic deformation (Brace and Kohlstedt, 1980; Meissner and Strehlau, 1982; Chen and Molnar, 1983; Sibson, 1983; Carter and Tsenn, 1987). Thus, at depth deformation should become more and more evenly distributed, which implies that faults do not extend into the lower crust. However, the occurrence of mylonitic shear zones shows that in mid-crustal levels deformation can still be localized in fault-like, but non-seismogenic, ductile zones (Tullis et al., 1982). Second, the change to ductile deformation creates structural disharmony at some depth, because in the brittle level the displacement varies laterally in a step-wise manner, whereas in the underlying zone it varies smoothly. Thus, some strain incompatibility should arise between the undeformed fault blocks and the underlying continuously deformed material. This will happen even if a part of the deformation at depth is localized in shear zones. These considerations help to discuss how shallow fault structures extend downward.

In areas of tilted block systems produced by normal faulting good evidence exists for signficant mid-crustal decoupling. In the Basin and Range province of western USA shallow seismic reflection (Smith and Bruhn, 1984) and COCORP profiles (Hague et al., 1987; Hauser et al, 1987) show that the offsets of shallow faults, reaching several km, and the tilting of the fault blocks, reaching tens of degrees, usually do not affect the shape of the top the reflective lower crust or its structure nor the shape of the remarkably flat Moho (Fig. 7A). These features show that the shallow fault-contolled structure is decoupled from the lower crust. Another pertinent case is the Gulf of Suez rift. Its entire width is underlain by a system of blocks which are mostly tilted by similar amounts and thus record a rather uniform extension along any traverse across the rift, yet the total subsidence varies across the rift (Fig. 7B; Moretti and Colletta, 1987). As subsidence depends on the amount of crustal stretching (McKenzie, 1978), this shows that the extension of the lower crust varies laterally, which indicates decoupling between the shallow and deep structures. Decoupling also appears to be common under continental margins where the shallow fault structure is not expressed at the level of the Moho (Fig. 7C), e.g. in the Bay of Biscay (Montadert et al., 1979) and off Newfoundland (Keen et al., 1987).

Thus, block systems formed by extensional faulting usually seem to be delimited by mid-crustal zones of decoupling or disharmony. In part this results from downward changes of rheology, as noted above. However, the tendency of displacement discontinuities to arise along boundaries of domains of rotated blocks (Fig. 2B, viewed as a vertical cross section) probably also promotes decoupling. This kinematic effect will augment any strain localizations that arise at mid-crustal levels, e.g. because of changes of rheology, and cause them to develop into a fault or a zone of concentrated strain, similar to the faults that often develop along boundaries of domains of strike slip faults. This is also expected to occur where normal faults have an oblique slip, i.e. have a strike slip component. The rotation of blocks on vertical axes which takes place in such situations (cf. Fig. 2C) probably also helps the development of decoupling. However, more must be known about such structures before this can be assessed.

The evidence regarding mid-crustal detachment zones and structural disharmony beneath systems of predominantly strike slip faults is less clear. Downward changes in fault geometry are indicated by secondary structures along bends of strike slip faults, e.g. rhomb-shaped pull aparts. As the margins of such basins are often little deformed, their volumes should equal the product of the lateral slip that produced them by the depth to which this slip extends, if slip is on vertical faults. Remarkably, such basins are often only a few kilometers deep. This shows that the geometry of the associated faults must change at quite shallow depths. The faults may become inclined or end on flat detachment surfaces, or else the lateral motion may be distributed over a wide zone not directly beneath the pull apart basins. Uplifts along compressional segments of strike slip faults (e.g. flower structures) are also often not very voluminous (Harding, 1985). This suggests, for the same reasons, that they too do not extend to great depths.

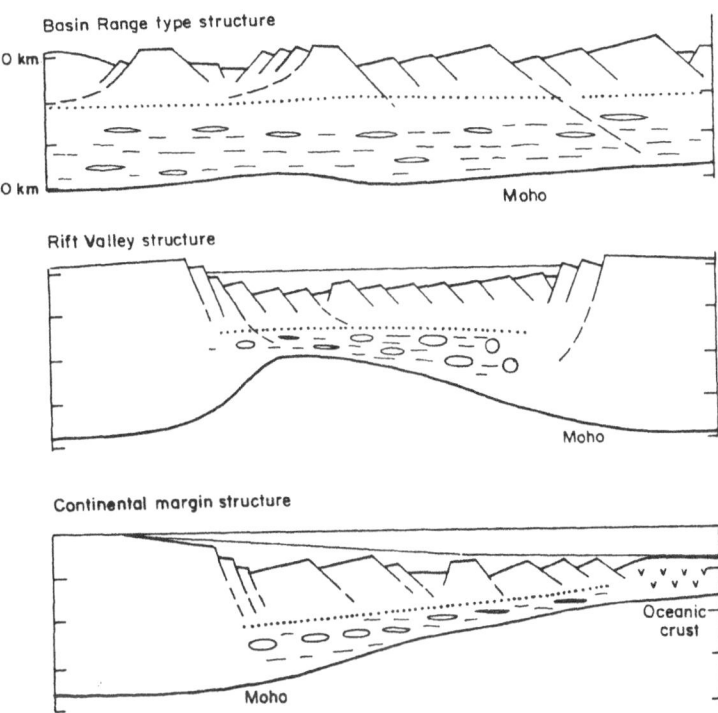

Fig. 7. The deep structure of areas of tilted blocks formed by normal faulting in different structural settings. Line of thick dots signifies level of disharmony and decoupling between shallow and deep structure.

Important evidence is given by the COCORP deep reflection profiles from the margin of the Mojave Desert (Cheadle et al., 1985). They show that the faults extend downward to less than 10 km, because flat-lying reflectors and this and at greater depths are generally not offset. Moreover, the Garlock fault does not displace a dipping refelctor at about 10 km depth, though the lateral offset of this fault exceeds 60 km where crossed by the COCORP profile. These relations indicate a rather shallow and flat decoupling zone. As the COCORP survey did not cross the central part of the Mojave Desert, where many long rotated strike slip faults exist (Garfunkel, 1974; Luyendyk et al., 1985; Hornafius et al., 1986), more profiles are needed to establish the generality of these relations. However, focal mechanisms which record slip on flat-lying faults under this and nearby regions (Webb and Kanamori, 1985; Nicholson et al., 1987) further support the existence of a mid-crustal disharmony beneath large areas of strike slip faults in southern California.

In fact, the downward change of rheology leading to distribution of deformation in large rock volumes should produce structural disharmonies beneath strike slip fault systems just as it does beneath normal faults. In order that shearing along shallow sub-vertical faults should become distributed at depth, then at some level shearing along horizontal or inclined surfaces should take place (Fig. 8). The evidence from southern California suggests that at least in some cases the the decoupling is very significant. On the other hand, the presence of steep mylonitic shear zones in terranes that were deformed in ductile conditions raises the possibility that some strike slip faults extend into mid-crustal levels (Sibson, 1983). However, there need not be a one-to-one correspondence between faults and deep shear zones, as even then some of the shallow deformation can become distributed downward in inclined zones of shearing (Fig. 8). It is also possible that strong decoupling occurs even where mid-crustal shear zones are present.

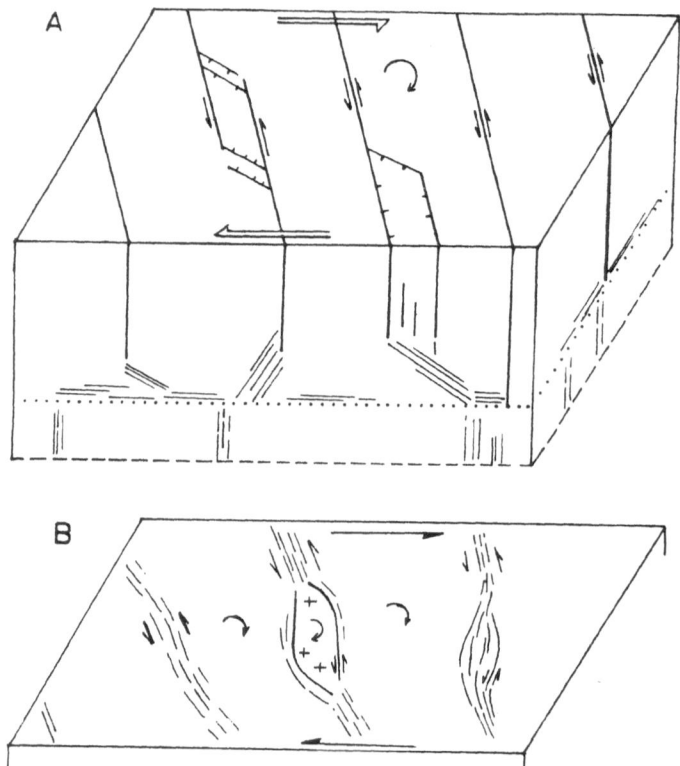

Fig. 8. The deep structure of areas of strike slip faulting. A - possible relations between shallow fault structure and deeper structure with shear zones. B - schematic structural patterns in deep levels seen in plan view.

Whatever the case, it appears that decoupling of some sort usually occurs at some depth under strike slip fault systems. Therefore the block rotations that occur in such settings need not continue to great depths. Perhaps block rotations even enhance decoupling, especially in multi-domain areas where rotations in opposite senses and by different amounts take place. It is not known if the overall deformations in the different levels are similar. However, where this is the case the deep deformation probably resembles the regional average of the shallow deformation. The heterogeneity of multi-domain areas can be mostly, if not entirely, attenuated in zones of detachment.

The deformation style of regions crossed by steep shear zones may bear some resemblance to that of fault blocks, regardless of the presence of shallower structural disharmonies. In such regions the shearing in plan view is unevenly distributed. A part is concentrated on the shear zones, the other part being distributed within the intervening ductile blocks. This was demonstrated, for example, in the Armorican Massif, France (Jagouzo, 1980; Percevaux and Cobbold, 1982). Clearly, all gradations can exist between such a situation and one in which the entire regional shearing is concentrated on faults. Therefore some of the deformation patterns existing in the latter case can be expected also where the deformation is unevenly distributed. Thus, regional shearing can be partly accommodated in zones whose shearing sense is the opposite of the regional shearing sense and whose strike is oblique to the direction of areal shearing, while the regions between these zones rotate (Fig. 8). This was suggested to have occured in the Hercynian basement complex of Extramadure, Spain, where less deformable granite plutons are embedded in more ductile metamorphics (Castro, 1985). The deformation may have even produced pull-apart-like features which guided granite emplacement. This interpretation still needs to be integrated into a wider picture which shows how it relates to nearby belts which were sheared in the oppsoite sense (e.g. Coimba-Cordoba shear zone, Burg et al., 1981). Nevertheless, it is worth to explore whether such deformation patterns occur in heterogeneous basement areas which include igneous plutons and were affected by overall shearing (e.g. as described by Brun and Pons, 1981; Davies, 1982).

It thus appears that the shallow structure of fault domains in which blocks rotate commonly changes at moderate depths, perhaps at the base of the seismogenic zone. It may end on flat decoupling zones, but a gradual change in zones where shearing on inclined surfaces occurs cannot be excluded. In any case the rotation of shallow blocks does not mirror rotations at mid-crustal levels. Much more needs to be learned about the manner in which the transition from the observable shallow structure to the underlying hidden structure is achieved. Only then will it be possible to place the fault controlled structures within the broader context of crustal- and lithospheric-wide deformation.

6. Concluding remarks

The foregoing considerations dealt with the character of shallow intracontinental deformation by block translation and rotation as well as with some outstanding problems relating to this structural style. Of particular importance is the recognition of the common occurrence of rotations on vertical axes. These can occur wherever fault slip has a strike-parallel component. The occurrence of rotations in different senses in domains of conjugate faults is considered to be a diagnostic feature of this deformation style. Much care is needed, however, to separate block rotation produced by fault motions from rotations having other causes.

Deformation by faulting is governed by mechanical and kinematic constraints. The latter depend on fault and domain geometry and are particularly important because of the disconticontinuous nature of the deformation. Because of these constraints rotation of blocks is an essential property of fault-controlled regional deformation. This allows fault domains to fit as closely as possible with each other and with their surroundings. However, these constraints can also produce misfits between domains and they may not be compatible with the overall boundary conditions that are externally imposed of multi-domain areas. These complications are probably resolved by formation of new structures and modification of the stress field within the faulted area. The mechanical constraints limit the amount of possible rotation of faults. However, their implications are difficult to analyze because the properties of faults and their depth variations, as well as the stress distribution in discontinuous media, are incompletely known. The block structures and the rotations in the brittle layer are partly or completely decoupled from the deeper structures, but the details still need study.

Block rotation was seen to be an essential manifestation of large-scale shearing in the brittle layer, rotations about vertical axes being associated with lateral motions. It now appears that such motions are quite common and occur in diverse tectonic settings. Strike slip fault systems are devloped along continental transform boundaries where igneous activity is scaerce, e.g. along the Dead Sea transform (Eyal et al., 1981; Ron et al., 1984) and in California, in orogenic belts with abundant magmatism, e.g. Japan (Research Group for Active Faulting in Japan, 1980), as well as in large areas undergoing late orogenic continental consolidation and collision, e.g. at the end of the Variscan Orogeny in Europe and nearby areas (Arthaud and Matte, 1977), at the end of the Pan African Orogeny in Arabia (Najd fault system: Moore, 1979) and in present-day central Asia (Molnar and Tapponier, 1975). In all cases domains of conjugate fault systems are present and block rotation is expected. From a wider point of view Woodcock (1985) showed that the present plate motions involve a component of shearing along most plate boundaries. This suggests that the extent of regional shearing in map view was greatly underestimated. Thus regional deformation by block translation and rotation in domains of conjugate faults is probably more common than hitherto recognized. However, the details of such structures and their relations with the structure of the underlying levels may vary according to the tectonic setting and mechanical-thermal state of the crust.

Acknowledgements

I am greatly indebted to Raphy Freund, my late teacher and collegue, who initiated some of the basic ideas leading to the present work. Warm thanks to C. Laj and C. Kissel for inviting me to the NATO workshop and to the participants of this workshop for many discussions which helped me to clarify the ideas presented in this paper.

Refecences

Anderson, E.M., 1951. The Dynamics of Faulting and Dyke Formation. Oliver and Byod, 206 p.
Angelier, J., Colletta, B., 1983, Tension features and extensional tectonics. *Nature*, v. **301**, p 49-51.

Arthaud, F., Matte, P., 1977. Late Paleozoic strike-slip faulting in southern Europe and northern Africa: Result of a right lateral shear zone between the Appalachians and the Urals. *Geol. Soc. America Bull.*, v. **88**, p. 1305-320.

Beck, M.E., 1980. Paleomagnetic record of plate-margin tectonic processes along the western edge of North America. *Jour. Geophys. Res.*, v. **85**, p. 7115-7131.

Brace, W.F., Kohlstedt, D.L., 1980. Limits on lithospheric stress imposed by laboratory experiments. *Jour. Geophy. Res.*, v. **85**, p. 6248-6252.

Brun, J.P., Pons, J., 1981. Strain pattern of pluton emplacement in a crust undergoing non-coaxial deformation, Sierra Morena, Southern Spain. Jour. *Structural Geology*, v. **3**, p. 219-229.

Burke, D.B., Hillhouse, L.W., McKee, E.H., Miller, S.T., Morton, J.L., 1982. Cenozoic rocks of the Barstow area of southern California - stratigraphic relations, radiometric ages and paleomagnetism. *U.S. Geol. Survey Bull.* **1529**-E, 16 p.

Burg, J.P., Iglesias, M., Laurent, Ph., Matte, Ph. Ribiero, A., 1981. Variscan intracontinental deformation: The Coimba-Cordoba shear zone (SW Iberian Peninsula). *Tectonophysics*, v. **78**, p. 161-177.

Byerlee, J.D., 1978. Friction in rocks. *Pure and Applied Geophysics*, v. **116**, p. 615-626.

Calderone, G., Butler, F.R., 1984. Paleomagnetism of Miocene volcanic rocks from southwestern Arizona. *Geology*, v. **12**, p. 627-630.

Carter, N.L., Tsenn, M.C., 1987. Flow properties of the continental lithosphere. *Tectonophysics*, v. **136**, p. 25-64.

Castro, A., 1985. The central Extramadura Batholith: geotectonic implications (European Hercynian) - an outline. *Tectonophysics*, v. **120**, p. 57-68.

Cheadle, M.J., and 7 others, 1986. The deep crustal structure of the Mojave Desert, California from COCORP reflection data. *Tectonics*, v. **5**, p. 293-320.

Chen, W.P., Molnar, P., 1983. Focal depths of intracontinental and intraplate earthquakes and their implications for the thermal and mechanical properties of the lithosphere. *Jour. Geophys. Res.*, v. **88**, p. 4183-4214.

Davies, F.B., 1982. Pan-African granite intrusion in response to tectonic volume changes in a ductile shear zone from northern Saudi Arabia. Jour. *Geology*, **90**, 467-483.

Dietrich, J.H., 1986. A model for the nucleation of earthquake slip. AGU Monograph 37 (Maurice Ewing series, v. 6), p. 37-47.

Eyal, M., Eyal, Y., Bartov, Y., Steinitz, G., 1981. The tectonic development of the western margin of the Gulf of Elat (Aqaba) Rift. *Tectonophysics*, v. **80**, p. 39-66.

Freund, R., 1980. Rotation of strike slip faults in Sistan, Iran. Jour. *Geology*, v. **78**, p. 188-200.

Freund, R., 1974. Kinematics of transform and transcurrent faults. *Tectonophysics*, v. **21**, p. 93-134.

Garfunkel, Z., 1974. Model for the Late Cenozoic tectonic history of the Mojave Desert, California, and for its relations to adjacent regions. *Geol. Soc. America, Bull.*, v. **85**, p. 1931-

Garfunkel, Z., Ron, H., 1985. Block rotation and deformation by strike slip faults 2. The properties of a type of macroscopic discontinuous deformation. *Jour. Geophys. Res.*, v. **90**, p. 8589-8602.

Granier, T., 1985. Origin, damping, and pattern of development of faults in granite. *Tectonics*, v. **4**, p. 721-737.

Hague, T.A. and 10 others, 1987. Crustal structure of western Nevada from COCORP deep seismic-reflection data. *Geol. Soc. America Bull.*, v. **98**, p. 320-329.

Harding, T.P., 1985. Seismic characristics and identification of negative flower structures, positive flower structures and positive structural inversion. *Amer. Assoc. Petrol. Geol. Bull.*, v. **69**, p. 582-600.

Hauser, E., and 9 others, 1987. Crustal structure of eastern Nevada from COCORP deep seismic-reflection data. *Geol. Soc. America Bull.*, v. **98**, p. 833-844.

Hancock, P.L., 1985. Brittle microtectonics: principles and practice. *Jour. Structural Geol.*, v. **7**, p. 437-457.

Handin,, J., 1969. On the Coulomb-Mohr failure criterion. *Jour. Geophys. Res.*, v. **84**, p. 5343-5348.

Hoeppener, R., Kalthoff, E., Schrader, P., 1969. Zur physikalischen Tektonik: Bruchbildung bei verschiedenen Deformationen im Experiment. *Geol. Rundschau*, **59**, 179-193.

Hornafius, J.S., Luyendyk, P.B., Terres, R.R., Kamerling, M.J., 1986. Timing and extent of Neogene tectonic rotation in the western Transverse *ranges, California. *Geol. Soc. America, Bull.*, v. **97**, p. 1476-1487.

Hudson, M.R., Geissman, W.J., 1987. Paleomagnetic and structural evidence for Middle Tertiary counterclockwise block rotation in the Dixie valley region, west-central Nevada. *Geology*, v. **15**, p.638-642.

Jaeger, J.C., Cook, N.G.W., 1979. Fundamentals of Rock Mechanics. John Wiley & Sons, New York, 593 p.

Jagouzo, P., 1980. The south Armorican shear zone. Jour. Structural *Geology*, v. **2**, p. 39-47.

Keen, C.E., Stockmal, G.S., Welsink, H., Quinlan, G., Mudford, B., 1987. Deep crustal structure and evolution of the rifted margin northeast of Newfoundland: results from LITHOPROBE East. *Canad. J. Earth Sci.*, v. **24**, p. 1537-1549.

Kissel, C., Laj, C., 1987. The Tertiary geodynamical evolution of the Aegean arc: a paleomagnetic reconstruction. *Tectonophysics*, v. **146**, p. 183-201.

Kissel, C., Laj, C., Sengor, A.M.C., Poisson, A., 1987. Paleomagnetic evidence for rotation in opposite senses of adjacent blocks in northeastern Aegea and western Anatolia. *Geophys. Res. Letters*, v. **14**, p. 907-910.

Lockwood, J.P., Moore, J.G., 1979. Regional deformation of the Sierra Nevada, California, on conjugate microfault sets. *Jour. Geophys. Res.*, v. **84**, p. 6041-6049.

Luyendyk, B.P., Kamerling, M.J., Terres, R.R., Hornafius, J.S., 1985. Simple shear of southern California during Neogene time suggested by paleomagnetic declinations. *Jour. Geophys. Res.*, v. **90**, p. 12454-12466.

Malvern, L.E., 1969. Introduction to the Mechanics of a Continuous Medium. Prentice-Hall, N.J., 713 p.

Mase, G.E., 1970. Continuum Mechanics. Schaum's Outline Series, McGraw- Hill, New York, 221 p.

McGarr, A., 1980. Some constraints on levels of shear stress in the crust from observations and theory. *Jour. Geophys. Res.* v. **85**, p. 6231-6238.

McKenzie, D., 1978. Some remarks on the development of sedimentary basins. *Earth Planet. Sci. Letters*, v. **40**, p. 25-32.

McKenzie, D., Jackson, J., 1983. The relationship between strain rates, crustal thickening, paleomagnetism, finite strain and fault movements within a deforming zone. *Earth. Planet. Sci. Letters*, v. **65**, p. 182-202.

McKenzie, D., Jackson, J., 1986. A block model of distributed deformation by faulting. *Jour. Geol. Soc. London*, v. **143**, p. 349-353.

Meissner, R., Strehlau, J., 1982. Limits of stresses in continental crust and their relation to the depth-frequency distribution of shallow earthquakes. *Tectonics*, v. **1**, p. 73-89.

Molnar, P, Tapponier, P., 1975. Cenozoic tectonics of Asia: Effects of a continental collision. *Science*, v. **189**, p. 419-426.

Montadert, L., Roberts, D.G., De Chapral, O., Guennoc, P., 1979. Riftingand subsidence of the northern continental margin of the Bay of Biscay. *Init. Reports DSDP*, v. **48**. p.1025-1060. US Gov't Printing Office, Washington D.C.

Moore, J.McM., 1979. Tectonics of the Najd transcurrent fault system, Saudi Arabia. *Jour. Geol. Soc. London*, v. **136**, p. 441-452.

Moretti, I., Colletta, B., 1987. Spatial and temporal evolution of the Suez rift subsidence. *Jour. Geodynamics*, v. **7**, p. 151-168.

Morrow, C.A., Shi, L.Q., Byerlee, J.D., 1982. Strain hardening and strength of clay-rich gouges. *Jour. Geophys. Res.*, v. **87**, p. 6771-6780.

Nicholson, C., Seeber, L., Williams, P., Sykes, L.R., 1986. Seismic evidence for conjugate slip and block rotation within the San Andreas fault system, southern California. *Tectonics*, v. **5**, p. 629-648.

Nur, A., Ron, H., Scotti, O., 1986. Fault mechanics and the kinematics of block rotations. *Geology*, v. **14**, p. 746-749.

Oertel, G., 1965. The mechanism of faulting in clay experiments. *Tectonophysics*, v. **2**, p. 343-393.

Paterson, M.S., 1978. Experimental rock deformation - the brittle field. Springer Verlag. 254 p.

Percevault, M.N., Cobbold, P.R., 1982. Mathematical removal of regional ductile strains in central Brittany: evidence for wrench tectonics.*Tectonophysics*, v. **82**, p. 317-328.

Proffett, J.M.Jr., 1977. Cenozoic geology of the Yerington district, Nevada, and implications for the nature and origin of Basin and Range faulting. *Geol. Soc. America Bull.*, v. **88**, p. 247-266.

Ransome, F.L., Emmons, W.H., Garrey, G.H., 1910. Geology and ore deposits of the Bullfrog district, *Nevada. U.S. Geol. Survey, Bull.* no. **407**, 130 p..

Reches, Z., Dietrich, J., 1983. Faulting of rocks in a three-dimensionalstrain field. I. Failure of rocks in polyaxial, servo-control experiments. *Tectonophysics*, v. **95**, p. 111-132.

Research Group for Active Faulting of Japan, 1980. Map of active faults in and around Japan, Scale 1:2 000 000. Univ. Tokyo Press.

Ron, H., Freund, R., Garfunkel, Z., Nur, A., 1984. Block rotation by strike slip faulting: structural and paleomagnetic evidence. *Jour. Geophys. Res.*, v. **89**, p. 6256-6270.

Ron, H., Aydin, A., Nur, A., 1986. Strike-slip faulting and block rotation in the Lake Mead fault system. *Geology*, v. **14**, p. 1020-1023.

Segall, P., Pollard, D.D., 1983. Nucleation and growth of strike slip faults in granite. *Jour. Geophys. Res.*, v. **88**, p. 555-568.

Sibson, R.H., 1983. Continental fault structure and the shallow earthquake source. *Jour. Geol. Soc. London*, v. **140**, p. 741-767.

Smith, R.B., Bruhn, R.L., 1984. Intraplate extensional tectonics of the eastern Basin-Range: Inferences on structural style from seismic reflection data, regional tectonics, and thermo-mechanical models of brittle-ductile deformation. *Jour. Geophys. Res.*, v. **89**, p. 5733-5762.

Stewart, J.H., 1980. Regional tilt patterns of Late Cenozoic Basin- Range blocks, western United States. *Geol. Soc. America Bull.*, v. **91** (part I), p. 460-464.

Thompson, G.A., 1960. Problem of the Late Cenozoic structure of the Basin Ranges. *Internat. Geol. Congress 21 (Copenhagen)*, v. **18**, p. 62-68.

Tullis, J.A., Snoke, A.W., Todd, V.R., 1982. Penrose conference report on significance and petrogenesis of mylonitic rocks. *Geology*, v. **10**, p. 227-230.

Webb, T.H., Kanamori, H., 1985. Earthquake focal mechanisms in the Eastern Transverse Ranges and San Emigdio Mountains, southern California, and evidence for a regional decollement. *Bull. Seismol. Soc. America*, v. **75**, p. 737-757.

Wells, R.E., Coe, R.S., 1985. Paleomagnetism and geology of Eocene volcanic rocks of southwest Washington, implications for mechanisms of tectonic rotations. *Jour. Geophys. Res.*, v. **90**, p. 1925-1947.

Wilcox, R.E., Harding, T.P., Seely, D.R., 1973. Basic wrench tectonics. *Amer. Assoc. Petroleum Gol. Bull*, v. **57**, p. 74-96.

Woodcock, N., 1986. The role of strike-slip fault systems at plate boundaries . *Phil. Trans. R. Soc. London*, v. **A347**, p. 13-29.

MECHANICS OF DISTRIBUTED FAULT AND BLOCK ROTATION

A. NUR[1], H. RON[2] and O. SCOTTI[1]
[1] Stanford Rock Physics Laboratory
Department of Geophysics
Stanford University
Stanford, California 94305,
U.S.A.

[2] Institute for Petroleum Research and Geophysics
P.O. Box 2286, Holon,
Israel

ABSTRACT. Most of the earth's crust is broken by dense *sets* of subparallel faults which are organized in *domains*. Joint analysis of structural and paleomagnetic data (Ron et al., 1984) demonstrate that these domains when subject to tectonic deform by distributed fault slip and block rotations, rather than by uniform straining. Many such domains have been recognized in the Western U.S., and in California and Nevada in particular. Precise mechanical considerations of stress at failure, strength and friction by Nur et al. (1986) reveal under what conditions new fault sets must form when these rotations are sufficiently large (25°-45°) leading to domains of *multiple sets*. Several domains of multiple fault sets have by now been recognized in California and Nevada.
Many past studies of rotations were limited to either structural or paleomagnetic data. However by combining paleomagnetic data, structural geology, and rock mechanics we are able to explore the validity of the block rotation concept and its significance in much greater detail than ever before. Our analysis is based on data from (1) Northern Israel, where fault slip and spacing is used to predict block rotation; (2) the Mojave Desert, with well-documented strike-slip fault sets, organized in at least three major domains. A new set of faults trending N-S may be in the process of formation here; and (3) the Lake Mead, Nevada, fault system with well-defined sets of strike-slip faults, which, in contrast with the Mojave region, are surrounded with domains of normal faults; and (4) the San Gabriel Mountains domain with a multiple set of strike-slip faults.
Block rotations can have profound influence on the interpretation of geodetic measurements and the inversion of geodetic data, especially the type collected in GPS surveys. Furthermore, block rotations and domain boundaries may be involved in creating the heterogeneities along active fault systems which may be responsible for the initiation and termination of earthquake rupture.

1. Blocks and fault rotations

1.1. KINEMATICS

It has long been recognized (Freund, 1970a) that fault blocks in strike-slip tectonic domains must progressively rotate on vertical axes as the overall strike-slip motion continues. Two direct consequences of this deformation mechanism are that (1) slip on each of the faults within a domain must be related to the rotation of the blocks bounded by these faults, and that (2) the faults

209

C. Kissel and C. Laj (eds.), Paleomagnetic Rotations and Continental Deformation, 209–228.
© *1989 by Kluwer Academic Publishers.*

themselves must also rotate because they are the boundaries of the blocks (Figure 1). These ideas were applied to strike-slip tectonics in Eastern Iran (Freund, 1970a), New Zealand (Freund, 1971), Mojave Desert, California (Garfunkel, 1974), Southern California (Luyendyk et al., 1980), and

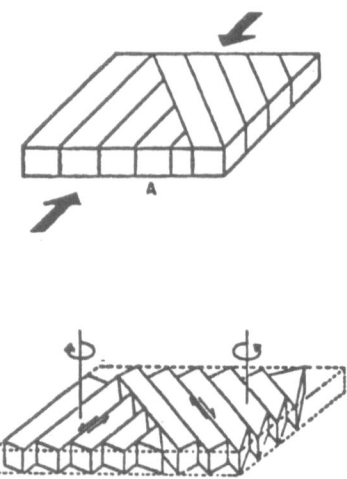

Fig. 1. A 2-D model which illustrates the simultaneous activity of strike-slip displacement and rotation of the faulted blocks. (a) The initial configuration; (b) after deformation. A set of left-lateral faults rotates clockwise and a set of right-lateral faults rotates counterclockwise.

Southeast Sinai (Frei, 1980). The geometric relations involved in this rotation process were analyzed by Freund (1974), Garfunkel (1974), and MacDonald (1980), who showed how the rotations of blocks and strike-slip displacement are interrelated and contemporaneous aspects of a single deformation process. Specifically if the blocks are rigid, then the model predicts a simple quantitative relationship between fault spacing, slip, and amount and sense of block rotation. Thus in areas where fault spacing and net slip can be determined geologically, paleomagnetic measurements may be used to test the sense and amount of rotation predicted by the model. One of the most basic aspects of the rigid fault and block rotation model (Freund, 1970a, b, 1974; Garfunkel, 1974) is that the sense of block rotation must be opposite to the sense of the fault slip, with left-handed slip associated with clockwise rotation, and right-handed slip with counterclockwise rotation.

The geometrical relation (Figure 2) between the displacement d along a fault (positive when right-lateral), the width w of the faulted block, the inital angle α between the faults and the boundary of the domain, and block rotation ϕ (positive when counterclockwise) is given by

$$d/w = \frac{\sin\phi}{\sin\alpha\sin(\alpha-\phi)} \qquad (1)$$

The relative elongation l/l_0 of the faulted domain parallel to its boundary, is given by

$$\lambda = l/l_0 = \frac{\sin\alpha}{\sin(\alpha-\phi)} \qquad (2)$$

The deformation within a single fault domain is simple shear.

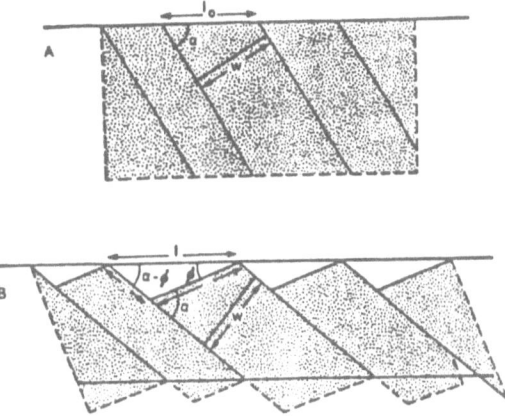

Fig. 2. Fault set kinematics. Geometrical relations among spacing between faults w, their initial orientation α, initial length l_0 to slip d, and rotation ϕ.

The block rotation model can be rigorously tested in areas where domains of strike-slip faults exist and where the blocks are fairly rigid and where paleomagnetic declinations can be measured to reveal possible horizontal rotation about vertical axes. If the model is correct, we expect to find agreement using equation (1) between structural rotations computed from the amount and sense of fault slip and fault spacing as measured in the field, and rotation amount and sense obtained from paleomagnetic declination measurements (Ron et al., 1984).

1.2. MECHANICS

In many published cases, the rotations inferred from paleomagnetic data are very large, exceeding 45°; some values approach 100°. A major question immediately arises: How much can a fault plane rotate away from the optimal direction of shearing before slip is inhibited by friction? If the magnitude of such rotation is restricted, then the kinematic model as described above must be modified accordingly.

An obvious constraint on the amount of fault rotation is imposed by the mechanical condition of faulting, namely that shear stress on the plane of the fault must exceed the shear resistance to slip along the fault. From extensive laboratory studies, it has been established that the shear stress τ

212

required for sliding on a fracture is controlled by the fracture cohesive strength S_1, and the normal effective stress σ_0 acting on the fracture.

$$\tau = S_1 \, \mu \sigma_0 = S_1 \, \mu(\sigma - p) \tag{3}$$

where μ is the coefficient of friction, σ is the normal stress, and p is pore pressure. If we consider now a fault originally formed at the optimal direction of failure ϕ_0 relative to the maximum stress as shown in the Mohr diagram of Figure 3, it follows that as deformation and rotation proceed, the shear stress acting on the fault plane decreases, and the normal stress increases. Beyond some critical angle of rotation ϕ_c it should become easier for a new fault to form in the rock mass with cohesive strength S_0 in direction ϕ_0, relative to the stationary direction of maximum stress, rather than continue sliding on the preexisting one, now oriented in the direction ϕ_c from ϕ_0.

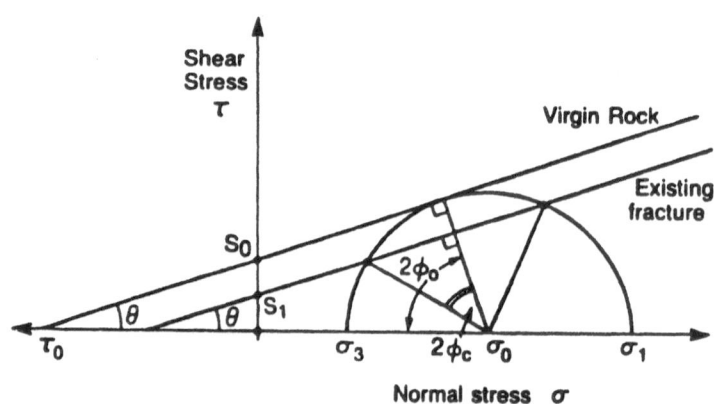

Fig. 3. Mohr circle representation of state of stress in rock with existing fracture: $\theta = \tan^{-1} \mu$ is angle of friction, S_0 and S_1 are cohesive strength of virgin rock and fracture, respectively, and ϕ_c is largest permissible angle between existing fracture and direction in which new shear fracture will develop if further rotation takes place.

The critical angle ϕ_c is given by (Nur et al., 1986)

$$\phi_c = \frac{1}{2} \cos^{-1} \left[1 - \frac{(1 - S_1 / S_0)}{(1 + \mu \sigma_0 / S_0)} \right] \tag{4}$$

where S_0 and S_1 are the cohesive strengths of the virgin rock mass and the preexisting fracture respectively, $\theta = \tan^{-1} \mu$, and σ_0 is overburden pressure.

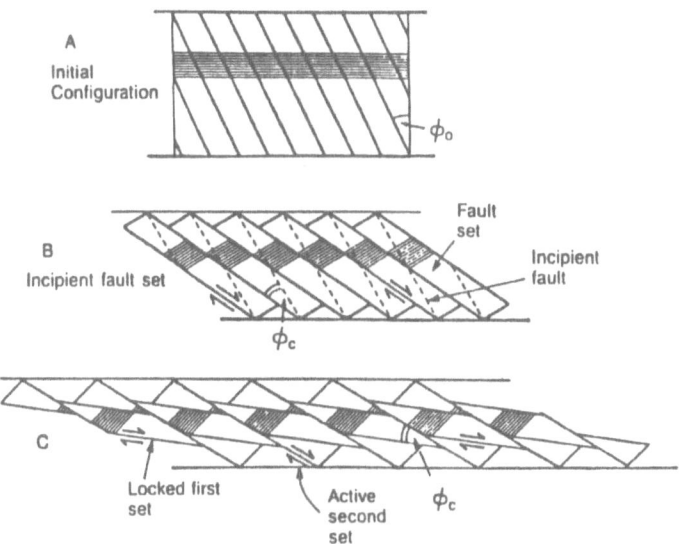

Fig. 4. Modified block model with critical angle given in Figure 5 showing new fault set required to accomodate block rotation greater than 45°. DST - Dead Sea transform.

Equation (4) shows that there is a mechanical limit on fault rotation under a stationary stress field. Apparently, the magnitude of the permissible rotation ϕ_c is dependent on the difference between the cohesive strength S_0 of the virgin rock and that of the preexisting faults S_1, the coefficient of friction $\mu = \tan\theta$, and effective overburden pressure σ_0 (or equivalently, depth). The results imply that the kinematic model must be modified, as shown in Figure 4, to include the appearance of a new set of shear faults when rotation is sufficiently large. The values of ϕ_c obtained from equation (2) show that the angle to which a fault set can rotate before a new set must appear to accommodate further block rotation is in the range of 25° to 45°. For the extreme cases $\mu = 0$ or σ_0 =0, we obtain the maximum possible angle of rotation of $\phi_c = 45°$, which is the absolute upper limit of block rottation which can be accommodated by one set of faults. In situ, the angle ϕ_c should be the angle between the direction of the faults in the older locked set, and the orientation of the faults in the newer set, which offsets the older one.

2. Field evidence

2.1. NORTHERN ISRAEL.

Probably the first suggestion that large spatial variation of paleomagnetic declinations may be due to tectonic rotations about vertical axes was made by Nur and Helsley (1971) as a result of a field study in Northern Israel in 1967. These results revealed large declination anomalies which have led to the idea of a rigorous test of the block tectonics model using paleomagnetic measurements. This test was carried out by Ron et al. (1984) in Northern Israel (Figure 5), where good structural data regarding fault spacing and slip were available and good paleomagnetic material yielded reliable and sufficiently accurate declination data.

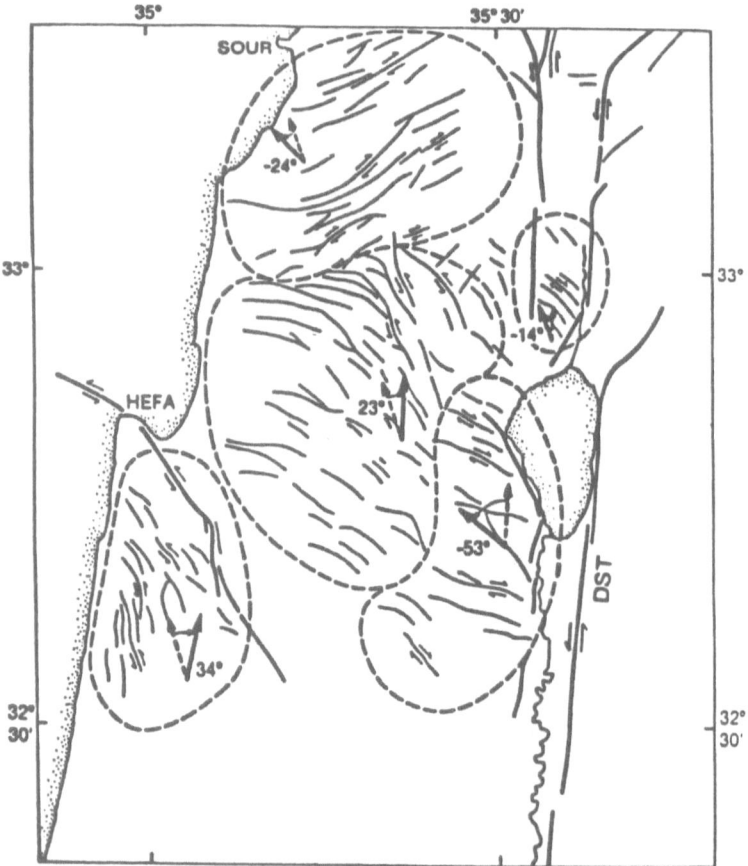

Fig. 5. Fault domains in northern Israel, showing fault sets, directions of slip, and paleomagnetically determined rotations in these domains.

The results of this study--both structural and paleomagnetic--revealed several domains of fault sets, some with clockwise and left-lateral fault slip, others with counterclockwise and right-lateral slip, as predicted by the model of Freund (1970). Figure 6 shows a summary of the structurally observed rotations vs. the paleomagnetically measured ones, with values of the rotations in the various domains ranging from 0° to 54°. The results clearly show that the magnitude and sense of block rotations as predicted from the structural data agree well with values obtained from the independent paleomagnetic determinations, suggesting that the rigid fault rotation process may thus be typical of such systems in general.

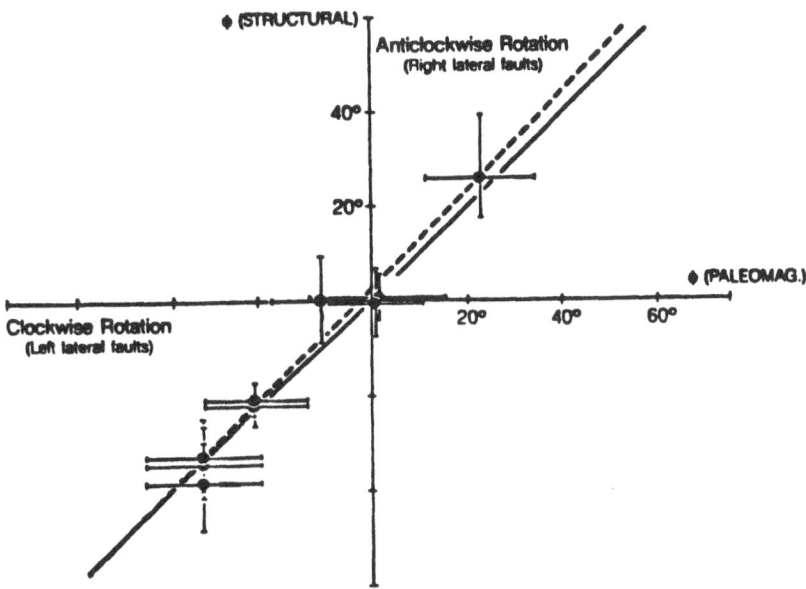

Fig. 6. Comparison between structurally derived and paleomagnetically determined rotations in northern Israel (see Figure 7). Solid line represents perfect agreement; dashed line is linear best fit to data.

2.2. MULTIPLE SETS

What about multiple sets? There are a few in situ cases in which the angle between the directions of faults in an older and a newer fault set within one domain have been determined: The values of these angles range between 25° to 45°, well within the predicted range of mechanically permissible fault rotations. These values are, however, significantly lower than many of the total block rotations determined paleomagnetically. According to our model block rotations larger than 45° must involve two or more sets of rotating faults; and rotations over 90° require three sets or more. One of the examples of table 1 was recognized by Freund (1970) in the Sistan, Iran, region, where two sets of strike-slip faults 40° to 45° apart were found to have accommodated the rotation of crustal blocks. Three sets may be involved in north-central Iceland (Young et al., 1985), and in the Snake Range, Nevada (Gans and Miller, 1983). Although the rotation in the Snake Range is associated with normal, not strike-slip, faulting, the model is similarly applicable. It is noteworthy that the angles between the sets as reported by Gans and Miller are again within the range of values permissible by the model. Proffett (1977) has similarly identified at least two multiple normal fault sets in the Yerington, Nevada, area with an angle between sets of around 40°.

216

The usefulness of the block tectonics approach has been evaluated by considering two areas-- the Mojave Desert, and the Lake Mead-Nevada fault zone--where comparisons between structurally inferred rotations, as derived from fault geometry and slip, and paleomagnetically derived rotations, as computed from some available declination measurements, can be made.

2.3. THE CENTRAL MOJAVE DOMAIN

The late Neogene crustal deformation in the Mojave Desert is predominantly accommodated by sets of NW-trending right-lateral strike-slip faults and E-W-trending left-lateral strike-slip faults (Figure 7) (Dibblee, 1961; Garfunkel, 1974; Dokka, 1983). On the basis of structural data, Garfunkel (1974) suggested that deformation in the Mojave is accommodated by contemporaneous right-lateral strike-slip faulting and counterclockwise (CCW) block rotation.

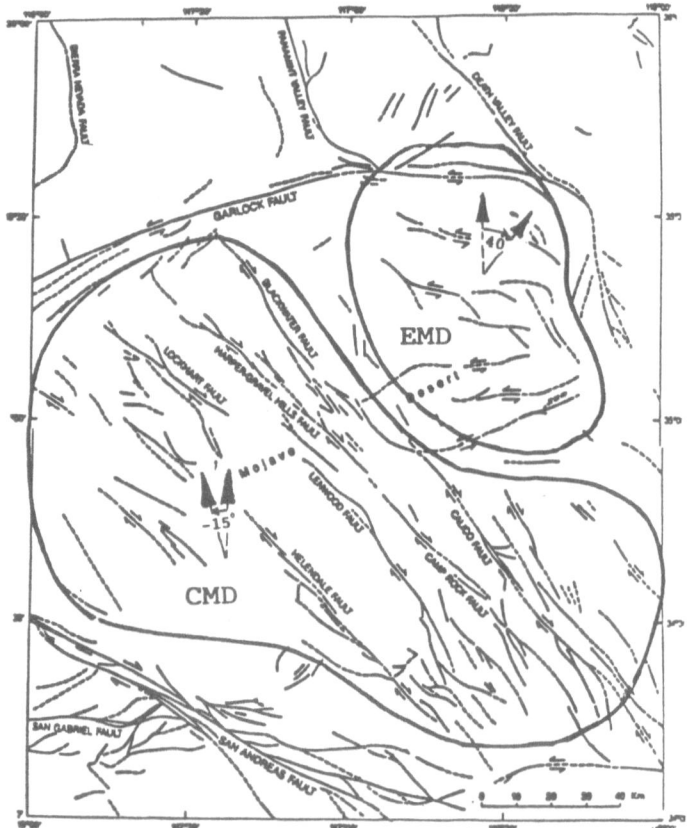

Fig. 7. Details of the central Mojave Desert (CDM) and eastern Mojave Desert (EMD). Each domain consists of roughly parallel fault sets of consistent sense of slip, which is expected to be directly associated with block and fault rotations: counterclockwise block rotation with right-lateral slip in the CMD, and clockwise block rotation with left-lateral slip in the EMD.

The anticipated rotation in the Central Mojave domain is counterclockwise. The magnitude of block rotation is estimated to be around 15° (CCW) using Dokka's (1983) fault slip values. Paleomagnetic data from this region (unpublished, Morton and Hillhouse, 1983) show about 15° CCW rotation of the Mojave blocks during the last 6 Ma.

Seismic data, active surface deformation, and geodetic data, may also be consistent with crustal deformation being accommodated by the block rotation model even at present. In 1975, the Galway Lake earthquake (M = 5.2) (Figure 8) was accompanied by a 6.8 km long surface rupture with

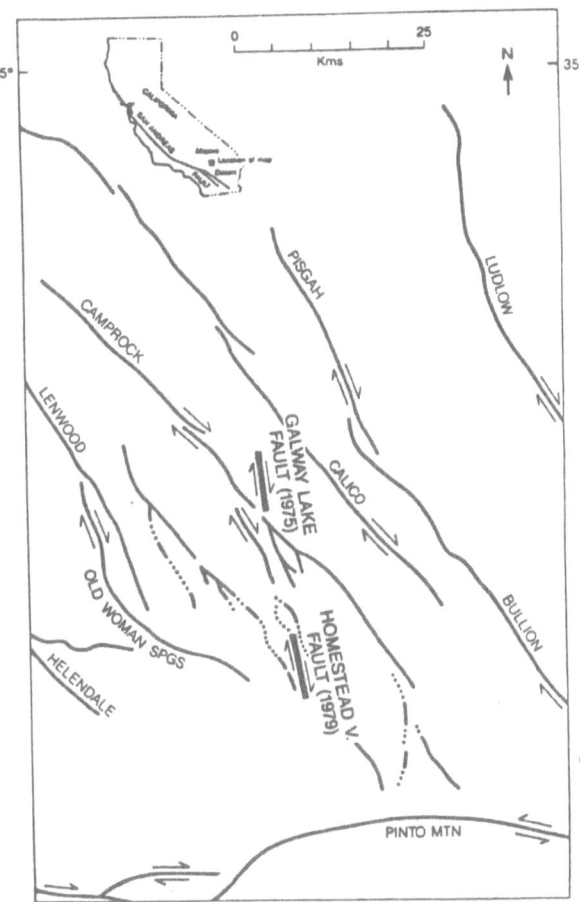

Fig. 8. The formation of a new fault set. The faulting directions associated with the 1975 Galway Lake earthquake and the 1979 Homestead Valley earthquake sequence. Note their azimuths which are distinctly different from the older, well-established central Mojave faults, and their alignement along a single trend. Shown also is the estimated direction of σ_{max}.

right-lateral displacement (Hill and Beeby, 1977) on a previously unmapped fault trending N20°W. Sauber et al. (1986) concluded that the preferred nodal planes for five out of nine recent earthquakes in this area are also oriented 20° to 35° away from the general strike of the Mojave faults, roughly in the direction of the Galway rupture direction, although they consider this direction of faulting to be "secondary." Furthermore, the fault plane solutions for the larger shocks of the 1979 Homestead Valley earthquake sequence (Hutton et al., 1980) show nodal planes trending in N4°W, again 30° to 40 ° away from the strike of the more developed, presumably older Mojave fault trend (Stein and Lisowski, 1983). It is remarkable that both ruptures together suggest a single fault line at least 25 km long.

We suggest that the direction of faulting in the central Mojave domain of the Galway Lake and Homestead Valley earthquakes might very well be part of a developing new fault set, which is gradually replacing the older, now rotated out of favor strike-slip faults here. This interpretation is consistent with the direction of the compressive stress in this area which is horizontal, at approximatley N30°E to N45°E as given by Zoback and Zoback (1980). Accordingly the older Mojave faults are thus unfavorably oriented relative to this stress direction, whereas the Galway Lake-Homestead Valley direction is mechanically quite favorable.

2.4. THE LAKE MEAD STRIKE-SLIP FAULT SYSTEM.

The Lake Mead fault system (LMFS) consists of a few major NE-trending left-lateral strike-slip faults. The accumulated offset across these faults is about 65 km, as obtained from offset of Late Neogene volcanic rocks (Anderson, 1973; Bohannon, 1979). Based on the ages of the displaced rocks of the Hamblin-Cleopatra volcano (Anderson et al., 1972) strike-slip faulting here post-dates 11 Ma.

A detailed study of the geometry and nature of faulting in the LMFS was carried out by Angelier et al. (1985), in the Hoover Dam area. They conclude that two sets of strike-slip faults are present: A set of major NE left-lateral faults, which are typically 50 to 150 km long, and a second set with faults trending roughly NW. The second set is divided into two fault groups: right-lateral strike-slip and normal dip-slip faults. The NW-trending right-lateral faults are typically 5 to 20 km long, and are roughly conjugate to the NE set.

Paleomagnetic data from the Hamblin-Cleopatra volcano were obtained (Ron et al., 1986) from 22 cooling units of volcanic rocks with K-Ar ages by Anderson et al. (1972) of 11.3 Ma and 12.7 Ma. The outward radial distribution of the lava flows' attitudes was used to control tectonic tilt. The paleomagnetic results reveal little if any rotations about the horizontal axis, corresponding to the lack of any vertical rotations associated with normal fault slip. In contrast, the declinations of the in situ field direction yielded large horizontal counterclockwise rotation of -29.4° ± 8.5° about the vertical axis. This counterclockwise sense of rotation is inconsistent, with the block rotation model with the left-lateral slip on the NE-trending major fault set. In contrast, this sense of rotation is to be expected if the associated slip took place on the shorter prevasive NW-trending right-lateral faults of the LMSZ. Our interpretation of the results in shown in Figure 9, in which the major NE fault coincides with the fixed boundary direction of this domain with internal block rotation associated only with the NW-trending faults.

Supporting evidence for this interpretation comes from the presently active faulting assuming active faulting is tectonically comparable with the late Miocene in the LMFS. According to Rogers and Lee (1976) the majority of eipicenters recorded are associated neither with the smaller NW-trending faults nor with the larger NE ones, but rather with younger and shorter, N-S-trending fault segments. Focal mechanisms for these earthquakes reveal right-lateral strike slip motion on these faults (Rogers and Lee, 1976).

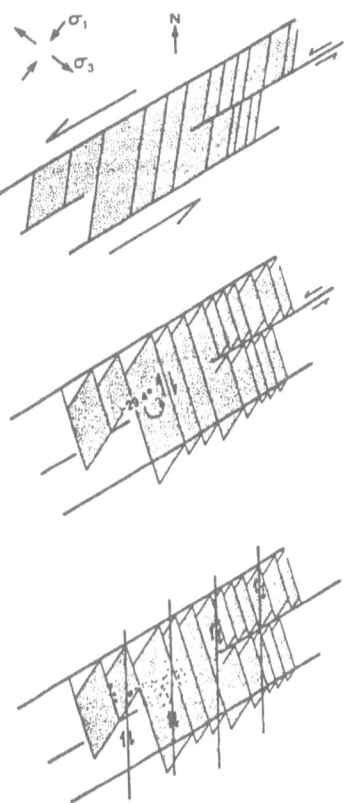

Fig. 9. Simplified geometric model for deformation of left-lateral Lake Mead fault system by right-lateral strike-slip faults and counterclockwise rotation. Direction of elongation is northwest so that the major NE-SW faults have *not* rotated. (a) Stage I. Slip on the major NE-trending faults; (b) Stage II. Slip and rotation on the shorter NW-trending fault; the original orientation of these faults, based on the measured rotation was N-S; (c) Stage III. Slip and seismicity on the active N-S-oriented faults. The slip was presumably transferred to these faults after the NW-trending set became locked due to their rotation.

Figure 9 presents a model for the fault geometry, the sense of horizontal slip, and the nature of block rotation in the LMFZ, beginning about 11 m.y. ago and still in progress now. Left-lateral shear in Miocene time caused local right-lateral horizontal displacements on a local set of faults initially trending N-S. These right-lateral displacements lead to an approximately 30° counterclockwise rotation of the blocks and their bounding faults, resulting in their present NW direction in which they became locked (Nur et al., 1986). Assuming that the stress orientation remained constant during the past 11 m.y. (Zoback and Zoback, 1980; Carr, 1984), subsequent deformation by right-lateral shear was accommodated by the newer set of more favorably oriented N-S faults which are active today.

This interpretation provides a link between the structural, paleomagnetic and seismic data and integrates them into a single continuous process of older and current crustal deformation by strike-slip and block rotation. Although the results are specific to the LMSZ, the model by analogy should be equally applicable to other Nevada strike-slip fault systems. Because this generalization could have profound impact on our understanding of the active tectonics of important portions of the Great Basin, much more work is needed in the future.

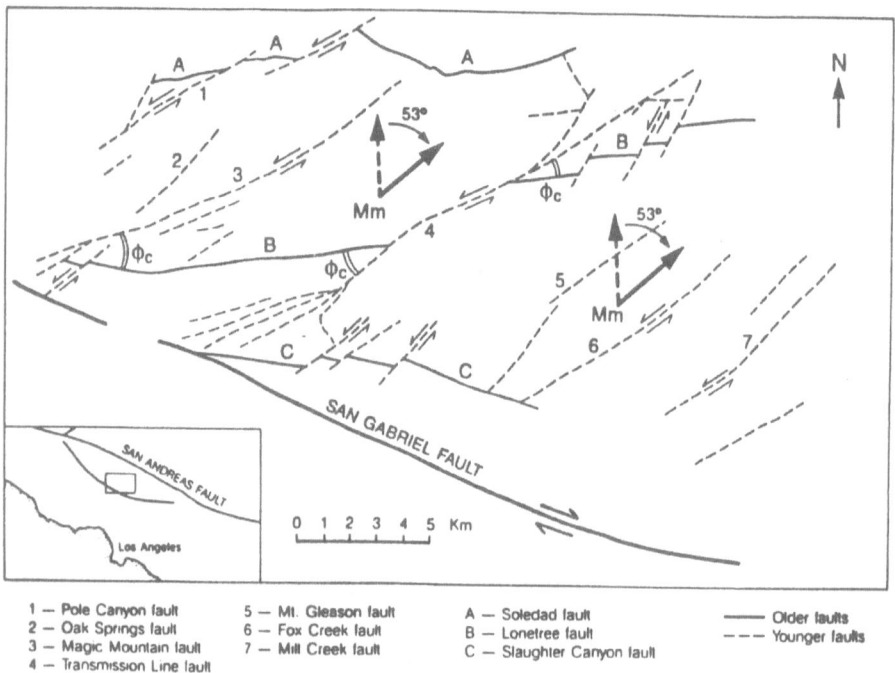

1 — Pole Canyon fault	5 — Mt. Gleason fault	A — Soledad fault
2 — Oak Springs fault	6 — Fox Creek fault	B — Lonetree fault
3 — Magic Mountain fault	7 — Mill Creek fault	C — Slaughter Canyon fault
4 — Transmission Line fault		

———— Older faults
– – – Younger faults

Fig. 10. Multiple strike-slip fault sets in the Gabriel Mountains (after Carter, 1982). Note the younger strike-slip NE-trending faults offsetting the older E-W-trending strike-slip. Both sets have the same left-handed sense of motion and are therefore not conjugate sets. The paleomagnetically determined clockwise rotation of 53° (Terres and Luyendyk, 1985) is consistent with the observed left-handed slip.

2.5. THE SAN GABRIEL MOUNTAIN DOMAIN.

The formation of multiple strike-slip fault sets, associated with clockwise rotation can explain some of the complexity of faulting in the San Gabriel Mountain domain (Figure 10). Dibblee (1961) long ago recognized that the set of subparallel left-handed strike-slip faults here imply clockwise rotation of both blocks and faults. Subsequent paleomagnetic results by Terres and Luyendyk (1985) revealed such rotation, with magnitudes reaching 50° to 55°. As suggested by the mechanical analysis of Nur et al. (1986) that such rotations may be larger than could be accommodated by a single fault set. As shown in Figure 10 (after Carter, 1982), there is good evidence for the presence here of an older strike-slip fault set, which is cut and offset by the younger set. According to the block rotation mechanics model, this older set, oriented now roughly NE, has rotated gradually clockwise, until it became locked due to the increasing normal stress and decreasing shear stress, with further deformation being accommodated by the new set of faults. The observed angles between the two sets is in the range of 30° to 40°, in good agreement with the predicted value of Nur et al. (1986).

3. Material rotation vs. stress field rotation

The kinematic interpretation of paleomagnetic declination anomalies in terms of rotation about vertical axes cannot be completed without structural information, as seen above specifically without the identification of the faults which accommodated the deformation, it is not possible to distinguish between terrane or microplate rotation to distributed fault and block rotation, as described here. However, as shown in the section on multiple sets, even when fault orientation and slip information are available, another ambiguity still exists, between the rotation of the rock mass itself and the rotation of the stress field which is presumably responsible for the creation and the slip on the faults involved. In Figure 11 we show an example from the Hoover Dam area, which tectonically is part of the LMFZ. On the basis of fault data collected and reported by Angelier et al. (1985), two least stress directions are inferred, an older one of N110°E and a younger direction N50°E. These directions can be interpreted to imply a 60° counterclockwise rotation of *stress field* in the Hoover Dam area. However this magnitude of rotation is inconsistent with the 30° stress field rotation derived by Zoback and Zoback (1980) for the entire basin and range province. Zoback and Zoback attribute this change in direction to a large-scale tectonic process, possibly the change in relative motion vectors between the Pacific and North American plates. It is therefore difficult to imagine how such plate motion could cause large local differences (from 30° to 60°) in stress field rotation. This inconsistency can be resolved when we include also the paleomagnetically derived block, or material rotation is described earlier, of about 30° CCW. By combining Zoback and Zoback's 30° stress field rotation with our 30° block rotation, we arrive at a total rotation of 60°, as inferred from fault data in the Hoover Dam area.

In Figure 11 we include also an illustration from Pavlides et al. (1988) in which a 30° rotation has been recognized. Without field data and knowledge about the stress field orientation, the interpretation of such results remains ambiguous. However, when paleomagnetic, structural and mechanical information is available, it becomes possible to make a detailed tectonic reconstruction of the history of crustal deformation.

222

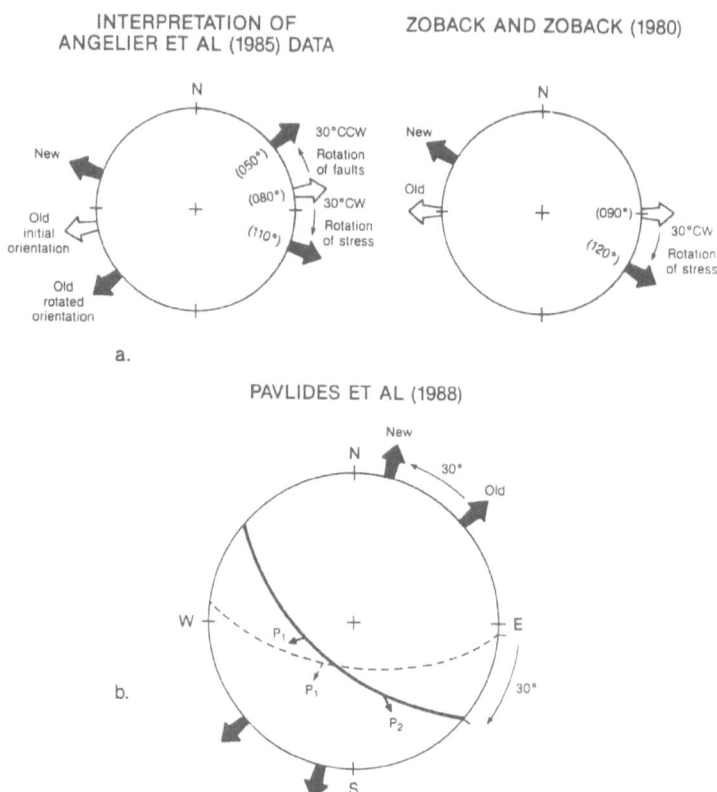

Fig. 11. Material rotation vs. stress field rotation: (a) Angelier's et al. (1985) data from the Hoover Dam, Nevada, area implies a 60° rotation of structural marker, which are probably due to ≈30° Basin and Range stress field rotation (Zoback and Zoback, 1980) *plus* ≈30° block rotation (Ron et al., 1986); (b) rotation inferred from structural and paleomagnetic data by Pavlides et al. (1988) for the Gmali fault system in Greece. The rotation here is probably a material rotation under constant direction stress field.

4. Relevance to geodynamics

4.1. SCALE DEPENDANCE AND MECHANICS.

The block rotation concept may be important for the interpretation of crustal deformation data e.g. in the Western United States, such as obtained from repeated geodetic measurements. The process of block rotation implies, for example, that strain, fault slip, and crustal deformation must generally be scale-dependent. On the scale of a few kilometers, deformation is controlled by fault spacing; on the scale of a few tens of kilometers, deformation is controlled by fault domains; and on a scale of a few hundreds of kilometeres, deformation is controlled by regional tectonics and plate motions. For example, a local domain can have left-lateral slip and clockwise rotation in an overall region with right-lateral and hence counterclockwise rotations. Furthermore faults must generally not be in directions of optimal failure and are therefore very ambiguous regional stress indicators.

In order to incorporate the roles of fault sets and domains in crustal deformation and evaluate their importance, their mechanical role must be understood in some detail. For example rock strength and frictional properties in two and three dimensions must be specifically considered in the formation of fault sets, domains, and domain boundaries. These properties may also control the formation of multiple fault sets of a given type (strike-slip, normal, thrust) or mixed types (strike-slip and thrust in the Transverse Ranges, strike-slip and normal in the Basin and Range). Consequently slip on a single fault, say during a single earthquake, such as the Galway Lake or the Homestead Valley events, does not reliably reveal the regional deformation field. Similarly, the deformation of a given domain does not mimic regional deformation. For example, the overall deformation of the Mojave region is the sum of the deformations of its separate domains. Clearly the strain of the central domain must differ markedly from the strain of the eastern domain.

4.2. STRESS-STRAIN RELATION FOR A SINGLE DOMAIN.

As slip on and rotation of the faults of a single domain set progress with time, the local shear stress required to strain the domain must increase, until, as shown earlier, it becomes more favorable to create a new set of faults. In the case of two-dimensional strike-slip faulting, analysis of Figure 4 yields the following relation between the shear stress τ required to cause fault slip, and block rotation

$$\tau = \tau_0 \frac{1}{\cos 2\phi} \tag{5}$$

where τ_0 is the shear stress for the initial, mechanically most favorable orientation of the fault set, and ϕ ($0 \leq \phi \leq \phi_c$) is the magnitude of block rotation.

The extensional strain ε associated with the stress τ is given by

$$\varepsilon = \frac{l - l_0}{l_0} = \frac{\sin \alpha - \sin(\alpha - \phi)}{\sin(\alpha - \phi)} \tag{6}$$

where l is the domain length, l_0 is the initial length (when rotation is $= \phi = 0$), and α is the initial orientation of the faults relative to the direction of shortening. This resulting stress-strain relation for a single domain can thus be used to relate local deformation to tectonic stress.

4.3. DEFORMATION OF A DOMAIN CLUSTER.

In Figures 12 we illustrate one of the problems which can arise in the interpretation of geodetic crustal deformation data when strike-slip-type block rotation is involved, especially when measurements of azimuth changes are concerned. Consider first two geodetic stations, A and B, within a single domain (Fig. 12a). As the domain deforms (Fig. 12a) each station undergoes rotation of the same sense, as would be the case for continuous uniform deformation within the domain. However if Station A is in one domain, and station B in a second (Fig. 12b), a major difference from the homogeneous case arises. Since station B, in the example rotates counterclockwise, whereas station A rotates clockwise. Consequently if the fact that the geodetic survey crosses a domain boundary, is not included in the analysis and the interpretation of the geodetic data, large and fundamental errors can be introduced. In the example here the two rotations would tend to cancel one another, leading to totally erroneous estimates of shear strain.

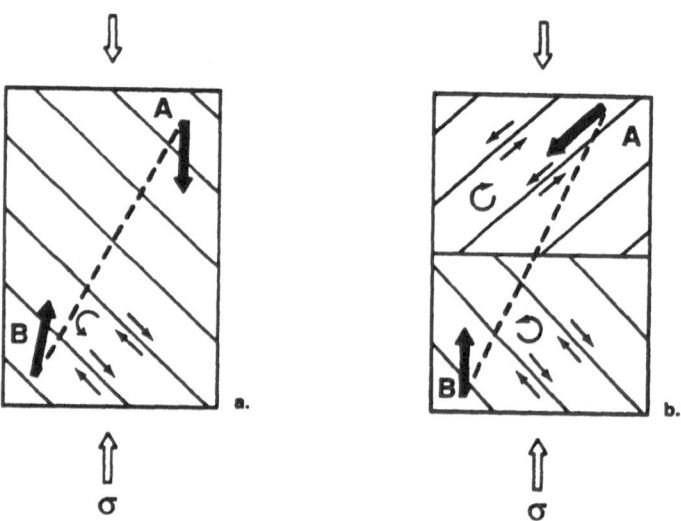

Fig.12. Sketch illustrating the influence of block rotation and domains on azimuthal changes, as determined from geodetic measurements. Heavy arrows indicate original lines connecting stations A and B. Light arrows indicate sense of fault slip.

5. Relevance to earthquake prediction.

Although the concept of block rotations is directly applicable to the understanding of distributed faulting, it is not immediately obvious whether and how it might be related to earthquake prediction. However two applications are worth exploring: the initiation and arrest of earthquake rupture, and the prediction of which fault will slip.

As suggested by a number of investigators, irregularities in the geometry of a fault or fault system may be the cause for either or both the initiation of earthquake ruptures and their arrest. Some models envision that rupture begins at a point of low strength or resistance to slip and is stopped at strong points, or barriers. Other models envision that rupture actually begins at strong points, and spreads into weak, or compliant regions where it is arrested. Regardless of the model, it is apparent that large heterogeneities of fault strength or normal and shear stresses are required, and that such hetereogeneities may be related to geometrical irregularities along faults and fault systems.

As pointed out by e.g. Freund and Merzer (1976) and Nicholson et al. (1986), the rotation of blocks within fault systems, and the rotation of the fault strands themselves may be directly related to the location and creation of these mechanical heterogeneities of the normal and shear stress along active faults. The study of the details of domains of block rotations and their development in time and space may consequently be significant for the understanding of both rupture initiation and stopping, and the prediction of sites for large future earthquakes.

The concept of blocks and their role may also be important in predicting the particular fault on which a future earthquake will occur. This question is of special interest because the most successful earthquake prediction to date--the Haicheng, China, earthquake of February 5, 1975-- involved the correct prediction of the time and the approximate location of the event, but failed to predict the fault on which this earthquake occurred. The epicenter instead of falling on the major through-going NW right-handed fault, actually fell on either a conjugate left slipping fault, or a fault in a conjugate set or domain. The stress required to rupture this fault may have been unusually high, due to the unfavorable orientation of the earthquake fault, so that the associated precursors were unusally strong.

A second example is the $M \approx 5.0$ Mount Lewis earthquake of 1985 east of the San Francisco Bay, California. Although the main strands of the San Andreas in this area, with trend NW-SE, and show right-handed slip, many active north-trending faults are present, possibly due to the complex of pull-aparts and pushups (Aydin and Page, 1984; Aydin and Nur, 1982). The Mt. Lewis earthquake fault plane solution indicates right-handed slip on one of these N-S-trending faults. The block rotation model (Fig. 3) provides an explanation for this puzzling earthquake, as an event on one of the faults in the N-S-trending strike-slip set of the east bay fault domain. This domain is bounded by the two major San Andreas strands, the Calaveras and Hayward faults.

6. Conclusion

The kinematics of block rotations by fault sets in domains, as first proposed by Freund (1970) provide an attractive model for understanding and analyzing crustal deformation when pervasive faulting is involved. The model is especially important because crustal deformation by pervasive faulting may very well be the rule.

Mechanical considerations of rock and fault strength and frictional properties suggest that new fault sets may form when older ones have rotated sufficiently away from the direction of optimal failure. Thus the combination of concept of block and fault rotation with fault mechanics provides a simple model which can explain rather complex fault patterns in situ.

The rotation of blocks and faults due to fault slip in a distributed fault system must be included in the analysis of geodetic and other crustal deformation data, because it can modify the interpretation of such data significantly. Finally, the mechanics of block and fault rotations cause spatial and temporal hetereogeneities at block and domain boundaries and may therefore be intrinsically linked with those heterogeneities of fault systems which are responsible for the initiation and arrest of rupture. This is true especially in distributed fault systems, where the prediction of an earthquake involves not only time and magnitude, but also the actual fault on which the event is to occur.

Acknowledgement

This study was supported by the Geodynamics Program of NASA, through grant no. NAG5-926.

References

Anderson, E.M., 1951, The dynamics of faulting: Oliver and Boyd, Edinburgh, 206 p.

Anderson, R.E., 1973, Large-magnitude late Tertiary strike-slip faulting north of Lake Mead, Nevada: *U.S. Geological Survey Professional* Paper 794.

Anderson, R.E., Longwell, R.C., Armstrong, R.L., and Marvin, R.F., 1972, Significance of K-Ar ages of Tertiary rocks from the Lake Mead region, Nevada-Arizona: *Geological Society of America Bulletin*, v. **83**, p. 273-288.

Angelier, J., Collette, B., and Anderson, E.R., 1985, Neogene paleostress changes in the Basin and Range: a case study at Hoover Dam, Nevada-Arizona: *Geological Society of America Bulletin*, v. **96**, p. 347-361.

Aydin, A., and Nur, A., 1982, Evolution of pull apart basins and their scale independence: *Tectonics*, v. **1**, no. 10, p. 91-105.

Aydin, A., and Page B., 1984, Diverse Pliocene-Quaternary tectonics in a transform evironment: San Francisco Bay region, California: *Geological Society of America Bulletin*, v. **95**, p. 1303-1317.

Billings, M.P., 1972, Structural Geology: Prentice-Hall, Englewood Cliff, N.J., 606 p.

Bohannon, R.G., 1979, Strike-slip faults of the Lake Mead region of southern Nevada, in Armentrout, Cole and TerBest, eds., *Cenozoic Paleogeography of the Western United States: Pacific Section, Society of Economic Paleontologists and Mineralogists*, p. 129-139.

Brown, L.L., and Golombek, M.P., 1985, Tectonic rotation within the Rio Grande rift: evidence from paleomagnetic studies: *Journal of Geophysical Research*, v. **90**, no. 1, p. 790-802.

Carr, W.J., 1984, Regional structural setting of Yucca Mountain, southwestern Nevada, and Late Cenozoic rates of tectonic activity in part of the southwestern Great Basin, Nevada and California: *U.S. Geological Survey* Open-File Report 84-854.

Carter, B.A., 1982, Geology and structural setting of the San Gabriel Mountains, Los Angeles County, California, in Geologic Excursions in the Transverse Ranges, a Guidebook, *Geological Society of America Cordilleran* Section, 78th Annual Meeting.

Dibblee, T.W., Jr., 1961, Evidence of strike-slip movement on northwest-trending faults in the western Mojave Desert, California: *U.S. Geological Survey* Professional Paper 424-B, p. B197-B199.

Dokka, R.K., 1983, Displacements on late Cenozoic strike-slip faults of the central Mojave Desert, California: *Geology*, v. **11**, p. 305-308.

Fagin, S.W., and Gose, W.A., 1983, Paleomagnetic data from the Redding section of the Eastern Klamath belt, northern California: *Geology*, v. **11**, p. 505-508.

Frei, L., 1980, Junction of two sets of strike slip faults in southeastern Sinai (in Hebrew with English abstract): M.Sc., thesis, Hebrew University, Jerusalem.

227

Freund, R., 1970a, Rotation of strike-slip faults in Sistan, southeastern Iran: *Journal of Geology*, v. **78**, p. 188-200.

Freund, R., 1970b, The geometry of faulting in the Galilee: *Israel Journal Earth Sciences*, v. **19**, p. 117-140.

Freund, R., 1971, The Hope fault, a strike slip fault in New Zealand: *New Zealand Geological Survey Bulletin*, v. **86**, 49 p.

Freund, R. 1974, Kinematics of transform and transcurrent faults: *Tectonophysics*, v. **21**, p. 93-134.

Freund, R., and Merzer, A.M., 1976, The formation of rift valleys and their zigzag fault patterns: *Geology Magazine*, v. **113**, no. 6, p. 561-568.

Gans, P.B., and Miller, 1983, Style of mid-Tertiary extension in east-central Nevada, in Guidebook, pt. 1, *Geological Society of America Rocky Mountain and Cordilleran Sections Meeting, Utah Geology and Mining Survey Special Studies*, v. **59**, p. 107-160.

Garfunkel, Z., 1970, The tectonics of the Western margins of the southern Arava: Ph.D. dissertation, Hebrew University, Jerusalem, 104 p.

Garfunkel, Z., 1974, Model for the late Cenozoic tectonic history of the Mojave Desert, California and for its relation to adjacent areas: *Geological Society of America Bulletin*, v. **85**, p. 1931-1944.

Greenhaus, M.R., and Cox, A., 1979, Paleomagnetism of Morro Rock-Islay Hill complex as evidence for crustal block rotation in central coastal California: *Journal of Geophysical Research*, v. **85**, no. 5, p. 2393-2400.

Hill, R.L., and Beeby, D.J., 1977, Surface faulting associated with the 5.2 magnitude Galway Lake earthquake of May 31, 1975: Mojave Desert, San Bernardino County, California: *Geological Society of America Bulletin*, v. **88**, p. 1378-1384.

Hoeppener, R., Kalthoff, E., and Schrader, P., 1969, Zur physikalischen Tektonik: Bruchbildung bei verschiedenen Deformationen im Experiment: *Geol. Rundsch.*, v. **59**, p. 179-193.

Hornafius, J.R., 1985, Neogene tectonic rotation of the Santa Ynez Range, western Transverse Ranges, California, suggested by paleomagnetic investigation of the Monterey Formation: *Journal of Geophysical Research*, v. **90**, no. 14, p. 12503-12522.

Hudson, M.R., and Geissman, W.J., 1985, Middle Miocene counterclockwise rotation of rocks from west-central Nevada: implication for Basin and Range extention: *Geological Society of America*, Abstracts and Program, Annual Meeting, Orlando, Florida, no. 59706, p. 615.

Hutton, L.K., Johnson, C.E., Pechmann, J.C., Ebel, J.E., Given, T.W., Cole, D.M., and German, P.T., 1979, Epicentral locations for the Homestead Valley earthquake sequences: California *Geology*, v. **33**, p. 110-164.

Kamerling, M.J., and Luyendyk, B.P., 1985, Paleomagnetism and Neogene tectonics of the Northern Channel Islands, California: *Journal of Geophysical Research*, v. **90**, no. 14, p. 12485-12502.

Luyendyk, B.P., Kamerling, M.J., and Terres, R., 1980, Geometric model for Neogene crustal rotations in southern California: *Geological Society of America Bulletin*, v. **91**, p. 211-217.

MacDonald, D.W., 1980, Net tectonic rotation, apparent tectonic rotation, and the structural tilt correction in paleomagnetic studies: *Journal of Geophysical Research*, v. **85**, p. 3659-3669.

Morton, J.L., and Hillhouse, J.W., 1983, Paleomagnetism and K-Ar ages of Miocene basaltic rocks in the western Mojave Desert, California: Unpublished manuscript.

Nadai, 1950, Theory of Fracture and Flow of Solids: McGraw-Hill, Longon, v. 1.

Nicholson, C., Seeber, L., Williams, P.L., and Sykes, L.R., 1986, Seismic deformation along the southern San Andreas fault, California: *Journal of Geophysical Research*, in press.

Nur, A., and Helsley, C.E., 1971, Paleomagnetism of Tertiary and recent lavas of Israel: *Earth and Planetary Science Letters*, v. **10**, p. 375-379.

Nur, A., Ron, H., and Scotti, O., 1986, Fault mechanics and the kinematics of block rotation: *Geology*, v. **14**, p. 746-749.

Pavlides, S.B., Kondopoulou, D.P., Kilias, A.A., and Westphal, M., 1988, Complex rotational deformations in the Sebro-Macedonian massif (North Greece): Structural and paleomagnetic evidence: *Tectonophysics*, v. **145**, p. 329-335.

Proffett, J.M., Jr., 1977, Cenozoic geology of the Yerington district, Nevada, and implication for the nature and origin of Basin and Range faulting: *Geological Society of America Bulletin*, v. **88**, p. 247-266.

Rogers, A.M., and Lee, W.H.K., 1976, Seismic study of earthquakes in the Lake Mead, Nevada-Arizona region: *Seismological Society of America Bulletin*, v. **66**, no. 5, p. 1631-1657.

Ron, H., Aydin, A., and Nur, A., 1986, The role of strike-slip faulting in the Late Cenozoic deformation of the Basin and range Province: *Geology*, in press.

Ron, H., Freund, R., Garfunkel, Z., and Nur, A., 1984, Block rotation by strike-slip faulting: Structural and paleomagnetic evidence: *Journal of Geophysical Research*, v. **89**, no. B7, p. 6256-6270.

Sauber, J., Thatcher, W., and Solomon, S.C., 1986, Geodetic measurement of deformation in the Central Mojave Desert, California: *Journal of Geophysical Research*.

Stein, R.S., and Lisowski, M., 1983, The 1979 Homestead Valley earthquake sequence. California: Control of aftershocks and postseismic deformation: *Journal of Geophysical Research*, v. **88**, no. B8, p. 6477-6490.

Terres, R.R., and Luyendyk, P.B., 1985, Neogene tectonic rotation of the San Gabriel region, California, suggested by paleomagnetic vectors: *Journal of Geophysical Research*, v. **90**, no. B7, p. 6256-6270.

Terres, R.R., and Sylvester, A.O., 1981, Kinematic analysis of rotated fractures and blocks in simple shear: *Seismological Society of America Bulletin*, v. **71**, no. 5, p. 1593-1605.

Wells, R.E., and Coe, R.S., 1985, Paleomagnetism and geology of Eocene volcanic rocks of southwest Washington, implication for mechanisms of tectonic rotation: *Journal of Geophysical Research*, v. **90**, no. 2, p. 1925-1947.

Young, K.D., Jancin, M., Vight, B., and Orkan, N.I., 1985, Tranform deformation of Tertiary rocks along the Tjornes fracture zone, north-central Iceland: *Journal of Geophysical Research*, v. **90**, no. 12, p 9986-10010.

Zoback, M.L. and Zoback, M.D., 1980, Faulting pattern in north-central Nevada and strength of the crust: *Journal of Geophysical Research*, v. **85**, no. B1, p. 275-284.

CRUSTAL ROTATION AND FAULT SLIP IN THE CONTINENTAL TRANSFORM ZONE IN SOUTHERN CALIFORNIA

B. P. LUYENDYK
Institute for Crustal Studies and Department of Geological Sciences,
University of California,
Santa Barbara, California 93106
U.S.A.

ABSTRACT. Computations from global plate circuits show that since the end of Oligocene time (24 Ma) the Pacific plate has moved northwestward relative to southern California 1072 ± 122 km (J. Stock and P. Molnar, 1988). During part of this period displacement across the San Andreas fault was 315 ± 15 km (5% error?). Paleomagnetic data infer that the Transverse Ranges crustal blocks rotated clockwise since 16 Ma, about 86 degrees west of the fault and 41 degrees on the east. These rotations were simultaneous with NW-SE dextral shear distributed on faults in non-rotating crust north and south of the Transverse Ranges. This shear is computed to be 265 ± 38 km west of the San Andreas and 110 ± 11 km east of it. The sum of the distributed shear and San Andreas slip is 690 ± 42 km. This sum is 382 ± 129 km less than the displacement between the plates. The difference is significant, and the missing displacement should be looked for in the eastern Mojave Desert (east of the Granite Mountain fault) and west of the western Transverse Ranges (west of the Hosgri fault). Because the timing of the rotations and slip on the San Andreas are also known, the partitioning of displacements across southern California can be described as a function of time. During Early Miocene time 45 km of dextral slip occurred on the northern San Andreas fault. An area of extended crust in the Mojave Desert also rotated 50 degrees clockwise during this time. During Middle Miocene time most of the shear occurred north and south of the western Transverse Ranges as they rotated almost 60 degrees. During this same interval the Tehachapi Mountains rotated clockwise about 41 degrees and the northern San Andreas fault may have slipped 30 km. During Late Miocene time the rotation rate decreased in the west. The northern San Andreas fault slipped about 55 km which may have been transferred to the San Gabriel fault and faults in the southern California borderland. Since Late Miocene time the western Transverse Ranges rotated 30 to 40 degrees and the San Andreas slipped 185 km north of the Transverse Ranges and 240 km in southern California. The eastern Transverse Ranges and portions of the central Mojave Desert rotated 41 degrees clockwise, presumably at the same time.

1. Introduction

Estimates of the amount and timing of slip on northwest-southeast trending Neogene dextral faults in southern California vary widely, even for the San Andreas fault. These estimates are based in part on field relations of piercing points in the best cases. Mostly the estimates are poorly constrained in both amount and timing.

Neogene dextral slip in southern California was accompanied by clockwise crustal rotation of large crustal domains (Luyendyk et al., 1980, 1985). These rotated domains are located within the broad simple shear zone between the Pacific and North American plates. In most cases the amount

229

C. Kissel and C. Laj (eds.), Paleomagnetic Rotations and Continental Deformation, 229–246.

of rotation and its error, and the timing of the rotation, are fairly well known. Additionally, the rotated domains extend across the shear zone and serve as markers which indicate the amount of dextral shear in the zones (Figure 1). By assuming rigidity of the domains during rotation and crustal dilation, the amount of distributed dextral shear and its error can be computed exactly (Hornafius, 1985; Luyendyk and Hornafius, 1987). Although the crust does not behave in precisely this fashion, this hypothesis permits upper bounds on distributed shear to be computed. Using the same assumptions, if it is known which faults in the shear zone were active during the rotation, the amount of slip on individual faults can be computed exactly.

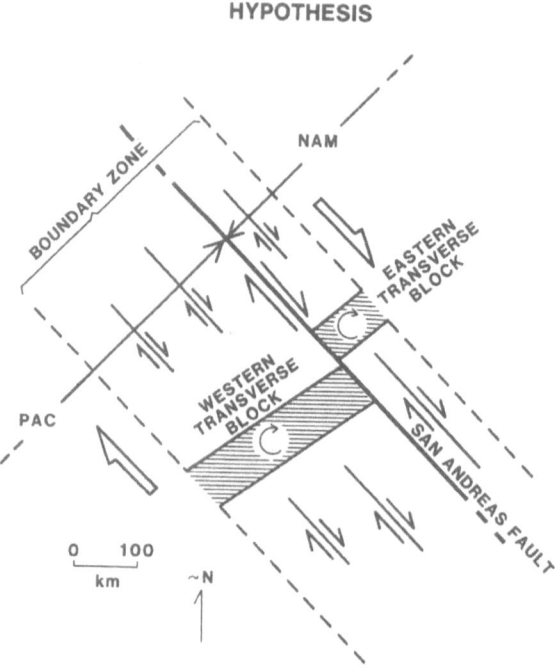

Figure 1. Representation of crustal rotations within the Pacific-North American plate boundary zone. The amount of rotation of the eastern and western Transverse Ranges indicates the amount of distributed shear across the zone.

2. Facts concerning deformation

The following facts on the amounts of crustal rotation and slip on the San Andreas fault, and the timing of these events were used to compute the distributed shear in southern California for Neogene and Quaternary time.

Near the end of Oligocene time large areas of southern California and Arizona crust were being extended, in a precursor to later Basin-and-Range extension. These include areas of the central Mojave Desert (Dokka, 1986), southeast California (Frost and Martin, 1982), and the central

Transverse Ranges (Bohannon, 1975). The Tehachapi Mountains region may have rotated about 20 degrees clockwise since 80 Ma and before Miocene time (Figure 2; McWilliams and Li, 1985).

BEGINNING MIOCENE
(Ca. 24 Ma)

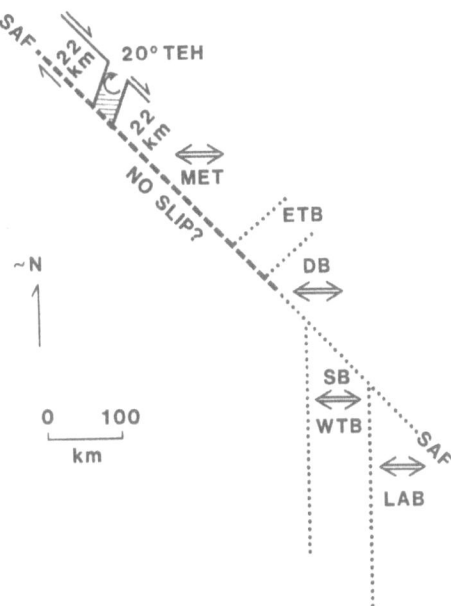

Figure 2. Tectonic elements in place at the beginning of Miocene time. Arrows indicate extension; circles rotation (or possible rotation?) occurring in pre-Miocene time. Twenty degrees of pre-Miocene clockwise rotation in the Tehachapi Mountains (TEH) causes 22 km of distributed shear to its north and south. The San Andreas fault (SAF) did not exist. ETR = eastern Transverse Ranges; WTR = western Transverse Ranges; MET = Mojave extensional terranes; DB = Diligencia Basin; SB = Soledad Basin; LAB = Los Angeles Basin.

During Early Miocene time (Figure 3) over 50 degrees and up to 120 degrees clockwise rotation occurred in the Central Mojave Desert Extensional Corridor (Ross, et al., 1987; 1988). Possibly at the same time, areas in the western Mojave (Golombeck and Brown, 1988) rotated about 40 to 50 degrees clockwise. On the northern San Andreas fault 45 km of right slip took place (Stanley, 1987). This slip stepped south to the California borderland and caused extension in the Los Angeles basin and offshore regions (LAB, Figure 3).

232

EARLY MIOCENE
(24 ⟶ 16)

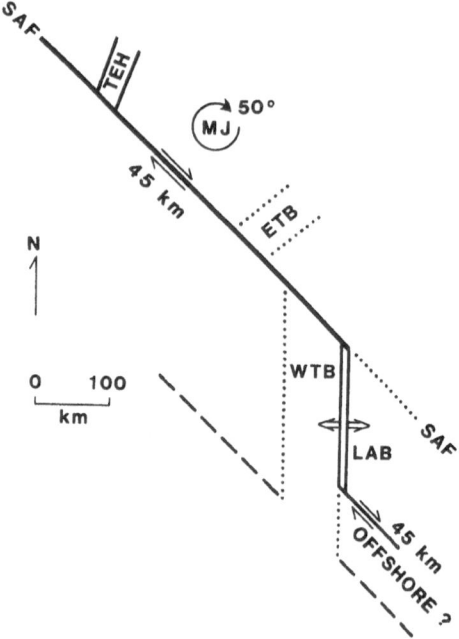

Figure 3. Geometry of the Middle Miocene tectonics in southern California. Slip on the northern San Andreas fault steps offshore and the Los Angeles Basin originates as a pull apart (Crowell, 1974).

During Middle Miocene time (Figure 4) the western Transverse Ranges and the San Gabriel Mountains (and possibly the Diligencia Basin) rotated clockwise 50 to 60 degrees (Crowell, 1975; Hornafius et al., 1986; Terres, 1984). About 40 degrees of additional clockwise rotation occurred in the Tehachapi Mountains. The northern San Andreas fault probably slipped 30 km during this time (Stanley, 1987). Also, right slip on the Fenner - San Francisquito - Clemens Well faults offset the Diligencia Basin in the Eastern Transverse Ranges and the Soledad Basin in the San Gabriel Mountains block during Middle Miocene time (Terres, 1984).

During Late Miocene time rotation slowed or stopped west of the northern San Andreas fault, but dextral slip on the fault continued and may have amounted to as much as 55 km (Hornafius et al., 1986) or as little as 15 km (Stanley, 1987). The San Gabriel fault initiated during this interval (Crowell, 1975; 1982) as did the Garlock fault (Burbank and Whistler, 1987). Slip on the San Andreas passed through to the San Gabriel and offshore faults. Additionally, the northwest-southeast fault pattern now present in the Mojave Desert was formed (Dokka, 1987).

MIDDLE MIOCENE
(16→11)

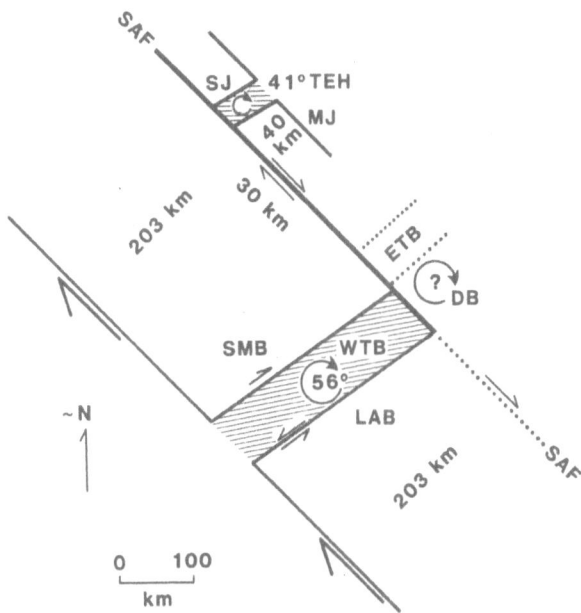

Figure 4. Geometry of the Middle Miocene tectonics in southern California. The western Transverse Ranges have rotated about 56 degrees clockwise and cause 203 km of distributed dextral shear to their north and south.

Following the Miocene, in the Pliocene and Quaternary (Figure 5), the eastern Transverse Ranges (Carter et al., 1987) and portions of the central Mojave Desert (Ross et al., 1987, 1988) rotated about 40 degrees clockwise. A counterclockwise rotation of about 15 degrees occurred in the western Mojave Desert on blocks bounded by northwest-southeast trending faults (Burke et al., 1982; Luyendyk et al., 1985). This rotation bent the San Andreas fault trace counterclockwise and also rotated the San Gabriel Mountains as they moved north into the bend. Following the Miocene and until present day the western Transverse Ranges rotated clockwise 30 to 40 degrees (Hornafius et al., 1986). The southern San Andreas initiated and slipped up to 240 km (Stanley, 1987; Ehlig et al., 1975; Crowell, 1962).

Major uncertainties still remain concerning the timing, amounts and locations of the crustal rotations. In particular, the timing and amounts of rotation in the eastern Transverse Ranges, Mojave Desert, and Tehachapi Mountains are unclear. Presumably these should be closely tied as these areas are geometrically connected. In the Mojave it is known that the clockwise rotations are Early

234

Miocene and in the eastern Transverse Ranges the rotation is post 10 Ma (Carter et al., 1987). Recent K-Ar dating suggests this rotation is post 6 Ma. Some evidence suggests that the central Mojave region may have rotated twice; 50 degrees clockwise in Early Miocene time and another 40 degrees in Late Miocene time (Ross et al., 1988). In the Tehachapi Mountains one-third of the rotation is pre-Early Miocene and the remainder is post-Early Miocene (Plescia et al., 1987). I have assumed that this rotation is Middle Miocene age because this is the time of the major deepening of the San Joaquin basin immediately north of the mountains (Crowell, 1987).

During the entire Neogene regions which underwent very little or no rotation and which absorbed dextral shear not taken up by the San Andreas fault, include the Coast Ranges and Salinian block north of the western Transverse Ranges and San Gabriel Mountains, the California Borderland south of the northern Channel Islands, the Peninsula Ranges in southern California, and southeast California east of the San Andreas fault and south of the eastern Transverse Ranges.

LATE MIOCENE - PRESENT

(11→0)

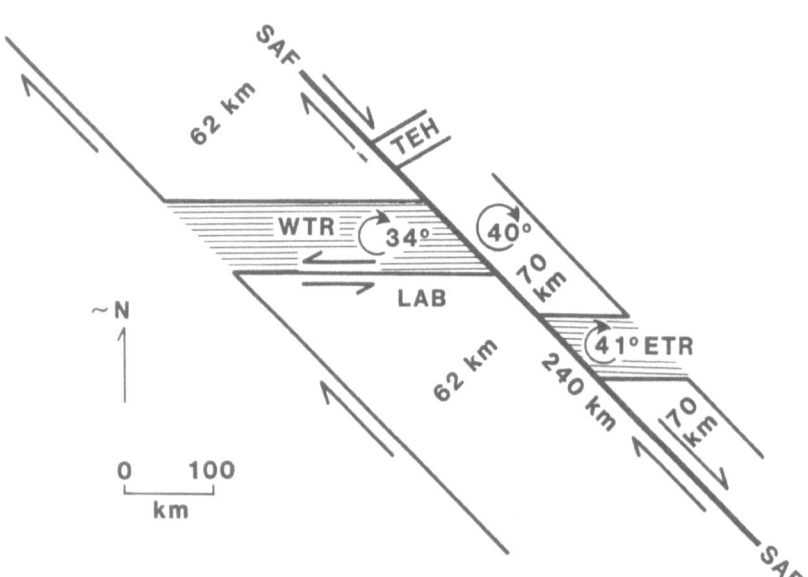

Figure 5. Present day geometry showing tectonic events in the period after 10 Ma. These included clockwise rotation of the western and eastern Transverse Ranges, breakthrough of the San Andreas fault into southern California, and counterclockwise rotations in most areas of the western Mojave Desert.

3. Amounts of dextral shear

The total shear plus slip value due to the rotation and the San Andreas offset for Miocene and later time is 690 ± 42 km (Table I). This value can be compared to estimates of Pacific - North American plate relative displacements for the same period. Stock and Molnar (1988) computed these displacements for a global plate circuit. They did this for sea floor spreading magnetic anomalies 3 (5.5 Ma), 5 (10.59 Ma), 6 (19.9 Ma) and 7 (25.82 Ma). A value of 1072 ± 22 km can be interpolated for the period 24 to 0 Ma, which is 382 ± 129 km more than the displacements computed for southern California due to rotation and slip on the San Andreas (Table II).This difference is significant.

SOUTHERN CALIFORNIA DEXTRAL SHEAR

Early Miocene (24-16 Ma) *Northern San Andreas fault[1] *Central Mojave extensional complex rotation 50 deg CW	45 km, ± 2? ?
Middle Miocene (16-11Ma) *Northern San Andreas fault[1] *Western Transverse Ranges rotation 56 deg CW *Tehachapi rotation 41 deg CW	30 km, ± 2? 203 ± 32 km 40 ± 05 km 273 ± 32 km
post Middle Miocene (11-0Ma) *San Andreas fault[1] *Western Transverse Ranges rotation 30 deg CW *Eastern Transverse Ranges rotation 41 deg CW (includes Central Mojave Block rotation)	240 km, ± 12? 62 ± 20 km 70 ± 10 km 372 ± 25 km
Total Miocene and younger shear *San Andreas fault[1] *distributed shear via rotation *sum	315 km, ± 15?[1] 375 ± 39 km 690 ± 42 km

1 Stanley, 1987

Table I. Partitioning of dextral shear in southern California into time intervals and between the San Andreas and distributed shear computed from crustal rotations

The total slip on the San Andreas fault during Miocene and later time is 330 km ± 15 (?) (Table I; Stanley, 1987) north of the juncture of the San Andreas and the San Juan-Chimineas (Figure 6) and about 315 km to the south. The distributed shear computed as due to rotation of the western Transverse Ranges and the eastern Transverse Ranges is 375 ± 39 km. These rotations are presumed to include post-Miocene rotations in the Mojave Desert which are within the limits of the shear zone defined by the width of the eastern Transverse Ranges. Slip on the San Gabriel fault is included in this figure as it may represent internal deformation of the western Transverse Ranges during rotation (see below) or slip relayed from the northern San Andreas to the California borderland. The distributed shear is computed using the method of Hornafius (1985) which assumes the shear zone expands in width during rotation to accommodate the length of the rotating blocks. The distributed shear is partitioned into 265 ± 38 km west of the San Andreas of which 203 ± 32 km occurred during Middle Miocene time and 62 ± 20 km occurred later. East of the San Andreas 70 ± 10 km occurred after the Middle Miocene and 40 ± 5 km during this time. This distributed shear is a minimum figure for Miocene time if in fact Early Miocene rotation occurred east of the San Andreas fault in the Mojave extensional terranes.

SOUTHERN CALIFORNIA CUMULATIVE DISPLACEMENTS

CONTINENT	PAC/NAM (STOCK & MOLNAR)	DIFFERENCE (PAC/NAM - CONTINENT)
post Miocene (0-6 Ma)		
312 ± 24 km	$325^1 \pm 45$ km	13 ± 51 km
post Middle Miocene (11- 0 Ma)		
372 ± 25 km	$635^2 \pm 70$ km	263 ± 79 km
post Early Miocene (16 - 0 Ma)		
645 ± 41 km	$774^3 \pm 70$ km	129 ± 81 km
post Oligocene (24 - 0 Ma)		
690 ± 42 km	$1072^4 \pm 122$ km	382 ± 129 km

1 at 5.5 Ma (anomaly 3)
2 at 10.59 Ma (anomaly 5)
3 interpolated between 19.0 (anomaly 6) and 10.59 Ma
4 interpolated between 25.82 (anomaly 7) 19.9 and 10.59 Ma

Table II. Cumulative displacements in southern California comparing the data from Table I with Pacific-North American plate separations computed by Stock and Molnar (1988).

The 382 ± 129 km of missing displacement must be searched for west of the Transverse Ranges, at the continental slope on faults like the Santa Lucia Bank fault zone, north and south of the Early Miocene Mojave extensional terranes, and/or east of the eastern Transverse Ranges; for example on the Furnace Creek - Death Valley - Granite Mountain fault system or to its east. This difference points out that the Pacific-North American plate boundary zone has been greater than 450 km wide during Neogene time.

To the best of my knowledge at present, dextral shear in southern California occurred in two intervals during Neogene time; during the Middle Miocene when the major rotation event took place, and during post-Miocene time when rotation of the Transverse Ranges and strike-slip on the San Andreas fault occurred. This is in contrast to a rather steady plate displacement at

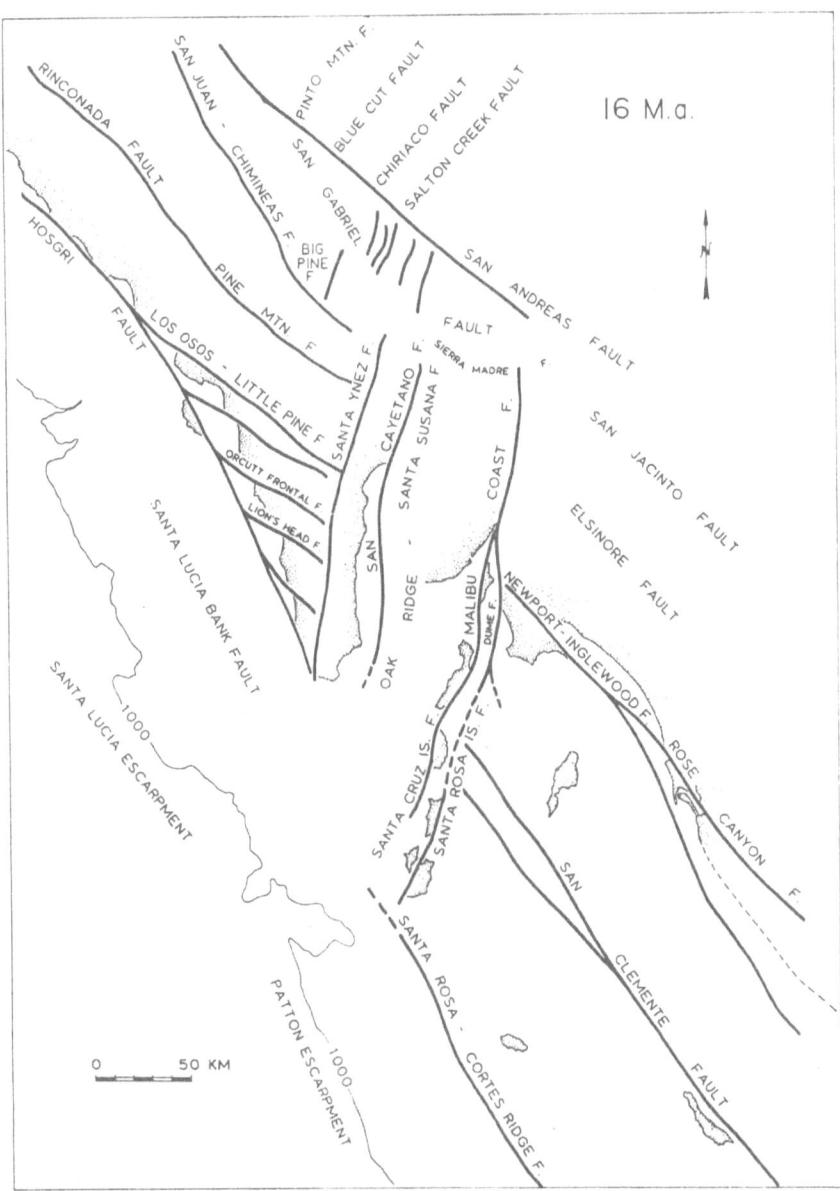

Figure 6. Palinspastic reconstruction for southern California at 16 Ma. Presently day shorelines are shown for reference (after Hornafius et al., 1986). Some fault names are shown near the location where they will appear in subsequent geologic time.

45 to 60 mm/yr during the same interval (Stock and Molnar, 1988). The data in Table II infer that most of the discrepancy between continent and plate displacement is prior to 6 Ma. Therefore, the plate boundary system on the continent is well described after 6 Ma but not before. During the Early Miocene and also during the Late Miocene, significant plate displacement occurred which is not obviously reflected on the continent. This includes about 250 km of missing displacement in each interval.

4. History of fault movements

Given the estimates of distributed shear from the rotations, the next questions to be asked are which faults accommodated the shear and how much of it? Field observations reveal which faults were active when; the amount of slip on the fault can be computed if the width of the block it bounds is known, the rotation is known, and the pre-rotation angle between the fault and the rotating block is known (Luyendyk et al., 1980; Luyendyk and Hornafius, 1987). Although the geometrical parameters are fairly well known, the ages of individual faults in the Pacific-North American plate boundary zone are poorly constrained.

At the end of Early Miocene time (16 Ma; Figure 6), prior to the major rotation event, it is known that the southern San Andreas fault plus the San Jacinto and Elsinore faults in southern California did not exist (Crowell, 1981; 1987; Crowell and Ramirez, 1979). Neither did the Oak Ridge - Santa Susana (or Malibu Coast?) faults exist in the Transverse Ranges (Yeats, 1983). The presently east-west trending faults in the eastern Transverse Ranges, such as the Pinto Mountain, Blue Cut and Chiriaco faults did not exist (Dibblee, 1975; Crowell and Ramirez, 1979). North of the western Transverse Ranges little is known of the age of the faults and the same can be said for faults in the borderland to the south (Howell and Vedder, 1981). Also, 45 km of Early Miocene right slip on the northern San Andreas fault (Stanley, 1987) had to be absorbed in a southward step to faults in the southern California borderland. This releasing step produced 45 km of extension in the region of the Los Angeles Basin. It is not clear which faults participated in this transfer or whether the San Andreas adjacent to the San Gabriel Mountains actually slipped.

By the end of Middle Miocene time (11 Ma) 203 ± 32 km of shear had to be distributed on faults north and south of the western Transverse Ranges to accommodate about 56 degrees of rotation (Figure 7; Table I). These faults probably included the Newport - Inglewood, San Clemente, and unknown offshore faults south of the western Transverse Ranges, and the Rinconada, Nacimiento, Hosgri, and unknown offshore faults north of the Transverse Ranges. South of the western Transverse Ranges slip was apparently equally distributed between three faults because each fault bounds a block of equal width. Therefore, the Newport - Inglewood, San Clemente, and Santa Rosa - Cortes fault zones may have slipped 68 km each during Middle Miocene time. If the Tehachapi Mountains block rotated 41 degrees during this period, then about 40 ± 5 km of dextral shear needs to be accommodated on unknown faults north and south of these mountains. The geometry of the rotation does not require a southern branch of the San Andreas fault south of the San Gabriel Mountains (Figure 8). Also, the slip on the northern San Andreas can be seen to be strictly a result of the rotation (Figure 8). Therefore, this slip figure should not be added to the total distributed shear (Table I) because the San Andreas slip is included in the shear due to the rotation. Figure 8 also shows that large gaps or basins may have formed as a result of the rotation event, such as the Los Angeles, Santa Monica, and Santa Maria Basins adjacent to the western Transverse Ranges. The San Joaquin Basin deepened radically due to

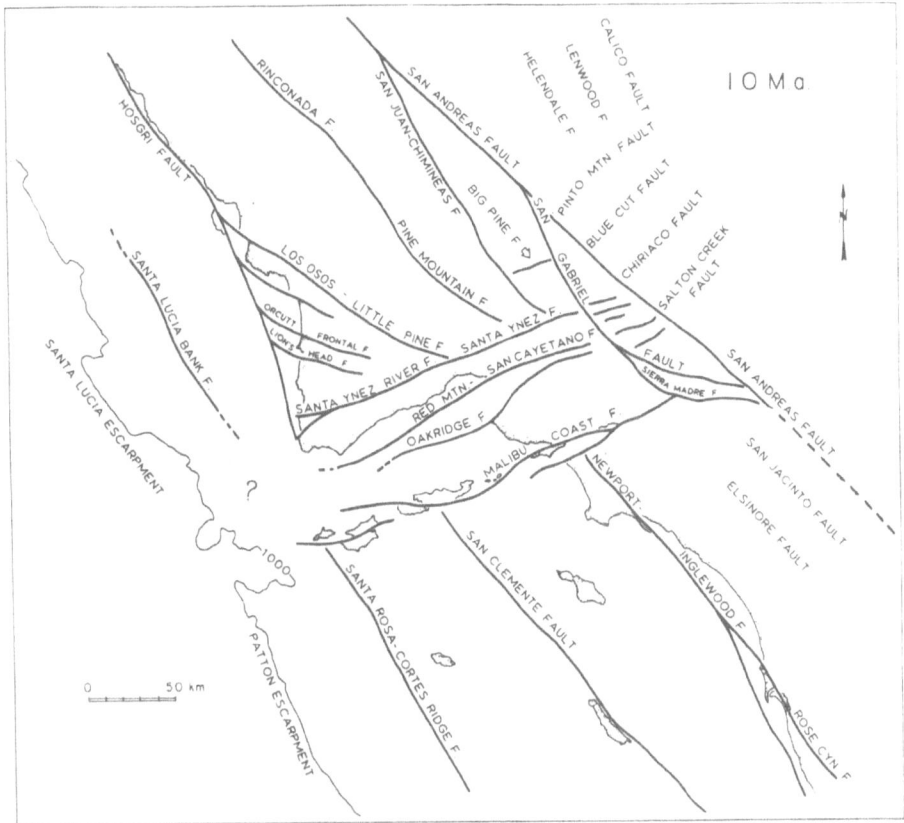

Figure 7. Palinspastic reconstruction for southern California at 10 Ma; after Hornafius et al. (1986).

rotation of the Tehachapi Mountains block. North of the western Transverse Ranges Hornafius (1985) estimated that 85 km of slip occurred on the Hosgri fault during Middle Miocene time. Because the Hosgri fault splays into Santa Maria Basin faults at its south end (Figure 7), dextral offset on the fault here will be less than to the north. If 30 km occurred on the northern San Andreas during the same time (Stanley, 1987), then 203 -(85 + 30) = 88 km of slip occurred on faults such as the Rinconada and others during this time period. However, an alternate interpretation is that about 60 km of slip occurred on the San Andreas (Hornafius et al., 1986) which leaves 58 km of Middle Miocene slip for the Rinconada and other faults.

The post-Miocene rotations were accommodated on the older faults and several newer ones such as the Elsinore and San Jacinto faults (Sharp, 1967), faults in the eastern Transverse Ranges (Dibblee, 1975), and faults in the Mojave Desert (Dokka, 1987). The southern San Andreas was also most active during this time period. Slip on the southern San Andreas is believed to total 240 km (Table I; Ehlig et al., 1975; Crowell, 1962). However recent discovery of ties between distinctive bodies of monzogranite across the central branch of the fault between the eastern and western Transverse Ranges limit the slip here to 160 km (Frizzell et al., 1986). The resolution of the apparent conflict in the offset estimates on the southern San Andreas awaits further field studies.

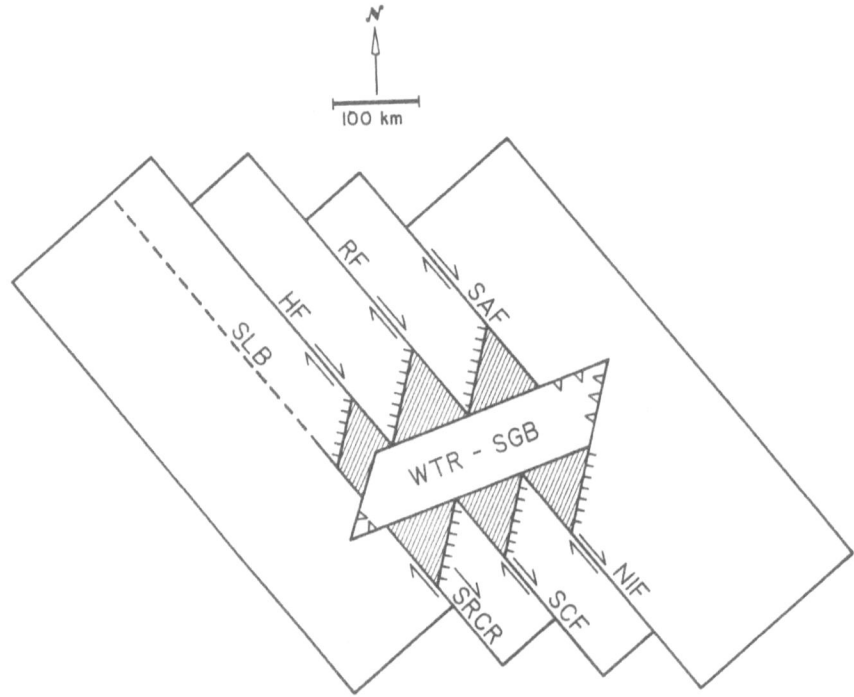

Figure 8. Geometric model of the Middle Miocene rotation in southern California. SAF = (northern) San Andreas fault, RF = Rinconada, H = Hosgri; SLB = Santa Lucia Bank, SRCR = Santa Rosa - Cortes Ridge, SCF = San Clemente, NIF = Newport - Inglewood, WTR-SGB = western Transverse Ranges - San Gabriel block. The displacement of the block edges depicts the idealized Middle Miocene displacement on the faults. Open teeth indicate crustal overlapping, shading and hatchures indicate crustal gapping and/or attenuation.

Because the western Transverse Ranges rotated 30 degrees clockwise towards the northern San Andreas fault since Miocene time, this section of the fault slips about 70 to 80 km more than the central branch, or 230 to 240 km compared to 160 km (Figures 8 and 9). Part of this northern slip could be partitioned into the San Juan - Chimineas fault. Rotation of the eastern Transverse Ranges can produce additional slip on the southern San Andreas immediately south of the ranges, and south of the monzogranite offset. This slip could be up to 70 km (Table I), but probably less if some of the slip was partitioned on unknown faults east of the southern San Andreas. Therefore, the rotation model predicts up to 230 km (160 ± 70) slip on the fault in southeast California, and gives offsets consistent with geologic data. The San Gabriel fault slips about 60 km right due to the fact that the western Transverse Ranges continue to rotate clockwise whereas the San Gabriel block does not (Figure 9).

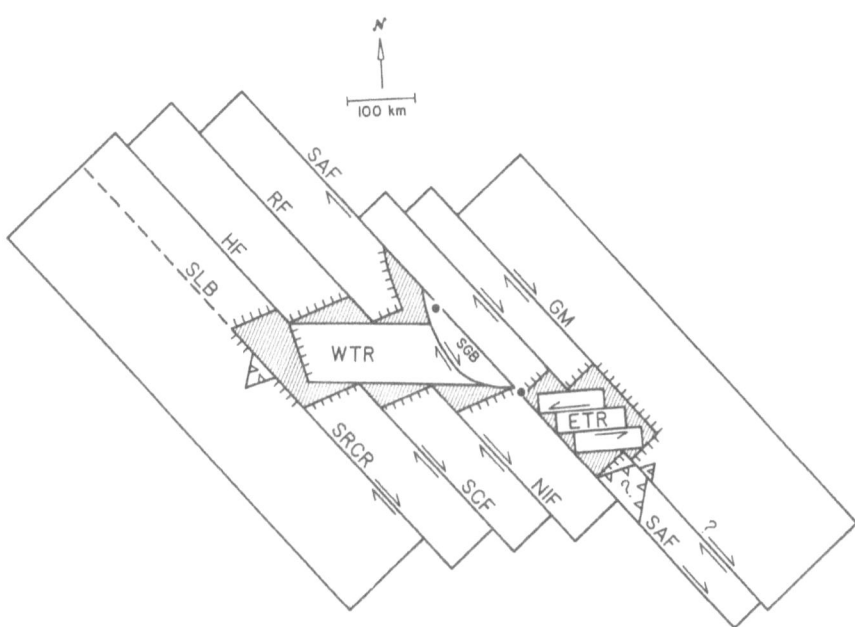

Figure 9. Geometric model of the post-Middle Miocene rotations. Legend in Figure 8. GM = Granite Mountain fault, ETR = eastern Transverse Ranges. The displacement of block edges indicate the idealized post-Middle Miocene displacement on faults. Two dots astride the central reaches of the San Andreas fault show 160 km of offset. Edges of the blocks indicate offsets on other faults.

Between 42 and 82 km of post-Miocene slip is partitioned on the Hosgri (and its splays), Rinconada, and other faults north of the western Transverse Ranges and on the San Jacinto, Elsinore, Newport - Inglewood, San Clemente, and Santa Rosa - Cortes Ridge faults south of the ranges. Between 60 and 80 km of slip are absorbed in the Mojave Desert and on the southern San Andreas and other southeast California faults. Slip on the Santa Rosa - Cortes Ridge fault and its possible continuation to the north to the Santa Lucia Bank fault is unconstrained by the rotation. However, the minimum case of slip which will accommodate the rotation is for the Santa Lucia Bank fault to be locked while the Santa Rosa - Cortes Ridge fault slips right an amount consistent with the required shear strain (Figures 8 and 9).

Because the shear computations and reconstructions assume that no crustal overlap occurs during rotation, extensive crustal stretching is implied with the formation of gaps (basins) in the geometric models. However, the Middle Miocene rotation produces overlap or thrusting at the ends of the Transverse Ranges (Figure 8) as the ranges extend out of the shear zone. In contrast, the post-Middle Miocene rotation causes the Transverse Ranges to retract back into the shear zone.This geometry sets up a broad regime of extension, in a roughly east-west sense, in the offshore Santa Maria Basin and westernmost Santa Barbara Channel, soon after the end of the Middle Miocene.

5. What has been learned?

This exercise of computing total dextral shear uses geometric assumptions relating to crustal rotations which yield the maximum estimates of shear if crustal blocks slipped and rotated in a rigid fashion and no overlapping is permitted between rotating and non-rotating blocks except at rotating block ends. If these assumptions are relaxed, then the shear estimates would be less. Nevertheless, 129 ± 81 km of Pacific - North American plate dextral shear is not accounted for since the end of Early Miocene time. The missing shear must be located west of the Hosgri fault or east of the Granite Mountain fault in the Mojave Desert.

The San Gabriel fault could have formed in its present curved shape and moved as a result of the western Transverse Ranges rotating about 30 degrees clockwise relative to the San Gabriel Mountains in post-Miocene time.

During Middle Miocene time the northern San Andreas fault slips in concert with clockwise rotation of the western Transverse Ranges and San Gabriel Mountains. Slip on this fault steps southward to southern California faults south of the rotating ranges. Since Middle Miocene time the San Andreas broke through to southern California and slipped about 160 km. Clockwise rotation of both the western and eastern Transverse Ranges produced an additional 70 to 80 km of slip on the San Andreas north and south of the ranges.

The location, amounts, and timing of southern California rotations agree in a large sense with the theoretical models of Sonder et al. (1986), who studied the deformation in the Pacific - North American plate boundary zone modeled as a thin sheet. The boundary zone lengthened in time as the Mendocino and Rivera triple junctions migrated apart. Points of agreement are, that the crustal rotations are larger and older in the west compared to the east because the plate boundary zone, or deformation zone, widens inland with time, and that the rate of rotation decreases with time.The predictive success of their calculations implies that the rotating blocks are pieces of a thin sheet, no more than about 35 km thick, and that the deformation is mainly an edge effect which extends no more than a few hundred kilometers inland.Their model also predicts that the amount of rotation increases to only slightly more than 90 degrees as the strike-slip plate boundary lengthens beyond 500 or 600 km.

6. Questions and problems remaining

1. Where is the missing shear ? A good possibility is slip on faults on or near the continental slope such as the Santa Rosa - Cortes Ridge fault. Slip on it is not constrained by the rotation. Lonsdale (1988) has documented from offshore magnetic anomalies that the oceanic plates adjacent to the California margin suffered extensive dextral shearing and clockwise rotation during Miocene time. A major part of the missing shear might be accommodated in the ocean basin. McKenzie and Jackson (1986) have pointed out that the vorticity of crustal flakes or microplates is one-half the vorticity of the distributed shear zone if the microplates are not

physically connected or pinned to the edges of the zone. This means that the amount of distributed shear via rotation (Table I) is 375 km x 2 = 750 km and the total continent shear is 1065 km compared to 1072 km for the plate displacement calculation. This is either an astonishing agreement or coincidence. If true, it would demand that no shear occurred on offshore faults or in the eastern Mojave Desert——an unlikely situation.

2. How much shear was caused by the Early Miocene clockwise rotations in the Mojave Desert extensional terranes ? Because the boundaries of the rotated region are not yet known, this cannot be estimated. However, Ross (1988) interprets this rotation to be related to a northeast-southwest trending dextral shear couple which is orthogonal to the Pacific - North American shear couple I am considering here.

3. What role does the Clemens Well - San Francisquito - Fenner fault system (Smith, 1977) play in the plate boundary shear ? Powell (1981) estimated about 100 km of Middle Miocene dextral shear on this fault system. However, it is not known if this fault zone was aligned parallel or transverse to Pacific - North American plate motion.

4. Why is the 70 ± 10 km of post-Middle Miocene slip in the Mojave Desert predicted by the rotation larger than the amount of 27 to 38 km of slip mapped (Dokka, 1983) ? This missing slip may have occurred on the Granite Mountain fault zone which is included in the rotation calculation but not in Dokka's estimate. Field mapping is needed here.

5. Do basins exist where the geometric model predicts them ? This is important to verify because the computation of dextral shear from the rotation requires no crustal overlap, only dilation. Some basins are known where predicted, such as the southern San Joaquin, Los Angeles, Santa Maria, offshore Santa Maria, and Santa Monica. However, no obvious crustal dilation has taken place immediately south of the eastern Transverse Ranges and east of the San Andreas, where it is predicted in Figure 9.

6. Is there evidence of Middle Miocene compression at the ends of the rotating blocks, particularly in the offshore west of the Santa Barbara Channel and/or in present southeast California south of the eastern Transverse Ranges and east of the San Andreas fault (Figure 9)? It is not known at which point the Transverse Ranges pivoted during the rotation, but if it assumed the shear zone maintained a constant width during the rotation, the ranges had to extend out of the zone at either or both ends, or their lengths become shortened by cross faulting.

7. The geometric models (Figures 8 and 9) constrain the San Gabriel fault to be a boundary for differential rotation of the western Transverse Ranges and San Gabriel Mountains. Paleomagnetic data constrain this differential rotation to be post-Miocene in age (Hornafius et al., 1986) yet geologic evidence suggests that slip on the fault was mostly Late Miocene in timing (Ehlig et al., 1975; Crowell, 1982). Unless the geologic data are in error, slip on this fault is apparently unrelated to crustal rotation. In this case the Late Miocene age slip would be connected from the northern San Andreas to the San Gabriel. A 16 degree counterclockwise rotation of the San Gabriel Mountains in post-Middle Miocene time (Terres and Luyendyk, 1985) could possibly cause some right slip on the fault at the proper time, but the geometry of this situation in unknown.

8. The computations used in Table I assume that the width of the shear zone is governed by the length of the rotating blocks (Hornafius, 1985). The geometric models assume the shear zone remains of constant width (Figures 8 and 9). Dextral shear computed under these assumptions is slightly different (Luyendyk, et al., 1980; Luyendyk and Hornafius, 1987), but within comparative error limits. It is known that the shear zone north of t he western Transverse

Ranges shortened in width during post-Miocene time (Crouch et al., 1984; Page, 1981; Namson and Davis, 1988).

Acknowledgements

Supported by National Science Foundation grant EAR85-18142. I want to thank John Crowell and Scott Hornafius for discussions and review of this manuscript.
May, 1988 NATO Conference on Crustal Rotations, Contribution of the Institute for Crustal Studies no.:0007-02TC

References

Bohannon, R. G., 1975. Mid-Tertiary conglomerates and their bearing on Transverse Ranges tectonics, southern California. *in* Crowell, J. C., ed., *The San Andreas Fault in Southern California, Spec. Rep. Calif. Div. Mines Geol.*, **118**, 75- 82.

Burbank, D.W., and Whistler, D. P., 1987. Temporally constrained tectonic rotations derived from magnetostratigraphic data: Implications for the initiation of the Garlock fault, California. *Geology*, **15**, 1172-1175.

Burke, D. B., Hillhouse, J. W., McKee, E. H., Miller, S. T., and Morton, J. L., 1982.Cenozoic rocks in the Barstow Basin area of southern California - stratigraphic relations, radiometric ages, and paleomagnetism. *U.S. Geol. Surv. Bull.* **1529-E**, 16 p.

Carter, J. N., Luyendyk, B. P., and Terres, R. R., 1987. Neogene clockwise rotation of the eastern Transverse Ranges, California, suggested by paleomagnetic data. *Geol. Soc. Am. Bull.*, **98**, 199-206.

Crouch, J. K., Bachman, S. B., and Shay, J. T., 1984,. Post-Miocene compressional tectonics along the central California margin. *In:* Crouch, J. K., and Bachman, S. B., eds., *Tectonics and Sedimentation along the California Margin:* Soc. Econ. Palentol. Mineral., Pac. Sec., Los Angeles, California, 37-54.

Crowell, J. C., 1962. Displacement along the San Andreas fault, California. *Geol. Soc. Am., Special Paper,* **71**, 61 p.

Crowell, J. C., 1974. Origin of the late Cenozoic basins in southern California. *Spec. Publ. Soc. Econ. Paleontol. Mineral.*, **22**, 190-204.

Crowell, J. C., 1975. The San Gabriel fault and Ridge Basin, southern California. *Spec. Rep. Calif. Div. Mines Geol.*, **118**, 208-219.

Crowell, J. C., 1981. An outline of the tectonic history of southeastern California. Chapter 18, *in* Ernst, G., ed., *The Geotectonic Development of California*, Rubey Volume I, (Prentice Hall, Englewood Cliffs, New Jersey), 583-600.

Crowell, J. C., 1982. The tectonics of Ridge Basin, southern California. *in* Crowell, J. C., and Link, M. H., eds., *Geologic History of Ridge Basin, Southern California, Soc. Econ. Paleon. Mineral., Pac. Section*, 25-42.

Crowell, J. C., 1987. Late Cenozoic basins of onshore southern California. Chapter 9, *in* Ingersoll, R., and Ernst, G., eds., *.Cenozoic Basin Development of Coastal California*, Rubey Volume VI, Prentice-Hall, Englewood Cliffs, New Jersey, 207-241.

245

Dibblee, T. W., Jr., 1975. Late Quaternary uplift of the San Bernardino Mountains on the San Andreas and related faults': *in* Crowell, J. C., ed., *The San Andreas Fault, Spec. Rep. Calif. Div. Mines*, **118**, 127-135.

Dokka, R. K., 1983. Displacements on late Cenozoic strike-slip faults of the central Mojave Desert, California. *Geology*, **11**, 305-308.

Dokka, R. K., 1986,. Patterns and modes of early Miocene crustal extension, central Mojave Desert, California': in Mayer, L., ed., Extensional Tectonics of the Southwestern United States - A Perspective on Processes and Kinematics: *Geol. Soc. Am. Special Paper* **208**, 75-95.

Dokka, R. K., 1987.New Perspectives on Late Cenozoic strike-slip faults of the Mojave Desert and their relationship to the Garlock fault. *Abstracts with Programs, Geol. Soc. Am. Ann. Meet.*, **19**, 645.

Ehlig, P. L. Ehler, K. W., and Crowe, B. M., 1975. Offset of the upper Miocene Caliente and Mint Canyon Formations along the San Gabriel and San Andreas faults. *Spec. Rep. Calif. Div. Mines Geol.*, **118**, 83-92.

Frizzell, V. A., Mattinson, J. M., and Matti, J. C., 1986. Distinctive Triassic megaporphyritic monzogranite: Evidence for only 160 km offset along the San Andreas fault, southern California. *J. Geophys. Res.*, **91**, 14080-14088.

Frost, E. G., and Martin, D. L., 1982. *Mesozoic - Cenozoic tectonic evolution of the Colorado River Region, California, Arizona, Nevada.* Cordilleran Pub., San Diego, California, 608 p.

Golombeck, M. P., and Brown, L. L., 1988. Clockwise rotation of the western Mojave Desertt. *Geology*, **16**, 126-130.

Hornafius, J. S., 1985. Neogene tectonic rotation of the Santa Ynez Range, western Transverse Ranges, California, suggested by paleomagnetic investigation of the Monterey Formation. *J. Geophys. Res.*, **90**, 12503-12522.

Hornafius, J. S., Luyendyk, B. P., Terres, R. R., and Kamerling, M. J., 1986. Timing and extent of Neogene tectonic rotation in the western Transverse Ranges, California. *Geol. Soc. Am. Bull.*, **97**, 1476-1487.

Lonsdale, P., 1988. Structural patterns of the Pacific floor offshore peninsular California. *Amer. Assoc. Petrol. Geol., Memoirs*, in press.

Luyendyk, B. P., and Hornafius, J. S., 1987. Neogene crustal rotations, fault slip, and basin development in southern California: Chapter 11, *in* Ingersoll, R., and Ernst, G., eds., *Cenozoic Basin Development of Coastal California*, Rubey Volume VI, (Pergammon Press, Elmsford, New York), 259-283.

Luyendyk, B. P., Kamerling, M. J., and Terres, R. R., 1980. Geometric model for Neogene crustal rotations in southern California. *Geol. Soc. Am. Bull.*, **91**, 211-217.

Luyendyk, B. P., Kamerling, M. J., Terres, R. R., and Hornafius, J. S., 1985. Simple shear of southern California during the Neogene. *J. Geophys. Res.*, **90**, 12454-12466.

McWilliams, M.O., and Li, Y., 1985. Oroclinal bending of the southern Sierra Nevada batholith. *Science*, **230**, 172-175.

Namson, J. S, and Davis, T. L., 1988. Seismically active fold and thrust belt in the San Joaquin Valley, central California. *Geol. Soc. Am. Bull.*, **100**, 257-273.

Page, B. M., 1981. The southern Coast Ranges': Chapter 13, *in* Ernst, G., ed., *The Geotectonic Development of California*, Rubey Volume I, (Prentice-Hall, Englewood Cliffs, New Jersey), 329-417.

Plescia, J. B., and Calderone, G. J., 1986. Paleomagnetic constraints on the timing and extent of rotation of the Tehachapi Mountains, California. *Abstracts with Programs, Geol. Soc. Am. Cordilleran Meet.*, **18**, 171.

Plescia, J. B., Calderone, G. J., and Snee, L. W., 1987. Paleomagnetic analysis of Miocene basalts from the Tehachapi Mountains, California,(Unpublished manuscript).

Powell, R. E., 1981. Geology of the crystalline basement complex, eastern Transverse Ranges, southern California: Constraints on regional tectonic interpretation. *Ph.D. dissertation* Calif. Inst. of Technol., Pasadena, California, 441 p.

Ross, T. M., 1988,. Neogene tectonic rotations in the central Mojave Desert, California, as indicated by paleomagnetic directions. *M. A. thesis,* University of California, Santa Barbara, 233p.

Ross, T. M., Luyendyk, B. P., and Haston, R. B., 1987. Neogene tectonic rotations in the Cady Mountains, Mojave Desert, California. *Abstracts with Programs, Geol. Soc. Am.,* **19,** 824.

Ross, T. M., Luyendyk, B. P., and Haston, R. B., 1988. Paleomagnetic evidence for Neogene tectonic rotations in the central Mojave Desert, California. *Abstracts with Programs, Geol. Soc. Am.,* **20,** 226.

Smith, D. P., 1977. San Juan - St. Francis fault: Hypothesized major Middle Tertiary right lateral fault in central and southern California. *Spec. Rep. Calif. Div. Mines Geol.,* **129,** 41-50.

Sonder, L. J., England, P C., and Houseman, G. A., 1986. Continuum calculations of continental deformation in transcurrent environments. *J. Geophys. Res.,* **91,** 4797-4810.

Stanley, R. G., 1987. New estimates of displacement along the San Andreas fault in central California based on paleobathymetry and paleogeography. *Geology,* **15,** 171-174.

Stock, J., and Molnar, P., 1988. Uncertainties and implications of the Late Cretaceous and Tertiary position of North America relative to the Farallon, Kula, and Pacific plates. (Unpublished manuscript).

Terres, R. R., 1984. Paleomagnetic and tectonics of the central and eastern Transverse Ranges, southern California. *Ph.D. dissertation,* University of California, Santa Barbara, California, 325p.

Terres, R. R., and Luyendyk, B. P., 1985. Neogene tectonic rotation of the San Gabriel region, California, suggested by paleomagnetic vectors. *J. Geophys. Res.,* **90,** 12467-12484.

Yeats, R. S., 1983. Large scale Quaternary detachments in Ventura basin, southern California. *J. Geophys. Res.,* **88,** 569-583.

EVIDENCE FOR CONTEMPORARY BLOCK ROTATION IN STRIKE-SLIP ENVIRONMENTS: EXAMPLES FROM THE SAN ANDREAS FAULT SYSTEM, SOUTHERN CALIFORNIA

C. NICHOLSON[1] and L. SEEBER[2]

[1] *Institute for Crustal Studies*
University of California
Santa Barbara California, 93106
U.S.A..

[2] *Lamont-Doherty Geological Observatory of Columbia University*
Palisades New York, 10964
U.S.A.

ABSTRACT. Recent moderate size earthquakes in southern California indicate seismic slip on both right-lateral faults trending north to northwest, and left-lateral faults trending northeast to east. These sets of faults are most prevalent in areas of local extension parallel to the regional strike of the San Andreas system, and are consistent with a model in which contemporary crustal deformation is partially accommodated by block rotation in response to distributed right-lateral shear. Block rotation by strike-slip faulting is inferred from the seismicity based on: sets of small crustal blocks defined by domains of active, nearly-parallel left-slip cross-faults; the unusual kinematic pattern of faulting exhibited by the seismicity around rotating block edges; and the identification at depth of seismic slip on low-angle detachments to decouple the blocks and permit rotational movement. As block corners rotate into or away from edge-bounding master faults, they increase or decrease normal stress and can therefore generate time-dependent asperities that control fault behavior, thus modifying the rupture patterns of large earthquakes. Where reliable data exists, there is often good agreement between seismicity and seismic reflection profiles, structural controls from geologic evidence, paleomagnetic results, and geodetic constraints for support of a rotating block model associated with the observed interacting left- and right-slip faults, distributed regional right-lateral shear, and basal detachments. Since edge-bounding secondary faults rotate with the adjacent block material, new faults must form as old faults rotate into positions unfavorable for further slip. Observations suggest that block rotation by strike-slip faulting is highly non-uniform in space and time, can occur on several scale lengths from millimeters to several tens of kilometers (requiring multiple zones of decoupling within the upper crust), and can accommodate both elastic and non-elastic strain accumulation. Identifying the conditions whereby block rotation is preferred over simple linear translation will be a critical element in understanding the processes controlling contemporary crustal deformation in strike-slip environments.

1. Introduction

Deformation of the earth's crust can often be partitioned into rotational and irrotational components. Paleo-magnetic data are particularly useful in identifying long-term accumulations of rotational strain, but are largely incapable of resolving how recent such observed tectonic rotations are, or whether certain types of volumetric changes and linear translations have also occurred. Conversely,

C. Kissel and C. Laj (eds.), Paleomagnetic Rotations and Continental Deformation, 247–280.
© *1989 by Kluwer Academic Publishers.*

geodetic data can resolve the extent of contemporary strain, but are incapable of identifying rigid-body rotations, especially at scale-lengths greater than the dimensions of the network or less than the spacing between grid points. In southern California, for example, geodetic data are typically reduced assuming no net rotations and minimizing the shear strains normal to the strike of the major regional faults [*e.g., King and Savage*, 1983]. Similarly, seismic data generated by earthquakes are inverted for dynamic slip histories and slip orientations assuming faults are planar and do not rotate during rupture (*i.e.*, angular momentum is conserved) [*Aki and Richards*, 1980]. Rotational strains, whether elastic or non-elastic, are thus difficult to define with standard seismological techniques [*Jackson*, this volume]. Yet if part of the large tectonic rotations observed in the past [*e.g.*, *Luyendyk et al.*, 1985] are also occurring in the present, then the distributed contemporary strain field in southern California must exhibit a net rotational component--a component that is largely ignored or assumed not to exist by most techniques used to resolve the present pattern of strain accumulation.

In this paper, we show that earthquake data can be used to identify contemporary rotations within strike-slip environments, by resolving the kinematic pattern of slip within individual earthquake sequences, and by the geometrical relationships defined by active fault structures. In southern California, these rotations are inferred to be largely the result of small crustal blocks rotating clockwise in response to distributed right-lateral shear [*Nicholson et al.*, 1986b]. The rotations are largely accommodated by left-slip on short, secondary northeast-trending faults, in a manner originally recognized and proposed by *Dibblee* [1954, 1977]. Other models may be equally valid, however, a simple block tectonic model is preferred based on: sets of blocks defined by domains of active, nearly-parallel left-slip cross-faults; the unusual kinematic pattern of faulting exhibited around rotating block edges; and the identification at depth of low-angle detachments to decouple the blocks and permit rotational movement. Independent evidence for the presence of the secondary left-slip cross faults, contemporary tectonic rotations, distributed shear strain, and midcrustal detachments is available from seismic reflection and refraction work, time-term analyses and regional velocity studies, gravity and geodetic data, paleomagnetic results, and local field mapping. Field observations indicate, however, that block rotations can occur on many different scale-lengths from millimeters to tens of kilometers [*Nicholson et al.*, 1986b], implying the presence of multiple detachment surfaces to decouple the blocks at various levels in the crust [*Brun et al.*, this volume]. This suggests a strong inter-relationship between the particular detachment surface used to accommodate any given rotation, the local mechanical properties and sizes of the rotating block material, the rotational strain rate, the width over which shear is distributed, and the initial boundary conditions, such as the local state of stress and fault orientations.

Of importance to geologists, geodesists, and seismologists is the implication that, at least in certain localized areas, components of the contemporary crustal deformation in southern California is being accommodated by local warping and small-scale block rotation. This has considerable significance in terms of understanding how certain small-scale tectonic basins may form in strike-slip environments [*e.g., Christie-Blick and Biddle*, 1985; *Karner and Dewey*, 1986], why slip rates may vary along strike of major wrench faults [*Salyards et al.*, 1987], the process of fault segmentation and the interaction between primary and secondary fault structures [*Nicholson*, 1988], earthquake nucleation [*Hudnut and Seeber*, 1987], and the proper interpretation of regional strain data. Recently, a series of moderate-sized earthquakes occurred in southern California, providing opportunities to test some of the structural, geometrical, and mechanical requirements of a rotating block model. In nearly each case, evidence in support of active left-slip faulting, detachment slip, and in some cases, net tectonic rotation of the material adjacent to the major bounding right-slip faults was provided. In the light of these results, re-examination of older

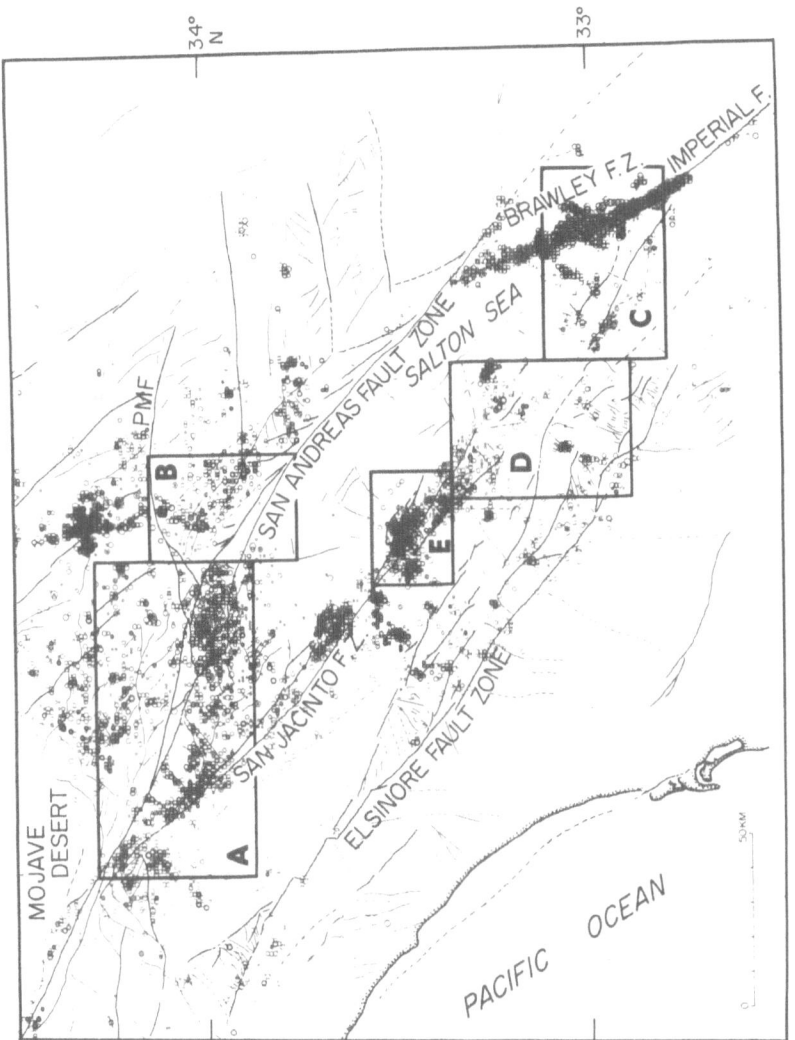

Fig. 1. Seismicity in southern California from 1975 to 1983. Only epicenters with horizontal errors in location of less than 2.5 km from the CIT-USGS catalog are shown. Faults are primarily from *Jennings et al.* [1975]. Dashed faults are inferred from seismicity or topography. Areas within boxes are examined in detail and discussed in the text. PMF = Pinto Mountain fault.

earthquake sequences provided additional new data in support of contemporary block rotation within the San Andreas fault system. In this paper, we attempt to review some of the latest available results from a number of different fields, and from a number of different investigators, on the contemporary crustal deformation of southern California. In some cases, the data are compelling, while in others they are only permissive of contemporary tectonic rotations largely because few experiments have been specifically designed to observe and measure net rotations within the contemporary strain field.

2. Data and interpretations

2.1 SEISMICITY AND THE KINEMATICS OF BLOCK ROTATION

One of the puzzling features of the microearthquake activity in southern California, is the lack of correspondence with major mapped surface traces of the San Andreas fault system [Richter, 1958; Allen et al., 1965; Brune and Allen, 1967; Allen, 1982]. Figure 1 shows earthquake epicenters from 8 years of continuous monitoring by Caltech and the U.S. Geological Survey.

Many of the earthquakes, however, define broad dense clusters (e.g., the Brawley seismic zone, and area E along the San Jacinto fault zone), as well as short linear segments extending up to several kilometers away from the major through-going fault strands. When the pattern of faulting along these short secondary structures was examined in detail, it was found that many of these features exhibit focal mechanisms consistent with predominantly left-slip motion on nearly vertical northeast-striking faults (E-H, Figure 2)[Nicholson et al., 1986b]. The nearly-parallel left-slip secondary faults were best defined near the intersection of major right-slip faults, such as the San Jacinto and San Andreas (Figure 2), thus describing a series of small crustal blocks. Further examination of the seismicity along the San Jacinto fault revealed an alternating pattern of normal (B and D) and reverse (C) faulting earthquakes. These earthquakes are located where a rotating block model predicts their occurrence, if the blocks were rotating clockwise as a result of distributed right-lateral shear (compare J with map, Figure 2). A similar pattern of alternating small-scale transtensional and transpressional features (sag ponds and normal faults, folds and thrust faults, etc.) is observed along the seismically quiescent San Andreas fault [Matti et al., 1985], but in opposite block corners, respectively, as expected in a "rigid-body" rotation model.

Other earthquakes that appear to exhibit left-slip on northeast-trending structures are found farther east between the Banning and Mission Creek faults (Figure 3)[Nicholson et al., 1986a]. These events exhibit a larger component of thrust faulting, indicating that the inferred pattern of deformation is more oblique. An interesting aspect of this seismicity is that all the earthquakes that exhibit left-slip on northeast-trending cross-faults typically occur at depths less than about 10 km (e.g., Figure 3). If the left-slip faults are related to block rotation, then they must be detached from the unrotated crust at depth [Terres and Sylvester, 1981]. The presence of a detachment to decouple the blocks and permit rotational movement suggests that either stress or strain at depth below the detachment must differ from that observed in the upper crust. In fact, earthquakes at depths greater than about 10-12 km, in about the same area as those in Figure 3, do exhibit a distinctly different style of deformation (Figure 4). The deeper seismicity appears to define a wedge-shape volume undergoing deformation on a complex combination of high-angle strike-slip faults and stacked low-angle thrust faults dipping to the north [Nicholson et al, 1986a]. Independent evidence for midcrustal detachments or ductile shear zones, of either local or regional

251

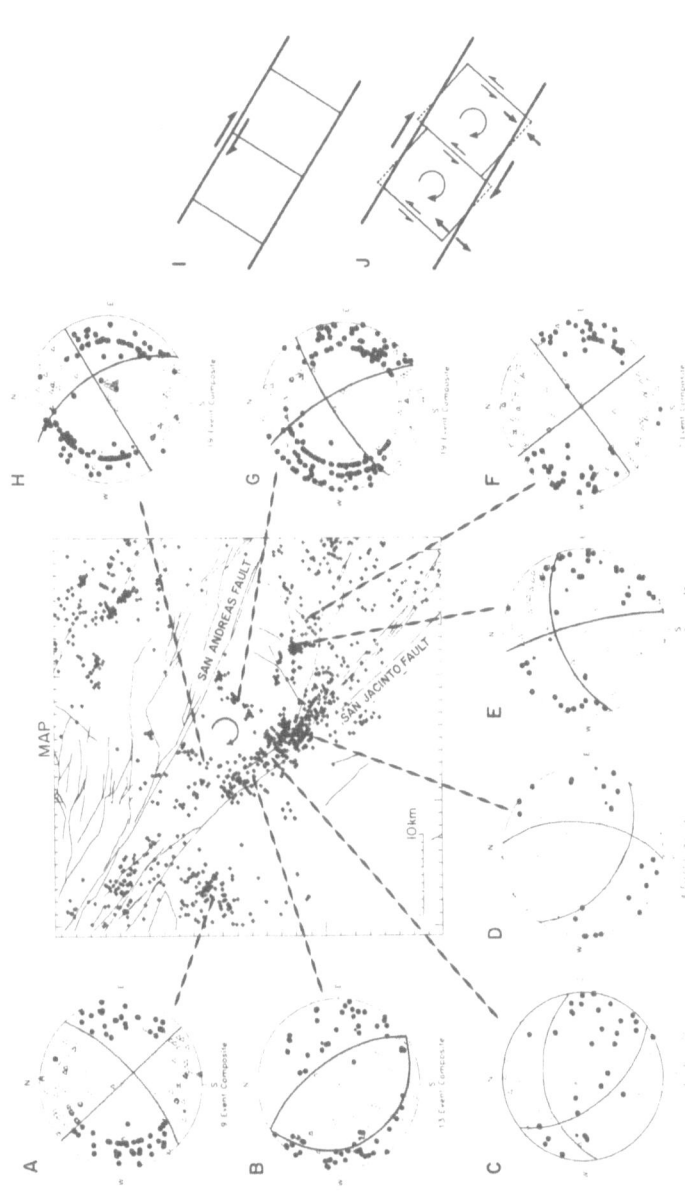

Fig. 2. Shallow seismicity used to define rotating blocks near the intersection of the San Andreas and San Jacinto faults (left side inset A, Figure 1)[*Nicholson et al.*, 1986b]. During a large earthquake, right-slip motion (*I*) occurs on one of the major bounding faults; however, during the interseismic period, the major faults become locked causing small blocks in between to rotate (*J*). This produces a pattern of northeast-striking left-slip faults (*E-H*), between which alternating groups of normal (*B* and *D*), and reverse (*C*) faulting earthquakes occur that match the particular pattern predicted by the model (compare *J* with map). Focal mechanism diagrams (*A-H*) are composite upper-hemisphere projections; solid symbols are compressions, open symbols are dilatations. Composite *A* (upper left) represents a set of subparallel left-slip faults previously identified by *Hadley and Combs* [1974].

252

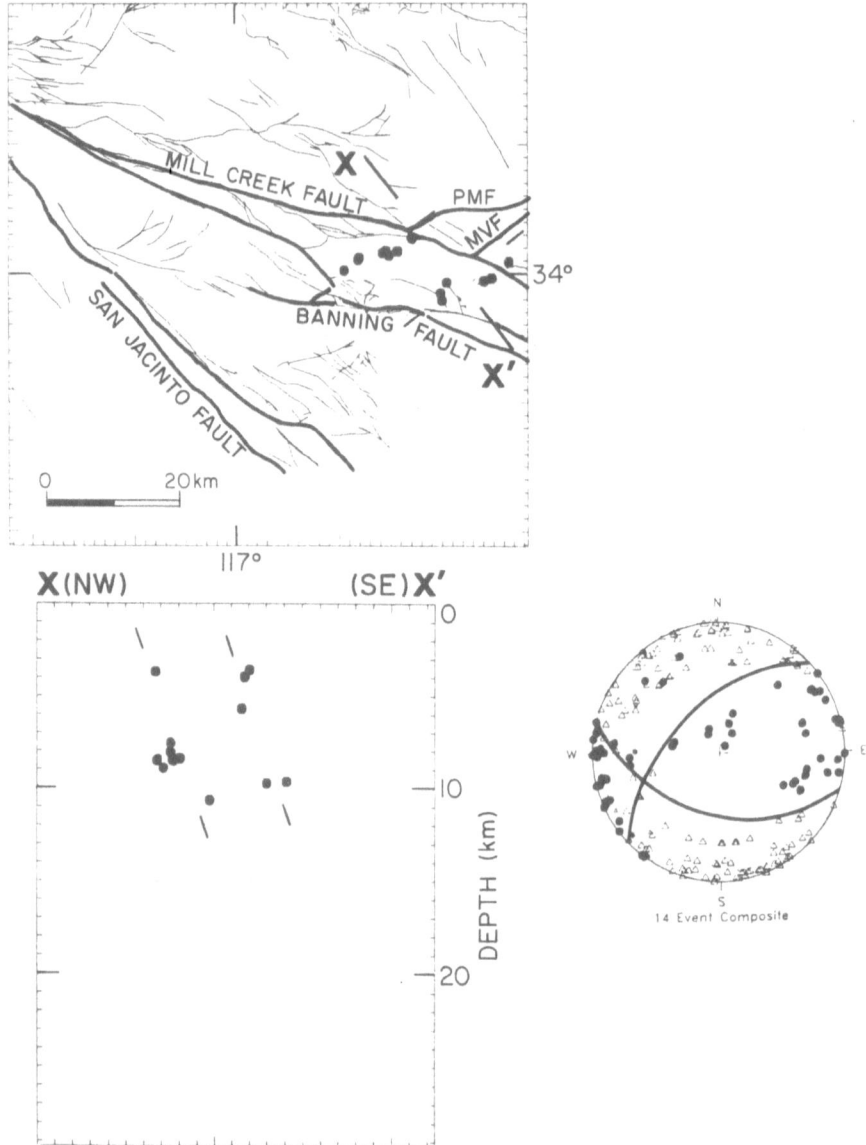

Fig. 3. Map and cross section of earthquake hypocenters above a depth of about 10 km between the Mission Creek and Banning faults (right side inset A, Figure 1) that exhibit left oblique-slip along two subsurface northeast-trending structures [*Nicholson et al.*, 1986b].

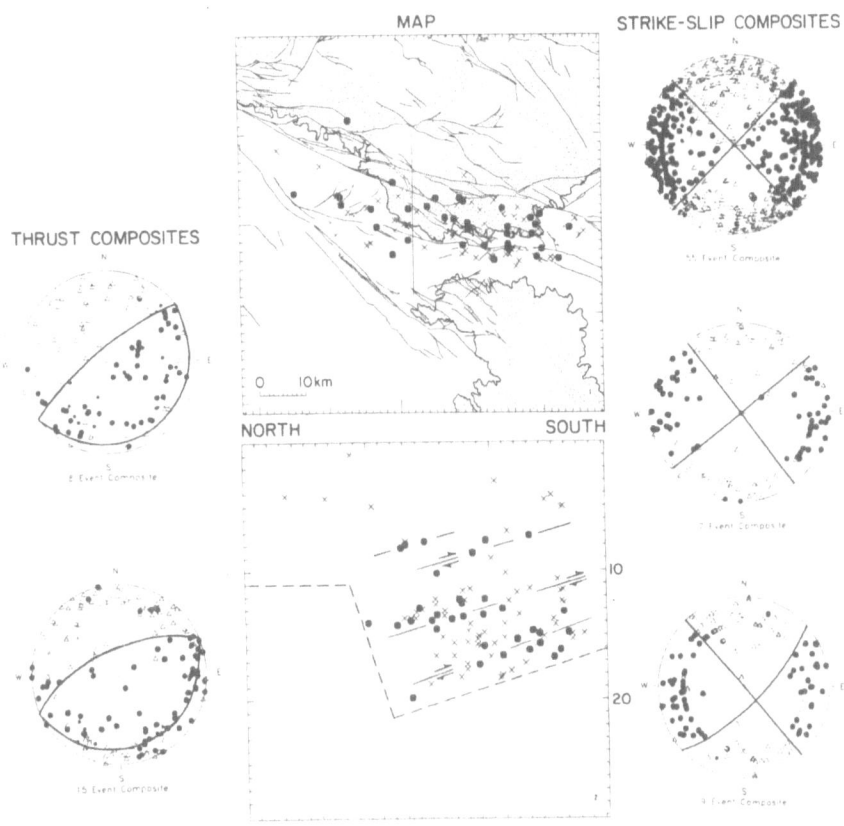

Fig. 4. Earthquake hypocenters at depths greater than about 10 km under San Gorgonio Pass that exhibit either strike-slip motion on high-angle faults or reverse slip on low-angle faults dipping to the north [*Nicholson et al.*, 1986b]. The change in deformation with depth from Figure 3 is interpreted as reflecting the presence of a detachment.

extent, adjacent to the San Andreas system and under the San Bernardino Mountains, is available from gravity and velocity studies [*Hadley and Kanamori*, 1977; *Hearn and Clayton*, 1986, *etc.*], geology and palinspastic reconstructions [*Meisling and Weldon*, 1986], and by the presence of moderate earthquakes at depth that exhibit low-angle nodal planes that parallel the shallow-dipping base of the seismogenic zone [*Webb and Kanamori*, 1985]. The seismicity data thus provide support for the three main elements of a rotating block model: (1) discrete crustal blocks defined by left-slip cross-faults; (2) characteristic patterns of deformation around block edges consistent with the kinematics of block rotation; and (3) detachment surfaces to permit rotational movement.

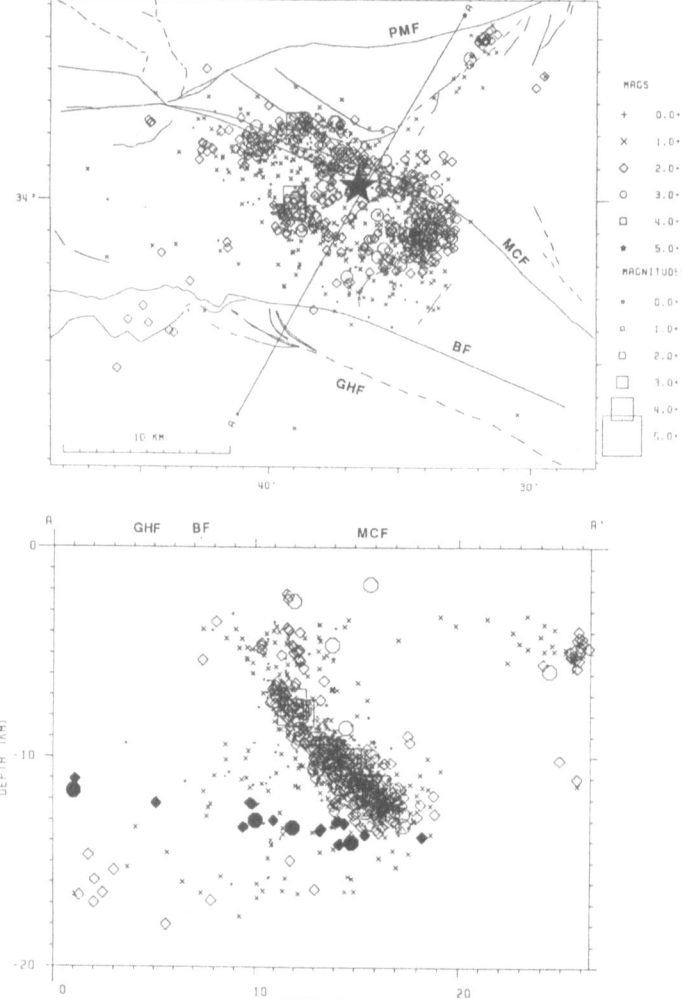

Fig. 5. Map and vertical cross section of well-located aftershocks of the 1986 North Palm Springs earthquake (star), located between the Banning (BF) and Mission Creek faults (MCF). Notice most of the hypocenters are confined between the two northeast-striking structures previously defined in Figure 3, are between 5 and 13 km depth, and define a northeast-dipping planar surface consistent with slip on the Banning fault. Off-fault aftershocks occur primarily as shallow left-slip events along the Morongo Valley fault to the northeast; or as deep (>10 km) events to the southwest, bounded above by a zone of larger aftershocks (solid symbols), many of which exhibit focal mechanisms with a low-angle nodal plane consistent with slip on a detachment.

2.2 RESULTS FROM THE 1986 NORTH PALM SPRINGS EARTHQUAKE

Unfortunately, few, if any, of the northeast-trending structures inferred from the seismicity in Figures 2 and 3 have much of a surface expression, except through the Crafton Hills where they are mapped as normal faults [*Matti et al.*, 1985]. However, in July of 1986, a moderate-sized earthquake (M_L=5.9) occurred along the Banning fault near North Palm Springs (Figure 5)[*Jones et al.*, 1986; *Nicholson et al.*, 1986c]. Aftershocks of the earthquake occurred primarily between depths of 5 and 13 km, and outlined a nearly planar surface dipping to the northeast at about 45° to 50°. Surprisingly, the limits of the rupture corresponded with the location of the two previously-defined subsurface faults in Figure 3. Focal mechanisms of the aftershocks at the ends of the rupture also agreed with earlier focal mechanism solutions.

The change in stress associated with slip during the mainshock triggered seismicity on adjacent secondary structures [e.g., *Das and Scholz*, 1981]. To the northeast, left-slip along the Morongo Valley fault occurred about 4 days later, but only at depths less than about 7 km (Figure 5). To the southwest, however, a complex pattern of faulting was triggered, but only at depths greater than about 10 km, similar in many respects with the pattern of deformation observed in Figure 4. The fact that the seismicity to the southwest was sharply bounded above by some of the larger aftershocks (solid symbols), and that many of these events exhibited a shallow-angle nodal plane in their focal mechanism solutions, strongly suggests that the upper limit of the seismicity corresponded with a detachment--the same detachment previously inferred based on the observed change in style of deformation with depth in the microseismicity (from Figure 3 to Figure 4). This detachment also apparently controlled the down-dip extent of the 1986 rupture, as defined by the aftershock activity (Figure 5).

2.3 BLOCK ROTATION AND INTERACTING LEFT-AND RIGHT LATERAL FAULTS NEAR THE SALTON SEA

Other areas in southern California that currently exhibit left-slip on sets of nearly parallel northeast-trending faults include the region east of the Mission Creek and south of the Pinto Mountain faults (Area B, Figure 1), as well as along the broad zone of activity that comprises the Brawley seismic zone south of the Salton Sea, and extending as far west as the Superstition Hills fault (Area C, Figure 1)[*Nicholson et al.*, 1986b]. Figure 6 shows a detailed map of this activity. Short northeast-striking lineations within the Brawley seismic zone and extending as far west as the Superstition Hills fault, are clearly evident and mirror in many ways the pattern of seismicity shown in Figure 2. Examination of individual earthquake swarms and sequences for well-defined planar features with consistent focal mechanism solutions reveals a complicated pattern of subsurface faulting (Figure 7)[*Johnson and Hadley*, 1976; *Fuis and Schnapp*, 1977; *Hutton and Johnson*, 1981; *Johnson and Hutton*, 1982; *Fuis et al.*, 1982; *Nicholson et al.*, 1985, 1986b; *Doser and Kanamori*, 1986b]. Interestingly, although this area would nominally be expected to exhibit normal faulting on northeast-trending structures (to accommodate the extension associated with a right-step in a right-lateral system), nearly all the northeast-trending faults exhibit predominantly left-slip motion as defined by the seismicity and the corresponding focal mechanism solutions. This pattern of faulting is consistent, however, with small-scale block rotation in response to distributed right-lateral shear.

In 1975, a particular set of earthquakes active during a specific earthquake swarm [*Johnson and Hadley*, 1976] demonstrated a pattern of faulting consistent with small-scale block rotation (Figure 8). The epicenters define two nearly-vertical, northeast-striking left-slip faults connected by a north-northwest-striking right-slip Brawley fault. Focal mechanisms of the earthquakes along the

116.00   115.25
33.50

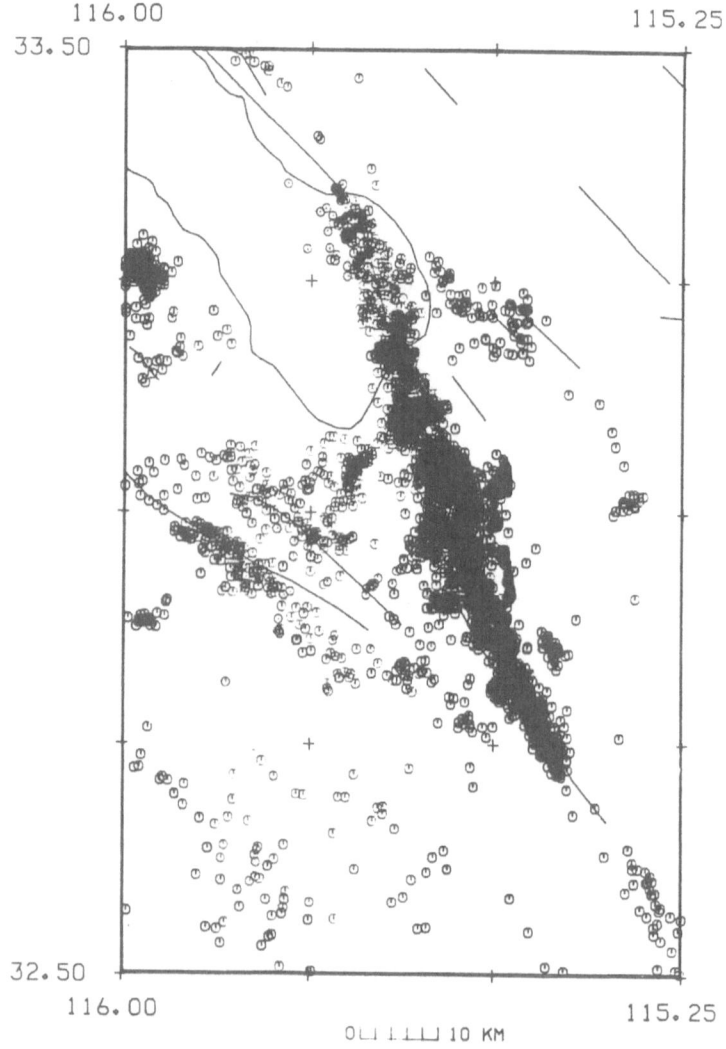

32.50
116.00   115.25

0 ⊔ I ⊥⊥⊔ 10 KM

Fig.6. Detailed map of earthquake epicenters along the Brawley seismic zone (inset C, Figure 1) at the southern end of the Salton Sea (upper left corner), 1975-1983. Earthquake locations define a broad zone several kilometers wide extending from the southern San Andreas to the northern Imperial fault, within which several well-defined short linear segments striking northeast to north are evident. Epicenters also extend as far west as the Superstition Hills fault on the west side of the Imperial Valley.

257

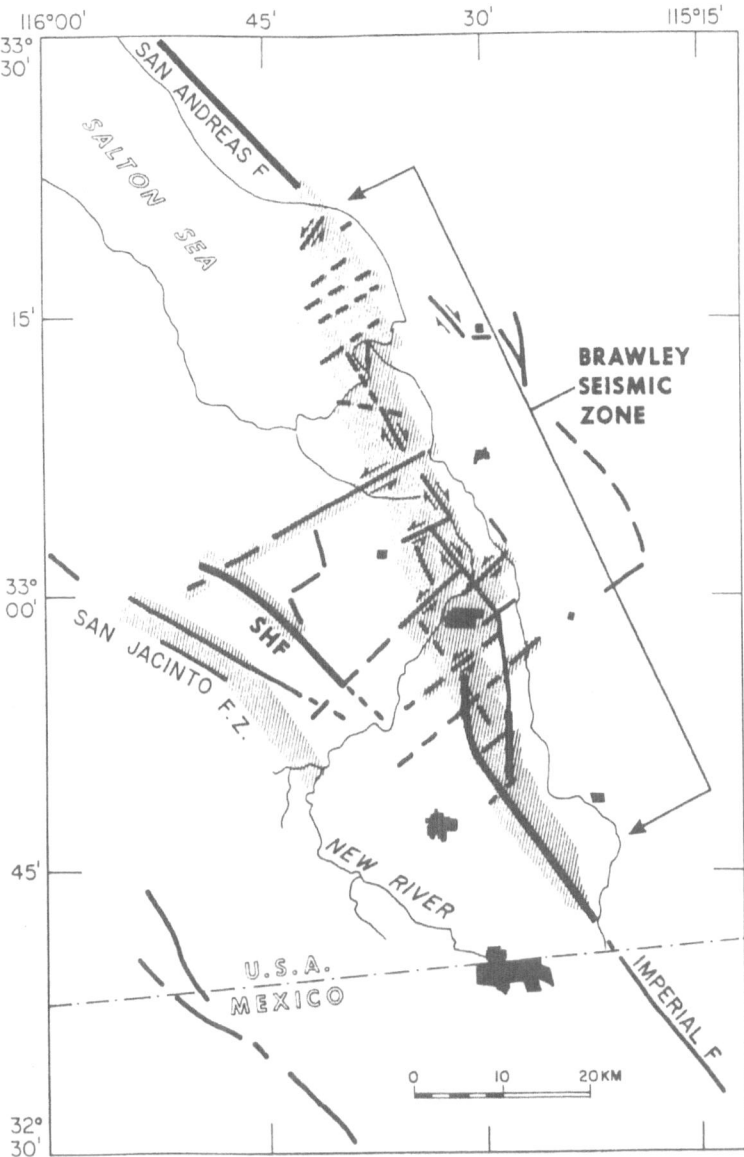

Fig.7. Inferred subsurface faults within the Brawley seismic zone defined by planar distributions of earthquake hypocenters, and consistent focal mechanism solutions of particular earthquake sequences and swarms [*Nicholson et al.*, 1985].

258

seismicity trend just south of the town of Brawley (Fault A), exhibit an unusual kinematic pattern of slip in space and time. At the corner where fault A intersects the Brawley fault, a series of reverse fault earthquakes occurred from January 23 to 25. Activity then migrated southwest, and the pattern

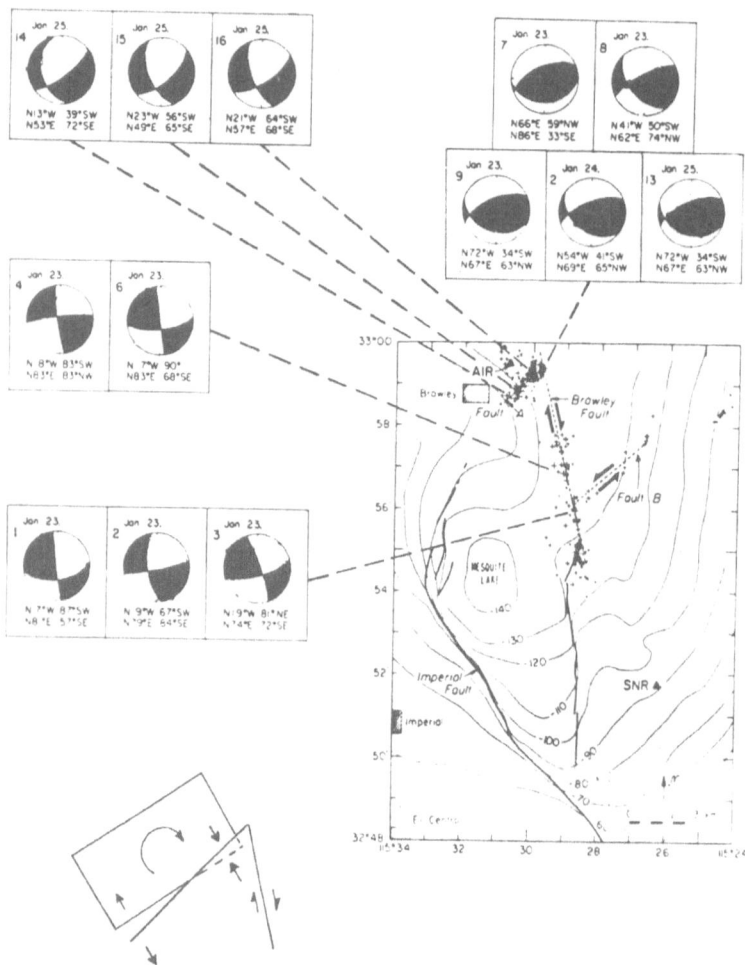

Fig.8. Seismicity and focal mechanisms associated with the 1975 Brawley swarm [*Johnson and Hadley*, 1976]. Earthquakes defined linear trends suggesting interacting left- and right-slip faults. Focal mechanisms are consistent with block rotation into (reverse faulting) and away from (normal faulting) the oblique left-slip northeast-striking fault (lower diagram) [after *Nicholson et al.*, 1986b].

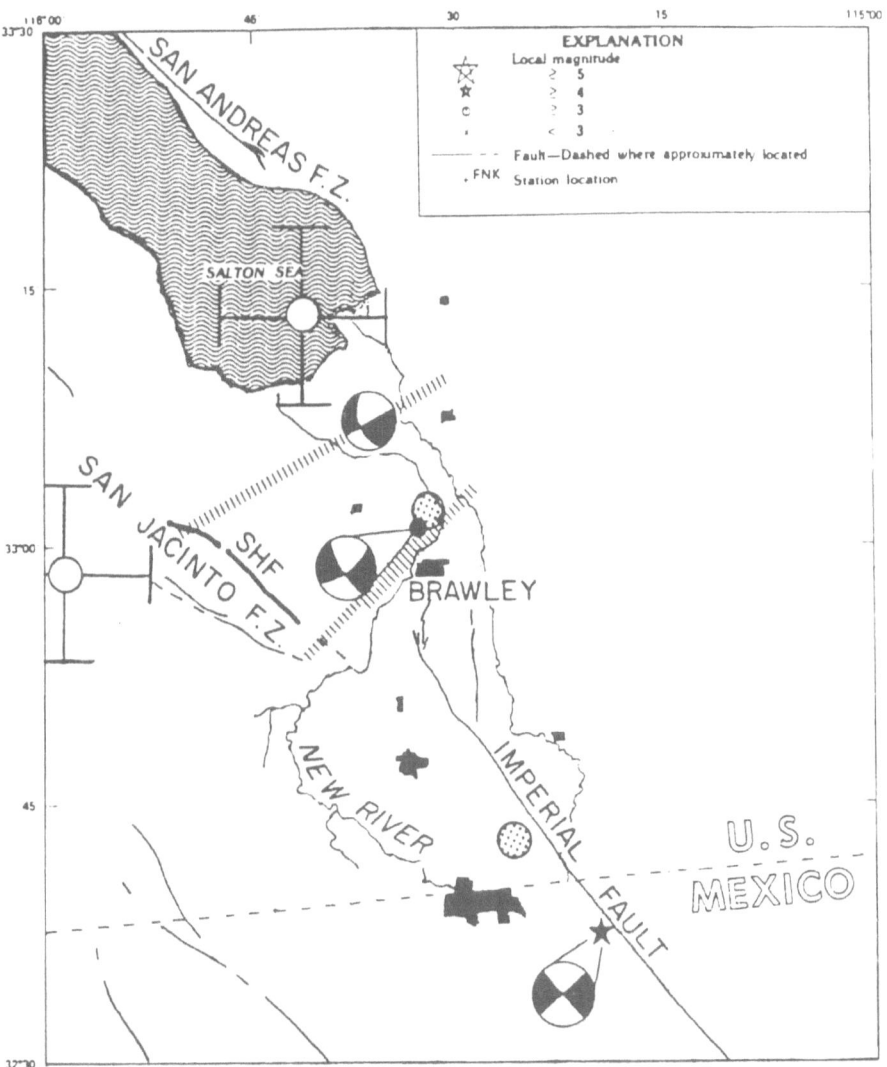

Fig. 9. Location of inferred principal left-slip cross faults (hachured zones) affecting rupture behavior of major earthquakes in the northern Imperial Valley. Shaded circles, 1940 Imperial Valley earthquake and large aftershock; solid symbols w/focal mechanisms, 1979 Imperial Valley and large left-slip aftershock; isolated focal mechanism, left-slip 1981 Westmorland mainshock; open symbols w/error bars, poorly located 1942 mainshock and large aftershock suggesting a northeast trend similar to the 1987 foreshock sequence (see Figure 10) [*Nicholson et al.*, 1985; *Doser and Kanamori*, 1986].

260

PRELIMINARY 1987 SUPERSTITION HILLS AFTERSHOCK LOCATIONS (NOVEMBER 25 - DECEMBER 4)

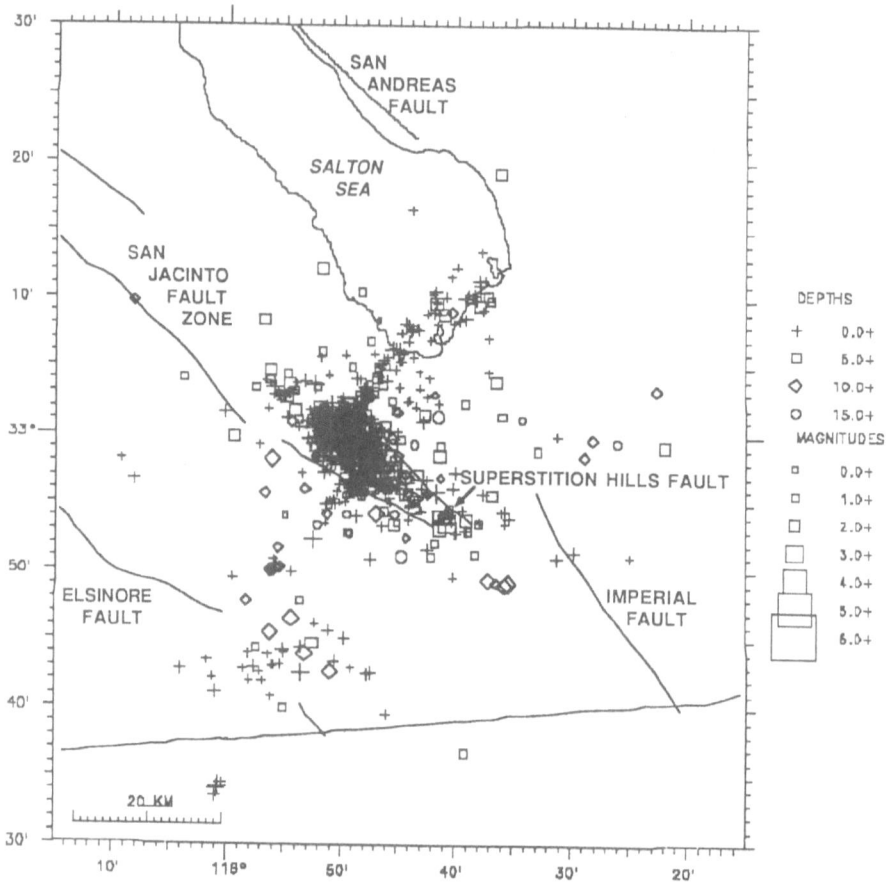

Fig. 10. Preliminary aftershock epicenters of the November 1987 Superstition Hills foreshock (M_S=6.2) and mainshock (M_S=6.6). Notice the well-defined northeast-striking zone extending into the Salton Sea, the concentration of aftershocks at the northern end of the Superstition Hills fault, and other short northeast-striking segments defined adjacent to the fault. All the events associated with the foreshock sequence occurred along the northeast trend that exhibited left-slip surface rupture (maximum 12.5 cm) and left-slip earthquake focal mechanism solutions. Surface rupture and focal mechanisms along the Superstition Hills fault following the mainshock exhibited predominantly right-slip (maximum 90 cm).

of faulting changed from reverse to one of increasing normal slip, in addition to the large component of left-slip motion along the northeast trend.

Models invoked to explain the deformation observed along the Brawley seismic zone usually involve pull-apart basins or small-scale spreading centers [*e.g.*, *Lomnitz et al.*, 1970; *Fuis et al.*, 1982].*Weaver and Hill* [1979] propose a model for continental pull-apart basins the involves oblique spreading and distributed dextral and sinistral slip. These irrotational models fail to explain, however, the presence of reverse faulting within the Brawley extensional environment, or account for the large amounts of left-slip along what many of the models would predict as primarily normal faults. Alternatively, this kinematic pattern of deformation is exactly what would be predicted by a block rotation model.*Johnson and Hutton* [1986] model the observed deformation within the Brawley area as the response of a fluid-saturated elastic layer, driven from below by a buried ridge-transform system, with basal shear distributed horizontally along an intervening detachment surface. This model accounts for a number of features characteristic of the Brawley area, including the presence and tectonic fabric of the Brawley seismic zone, the location of larger, plate-rupturing right-slip earthquakes, the occurrence of creep and swarms of shallow microearthquakes, and the clockwise rotation of crustal microblocks.

Although the amount of plate motion accommodated by secondary left-slip faults is small compared to the total slip on major right-slip faults, an important feature of a rotating block model is the possible interaction between left- and right-slip faults, and the effect on fault behavior as rotations alter the local stress conditions and relative fault geometry. Figure 9 shows the orientation and position of two of the major northeast-striking left-slip faults (hachured zones) connecting the Brawley seismic zone with the Superstition Hills fault (Figure 7)[*Nicholson et al.*, 1985; *Doser and Kanamori*, 1986a]. Following the 1940 and 1979 Imperial Valley earthquakes, large left-slip aftershocks were triggered along the fault located near the town of Brawley [*Johnson and Hutton*, 1982; *Nicholson et al.*, 1986b]. This cross-fault is also apparently responsible for the deflection of the New River to the northeast, and the southern termination of where triggered right-slip occurred on the Superstition Hills fault following the 1968 Borrego Mountain and 1979 Imperial Valley earthquakes [*Allen et al.*, 1972; *Fuis*, 1982]. The other major northeast-striking cross-fault was responsible for the left-slip 1981 Westmorland mainshock, which triggered right-lateral sympathetic slip along the Superstition Hills and Imperial Valley faults [*Sharp et al.*, 1986]. The presence of even yet another major northeast-trending cross-fault, located farther north and extending out into the Salton Sea, was suggested by the occurrence of the 1942 earthquake and its largest aftershock [*Nicholson et al.*, 1986b; *Sanders et al.*, 1986].

2.4 RESULTS FROM THE 1987 SUPERSTITION HILLS EARTHQUAKE SEQUENCE

Confirmation of active left-slip cross-faults in the northern Imperial Valley, and the importance of fault interaction was provided by the November, 1987 Superstition Hills foreshock-mainshock-aftershock sequence (Figure 10). This series of events began with a M_S=6.2 foreshock along a northeast-striking left-slip cross-fault and culminated 12 hours later with a M_S=6.6 mainshock along the northwest-striking right-slip Superstition Hills fault. Prior to the 1987 earthquake, the left-slip cross-fault responsible for the foreshock sequence exhibited little or no microearthquake activity (see Figure 6), but had been previously mapped as a major left-slip fault deforming the sediments of the Imperial Valley [*Dibblee*, 1954], and had generated a series of *en échelon* folds characteristic of left-slip faults [*T. Rockwell*, pers. comm., 1988]. Following the mainshock, left-slip surface rupture was observed along several northeast-trending fault strands extending towards

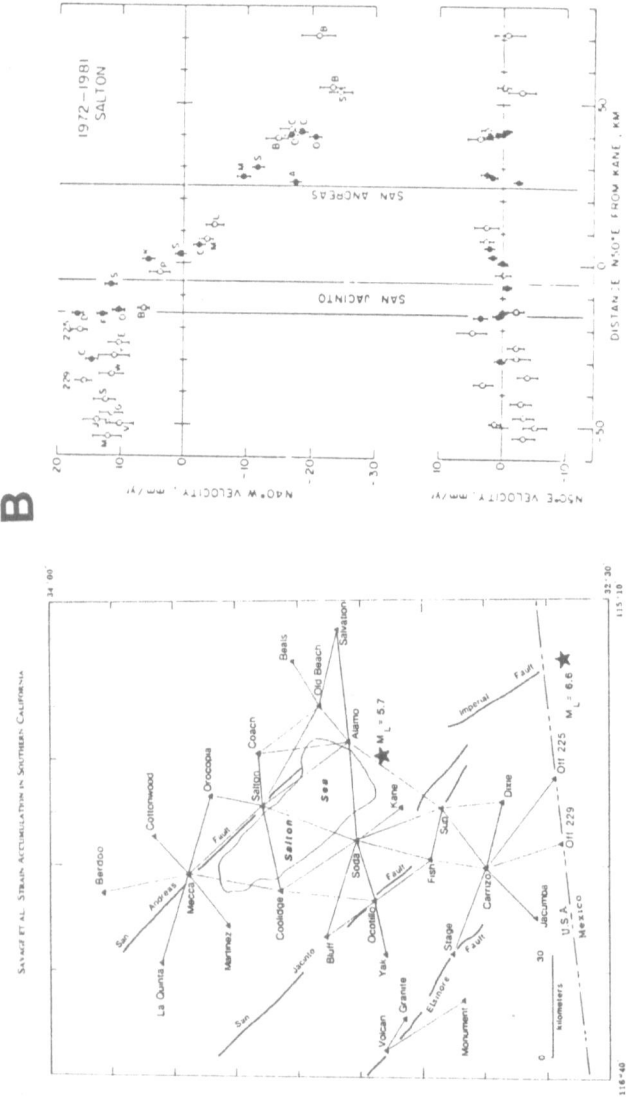

Fig. 11. (*a*) Map of the Salton trilateration network. The two stars are the epicenters of the 1981 Westmorland and 1979 Imperial Valley earthquakes [*Savage et al.*, 1986]. (*b*) (*top*) Velocity in mm/yr parallel to N40°W as a function of distance from station KANE along a line N50°E across the northern Imperial Valley; (*bottom*) Velocity parallel to N50°E [*Savage et al.*, 1981]. Shear strain is not concentrated within discrete zones associated with major faults, but rather is distributed between the Superstition Hills (San Jacinto Fault Zone) and southern San Andreas faults.

the Salton Sea, and located directly above the northeast-trending zone of aftershocks (Figure 10)[*K. Hudnut*, pers. comm., 1987; *Rockwell et al.*, 1988; *Budding and Sharp*, 1988].

The 1987 mainshock initiated at the northern end of the Superstition Hills fault, where the two conjugate systems intersect, and then ruptured unilaterally to the southeast, consistent with the reduction in normal stress along the Superstition Hills fault caused by the foreshock sequence. Other short northeast trends are evident within the aftershock distribution farther along strike of the Superstition Hills fault (Figure 10). Many of these secondary cross-faults have orientations and positions similar to previously inferred subsurface faults based on seismicity (*e.g.*, Figure 7). The concentration of aftershocks at the northern end of the Superstition Hills fault is clearly controlled by the relative position of secondary cross-faults, as well as the sudden decrease in aftershock activity, about a third of the way farther southeast along strike, where short northeast trends in aftershocks are recognizable to either side. This drop-off in aftershock activity corresponds with a mapped discontinuity and change in strike of the surface trace of the Superstition Hills fault [*Sharp et al.*, 1986], the intersection of the Superstition Hills fault with a subsurface fault in basement topography inferred from seismic refraction studies [*Fuis et al.*, 1982] and time-term analyses [*Kohler and Fuis*, 1986], and the likely intersection of the Superstition Hills fault with the fault responsible for the 1981 Westmorland mainshock near the Brawley seismic zone (see Figure 9).

No surface rupture was noted on the Superstition Hills fault following the foreshock and before the mainshock [*P. Williams*, pers. comm., 1987], but extensive right-slip surface faulting did appear after the mainshock [*Sharp*, 1987; *Magistrale and Williams*, 1987]. A considerable amount of post-seismic creep or afterslip was observed, with typical amounts of right-slip along the Superstition Hills fault varying from about 10-20 cm shortly after the mainshock, to more than 70 cm at the end of a week [*Sharp*, 1987; *Magistrale and Williams*, 1987]. Subsequent re-surveying of the Salton trilateration network (Figure 11a) following the 1987 earthquake sequence indicated about 1.2 m of right-slip had occurred on the Superstition Hills fault, and about 0.4 m of left-slip on the secondary cross-fault responsible for the foreshock sequence [Lisowski and Savage, 1988]. Prior to the 1987 mainshock, the reduction in normal stress and the triggering of major right-slip earthquakes by left-slip on cross-faults was predicted based on a rotating block model [*Nicholson et al.*, 1986b; *Hudnut and Seeber*, 1987].

If the northeast-striking left-slip cross-faults in the northern Imperial Valley are indeed responding to block rotation, then dextral shear related to plate boundary motion must be distributed over the entire area between the Superstition Hills fault and the Brawley seismic zone. Geodetic evidence suggests that this is true. Figure 11b shows the rate of plate velocity in mm/yr across the Salton network, resolved along a line perpendicular to the relative plate-motion vector (N40°W) during the time interval 1972-1981. The data suggest that, at least in the northern Imperial Valley, dextral shear is not concentrated in discrete narrow fault zones as it is farther north [*King and Savage*, 1983], but rather is distributed across the entire width of the zone between the major bounding faults, consistent with a rotating block model.

2.5 RE-INTERPRETATION OF OLDER EARTHQUAKE SEQUENCES ALONG THE SAN JACINTO FAULT

The interaction between left- and right-slip faults during the 1979 Imperial Valley, 1981 Westmorland, and 1987 earthquakes, suggests that older earthquake sequences may have also involved combinations of left- and right-slip faults, detachment slip, and possible block rotation. Consequently, the seismic deformation associated with the 1968 (M_L=6.5) Borrego Mountain and 1969 (M_L=5.8) Coyote Mountain (Area D, Figure 1) earthquakes was recently re-examined.

264

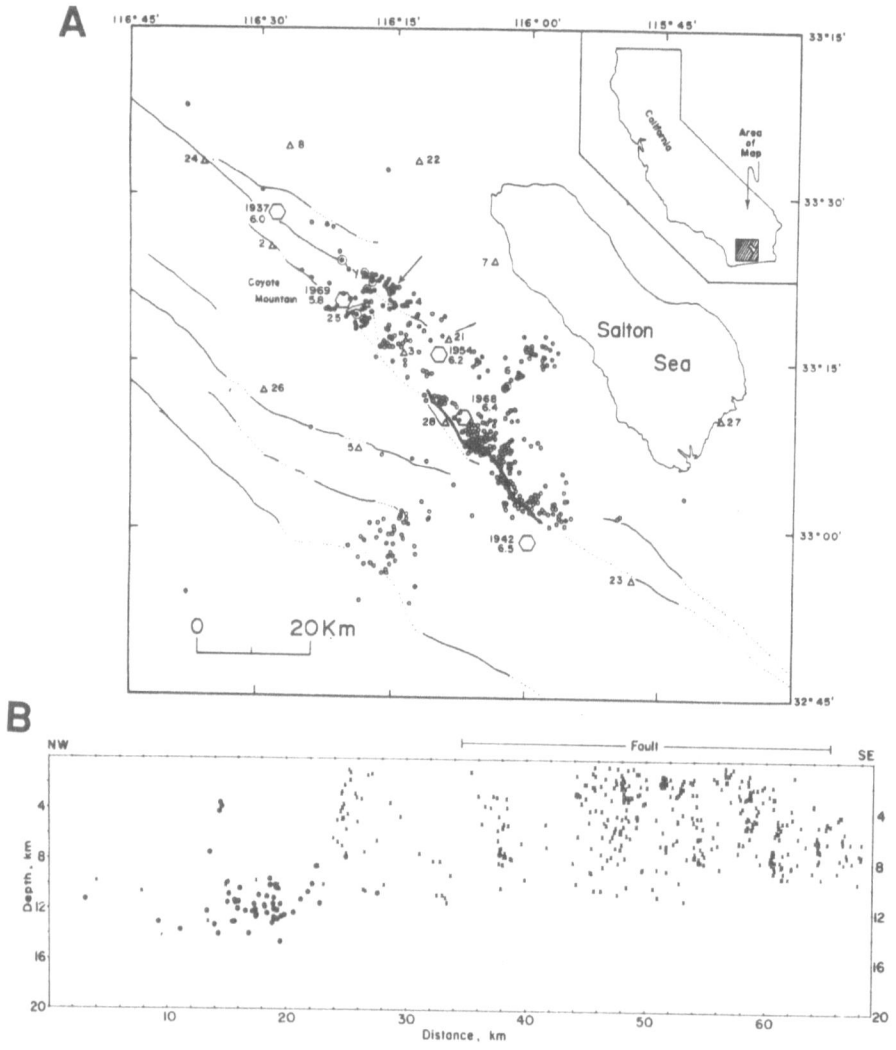

Fig.12. (a) Map showing aftershock epicenters associated with both the 1968 Borrego Mountain (M$_L$=6.5) (open symbols) and 1969 Coyote Mountain (M$_L$=5.8) (closed circles) earthquakes along the San Jacinto fault zone (insets D and E, Figure 1)[*Thatcher and Hamilton*, 1973]. Hexagons represent locations of large regional events; triangles are stations. Circled dots are Coyote Mountain afteshocks used in composite focal mechanism shown in 12c (left); arrow points to cluster of aftershocks used in composite 12c (right).The complex 1968 rupture involved two subevents: a larger pure right-slip event and a smaller low-angle thrust at a depth of 11 km [*Petersen et al.*, 1987]. (b) Vertical cross section of Coyote Mountain (solid circles) and Borrego Mountian (crosses) aftershock hypocenters [*Thatcher and Hamilton*, 1973].

265

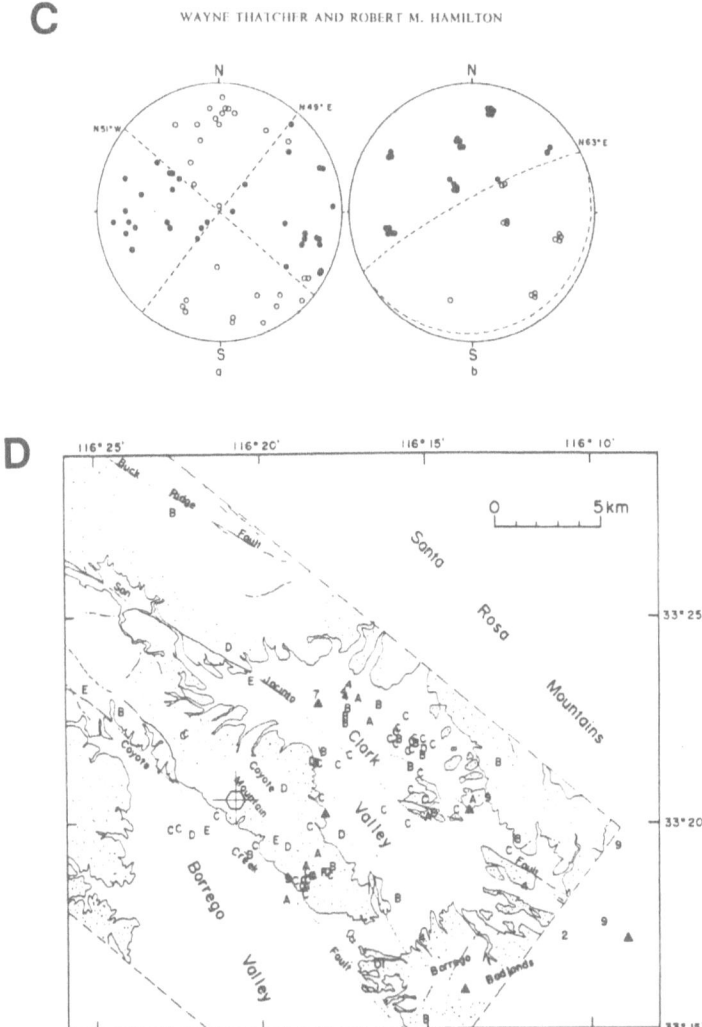

Fig. 12 cont. (*c*) Composite focal mechanism solutions of earthquakes identified in 12a [*Thatcher and Hamilton*, 1973]. (*d*) Detailed epicenter map of the Coyote Mountain (mainshock = hexagon) aftershocks [*Thatcher and Hamilton*, 1973]. Symbol plotted at epicenter indicates focal depth (*i.e.*, 1=1 km, 2=2 km, ..., A=10 km, B=11 km...). Slip in 1969 is believed to have involved both left-slip on a northeast-striking secondary fault [*Petersen et al.*, 1987], and detachment slip at depth on a low-angle structure, as indicated by the aftershock distributions and focal mechanisms, causing a net rotation of Clark Valley [*Hudnut and Seeber*, 1987].

Aftershock locations of the 1968 event were found to extend up to several kilometers away from the main surface fault trace, and in many cases, defined short linear trends with northeast orientations (Figure 12a)[*Hamilton, 1972; Nicholson et al., 1986b*]. Focal mechanisms of several of the aftershocks suggest that slip along these northeast trends was left-lateral. Body-wave inversion for the source characteristics of the rupture indicates that the 1968 earthquake sequence was largely composed of two sub-events: the first and largest involving nearly-pure right-slip along

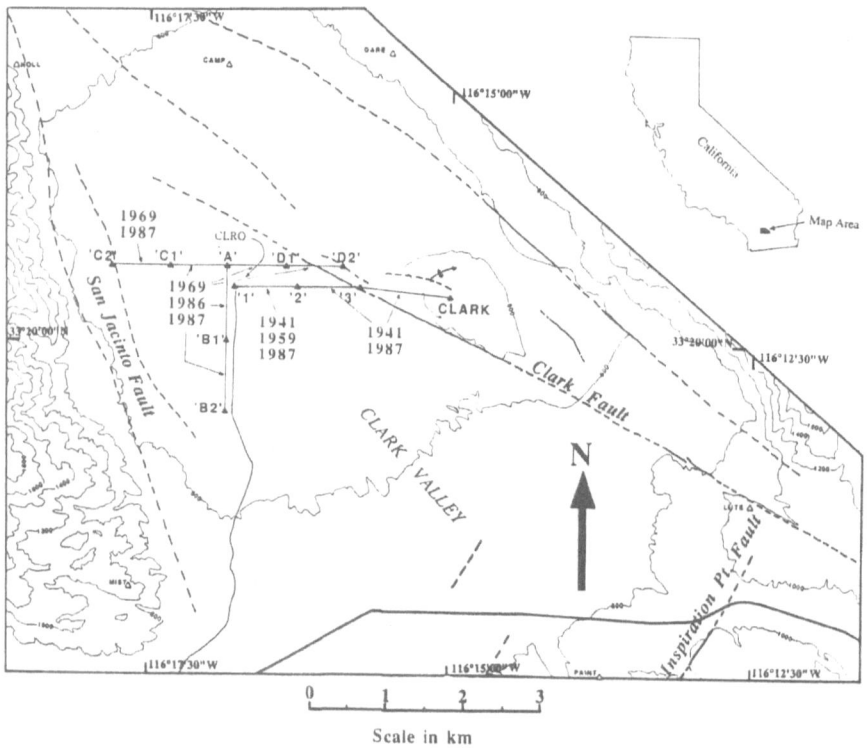

Scale in km

Fig. 13. Map of the Clark Lake Radio Observatory (CLRO), directly above the 1969 rupture (Figure 12d), and adjacent major faults [*Hudnut and Seeber, 1987*]. Sediments in the triangular basin between the Clark and San Jacinto faults are expected to rotate, if slip in 1969 involved a left-slip secondary fault. Dates of geodetic surveys of the CLRO radio antenna are indicated by arrows to the specific lines. Recently re-surveyed benchmarks following the 1969 earthquake are shown as triangles. The 1986 survey along the "T-array" indicated ~4 μrad/yr of net clockwise rotation since 1969 relative to the astro-azimuth of Polaris, on baselines A-C1, A-B1, A-B2, and A-D1 [*Hudnut and Seeber, 1987*]. Benchmarks C2 and D2 were both found to have translated relative to the others, indicating the accumulation of either slip or strain across the San Jacinto and Clark faults.

the Coyote Creek fault; the second occurring about 11 s later and involving reverse faulting on a low-angle detachment at a depth of about 11 km [*Petersen et al.*, 1987].

Almost a year later, the 1969 Coyote Mountain earthquake occurred at the northern end of the 1968 rupture (Figure 12a, solid circles). Most of the aftershocks of this later event were restricted to depths between about 10-14 km (Figure 12b), and defined an elongate pattern whose major axis was oriented northeasterly [*Thatcher and Hamilton*, 1973]. *Petersen et al.* [1987] model the rupture source of the earthquake as occurring on a high-angle left-slip cross-fault. Some of the aftershock focal mechanisms were consistent with left-slip on a northeast-trending structure (Figure 12c, left), however, several aftershocks also exhibited focal mechanisms with a low-angle nodal plane (Figure 12c, right)[*Thatcher and Hamilton*, 1973], consistent with seismic slip on a possible basal detachment. If the pattern of faulting involved in the 1969 earthquake was related to block rotation, then the sediments in the triangular basin that forms Clark Valley above the earthquake rupture (Figures 12d and 13) should have rotated. In 1986, re-survey of the benchmarks installed prior to 1969, as part of the Clark Lake Radio Observatory (CLRO), indicate that part of the radio antenna had indeed rotated relative to the astroazimuth of Polaris following the earthquake [*Hudnut and Seeber*, 1987]. The rotation is believed to have occurred as a rigid block, and not by distributed simple shear, since the area defined by the 5 rotated benchmarks (A, B1, B2, C1, and D1, Figure 13) exhibit no internal deformation within the resolution of the survey. The amount of rotation measured corresponded with a rate of rotation of 4.2 ± 1.0 μrad/yr, or about an order-of-magnitude greater than the rate determined paleomagnetically since the mid-Pleistocene (Figure 14), adjacent to the left-slip Inspiration Point fault located farther south (Figure 13)[*Bogen and Seeber*, 1986].

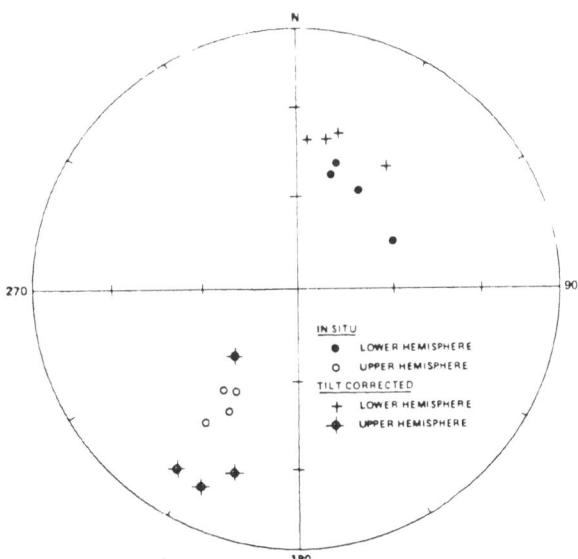

Fig. 14. Paleomagnetic declinations of sample sites next to the left-slip Inspiration Point fault (Figure 13) [*Bogen and Seeber*, 1986]. Data suggests ~30° of clockwise rotation in the last 700,000 years, or about 0.75 μrad/yr.

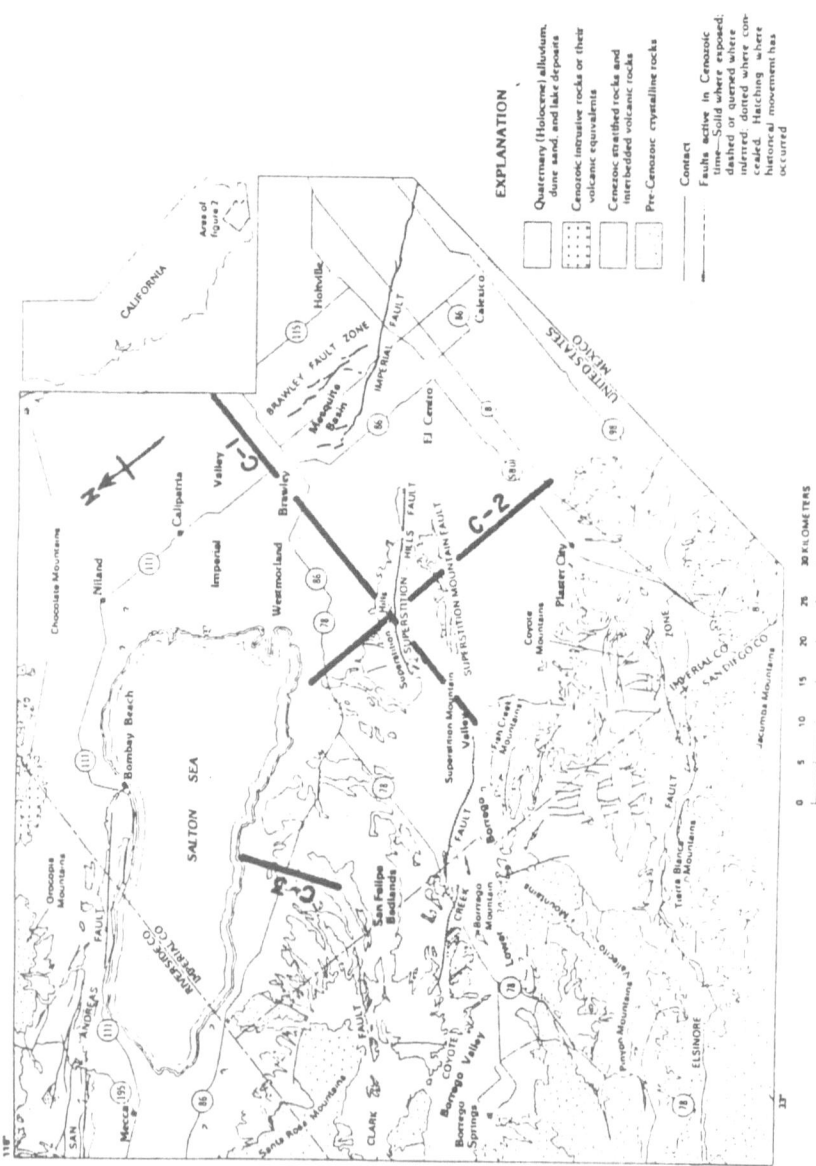

Fig. 15. Generalized geologic map of the western Imperial Valley [modified from *Sharp*, 1982] showing the location of reprocessed seismic reflection profiles of *Severson* [1987]. Line C-3 is shown in Figure 16.

2.6 ADDITIONAL EVIDENCE FOR LEFT-SLIP FAULTS, DETACHEMENTS, AND BLOCK ROTATION

Detailed mapping of existing geologic structures adjacent to major fault strands of the San Andreas system reveals a number of low-angle detachment surfaces and high-angle left-slip faults, much like those suggested by the seismicity [*Dibblee*, 1954, 1977, 1984; *Crowell*, 1974; *Yeats*, 1981; *Powell*, 1981; *Engel and Schultejann*, 1984; *May*, 1985; *Isaacs et al.*, 1986]. Other studies based on well data, seismic reflection and refraction data, and reduced travel-time anomalies imply similar northeast-striking structures in the basement topography surrounding the northern Imperial Valley region [*Fuis et al.*, 1982; *Kohler and Fuis*, 1986]. Several deep-crustal seismic profiles have successfully imaged basal detachments [*Wallace and English*, 1982; *Frost and Okaya*, 1985; *Cheadle et al.*, 1986; *Frost et al.*, 1987; *Severson*, 1987], some of which truncate high-angle strike-slip faults and locally crop out along known dip-slip faults. Figure 15 shows a map of some recently re-processed seismic reflection profiles in the Imperial Valley region [*Severson*, 1987]. Line C-3 is shown in Figure 16. The data have been interpreted as reflecting a shallow detachment at about 1.5 s two-way travel time [*Severson*, 1987]. This structure is located adjacent to the extrapolated position of the recently active right-slip Clark fault (Figure 13), at the southern end of the uplifted Santa Rosa Mountains, where similar detachment structures have been previously imaged [*Wallace and English*, 1982] and mapped in outcrop [*Dibblee*, 1984]. At the southern end of line C-2 (Figure 15), a similar reflective horizon corresponds to a highly-altered zone at the top of the basement, as indicated by well cores, and is also believed to represent a detachment surface at a depth of about 2.4 km [*Severson*, 1987].

Empirical evidence now exists that block rotation by strike-slip faulting can occur on many different scale-lengths, from millimeters to kilometers (Figure 17)[*Nicholson et al.*, 1986b]. Blocks of even larger dimensions have been used to account for the counterclockwise rotation of the Mojave [*Garfunkel*, 1974], the overall deformation of southern California [*Luyendyk et al.*, 1980], and the clockwise rotation of large massifs along the Aleutian arc [*Geist et al.*, 1988]. On the scale of meters to kilometers, *Young et al.* [1985] describe the development of Tertiary rotational features and the evolution of an extensive rotating block system along transform faults in Iceland. Each of these examples requires a detachment surface at depth to decouple the blocks. Minimum energy requirements and the maximum strength of materials suggest that blocks cannot exceed certain dimensions of length, width, and thickness before they will deform and breakup into smaller pieces.Because some block sizes are controlled by a preexisting fabric, this implies that multiple detachment surfaces exist at appropriate levels in the crust to correspond with different block sizes.Thus, in the case of Figure 17a, blocks are defined by crystal cleavage planes [*Simpson and Schmid*, 1983]; whereas, in Figure 17b, deformation of the hard surface layer, caused by the mole-track of the 1979 Imperial Valley earthquake, was controlled by evenly spaced furrows of a plowed field, and the depth to a more ductile, moist sublayer [*Terres and Sylvester*, 1981]. A block rotation model can also account for the observed uplift, and rotated material along Coyote Ridge (Figure 17c)[*Nicholson et al.*, 1986b], in an area the otherwise would be expected to experience extension and subsidence [*Rodgers*, 1980].

Geodetic data are still largely interpreted assuming irrotational models, however, re-interpretation of the vertical deformation associated with strike-slip faults [i.e., *Chinnery*, 1963], along the Brawley seismic zone suggests that northeast-striking left-slip faults are actively involved in the process of strain accumulation [*Larsen and Reilinger*, 1986]. More and more paleomagnetic data provide constraints on the amount, timing and evolution of local shear rotations, and show that tectonic rotations can be quite rapid in some cases, particularly in areas where blocks are small.

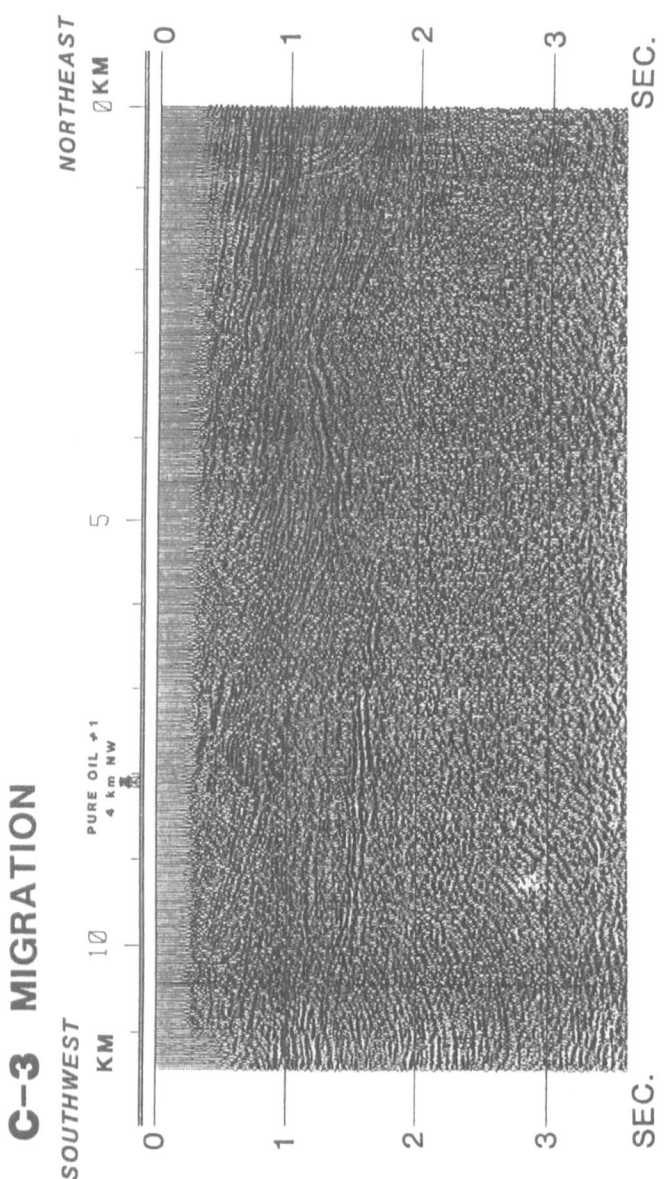

MIGRATED SECTION OF LINE C-3.

Fig. 16. Migrated seismic reflection profile C-3 at the southern end of the Santa Rosa Mountains, extending from the Clark fault to the Salton Sea [*Severson*, 1987]. Data suggests the presence of a possible detachment at a depth corresponding to about 1.5 s two-way travel time, adjacent to a possible extension of a dipping Clark fault.

271

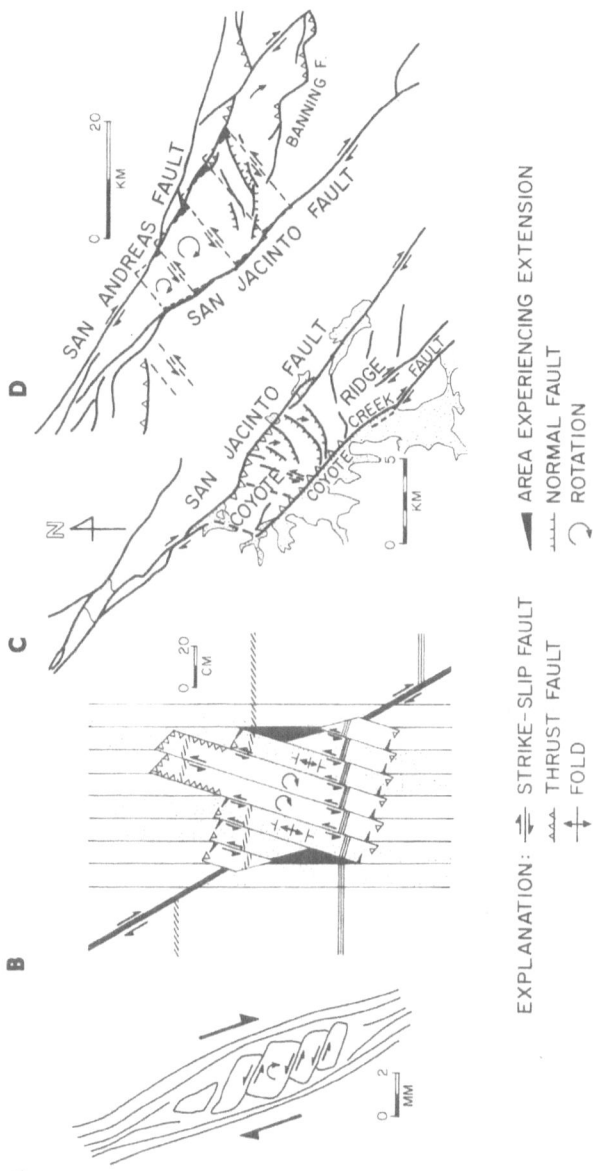

Fig. 17. Geologic examples of block rotation by strike-slip faulting on scales from millimeters to kilometers [*Nicholson et al.*, 1986b]. (*A*) Fracture and rotation of a feldspar crystal along cleavage planes in a ductile matrix [*Simpson and Schmid*, 1983]; (*B*) rotation of a hard desiccated surface layer caused by the mole track of the 1979 Imperial Valley earthquake and controlled by evenly-spaced furrows of a plowed field (ruled lines) [*Terres and Sylvester*, 1981]; (*C*) rotating blocks defined by secondary cross faults between an overlapping right-step from the Coyote Creek fault to the San Jacinto fault (inset E, Figure 1) [*Seeber and Nicholson*, 1986]; (*D*) block model for shear rotation near the intersection of the San Jacinto and San Andreas faults inferred from geology and seismicity (Figure 2)[*Matti et al.*, 1985; *Nicholson et al.*, 1986a].

Johnson et al. [1983] report on the Quaternary deformation of the Vallecito-Fish Creek basin adjacent to the Elsinore fault, southwest of the 1968 Borrego Mountain rupture (Figure 12a).This basin underwent 35° of clockwise rotation in less than 0.9 Ma, after sedimentation ceased and the basin uplifted in a manner that may have been similar to the proposed history of Coyote Ridge (Figure 17c). This rate of rotation would correspond to an average slip rate along the Elsinore fault of 5-7 mm/yr. Although this rate of fault slip is high relative to some earlier estimates [e.g., *Bird and Rosenstock*, 1984], it is within the range of recent Quaternary slip measurements based on trenching [*Rockwell et al.*, 1985]. Care must be taken, however, in averaging sample sites.*Rymer et al.* [1987] describe measured rotations in mid- to early Pleistocene sediments adjacent to the southern San Andreas fault in the Indio Hills. They find small discrete fault-bounded domains of material rotated both clockwise and counterclockwise by as much as 15°. Farther southeast, however, *Chang et al.* [1987] find no net tectonic rotation in sediments located within the southern Mecca Hills, however, mapping by *Sylvester and Smith* [1976] indicate that the sample sites are not within the main zone of strike-slip deformation. These data imply that periods of shear rotation, like displacements, can be episodic and inhomogeneous, but that once conditions are favorable, rotations do occur and are closely coupled to the strike-slip tectonics of the San Andreas System.

3.Discussion

These results and observations imply that block rotation by strike-slip faulting is a contemporary component of the plate boundary deformation in southern California. Thus, left-slip cross-faults, low-angle detachments, and crustal rotations must all be considered when examining the present pattern of strain accumulation. Classical interpretations of small-scale tectonic features along major strike-slip faults that assume irrotational models may need to be modified to allow for various degrees of block rotation, particularly between fault zones or fault systems of finite width. Since block rotation requires detachment surfaces at depth to decouple the blocks, and blocks of different sizes are observed to rotate, this necessarily implies that multiple detachments exist at various levels throughout the crust to accommodate appropriate rotating block dimensions.

A block nature of the crust implies that not only will strains be inhomogeneous and concentrated along edge-bounding weak faults [*Bilham and Beavan*, 1979; *Garfunkel and Ron*, 1985], but that both the elastic and non-elastic behavior of the crust will strongly depend on the nature of any preexisting fabric and the depth to either a detachment or ductile shear zone. Local stress orientations will largely be responding to local kinematic constraints of block rotation and fault interaction. As blocks rotate, the level of normal stress may increase or decrease along strike as block corners rotate into or away from the sides of major bounding faults.This explains how motion on left-slip cross-faults can trigger right-slip (earthquakes) on adjacent major faults (*e.g.*, the 1987 Superstition Hills sequence).Thus, although the total slip accommodated by secondary faulting and block rotation is small compared to the long-term plate motion, these secondary structures can have a significant effect in terms of controlling the rupture behavior of major fault strands during large earthquakes. Nucleation points, characteristic rupture lengths, and the distribution of seismic slip may all be strongly affected by the prior history, position, and fault geometry of secondary structures that are responding to, or a reflection of, varying amounts of block rotation and the local state of stress.

Block rotation will also change the initial angle between left- and right-slip faults, as well as their orientation relative to the regional stress field [*Garfunkel and Ron*, 1985; *Nur et al.*, 1986]. This explains how changes in strike of major strike-slip faults continue to develop, why secondary

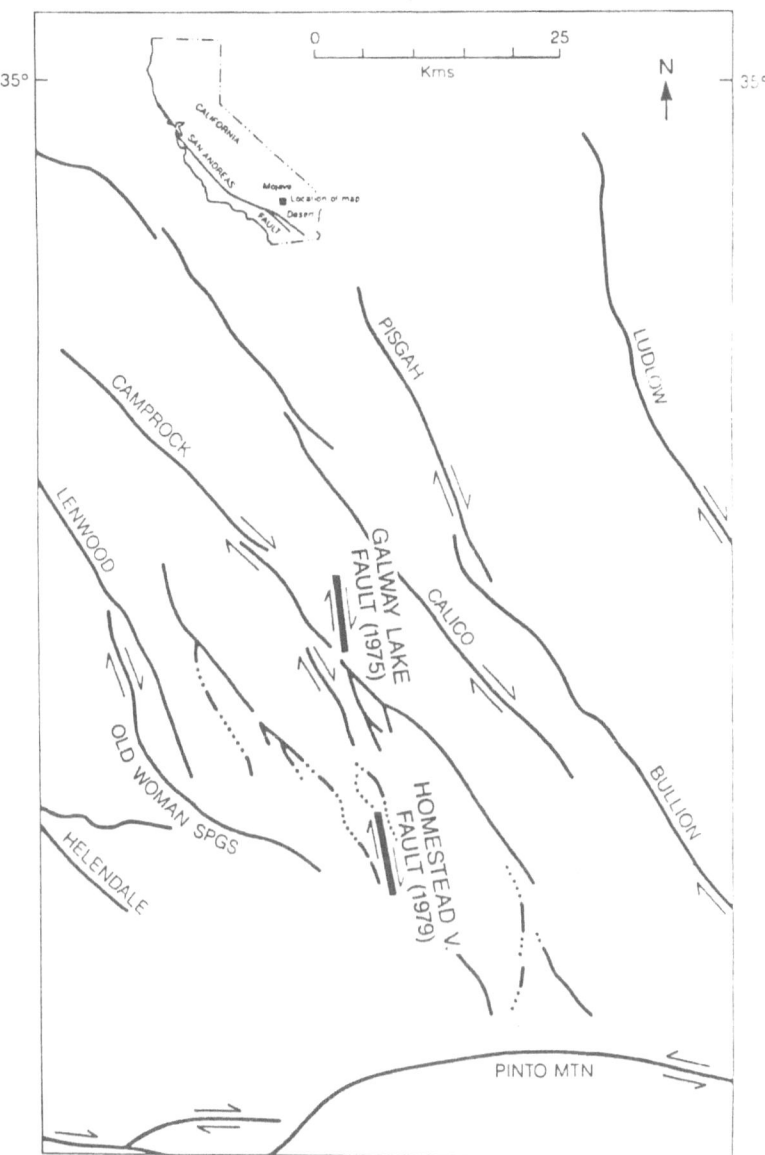

Fig. 18. Possible formation of a new fault set in the Mojave Desert as indicated by the faulting directions associated with the 1975 Galway Lake and 1979 Homestead Valley earthquake sequences [*Nur and Ron*, 1987]. Note their azimuths are distinctly different from the older, rotated central Mojave faults, and their alignment along a single trend.

274

faults may change their sense of slip with time, why rotations eventually cease or reverse, and why new faults may form as old faults are rotated into positions unfavorable for further slip. Figure 18 shows the possible development of a new fault set (as reflected by recent earthquake sequences in 1975 and 1979) in the Mojave Desert [*Nur and Ron*, 1987]. The older regional trend has been rotated some 23° counterclockwise [*Valentine et al.*, 1987], as predicted by *Garfunkel* [1974], and is now no longer favorably oriented relative to the current regional stress field [*Mount and Suppe*, 1987; *Zoback et al.*, 1987].

A major question regarding tectonic rotations in strike-slip environments is the specification of the conditions under which block rotation is preferred over simple shear. In southern California, contemporary block rotation appears to be most prevalent in areas near the intersection of major right-slip faults (*e.g.*, the intersection of the San Jacinto and San Andreas faults, or the intersection of the Brawley seismic zone and the Superstition Hills fault), where shear is distributed and the area is experiencing localized extension parallel to the trends of the major bounding faults (*e.g.*, Figure 11b). Most of these areas also exhibit evidence of a detachment at depth to decouple the blocks. What is uncertain is whether block rotation is driven from below by basal shear in response to ductile flow in the lower crust (as proposed by *Johnson and Hutton* [1986]), or from the sides by an applied horizontal shear couple.

In New Zealand, triangulation surveys along the plate boundary zone exhibit relatively high rates of net tectonic rotation, on the order of 8°/Ma [*Walcott*, 1984]. The short-term geodetic values are consistent with longer-term rates measured paleomagnetically, which show a gradual increase in the amount of plate boundary deformation taken up by shear rotation since the orientation of the plate motion vector became more oblique to the plate boundary fault system about 5 Ma ago [*Walcott*, 1981;*Walcott*, 1984]. This suggests that a certain degree of coupling perpendicular to the fault zone is required before rotations are observed. The fact that the stress field in California is now considered to be more oblique (even perpendicular) to the regional strike of the San Andreas system than originally thought [*e.g.*, *Zoback et al.*, 1987], suggests that coupling from the sides may be a necessary condition. Obviously, more information on the timing and spatial distribution of rotations, as well as the correlation between short-term geodetic measurements, and long-term cumulative effects determined geologically or paleomagnetically, are needed to better understand the process of how strain is partitioned between both elastic and non-elastic deformation, and the extent to which tectonic rotations accommodate shear within strike-slip environments.

Acknowledgements

The authors are indebted to a number of individuals who provided figures, photographs, and discussions of their work on the contemporary crustal deformation in southern California.The authors especially like to thank Tom Rockwell, Art Sylvester, Lupe Severson, Ken Hudnut, Nick Bogen, and Amos Nur for providing material used in the presentation at the conference and in the manuscript, much of which was made available prior to its publication. Eric Frost brought to our attention the recent results from CALCRUST on detachments in the Imperial Valley/Salton Trough area. One of the authors (C. N.) acknowledges the generous support of NATO to attend the workshop, and of ICS in the preparation of the paper. We thank Dick Walcott and Zvi Garfunkel for their reviews of the original manuscript. Institute for Crustal Studies Contribution No. 0008-03TC-02EQ.

References

Aki, K. and P.G. Richards, 1980.*Quantitative Seismology: Theory and Methods*, W.H. Freeman and Co., San Francisco, Calif.

Allen, C. R., P. St. Armand, C. F. Richter, and J. M. Nordquist, 1965. Relationship between seismicity and geologic structure in the southern California region, *Bull. Seismol. Soc. Am.*, **55**, 753-797.

Allen, C. R., 1957. San Andreas fault zone in San Gorgonio Pass, southern California, *Geol. Soc. Am. Bull.*, **68**, 315-350.

Allen, C. R., 1982. The modern San Andreas fault, in *The Geotectonic Development of California*, W.G. Ernst, ed., Prentice-Hall, Inc., Englewood Cliffs, New Jersey, 511--534 .

Allen, C.R., M. Wyss, J.N. Brune, A. Grantz, and R.E. Wallace, 1972. Displacements on the Imperial, Superstition Hills, and San Andreas faults triggered by the Borrego Mountain earthquake, in The Borrego Mountain Earthquake of April 9, 1968, *U.S. Geol. Surv. Prof. Paper 787*, 87--104

Angelier, J., B. Colletta, J. Chorowicz, L. Ortlizb, and C. Rangin, 1981. Fault tectonics of the Baja California Peninsula and the opening of the Sea of Cortez, Mexico, *J. Struct. Geol.*, **3**, 347-359 .

Beck, M. E., Jr., 1976. Discordant paleomagnetic pole positions as evidence of regional shear in the Western Cordillera of North America, *Am. J. Sci.*, **276**, 694-712.

Beck, M. E., Jr., 1980. Paleomagnetic record of plate margin tectonic processes along the western edge of North America, *J. Geophys. Res.*, **85**, 7115-7131.

Bilham, R. and P. Williams, 1985. Sawtooth segmentation of the southern San Andreas fault, California, *Geophys. Res. Lett.*, **12**, 557--560 .

Bilham, R. G., and R. J. Beavan, 1979. Strains and tilts on crustal blocks, *Tectonophysics*, **52**, 121-138.

Bird, P., and R. W. 1984. Rosenstock, Kinematics of present crust and mantle flow in southern California, *Geol. Soc. Am. Bull.*, **95**, 946-957.

Bogen, N. L., and L. Seeber, 1986. Neotectonics of rotating blocks within the San Jacinto fault zone, southern California (abstract), *Eos Trans. AGU*, **67**, 1200 .

Brun, J.P., P.R. Cobbold, and P. Davy, 1989. Experimental insights into mechanisms of block rotation, this volume.

Brune, J. N., and C. R. Allen, 1967. A microearthquake survey of the San Andreas fault system in southern California, *Bull. Seismol. Soc. Am.*, **57**, 277-296.

Budding, K.E. and R.V. Sharp, 1988. Surface faulting associated with the Elmore Desert Ranch and Superstition Hills, California, earthquakes of 24 November 1987 (abstract), *Seismol. Res. Lett.*, **59**, 49.

Carey, S. W., 1958. A tectonic approach to continental drift, in *Continental Drift:* A Symposium convened by S. W. Carey, 177-355, University of Tasmania, Hobart .

Chang, S-B R., C.R. Allen, and J. L. Kirschvink, 1987. Magnetic stratigraphy and a test for block rotation of sedimentary rocks within the San Andreas fault zone, Mecca Hills, southern California, *Quat. Res.*, **27**, 30-40.

Cheadle, M. J., B. L. Czuchra, T. Bryne, C. J. Ando, J. E. Oliver, L. D. Brown, S. Kaufman, P. E. Malin, and R. A. Phinney, 1986. The deep crustal structure of the Mojave desert, California, from COCORP seismic reflection data, *Tectonics*, **5**, 293-320.

Chinnery, M.A., 1963. The stress changes that accompany strike-slip faulting, *Bull. Seismol. Soc. Am.*, **53**, 921--932.

Christie-Blick, N., and K. T. Biddle, 1985. Deformation and basin formation along strike-slip faults, in *Strike-slip Deformation, Basin Formation, and Sedimentation*, edited by K. T. Biddle and N. Christie-Blick, *Soc. Econ. Paleontol. Mineral. Spec. Publ.*, **37**, 1-34.

Clark, M.M., 1984. Map showing recently active breaks along the San Andreas fault and associated faults between Salton Sea and Whitewater River--Mission Creek, California, *U.S. Geol. Surv. Misc. Investigations Map*, I-**1483**, scale 1:24,000.

Cloos, E., 1955. Experimental analysis of fracture patterns, *Geol. Soc. Am. Bull.*, **66**, 241-256.

Cox, A., 1980. Rotation of microplates in western North America, in *The Continental Crust and its Mineral Deposits*, edited by D. W. Strangeway, *Geol. Assoc. Can. Spec. Pap.*, **20**, 305-321.

Crowell, J. C., 1974. Origin of late Cenozoic basins in southern California, in *Tectonics and Sedimentation*, edited by D. G. Howell, *Soc. Econ. Paleontol. Mineral. Spec. Publ.*, **22**, 190-204

Crowell, J. C., 1975. Geologic sketch of the Orocopia Mountains, southeastern California, *Calif. Div. Mines Geol. Spec. Rep.*, **118**, 97-110.

Das, S., and C. Scholz, 1981. Off-fault aftershock clusters caused by a shear stress increase?, *Bull. Seismol. Soc. Am.*, **71**, 1669-1675.

Dibblee, T. W., Jr., 1954. Geology of the Imperial Valley region, California, in *Geology of Southern California*, edited by R. H. Jahns, *Calif. Div. Mines Bull.*, **170**, 21-28 .

Dibblee, T. W., Jr., 1977. Strike-slip tectonics of the San Andreas fault and its role in Cenozoic basin evolution, in *Late Mesozoic and Cenozoic Sedimentation and Tectonics in California*, edited by T. H. Nilsen, 26-38, San Joaquin Geological Society, Bakersfield, Calif.

Dibblee, T.W., Jr. 1984. Stratigraphy and tectonics of the San Felipe Hills, Borrego Badlands, Superstition Hills, and vicinity. In*The Imperial Basin: Tectonics and thermal aspects*, C.A. Rigsby, Ed., 31-44, Soc. Econ. Paleontol. and Mineral., Pacific Section, Sacramento, Calif.

Doser, D.I., and H. Kanamori, 1986a. Depth of seismicity in the Imperial Valley region (1977-1983), and its relationship to heat, flow, crustal structure, and the October 15, 1979 earthquake, *J. Geophys. Res.*, **91**, 675--688 .

Doser, D.I., and H. Kanamori, 1986b. Spatial and temporal variations in seismicity in the Imperial Valley (1902-1984), *Bull. Seismol. Soc. Am.*, **76**, 421--438.

Engel, A. E. J., and P. A. Schultejann, 1984. Late Mesozoic and Cenozoic tectonic history of south central California,*Tectonics*, **3**, 659-676.

Freund, R., 1974. Kinematics of transform and transcurrent faults, *Tectonophysics*, **21**, 93-134.

Freund, R., 1970. Rotation of strike-slip faults in Sistan, southeast Iran, *J. Geol.*, **78**, 188-200 .

Frost, E. G., and D. A. Okaya, 1985. Geometry of detachment faulting in the Old Woman-Turtle-Sacramento-Chenehuevi Mountains region of SE California (abstract), *EOS Trans. AGU*, **66**, 978.

Fuis, G.S., and M.R. Schnapp, 1977. The November-December 1976 earthquake swarms in the northrn Imperial Valley, CAlifornia: Seismicity on the Brawley fault and related structures (abstract), *Eos Trans. AGU*, **58**, 1188 .

Fuis, G. S., W. D. Mooney, J. H. Healey, G. A. McMechan, and W. J. Lutter, 1982. Crustal structure of the Imperial Valley region, *U.S. Geol. Surv. Prof. Paper*, **1254**, 25.80.

Fuis, G. S., 1982. Displacement on the Superstition Hills fault triggered by the earthquake, *U.S. Geol. Surv. Prof. Paper*, **1254**, 145-154 .

Garfunkel, Z., 1974. Model for the late Cenozoic tectonic history of the Mojave Desert, California, and its relation to adjacent regions, *Geol. Soc. Am. Bull.*, **85**, 1931-1944.

Garfunkel, Z., and H. Ron, 1985. Block rotation and deformation by strike-slip faults, 2: The properties of a type of macroscopic discontinuous deformation, *J. Geophys. Res.*, **90**, 8589-8602 .

Geist, E. L., J. R. Childs, and D. W. Scholl, 1988. The origin of summit basins of the Aleutian Ridge: Implications for block rotation of an arc massif, *Tectonics*, **7**, 327-341.

Hadley, D. and J. Combs, 1974. Microearthquake distribution and mechanisms of faulting in the Fontana-San Bernardino area of southern California, *Bull. Seismol. Soc. Am.*, **64**, 477-1499.

Hadley, D. and H. Kanamori, 1977. Seismic structure of the Transverse Ranges, California, *Geol. Soc. Am. Bull.*, **88**, 1469-1478.

Hamilton, R. M., 1972. Aftershocks of the Borrego Mountain earthquake from April 12 to June 12, 1968, *U.S. Geol. Surv. Prof. Paper*, **787**, 31-54 .

Hearn, T. M. and R. W. Clayton, 1986. Lateral velocity variations in southern California: I-Results for the upper crust from Pg waves. *Seism. Soc. Am. Bull.*, **76**, 495-509.

Hempton, H. R., and K. E. Neher, 1986 . Experimental fracture, strain and subsidence patterns over échelon srike-slip faults: Implications for the structural evolution of pull-apart basins. *J. Struc. Geol.*, **8**, 597-605.

Hudnut, K.W. and L. Seeber, 1987. Astroazimuthal geodetic measurement of block rotation in the southern San Jacinto fault zone, California (abstract), *EOS Trans. AGU*, **68,** 287.

Hutton, L. K., and C. E. Johnson, 1981. Preliminary study of the Westmorland, California earthquake swarm (abstract), *Eos Trans. AGU*, **62,** 957.

Hutton, L. K., C. E. Johnson, J. C. Pechman, J. E. Ebel, J. W. Given, D. M. Cole, and P. T. German, 1980. Epicentral locations for the Homestead Valley earthquake sequence, March 15, 1979, *Calif. Geol.*, **33**, 110-114.

Isaac, S., T.K. Rockwell, and G. Gastil, 1986. Plio-Pleistocene detachment faulting, Yuha Desert region, western Salton Trough, northern Baja California (abstract), *Geol. Soc. Am. Abstr. Programs*, **18**, 120.

Jackson, J. A., 1989. Seismicity and rotations in areas of active distributed continental tectonics, *this volume.*

Jennings, C. W., et al., 1975. *Fault map of California*, Calif. Div. of Mines and Geol., Sacramento.

Johnson, C. E., and D. M. Hadley, 1976. Tectonic implication of the Brawley earthquake swarm, Imperial Valley, California, January 1975, *Bull. Seismol. Soc. Am.*, **66**, 1133-1144.

Johnson, C. E., 1977. Swarm tectonics of the Imperial and Brawley faults in southern California (abstract), *Eos Trans. AGU*, **58**, 1188.

Johnson, C.E., 1979. I. CEDAR--An approach to the computer automation of short-period local seismic networks, and II. Seismotectonics of the Imperial Valley of Southern California, *Ph.D. Dissertation*, Calif. Inst. of Tech., Pasadena, California, 332 p.

Johnson, C. E., and D. P. Hill, 1982. Seismicity of the Imperial Valley region, *U.S. Geol. Surv. Prof. Paper*, **1254**, 5-14

Johnson, C. E., and L. K. Hutton, 1982. Aftershocks and preearthquake seismicity, *U.S. Geol. Surv. Prof. Paper*, **1254**, 59-76.

Johnson, C. E., and L. K. Hutton, 1986. A tectonic model for the Imperial Valley and its relation to seismic risk on the southern San Andreas fault (abstract), *Eos Trans. AGU*, **67,** 1200.

Johnson, N. M., C. B. Officer, N. D. Opdyke, G. D. Woodard, P. K. Zeitler, and E. H. Lindsay, 1983. Rates of late Cenozoic tectonism in the Vallecito-Fish Creek basin, western Imperial Valley, California, *Geology*, **11**, 664-667.

278

Jones, L.M., L.K. Hutton, D.D. Given and C.R. Allen, 1986. The North Palm Springs, California, earthquake sequence of July 1986, *Bull. Seismol. Soc. Am.*, **76**, 1830-1837.

Karner, G. D., and J. F. Dewey, 1986. Rifting: Lithospheric versus crustal extension as applied to the Ridge basin of southern California, in *Future Petroleum Provinces of the World*, M. T. Halbouty, ed., *Amer. Assoc. Petrol. Geol. Memoir*, **40**, 317-337.

King, G. C. P., and J. Nabelek, 1985. The role of bends in faults in the initiation and termination of earthquake rupture: implications for earthquake prediction, *Science*, **228**, 984-987.

King, N. E. and J. C. Savage, 1983. Strain rate profile across the Elsinore, San Jacinto and San Andreas faults near Palm Springs, California 1973-1981, *Geophys. Res. Lett.*, **10**, 55-57 .

Kohler, W. M., and G. S. Fuis, 1986. Travel-time, time-term, and basement depth maps for the Imperial Valley Region, California, from explosions, *Bull. Seismol. Soc. Amer.*, **76**, 1289-1303.

Larsen, S., and R. Reilinger, 1986. Fault behavior in the Imperial Valley, California: Evidence from the 1940 and 1979 earthquakes (abstract), *EOS Trans. AGU*, **67**, 1200.

Lisowski, M. and J.C. Savage, 1988. Deformation associated with the Superstition Hills, California, earthquakes of November 1987 (abstract), *Seismol. Res. Lett.*, **59**, 35.

Lomnitz, C., F. Mooser, C. R. Allen, and W. Thatcher, 1970. Seismicity and tectonics of the northern Gulf of California region, Mexico: preliminary results, *Geofis. Int.*, **10**, 27-48.

Luyendyk, B. P., M. J. Kamerling, and R. Terres, 1980. Geometric model for Neogene crustal rotations in southern California, *Geol. Soc. Am. Bull.*, **91**, 211-217.

Luyendyk, B. P., M. J. Kamerling, R. R. Terres, and J. S. Hornafius, 1985. Simple shear of southern California during Neogene time suggested by paleomagnetic declinations, *J. Geophys. Res.*, **90**, 12,455-12,466.

Magistrale, H. and P. Williams, 1987.Earthquakes and Afterslip of the 1987 Superstition Hills earthquake sequence (abstract), *Fall AGU meeting*, December.

Matti, J. C., D. M. Morton, and B. F. Cox, 1985. Distribution and geologic relations of fault systems in the vicinity of the Central Transverse Ranges, southern California, *U.S. Geol. Surv. Open File Report* **85-365**, 27 pp.

May, D. J., 1985. Mylonite belts in the southeastern San Gabriel Mts., California: remnants of a late Cretaceous sinistral transcurrent shear zone (abstract), *Geol. Soc. Am. Abstr. Programs*, **17**, 368.

Meisling, K. E., and R. J. Weldon, 1986. Cenozoic uplift of the San Bernardino Mountains: Possible thrusting across the San Andreas fault (abstract), *Geol. Soc. Am. Abstr. Programs*, **18**, 157.

Mount, V.S. and J. Suppe, 1987.State of stress near the San Andreas fault: Implications for wrench tectonics, *Geology*, **15**, 1143-1146.

Nicholson, C., P. Williams, L. Seeber, and L. R. Sykes, 1985. Seismicity and fault kinematics along the Brawley seismic zone and adjacent regions (abstract), *EOS Trans. AGU*, **66**, 953.

Nicholson, C., L. Seeber, P. Williams and L.R. Sykes, 1986a. Seismicity and fault kinematics through the eastern Transverse Ranges, California: block rotations, strike-slip faulting and low-angle thrusts, *J. Geophys. Res.*, **91**, 4891-4908.

Nicholson, C., L. Seeber, P. Williams and L.R. Sykes, 1986b. Seismic evidence for conjugate slip and block rotation within the San Andreas fault system, southern California, *Tectonics*, **5**, 629-648.

Nicholson, C., R.L. Wesson, D. Given, J. Boatwright and C.R. Allen, 1986c. Aftershocks of the 1986 North Palm Springs earthquake and relocation of the 1948 Desert Hot Springs earthquake sequence (abstract), *EOS Trans. AGU*; **67**, 1089--1090.

Nicholson, C., 1988. Fault interaction and segmentation along the San Andreas fault system, southern California, *USGS Workshop: Fault Segmentation and Controls of Rupture Initiation and Termination*, Palm Springs, Calif., March 1988.

Nur, A., H. Ron, and O. Scotti, 1986. Fault mechanics and the kinematics of block rotations, *Geology*, **14**, 746-749.

Nur, A., and H. Ron, 1987. Kinematics and mechanics of tectonic block rotations: A review, *U.S. Geol. Surv. Open-file Report* **87-591**, 797-823.

Petersen, M.D., L. Seeber, and K. Hudnut, 1987. Importance of transverse features along the southern San Jacinto fault zone, California (abstract), *Eos Trans. AGU*, **68**, 1507.

Richter, C. F., 1958. *Elementary Seismology*, W. H. Freeman, San Francisco, Calif., 768 pp.

Rockwell, T. K., D. L. Lamar, R. S. McElwain, and D. E. Millman, 1985. Late Holocene recurrent faulting on the Glen Ivy north strand of the Elsinore fault, southern California (abstract), *Geol. Soc. Am. Abstr. Programs*, **17**, 404.

Rodgers, D. A., 1980. Analysis of basin development produced by en échelon strike slip faults, in *Sedimentation at Oblique Slip Margins*, edited by P. F. Ballance and H. G. Reading, *Int. Assoc. Sedimentol. Spec. Publ.*, **4**, 27-41.

Ron, H., R. Freund, Z. Garfunkel, and A. Nur, 1984. Block rotation by strike-slip faulting: Structural and paleomagnetic evidence, *J. Geophys. Res.*, **89**, 6256-6270.

Rymer, M. J., J. Boley, and R. Weldon, 1987. Nonuniform rotation (Pleistocene) along the San Andreas fault in the Indio Hills, southern California (abstract), *Eos Trans. AGU*, **68**, 1507.

Salyards, S. L., K. E. Sieh, and J. L. Kirschvink, 1987. Paleomagnetic measurement of dextral warping during the past three large earthquakes at Pallett Creek, southern California (abstract), *Geol. Soc. Amer. Abstr. Programs*, **19**, 828.

Sanders, C., H. Magistrale, and H. Kanamori, 1986. Rupture patterns and preshocks of large earthquakes in the southern San Jacinto fault zone, *Bull. Seismol. Soc. Am.*, **76**, 1187--1206.

Savage, J.C., W.H. Prescott, M. Lisowski, and N.E. King, 1979. Deformation across the Salton Trough, California, 1973-1977. *J. Geophys. Res.*, **84**, 3069--3080.

Savage, J.C., W.H. Prescott, and G. Gu, 1986. Strain accumulation in southern California, 1973--1984, *J. Geophys. Res.*, **91**, 7455-7473.

Savage, J.C., W.H. Prescott, M. Lisowski, and N.E. King, 1981. Strain accumulation in southern California, 1973-1980, *J. Geophys. Res.*, **86**, 6991-7001.

Seeber, L., and C. Nicholson, 1986.Block/fault rotation in geologic and interseismic deformation, in National Earthquake Prediction Council Special Report 1: Workshop on Special Study Areas in Southern California, edited by C. Shearer, *U.S. Geol. Surv. Open File* Report **6-580**, 185-203.

Seeber, L., and N. L. Bogen, 1985. Block/fault rotation along the southern San Jacinto fault zone (abstract), *Eos Trans. AGU*, **66**, 953.

Seeber, L., J.G. Armbruster, P. Williams, and L.R. Sykes, 1986. Faults antithetic to the San Andreas delineated by seismicity in the San Gorgonio Pass area of southern California, *EOS Trans. AGU*, **67**, 1200.

Severson, L.K., 1987. Interpretation of shallow crustal structure of the Imperial Valley, California, from seismic reflection profiles, *M.S. Thesis, Lawrence Berkeley Laboratory* Report - **LBL-23888**, University of California, Berkeley, 68 pp.

Sharp, R. V., San Jacinto fault zone in the Peninsular Ranges of southern California, *Geol. Soc. Am. Bull.*, **78**, 705-730 (1967).

Sharp, R.V., M.J. Rymer, and J.J. Lienkaemper, 1986. Surface displacement on the Imperial and Superstition Hills faults triggered by the Westmorland, California, earthquake of 26 April 1981, *Bull. Seismol. Soc. Am.;* **76**, 949-965.

Sharp, R.V., 1987. Surface rupture associated with the 1987 Superstition Hills earthquake sequence (abstract), Fall AGU meeting, December 1987.

Simpson, C., and S. M. Schmid, 1983. An evaluation of criteria to deduce the sense of movement in sheared rocks, *Geol. Soc. Am. Bull.*, **94**, 1281-1288.

Stein, R.S., and M. Lisowski, 1983. The 1979 Homestead Valley earthquake sequence, California: Control of aftershocks and postseismic deformation, *J. Geophys. Res.*, **88**, 6477-6490.

Sylvester, A. G., and Smith, R. R., 1976. Tectonic transpression and basement-controlled deformation in the San Andreas fault zone, Salton Trough, California, *Am. Assoc. Petrol. Geol. Bull.*, **60** (12), 2081-2102.

Tchalenko, J. S., 1970. Similarities between shear zones of different magnitudes, *Geol. Soc. Am. Bull.*, **81**, 1625-1640.

Terres, R. R., and A. G. Sylvester, 1981. Kinematic analysis of rotated fractures and blocks in simple shear, *Bull. Seismol. Soc. Am.*, **71**, 1593-1605.

Terres, R. R., and B. P. Luyendyk, 1985. Neogene tectonic rotation of the San Gabriel region, California, suggested by paleomagnetic vectors, *J. Geophys. Res.*, **90**, 12,467-12,484.

Thatcher, W. and R. M. Hamilton, 1973. Aftershocks and source characteristics of the 1969 Coyote Mountain earthquake, San Jacinto fault zone, California, *Bull. Seismol. Soc. Am.*, **63**, 647-661.

Valentine, M. J., L. Brown, and M. P. Golombek, 1987. Tectonic rotation of the Barstow area, central Mojave desert, California, inferred from paleomagnetic data (abstract), *Eos Trans. AGU*, **68**, 1254.

Walcott, R. I., D. A. Christoffel, and T. C. Mumme, 1981. Bending within the axial tectonic belt of New Zealand in the last 9 my from paleomagnetic data, *Earth Planet. Sci. Lett.*, **52**, 427-434.

Walcott, R. I., 1984. The kinematics of the plate boundary zone through New Zealand: A comparison of short- and long-term deformation, *Geophys. J. R. Astr. Soc.*, **79**, 613-633.

Wallace, R.D., and D.J. English, 1982. Evaluation of possible detachment faulting west of the San Andreas, southern Santa Rosa Mountains, California, in *Mesozoic-Cenozoic Tectonic Evolution of the Colorado River Region, California, Arizona, and Nevada*, E.G. Frost and D.L. Martin, eds., Geol. Soc. Amer., Boulder, Colorado, 502-510.

Weaver, C. S., and D. P. Hill, 1979. Earthquake swarms and local crustal spreading along major strike-slip faults in California, *Pure Appl. Geophys.*, **117**, 51-64.

Webb, T. H., and H. Kanamori, 1985. Earthquake focal mechanisms in the Eastern Transverse Ranges and San Emigdio Mountains, southern California, and evidence for a regional decollement, *Bull. Seismol. Soc. Am.*, **75**, 737-757.

Wilcox, R. E., T. P. Harding, and D. R. Seely, 1973. Basic wrench tectonics, *Am. Assoc. Pet. Geol. Bull.*, **57**, 74-96.

Yeats, R. S., 1981. Quaternary flake tectonics of the California Transverse Ranges, *Geology*, 9, 16-20.

Young, K. D., M. Jancin, B. Voight, and N. I. Orkan, 1985. Transform deformation of Tertiary rocks along the Tjornes fracture zone, north central Iceland, *J. Geophys. Res.*, **90**, 9986-10,010.

Zoback, M.D. et al., 1987. New evidence on the state of stress of the San Andreas fault system, *Science*, **238**, 1105-1111.

THE IMPORTANCE OF MAGNETOSTRATIGRAPHY FOR STUDIES OF TECTONIC ROTATIONS: EXAMPLES FROM THE MIO-PLIOCENE OF CALIFORNIA

K. L. VEROSUB and E. J. HOLM
Department of Geology
University of California
Davis, California USA 95616

ABSTRACT. Magnetostratigraphy provides a means of identifying paleomagnetic samples with a strong secondary overprint as well as a means of determining whether the time-averaged field can be assumed to be a geocentric axial dipole. Two Neogene sedimentary sequences in California provide examples of the application of magnetostratigraphy to studies of tectonic rotations. In the case of the Ridge Basin sediments of southern California, the magnetostratigraphic analysis leads to a new interpretation of previously published data. In the case of the Purisima Formation of coastal central California, it provides evidence that an observed rotation of 50° in less than 3.4 million years is due to tectonic rather than geomagnetic or paleomagnetic factors. The rate of rotation of the Purisima Formation is among the highest rates ever documented and demonstrates that for older rocks it may be impossible to distinguish clockwise rotations from counterclockwise rotations.

1. Introduction

1.1. FUNDAMENTAL ASSUMPTIONS

Determination of apparent polar wander paths for major tectonic blocks usually involves data from many geologic formations spanning a long interval of time. In contrast, studies of tectonic rotations usually involve small structural units which contain a limited number of formations only some of which are suitable for paleomagnetic study. The smaller spatial and temporal extent of these structural units makes it more difficult to satisfy two of the fundamental assumptions involved when paleomagnetic data are interpreted in terms of tectonic rotations. These assumptions are that 1) the paleomagnetic sampling represents sufficient time that the geomagnetic field has averaged to a geocentric axial dipole and 2) the observed paleomagnetic directions represent the original primary directions with no secondary overprints. For small structural units, it is often difficult to be certain that either of these conditions has been met.

Although originally it was assumed that the field would average to a geocentric axial dipole in as little as 5,000 years, new evidence suggests that non-dipole features in the field can persist for periods of at least a hundred thousand and perhaps even a million years. In a study of sediments from a now-drained middle Pleistocene lake in Nevada, Negrini et al. (1987) examined the validity of the geocentric axial dipole hypothesis for time periods that were different from and considerably longer than the last 10,000 years, the period on which most conclusions about the geocentric axial dipole hypothesis have been based. They were able to document that mean offsets of 19° in declination and 22° in inclination had existed over a time period estimated to be at least 115,000

281

C. Kissel and C. Laj (eds.), Paleomagnetic Rotations and Continental Deformation, 281–292.

years. It is not yet known whether the time interval of one hundred thousand years is near the upper limit for the lifetime of such large-scale non-dipole offsets. Schneider and Kent (1986) concluded from an analysis of paleomagnetic studies of marine sediments that non-geocentric axial components might exist for much longer periods of time although the effect on the geocentric axial dipole direction would be relatively small and would primarily involve the inclination rather than the declination. It is also not yet known whether the interval studied by Negrini et al. should be considered typical or anomalous. That question can only be answered by studying sedimentary sections of different age but similar duration. Until these studies are completed, we must consider several hundred thousand years to be the minimum amount of time necessary for the field to average to a geocentric axial dipole. In order to demonstrate that this condition has been met, any study of tectonic rotations must include some estimate of the time interval encompassed by the paleomagnetic sampling.

With regard to the second assumption, thermal demagnetization or alternating field demagnetization are not always fully effective in removing secondary components of magnetization, particularly for sedimentary rocks. When data from many different formations are averaged together, the incomplete removal of secondary components is a relatively minor problem. When only a single formation is involved, this problem can be considerably more important. In particular, Late Cretaceous and Tertiary pole positions are usually not substantially different from the present pole. For a small structural unit which has not been rotated, the succession of normal and reversed dipole orientations to which the rocks have been exposed will be essentially antipodal. Any secondary components are likely to be randomly distributed around the original direction or its antipode. The incomplete removal of these secondary components during demagnetization will increase the α_{95} but will not change the mean direction. On the other hand, as a unit is rotated tectonically, the succession of dipole directions forms an arc between the original and final geocentric axial dipole direction. Secondary components have a high probability of lying on or near this arc or its antipode. In this case, the effects of the incomplete removal of the secondary components will be coherent, and the mean paleomagnetic direction will lie on the great circle that includes the arc. The measured declination will be shifted from the actual primary declination leading to the incorrect determination of the amount of rotation. This problem is most severe when the secondary component is of opposite polarity to the primary component. Therefore it is important to be able to identify and eliminate samples which are strongly overprinted with components of the opposite polarity.

1.2. ROLE OF MAGNETOSTRATIGRAPHY

Magnetostratigraphy provides a means of addressing both of the problems described above. The first step in developing the magnetostratigraphy of a section is to establish the magnetic polarity zonation, that is, the sequence of normal and reversed polarity intervals contained within the section. Radiometric or biostratigraphic data are then used to correlate the magnetic polarity zonation to the Magnetic Polarity Time Scale. The magnetic polarity zonation is particularly useful in identifying samples of the "wrong" polarity within a given zone. For example, without the magnetic polarity zonation, reversed samples with strong normal overprints might incorrectly be grouped with normal polarity samples thus biasing the mean direction for the normal polarity samples. When the magnetic polarity zonation has been correlated to the Magnetic Polarity Time Scale, it is possible to determine the duration of each polarity interval used in the study. Although we do not yet know how much time is "enough time," knowledge of the actual time involved allows us to make a relative assessment of the reliability the results of a study. Two examples of the advantages and advantages and limitations of magnetostratigraphy in the study of tectonic rotations are provided by sediments of the Ridge Basin of southern California and the Purisima Formation of central coastal California.

2. Ridge Basin

2.1. GEOLOGIC SETTING

The Ridge Basin is an uplifted, 11,000 meter sequence of interbedded sandstones and shales comprising the Castaic, Ridge Route, Peace Valley, and Hungry Valley formations (Crowell, 1975). These sediments were deposited during the Late Miocene and Early Pliocene in an estuarine basin located at the juncture of two of the major active faults in California, the San Andreas Fault and the Garlock Fault (Figure 1). The two faults intersect at an angle of about 60°, and the Ridge Basin developed as a result of the interaction between the two faults. In fact, what is now the San Gabriel Fault is believed to have been the ancestral San Andreas Fault. In the Miocene, left-lateral strike-slip motion on the Garlock Fault produced a zone of compression north of the intersection and a zone of extension (the Ridge Basin) south of it. Various models for the tectonic evolution of southern California during the Neogene imply some degree of rotation for the sediments of the Ridge Basin. For example, Garfunkel (1974) has proposed a counterclockwise rotation of 30° during the Late Miocene and Pliocene while Luyendyk et al. (1980) have proposed that there was a clockwise rotation of 75° at some time in the Miocene.

2.2 PALEOMAGNETIC PROCEDURES

Almost 2,000 paleomagnetic samples were collected from 670 horizons in the lower 6,000 meters of the Ridge Basin sequence, representing the Castaic, Ridge Route and Peace Valley formations (Ensley and Verosub, 1982). Because the shales were extremely fissile, samples were collected only from the sandstone layers, using a portable diamond core drill. Studies of the acquisition and demagnetization of a saturation isothermal remanent magnetization demonstrated that the principal magnetic carrier was magnetite. A fold test showed that the magnetization of the sediments had been acquired prior to regional deformation.

All of the samples were subjected to step-wise alternating field demagnetization, and median destructive fields ranged from 15 to 30 mT. Although most of the samples reached stable endpoints, in many cases the final directions of magnetization were of intermediate polarity. In addition, very few samples exhibited fully reversed final directions. These observations were taken as evidence for the presence of a high-coercivity normal overprint at least on the reversely magnetized samples.

2.3 MAGNETIC POLARITY ZONATION

In order to overcome this problem, it was necessary to examine the behavior of the virtual geomagnetic pole (VGP) of each sample during demagnetization. In general the samples could be classified in one of three ways. Normal samples had VGP latitudes that were within 40° of the north geographic pole or showed more than 10° of movement toward that pole. Reversed samples had VGP latitudes that were within 40° of the south geographic pole or showed more than 10° of movement toward that pole. Intermediate samples had VGP latitudes that were more than 40° from

284

either pole or showed less than 10° of movement toward a pole. The polarity of an horizon was based on the polarities of the samples. An horizon was considered normal if it contained two or more normal samples and no reversed samples, reversed if it contained one or more reversed samples, and intermediate if it did not fall into either of the above categories. The magnetic polarity zones were identified by the consecutive occurrence of horizons of the same polarity (Ensley and Verosub, 1982).

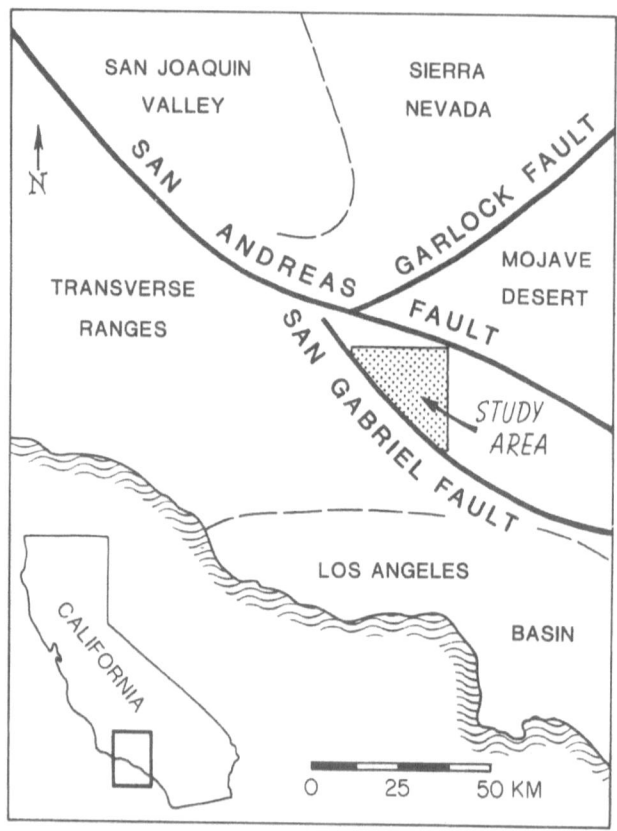

Figure 1. Geologic setting of the Ridge Basin.

This approach resulted in the development of a magnetic polarity zonation containing seven normal magnetozones and six reversed magnetozones. The zonation was correlated to the Magnetic Polarity Time Scale using biostratigraphic information which was available for the bottom and top of the entire sequence of Ridge Basin sediments. Near the bottom, just below the lowest sampling horizon, was a marine invertebrate fauna that had been identified as being earliest Jacalitos, a provincial Californian molluscan stage which lasted from about 9 Ma to 6 Ma. At the top of the sequence, 5,000 meters above the highest paleomagnetic horizon, was a terrestrial vertebrate fauna identified as late Hemphillian and having an age of 5.5 Ma. Thus the magnetic polarity zonation had to represent a time interval which was younger than 9 Ma but sufficiently older than 5.5 Ma to permit the deposition of the upper half of the sedimentary sequence. The correlation which best fits these criteria is shown in Figure 2.

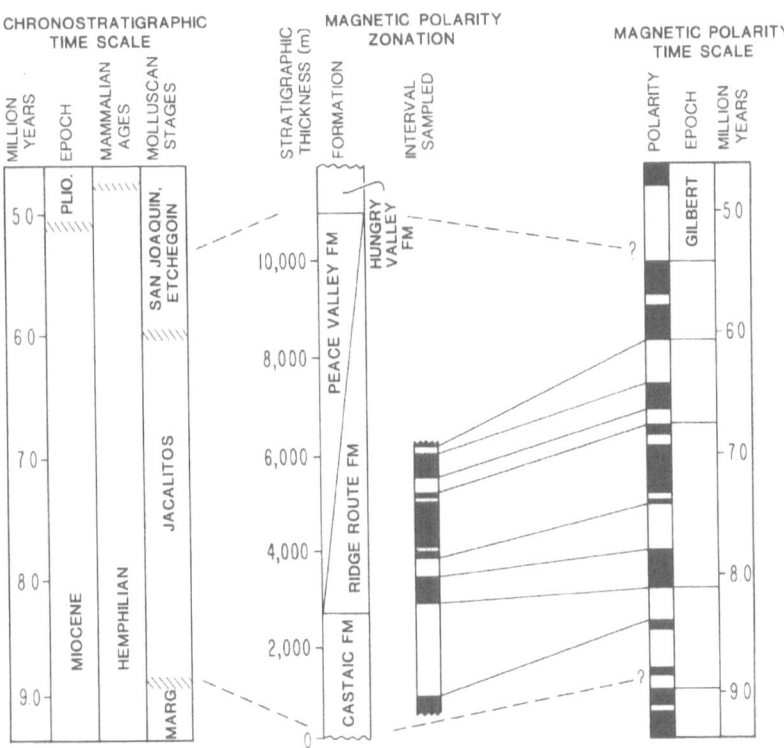

Figure 2. Magnetic polarity zonation (MPZ) for the sediments of the Ridge Basin and its correlation to the Magnetic Polarity Time Scale (MPTS).

2.4. TECTONIC ROTATION

With this correlation it was possible to analyze the data for possible tectonic rotations. Mean directions were computed from each of the normal magnetozones which are designated from oldest to youngest as N1 to N7. The existence of the magnetic polarity zonation for this sequence insured that only samples of normal polarity were included in the computation of these mean directions. Mean directions for the reversed magnetozones were not computed because so few samples exhibited fully reversed directions. The declinations corresponding to the mean directions of the normal magnetozones are shown in Figure 3 along with the range in declination corresponding to the values of α_{95} for each magnetozone.

The data in Figure 3 were originally interpreted as representing first a clockwise and then a counterclockwise "jostling" of the basin (Enslay and Verosub, 1982). Such behaviour would not be

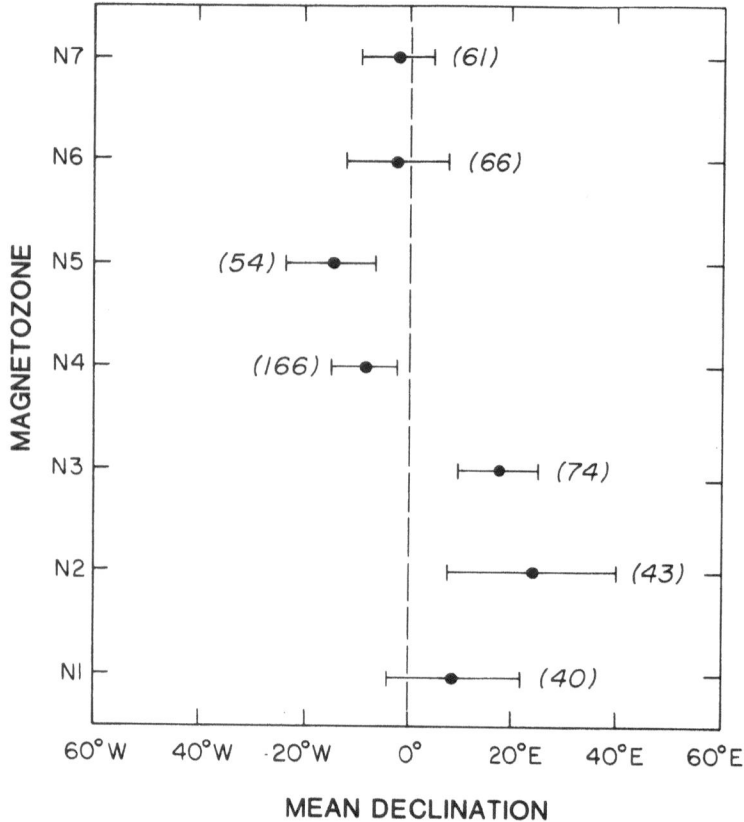

Figure 3. Mean declination of samples from each normal polarity zone of the sediments of the Ridge Basin. The ranges are derived from the α_{95} of the mean directions.

surprising in such a tectonically active area. Furthermore, because samples in the reversed magnetozones were so strongly affected by a normal overprint, one could not exclude the possibility that the mean normal directions had been influenced by the same overprint. As there was evidence that the overprint was considerably younger than the original magnetization, the implication was that the secondary components would have been rotated less than the primary components. Thus the rotations shown in Figure 3 were considered minimum values. However, using the magnetostratigraphy, we have computed the amount of time actually represented by the magnetozones N2-N6. (This can not be done for magnetozones N1 and N7 because we do not know how much is missing from the bottom and top, respectively, of these zones.) According to the Magnetic Polarity Time Scale (Harland et al., 1982), magnetozones N2, N4 and N6 represent time intervals of .37, .40, and .13 million years, respectively, but magnetozones N3 and N5 represent only .05 and .09 million years, respectively. In view of the higher reliability associated with results from longer time intervals, the original interpretation should be revised. Although jostling of the basin is still a possibility, an alternate interpretation of the results is that there was clockwise rotation of the Ridge Basin prior to about 7.5 million years but probably little or no rotation after that.

3. Purisima Formation

3.1 GEOLOGIC SETTING

The Mio-Pliocene Purisima Formation is a shallow-water marine siltstone and sandstone sequence found in central coastal California on several tectonic blocks bounded by faults of the San Andreas

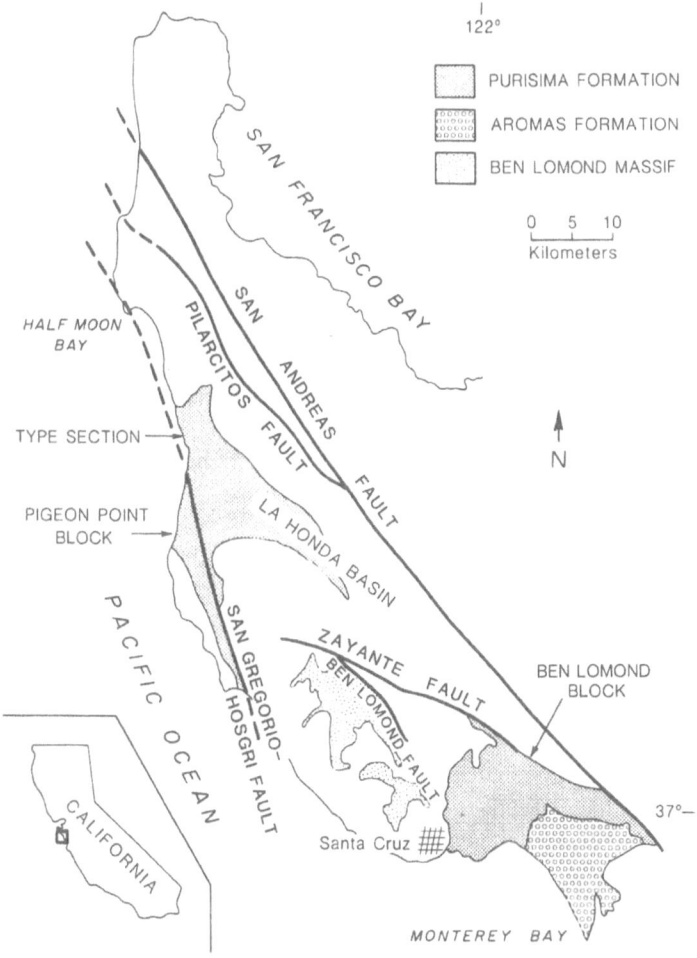

Figure 4. Geologic setting of the Purisima Formation.

Fault System (Cummings et al., 1962). The San Gregorio-Hosgri Fault is generally regarded as the westernmost major fault of this system. The fault is nearly 400 kilometers long with most of its exposure lying offshore. It is a zone of active faulting and is responsible for both the 1927 Lompoc and the 1926 Monterey earthquakes. Estimates for the right lateral Neogene offset along this fault zone range from 80 to 115 kilometers (Graham and Dickinson, 1978). One of the tectonic blocks on which the Purisima Formation is found is the Pigeon Point block which is bounded on the east by the San Gregorio Fault and on the west by the Pacific Ocean (Figure 4).

3.2 PALEOMAGNETIC PROCEDURES

Paleomagnetic samples were collected from a nearly-continuous 300 meter coastal section within the Pigeon Point block. The predominate lithologies are fine grained sandstones, siltstones and claystones interbedded with several tuffs and tuffaceous mudstones. At least three samples were collected from each of 102 sampling horizons. The typical vertical separation between horizons was 3 meters. Samples were collected by carving a small pedestal of sediment in the outcrop using stainless steel tools. A plastic box (2.5 cm x 2.5 cm x 1.8 cm) was then placed over the pedestal, and the full orientation of the box was recorded before the box and its sample were removed from the outcrop.

One-third of the samples were measured with a spinner magnetometer using a six spin procedure. The remainder of the samples were measured with a 3-axis cryogenic magnetometer.

In a pilot study samples from several horizons were subjected to thermal demagnetization in 50°C increments from 0° to 300° C and then 100° C increments to a maximum temperature of 600°C. Large changes in magnetic susceptibility above 150° C, erratic changes in direction, and a marked reddening of the samples provided evidence that chemical alteration was occurring that would preclude the use of thermal demagnetization for these samples. Consequently, all samples were subjected to alternating field demagnetization in 5 mT increments from 0 to 30 mT then in 10 mT increments to a maximum applied field of 60 mT. A variety of rock magnetic studies demonstrated that the magnetic carriers were single domain and pseudo-single domain magnetite.

3.3. MAGNETIC POLARITY ZONATION

Initially we attempted to determine the magnetic polarity zonation for this section using the methods described above for the Ridge Basin study. However, when the latitudes of the virtual geomagnetic poles were plotted as a function of stratigraphic position, no clear polarity pattern emerged. Examination of Zijderveld plots and plots of the normalized intensity allowed us to identify a number of samples whose behavior indicated that they were probably of normal polarity. However, the mean direction of these samples corresponded to a declination of 50°E, implying that the entire section had been rotated. Data from all of the samples were then rotated counterclockwise by 50° and the magnetostratigraphic analysis was repeated. This time five normal and five reversed polarity intervals could be recognized.

The magnetic polarity zonation was correlated to the Magnetic Polarity Time Scale using a tephra layer and a biostratigraphic constraint. The tephra layer was located eighteen meters below the top of the sampled section, and its trace element geochemistry is very similar to that of the Nomlaki Tuff which has an age of 3.4 Ma (A. Sarna-Wojcicki, oral communication, 1987). The biostratigraphic constraint was provided by the presence of the diatoms *Thalassiosira oestrupii* and *Thalassiosira praeoestrupii* about 15 meters above the base of the section (J. D. Barron, written communication, 1987). According to the North Pacific Diatom Zonation (Barron, 1986), the occurrence of both of these species dates this horizon as about 5.2 Ma. The only possible correlation of the magnetic polarity zonation with the Magnetic Polarity Time Scale that satisfies these age constraints is shown in figure 5. This correlation implies that this 300 meter section of the Purisima Formation was deposited between about 5.2 Ma and 3.4 Ma.

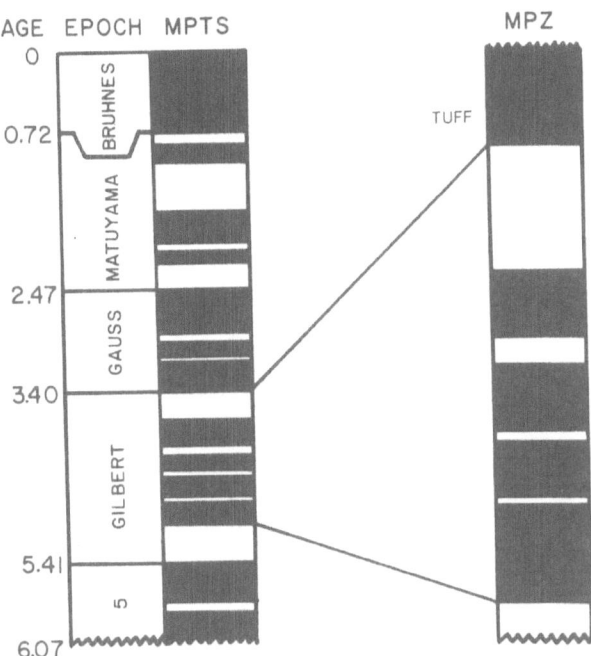

Figure 5. Magnetic polarity zonation for the sediments of Pigeon Point block of the Purisima Formation and its correlation to the Magnetic Polarity Time Scale.

3.4. TECTONIC ROTATION

With the magnetic polarity zonation, it is possible to return to the original paleomagnetic database and determine the tectonic rotation. For all of the normal samples the mean direction corresponds to a clockwise rotation of $49°\pm3°$ (Figure 6). As in the study of the samples from the Ridge Basin, the strong secondary overprint on the reversed samples precludes their use for the determination of a mean direction. The possibility that a secondary overprint is also present in the normal samples means that 49° must be regarded as a minimum amount of rotation. However, the magnitude of the rotation and the fact that it has been determined from samples that represent time-averaging over a period of 1.8 Ma gives us considerable confidence in the existence of such a large rotation.

This result for the Pigeon Point block of the Purisima Formation can be compared to a result from the Ben Lomond block of the Purisima Formation. The Pigeon Point block is now located northwest of the Ben Lomond block (Figure 1). It has been suggested that the Pigeon Point block is a deeper water equivalent of the Ben Lomond block and that motion on the San Gregorio-Hosgri Fault has translated the Pigeon Point block from a position southwest of the Ben Lomond block to its present position (Greene and Clark, 1978). A magnetostratigraphic study of 240 samples from 80 horizons of a 225 meter section of the Purisima Formation on the Ben Lomond block showed that these sediments had been deposited between 6.1 Ma and 2.5 Ma with a hiatus between 4.5 Ma and 3.5 Ma (Madrid et al., 1986). Our compilation of the data from the normal polarity intervals of that study yields a mean direction corresponding to a rotation of 4°±3° (Figure 6).

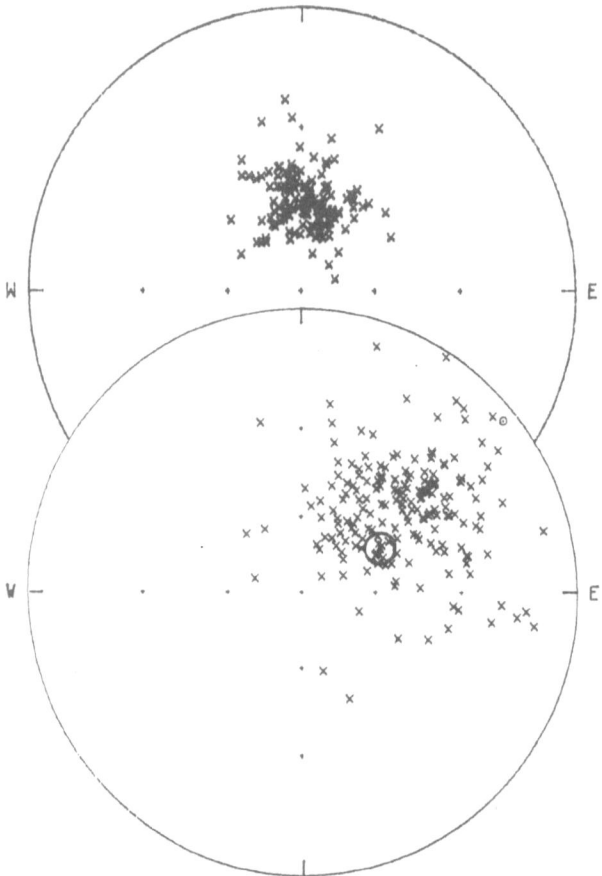

Figure 6. Stereographic projection of stable endpoints of samples from normal polarity zones of the Purisima Formation. Bottom: Pigeon Point block; top: Ben Lomond block.

3.5. IMPLICATIONS

Recognition of 49° of clockwise rotation on the Pigeon Point block west of the San Gregorio-Hosgri Fault and essentially no rotation on the Ben Lomond block east of it implies that the interaction between the Pacific Plate and the North American Plate is more complex than the simple model of a transform fault boundary would suggest. The fact that the Pigeon Point block has been rotated is strong evidence for the existence of another major active fault offshore to the west of the

San Gregorio-Hosgri Fault. Such a fault has previously been proposed as a means of explaining the discrepancy between the rate of slip on the San Andreas Fault as determined from geologic and geodetic data and the rate as determined from rigid plate models (Minster and Jordan, 1987). According to a simple "ball-bearing" model (Beck, 1976), right-lateral strike-slip motion on this offshore fault and on the San Gregorio-Hosgri Fault would produce the clockwise rotation observed in the Pigeon Point block. However, the block as presently exposed is a long sliver along the San Gregorio-Hosgri Fault. A palinspastic restoration based on this geometry places the block across the very fault which is presumably responsible for its rotation. In addition, if the rotation of the Pigeon Point block is attributed to its location between two active strike-slip faults, one would expect significant rotation of the Ben Lomond block which is located between two known, active faults. Furthermore, the San Gregorio-Hosgri Fault is usually considered subsidiary to the San Andreas Fault but the observed rotations imply that the former has possibly been a more important tectonic element in central coastal California than the latter. The questions raised by these studies can only be resolved by addition work on the Purisima and equivalent formations in the region.

Whatever the resolution of these problems, a 49° rotation of the Pigeon Point block in the last 3.4 million years implies a minimum rate of rotation on the order to 15°/my. This is one of the highest rates ever documented and provides graphic evidence of the extreme mobility of small crustal plates. Assuming that the tectonic regime does not change, the Pigeon Point block would rotate 360° in 30 million years. One implication of this observation is that for rocks that are Paleogene in age or older, it is not possible to determine whether a tectonic rotation has been clockwise or counterclockwise because rotations of more than 180° would be possible over this interval of time.

4. Conclusion

Although magnetostratigraphy can not solve all of the problems arising from the limited spatial and temporal extent of small structural blocks, it can provide a chronostratigraphic framework for samples from such blocks. This framework serves as a means of estimating the degree to which the geocentric axial dipole hypothesis is valid as well as identifying samples that have been incorrectly included in a data set. In some cases, the framework can lead to more subtle and more complex interpretations of the tectonic evolution of a block, but in other cases the magnetostratigraphy will simply bring out the limitations of the paleomagnetic method.

Acknowledgements

This research was supported by the donors of the Petroleum Research Fund, administered by the American Chemical Society.

References

Barron, J.A., 1986,Paleoceanographic and tectonic controls on deposition of the Monterey Formation and related siliceous rocks in California. *Paleo. Paleo. Paleo*, **53,** 27-45.

Beck, M.E., 1976,Discordant paleomagnetic pole positions as evidence of regional shear in the western Cordillera of North America. *Amer. J. Sci.*, **276**, 694-712 .

292

Crowell, J.C., 1975, The San Gabriel Fault and the Ridge Basin, southern California, in *The San Andreas Fault in Southern California*, edited by J.C. Crowell, Calif. Div. Mines Geol. Spec. Publ. 2, 208-233.

Cummings, J.C., R.M. Touring and E.E. Brabb, 1962, Geology of the northern Santa Cruz Mountains, California, in *Geologic Guide to the Gas and Oil Fields of California*, edited by O.E. Bowen, Calif. Div. Mines. Geol. Bull. 181, 179-220.

Ensley, R.A., and K.L. Verosub, 1982, A magnetostratigraphic study of the sediments of the Ridge Basin, southern California, and its tectonic and sedimentologic implications. *Earth Planet Sci. Lett.*, **59**, 192-207.

Garfunkel, Z., 1974, Model for the late Cenozoic tectonic history of the Mojave Desert, California, and for its relation to adjacent regions. *Geol. Soc. Amer. Bull.*, **85**, 1931-1944.

Graham, S.A., and W.R. Dickinson, 1978, Evidence for 115 kilometers of right slip on the San Gregorio-Hosgri fault trend. *Science*, **199**, 179-181.

Greene, H.G., and J.C. Clark, 1978, Neogene paleogeography of the Monterey Bay area, California, in *Pacific Coast Paleogeography of the Western United States*, edited by D.G. Howell and K.A. McDougall, Pacific Section, SEPM, 277-296.

Harland, W.B., A.V. Cox, P.G. Llewellyn, C.A.G. Pickton, A.G. Smith, and R. Walters, 1982, *A Geologic Time Scale*, Cambridge Univ. Press, Cambridge, 131 pp.

Luyendyk, B.P., M.J. Kamerling, and R. Terres, 1980, Geometric model for Neogene crustal rotations in southern California. *Geol. Soc. Amer. Bull.*, **91**, 211-217.

Madrid, V.M., R.M. Stuart, and K.L. Verosub, 1986, Magnetostratigraphy of the Late Neogene Purisima Formation, Santa Cruz County, California, *Earth Planet Sci. Lett.*, **79**, 431-440.

Minster, J.B., and T.H. Jordan, 1987, Vector constraints on western U.S. deformation from space geodesy, neotectonics, and plate motions. *J. Geophys. Res.*, **92**, 4798-4804.

Negrini, R.M., K.L. Verosub, and J.O. Davis, 1987, Long-term, non-geocentric axial dipole directions and a geomagnetic excursion from the Middle Pleistocene sediments of the Humboldt River Canyon, Pershing County, Nevada, U.S.A., *Jour. Geophys. Res.*, **92**, 10617-10628.

Schneider, D.A., and D.V. Kent, 1986, Influence of non-dipole field on determination of Plio-Pleistocene true polar wander, *Geophys. Res. Lett.*, **10**, 471-474.

THE APPLICATION OF PALAEOMAGNETISM TO EXTENSIONAL TECTONICS: A PALAEOMAGNETIC STUDY OF THE PARKER DISTRICT, BASIN AND RANGE PROVINCE, ARIZONA.

P.DAGLEY and J.D.A. PIPER
Dept. of Earth Sciences,
University of Liverpool,
PO Box 147,
U.K.

ABSTRACT. A palaeomagnetic study of metamorphic core complexes within the Basin and Range province of North America has been started with the aim of testing various models of extensional tectonics. Here we report the first results from the Whipple-Buckskin-Rawhide Core complex situated across the Californian-Arizona boundary near Parker, Arizona. Samples were collected from both the metamorphic core and the Tertiary cover rocks. Normal and reversed directions obtained for the latter do not differ significantly from directions predicted for stable North America at that time. The direction of the remanence of the rocks from below the detachment zone does not correspond to any identifiable period. We suggest that the remanence was acquired during the uplift of the core and has been rotated by about 30°. Some possible interpretations are discussed.

1. Introduction

The failure of continental crust in an extensional stress regime can be usefully considered in the context of two limiting situations defined in simplistic terms as "simple shear" and "pure shear" models (Figure 1). The simple shear model has been developed notably by Wernicke (1985)and postulates failure with a uniform sense of slip along a discrete low angle detachment zone extending through the thickness of the continental crust. Wernicke's "simple shear" model is an asymmetric one and it accommodates the observation that lower plate ductile deformationin some extensional provinces such as the Basin and Range (and as indicated by K-Ar mineral ages) pre-dates large amounts of upper plate extension.

Wernicke's model can be considered to evolve in four stages (Figure 2):

(a) A shallowly inclined fault merging into a shear zone at depth develops through the thickness of the crust as a detachment zone;

(b) as the fault moves a sedimentary basin grows and the earliest-formed mylonite beings its ascent;

(c) after a substantial amount of movement has occurred on the fault, penetrative brittle deformation begins to affect the sedimentary basin and its basement. At this point, upper-lithosphere thinning has been sufficient to move the mylonites into brittle conditions;

(d) as shearing continues, a portion of the upper lithosphere may begin to extend preferentially relative to the surrounding areas and if the extension is large enough, folds in the detachment surface may develop by isostatic rebound of the unloaded terrane.This produces surface exposures

293

C. Kissel and C. Laj (eds.), Paleomagnetic Rotations and Continental Deformation, 293–311.
© *1989 by Kluwer Academic Publishers.*

of deep crustal rocks referred to as *metamorphic core complexes*.

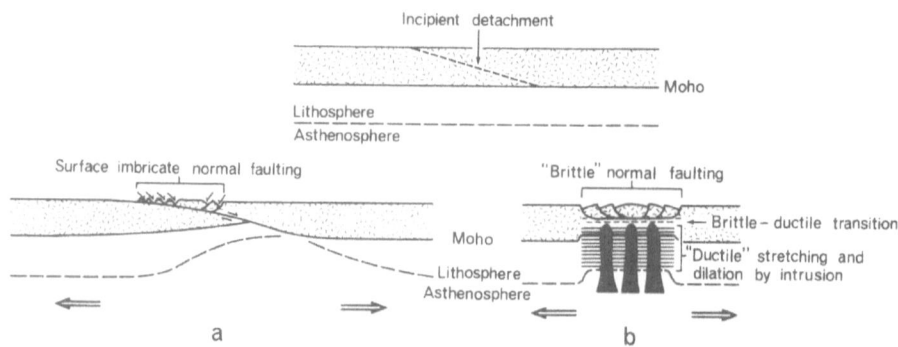

Figure 1. The end member models for strain geometry in rifted continental lithosphere. In the "pure shear" model the crust and lithosphere are attenuated uniformly along any given vertical reference line, while in the "simple shear" model the relative extension of crust and mantle lithosphere along any given vertical line is non-uniform. After Wernicke (1985).

The "simple shear" model therefore requires that (a) failure is concentrated so that large volumes of lower crust and mantle lithosphere within the divergent zone may move but remain internally undeformed and (b) that shear of the lithosphere presumably occurs on a low strength zone at a low angle to the axis of minimum principal stress.

The alternative interpretation of crustal failure during extension considers that it is accommodated by the mechanism of pure shear and the model must at least commence with a symmetrical form (McKenzie, 1978 and Figure 1). This model assumes that the thin-skinned normal faulting at high crustal levels is accommodated in the middle and lower crust by stretching and igneous dilation. This view regards the detachment zones, which demonstrably underlie the normal fault systems in many areas, as brittle-ductile transition zones. Within this model the attenuation and uplift of the middle and lower crust are essentially contemporaneous with the development of the high level fault systems (Figure 1).

The practical application of these concepts to the real crust will depend on the strength and structure of the lithosphere and on the magnitude of intraplate stresses. Thus creep in the lower ductile lithosphere causes the stress to decay and leads to amplification of the stress in the upper lithosphere; stresses there are increased and ultimately fracture may occur. If the whole of the brittle upper lithosphere fractures, stress is transferred to the lower ductile lithosphere, which in turn transfers the stress back to the upper lithosphere producing furtherfracture (Kusznir and Park, 1982). Rifting and extension are then part of a cyclic process of upper lithosphere fracture and lower lithosphere creep; they lead to a state of complete failure throughout the brittle lithosphere which is referred to as whole lithosphere failure (WLF). The stress transfer, and the consequent lithosphere deformation, are controlled by rheology; the most important parameters are therefore the temperature structure of the lithosphere and its composition. By evaluating a crustal model with the behaviours of upper crust, lower crust and mantle controlled by the dislocation creep in quartz, plagioclase andolivine respectively, Kusznir and Park (1987) show that significant extensional deformation will occur in regions with surface heat flows inexcess of 65 mW/sq.m subjected to a tensile force of greater than 3×10^{12}N/m. The positions of the low strength zones, which in turn control the location of the detachment zones along which the crustal extension occurs, are controlled principally by the major compositional boundaries and geothermal gradients: high geothermal gradients favour the development of shallow detachment zones whereas low geothermal gradients favour the development of deep detachment zones especially located near the base of the crust where there is a low stress-low strength region caused by the contrasting plagioclase and olivine rheologies at intermediate geothermal gradients (Kusznir and Park 1987).

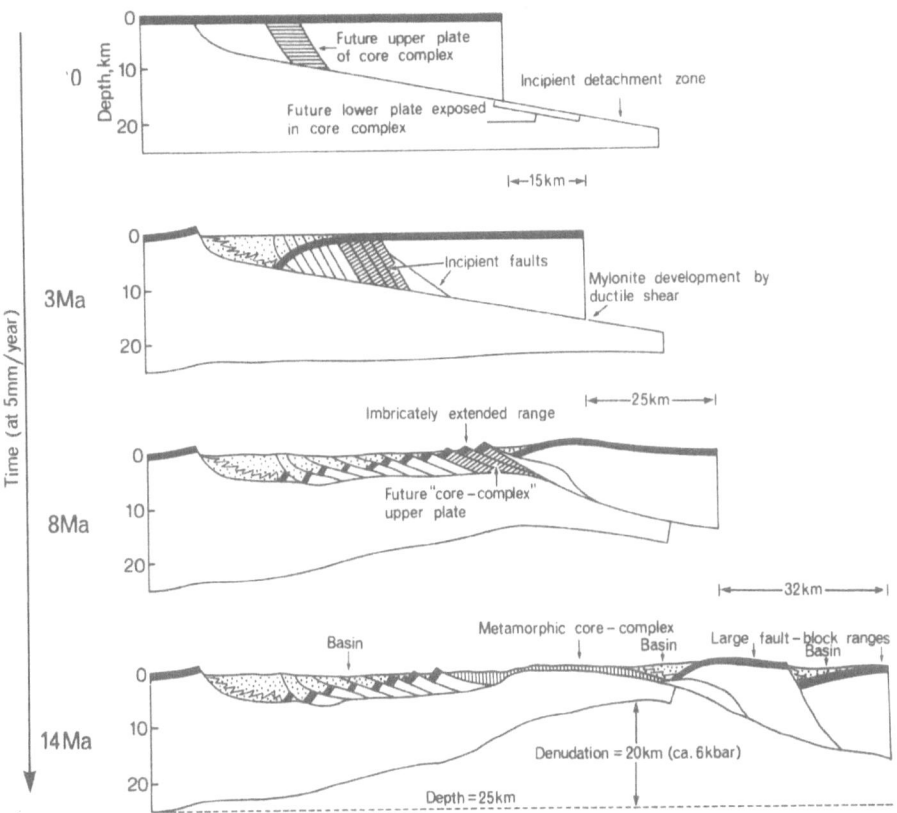

Figure 2. The development of a detachment zone in the continental crust under an extensional shear regime after Wernicke (1985). The four stages at time intervals representative of be Basin and Range Province illustrate the development (b) of a mid-crustal mylonite zone under greenschist to amphibolite facies metamorphic conditions followed by uplift resulting in isotopic closure (c) and brittle reworking, and adjustment of the lithosphere section by sedimentary infilling, imbricate distention and subaerial denudation ((c) and (d)).

The other important implication to come from modelling studies is that the lateral distribution of the rifting of the lithosphere is dependent on strain rate. Under slow strain rates the lithosphere strengthens (or "hardens") with extension and the deformation will be transferred laterally to undeformed areas progressively widening the zone of active deformation (Kusznir and Park op.cit.). For fast strain rates the lithosphere weakens (or "softens") with progressive extension so that an intense but localised zone of extension is produced characterised by highheat flow and high lithosphere stretching factors (β, McKenzie 1978).

Palaeomagnetic studies can potentially make an important contribution to the study of extensional tectonics because they are able to provide a reference vector for comparison with the stable cratonic frame, and thereby to constrain relative movements between the footwall of the detachment zone, the cover rocks and the stable reference crust bordering the zone of extension. Considering the limiting "pure shear" and "simple shear" models shown in Figure 1, we may expect them to influence the magnetic remanence record in the following contrasting ways:

(i) The footwall comprising the metamorphic complex should possess an uplift and cooling remanence which is somewhat older (simple shear) or about the same age (pure shear) as the normal faulting in the upper plate. Since the core complex is detached from the stable cratonic crust on one side or the other in the simple shear solution, magnetic declinations will not necessarily accord with contemporaneous declinations from the reference craton. No detachment is present in the pure shear solution and the declinations should therefore, broadly agree although the remanence may have been influenced by the bulk strain.

(ii) The lateral transport of the imbricate fault blocks in the upper plate may be up to an order greater in the simple shear model compared with the pure shear model. Accordingly, a much greater dispersion of magnetic declinations is possible in the cover rocks with the simple shear model than is permissible with the pure shear model. In either case a greater dispersion of declinations is to be expected from the cover rocks within the province than will be observed between the stable reference blocks on either side.

(iii) The detachment zone will initially have developed at shallow (high geothermal gradient) to deep (lower geothermal gradient) depths in the continental crust and the uplift to form the surface core complex will have involved rotation about a horizontal axis in the simple shear situation. The scale of uplift and consequent rotation may be sensitive to palaeomagnetic resolution.

To test these predictions, and thereby to evaluate the effective contribution of palaeomagnetism to extensional tectonics, we have commenced studies on the basement and cover sequences in zones of extensional activity. The Basin and Range province of western North America is an obvious starting point both because the detachment footwall zones are well developed in metamorphic core complexes, and because the palaeomagnetic studies of the part of North America essentially uninfluenced by Mesozoic and Cenozoic tectonism provide a relatively well-defined reference frame with which the data can be compared. In this paper we report results from core rocks and cover sequences in the Parker District of the western part of the province (Figure 3).

2. Geology and sampling

2.1 REGIONAL OUTLINE

The Basin and Range Province is a broad zone of extensional faulting extending for more than 1000km along the western Cordillera of North America. In general terms the structure is linked to block tilting along downward-flattening (listric) faults with the upthrown blocks forming themountain ranges and the downslope parts forming the valleys. About two thirds of the tilts (ranging from a few degrees up to about 30 degrees) are to the east, and one third are to the west: the predominance of westward tilts in the western part of the province and eastward tilts in the eastern part contributes to an overall bilateral symmetry which is expressed in regional topography, Bouguer gravity and lithosphere thickness (Stewart, 1978). The region is characterised by a thin crust (20-35km), high heat flow, and an anomalous upper mantle (Thompson and Burke, 1974).

There are two contrasting geological elements within the Basin and Range Province:

(i) Metamorphic Core Complexes comprise a basement of metamorphicand plutonic igneous rocks ranging from Precambrian to mid-Tertiary in age. These rocks have all been overprinted by a low angle foliation which invariably has a low angle of dip conforming to the overall domal or archlike shape of outcrop of the complex; a mineral lineation within this foliation is remarkably constant in trend and seems to reflect the regional nature of the stress field responsible for ·emplacing the complexes.The igneous-metamorphic basement passes upwards through a zone of steep metamorphic gradient, which is typically strongly mylonitised, into a low-angle decollement (detachment) zone. Themylonitic facies are generally regarded as the deeper seated manifestations of low angle normal faulting (eg. Davis 1983), and both the plutonic fabrics and the mylonites become thoroughly brecciated into the zones of decollement (Coney 1980). More than 25 metamorphic core complexes have now been identified following an en echelon pattern along a narrow sinuous belt between Southern Canada and NW Mexico. They lie entirely within the cratonic basement of North America, although they are all found within 300km of the western margin of this basement (as defined by 87Sr : 86Sr ratios) and close to the zone of suspect terranes accreted against the Cordillera margin by progressive subduction of Pacific Ocean crust (Jones et al 1982).

(ii) The Unmetamorphosed Cover Terranes comprise rocks of a wide range of ages from Late Precambrian to Mid-Tertiary which have moved relative to the core complexes along the zones of decollement. These cover rocks are attenuated and sliced by closely-spaced subhorizontal normal faults. The most typical component of the cover sequence is a Lower to Middle Tertiary continental sedimentary succession which is often associated with, or succeeded by, Mid-Tertiary volcanic rocks; the latter record a widespread episode of Oligocene-Miocene igneous activity and include extensive ash-flow deposits linked to widespread ignimbrite eruption within the Basin and Range Province during the Oligocene to early Miocene interval.

The spatial association of core complexes and dissected cover rocks has formerly been interpreted in terms of both tensional and compressional models.The interpretation of the decollement zones has been critical to the development of these models: they have been regarded (see models of Figure 1 and section 1) both as major detachment faults defining zones of large lateral movement, or as exhumed brittle to ductile transition zones in the crust. It is now recognised that many decollements are zones of major lateral transport: a comparison of hanging wall and footwall successionson either side of the Bullard Fault in west central Arizona, forexample, suggests that at least 50km of lateral movement to the NE has taken place here (Reynolds and Spencer, 1985). The deeper levels of the fault almost certainly moved in a ductile regime at the time of major translation in mid to late Tertiary times; this fault appears to have had an original dip to the NE comparable with the low angle normal detachment form advocated by Wernicke (1981), Bosworth (1985) and others.

The evidence for earlier compression succeeded by later tension in the Basin and Range Province has been integrated with the geological and geophysical evidence by Coney and Harms (1984). They propose that crustal telescoping along the western margin of North America linked to the Laramide orogeny in Mesozoic and early Tertiary times produced an over-thickened crustal

298

welt. Subsequently, this welt has collapsed in a tensional regime, motivated both by the
gravitational instability of the crustal section, and by the thermal consequence of mid-Tertiary
magmatic activity which has acted to reduce the crustal viscosity.The high geothermal gradients of
these times would have been conducive to failure along shallow crustal detachment zones. In
addition to the applications noted in section 1, palaeomagnetic studies of this region are of interest
for another reason: aeromagnetic surveys over the Basin and Range Province have largely failed to
identify a signature here which can be attributed to rocks beneath the Phanerozoic cover (Mabey et
al., 1981).This could be the consequence of several factors including the composition of the
basement rocks, the conversion of magnetic minerals during metamorphism, and a thin magnetic
basement; the latter effect is certainly a contributory factor because the high heat flow here would
be responsible for elevating the Curie point isotherm, but it is evidently not a complete explanation
(Shuey et al., 1973).

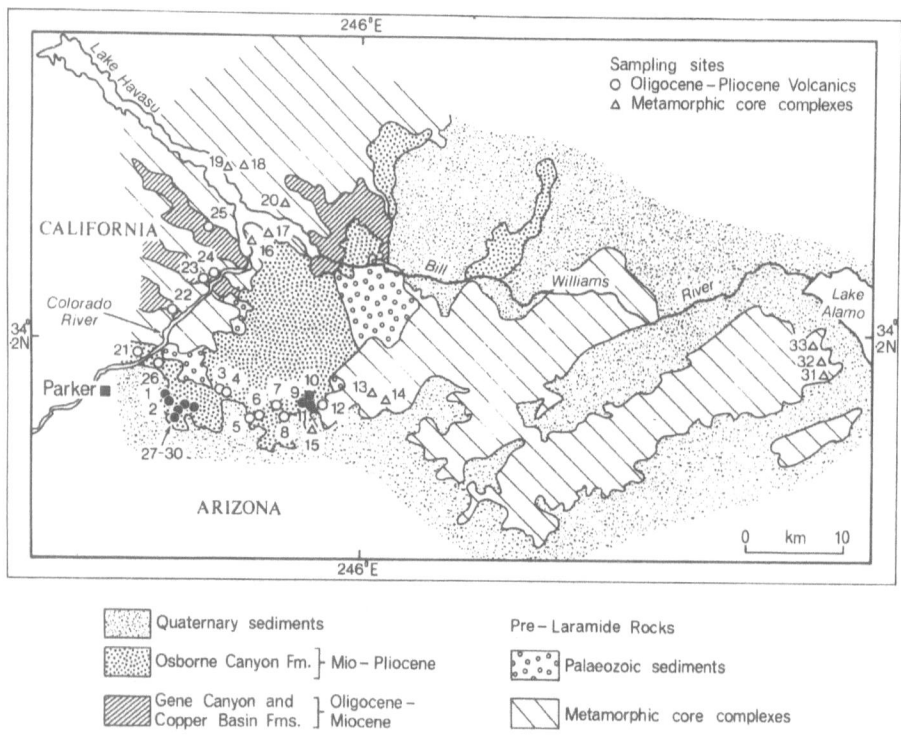

Figure 3. Outline geology (with palaeomagnetic sampling sites of this study) of the sector of the
Basin and Range Province adjoining the metamorphic Whipple Mountains of eastern California and
the Buckskin Mountains of western Arizona. The white circles refer to normally-magnetised sites in
the lavas and the black circles refer to reversely-magnetised sites; the sites in the core complex are
indicated by triangles.

2.2 LOCAL GEOLOGY AND SAMPLING

The segment studied here comprises the outcrop between the metamorphic core complexes comprising the Whipple Mountains and the Buckskin and Rawhide Mountains straddling the Colorado River near Parker, Arizona (Figure 3). The core complexes (sites 13-20, 22 and 31-33) below the detachment zone comprise a range of quartzo-feldspathic gneisses and amphibolites (Shackelford, 1977, Rehrig and Reynolds, 1980) with low but variable angles of foliation, which become progressively brecciated with the development of low grade assemblages including chlorite and epidote upwards towards the detachment surface. Age data from these rocks are sparse, but K-Ar cooling ages from the Whipple Complex are in the range 30-18Ma (Davis et al., 1982).

The deposition of cover rocks in mid-Tertiary times followed the Laramide orogeny and post-Laramide peneplanation. They include thick fluviatile clastic sequences and rhyolitic to basaltic lavas and tuffs. The latter (sites 23-25) record a voluminous episode of volcanism in western Arizona which spanned the interval 35 to 12Ma (Shafiqullah et al., 1980) incorporating the transition (24-12Ma) from mid-Tertiary orogeny to Basin and Range tensional faulting. In the Whipple Mountains these formations are subdivided into the Gene Canyon and Copper Basin Formations (Davis et al., 1980). In the western part of the Buckskin rangea ca 2000m thick succession of immature clastic deposits rests on a detachment surface above mylonitic gneisses (Wodzicki et al., 1982). Referred to locally as the "Osborne redbeds", they include numerous andesite flows dated (20-16Ma) and pass upwards via local unconformities into poorly consolidated fluvioclastic sediments and thence into marine sediments of the Pliocene Formation deposited in an embayment of the Gulf of California.The palaeomagnetic sites in lava flows of the mid-Upper Tertiary cover rocks (1-12, 21 and 26 Figure 3) are concentrated within units erupted during the ca 19-12Ma peak of volcanic activity in this area, but include some young flows erupted during the waning stages of activity in this area between 9 and 4Ma (Shafiqullah et al 1980, Wodzicki et al., op.cit).

3. Paleomagnetic results

Samples were collected using a portable drill and individually oriented using a sun-compass. Measurements were made using parastatic and ring-fluxgate (Minispin) spinner magnetometers. Each sample was demagnetised either by alternating fields in steps of 5 mT. to 40 mT. or 200 mT. as required or by heating to at least 600°C in steps of 50 or 100° in a Magnetic Measurements thermal demagnetiser. Thermal demagnetisation was generally found to define the same vectors asalternating field treatment (Figure 4). Site mean directions calculatedusing the method of Dagley and Ade Hall (1970) are given in Table I for the a.f. data.

3.1 COVER SUCCESSIONS

Osborne Canyon Formation:

The sites in this formation comprise a range of andesitic and basaltic lava flows sampled at the locations shown in Figure 3. They are variably deformed and have tilts ranging from horizontal up to 30°. Normally magnetised sites are, (21, 26, 3, 4, 5, 6, 7, 8, 12): Site 21 (Figure 5), before adjustment for its steep tilt actually has -ve inclination; its mean direction is shallower and more westerly than the main group and has been excluded from the group mean.

300

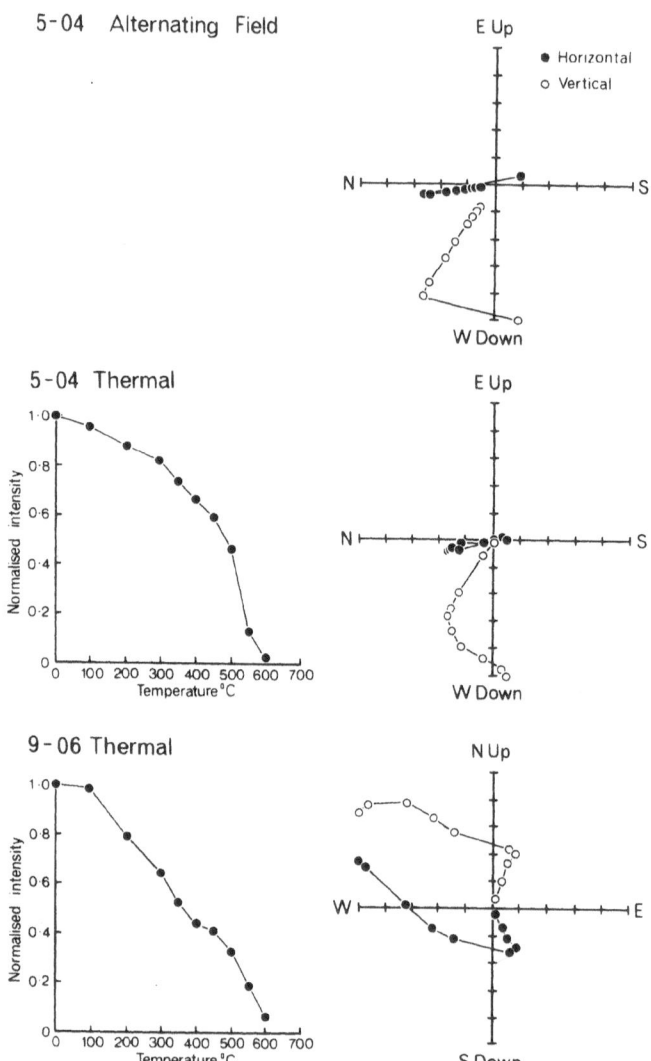

Figure 4. Representative thermal and a.f. demagnetisation results from: a) normally-magnetised lavas of the Osborne Canyon Formation; b) Reversely-magnetised lavas of the Osborne Canyon Formation. Fields of treatment are indicated in milli Tesa (mT.) and °C; open symbols are plots onto the horizontal plane. The axes are scaled in x10^{-5} Am^{2}kg^{-1}.

301

TABLE 1 : Site and group mean palaeomagnetic results from the cover successions

Site No.	N/n	R	α95	In situ D	In situ I	Tilt Adjusted D	Tilt Adjusted I	Polarity
(i) Osbourne Canyon Formation (Mio-Pliocene)								
1	8/8	7.98	2.3	183	-24	176	-29	R
2	4/4	3.99	4.8	177	-34	167	-38	R
3	4/4	3.99	1.5	0	69	353	62	N
4	4/4	3.99	2.3	341	72	338	64	N
5	4/3	2.99	8.5	355	54	354	59	N
6	4/4	3.99	4.4	339	50	337	55	N
7	4/4	3.99	4.9	351	33	348	37	N
8	4/4	3.98	6.7	333	38	326	42	N
9	4/4	3.99	5.0	215	-41	215	-54	R
10	4/4	3.86	19.9	148	-54	155	-44	R
11	4/3	2.99	9.7	187	-48	200	-52	R
12	4/4	3.80	24.7	354	69	344	68	N
21	4/3	2.96	178	307	-30	307	20	I
26	4/4	3.96	11.0	342	43	342	43	N
27	4/4	3.98	8.1	192	-21	187	-39	R
28	4/4	3.99	5.2	181	-35	169	-51	R
29	4/4	3.99	2.4	178	-28	168	-43	R
30	4/4	3.13	58.6	186	-39	173	-56	R
(ii) Gene Canyon Formation (Oligo-Miocene)								
23	4/3	2.98	12.0	244	39	283	41	I
24	4/3	2.97	15.5	22	16	22	-16	I
25	4/3	2.99	9.0	22	14	21	45	N

Osbourne Canyon, overall mean calculation[+]

	N(sites)	R	α95	D	I
In situ	15	14.22	9.26	355	44
Tilt adjusted	15	14.51	7.3(2)	351	49

+ excluding sites 38, 56 because $\alpha_{95} > 20°$ and site 47 because its direction is intermediate

Although site 12 has a mean direction close to thoseof nearby sites 3, 4 its precision is poor (Table I); the strength of magnetisation and scatter between cores may indicate lightning-induced IRMs and it has also been excluded from the group mean. Thedemagnetisation behaviour of the others is straight forward and defines single convergent vectors, (Figure 4). There appears to be a progression (excepting 12) from somewhat steeper directions in the west (3, 4) to directions 20-30 degrees shallower in the east (7, 8 and 26), of uncertain significance. Geographically there are two groups (3, 4) and (5, 6, 7, 8). Within these, the pairs (3, 4), (5, 6) and (7, 8) have very similar tilt adjustments and only for 7, and 8 does the adjustment improve the precision of the mean direction. The group (5, 6, 7, 8) clusters less tightly. Except for 21, none of the dips are very large and the change in direction has a very small affect on the agreement with the predicted Miocene direction. Following tilt adjustment, the clustering of the group is slightly improved (angular std. dev. 11.6 cf 14.6 deg.) and the change in site 12 is in the same sense.

Reversely Magnetised Sites form two groups (1, 2, 27, 28, 29, 30) and (9, 10, 11) approximately 15 km apart (Figure 3). All except 10 and 30 have well defined site mean directions (Table I). In general the demagnetisation behaviour is straightforward (Figure 4).

Each core of 10 is quite stable but the core directions are dispersed. The NRMs of the cores from site 30 are very strong and directions are discordant; since the site is at the summit of Black

302

Peak a lightning strike is suspected, although after a.f. demagnetisation at high fields (200 mT.) 2 cores agree amongst themselves and with the remainder of the group.

The tilts are significant (24 and 20 degrees) but the group mean of the first cluster is slightly less well defined after adjustment than before (ang. std. dev. 7.3 and 9.2 degrees): the tilts measured for the second group (9, 10, 11) are ca. 12 degrees in different directions and there is some improvement after adjustment although they move away from the other group.

The overall mean direction improves on making the tilt adjustments (angular std. dev. 14.7 and 16.6 degrees) and, as there are no grounds on which to reject any sites, the mean is presumed to have averaged secular variation and imprecise tilt measurements. This direction, when inverted, lies closer to the normal group mean and to the predicted direction than the uncorrected mean (Figure 5).

We do not yet have enough data from the Gene Canyon and Copper Basin Formations to make any valid comparisons but the directions are normal and generally lie to the east of the predicted field for Oligo-Miocene times.

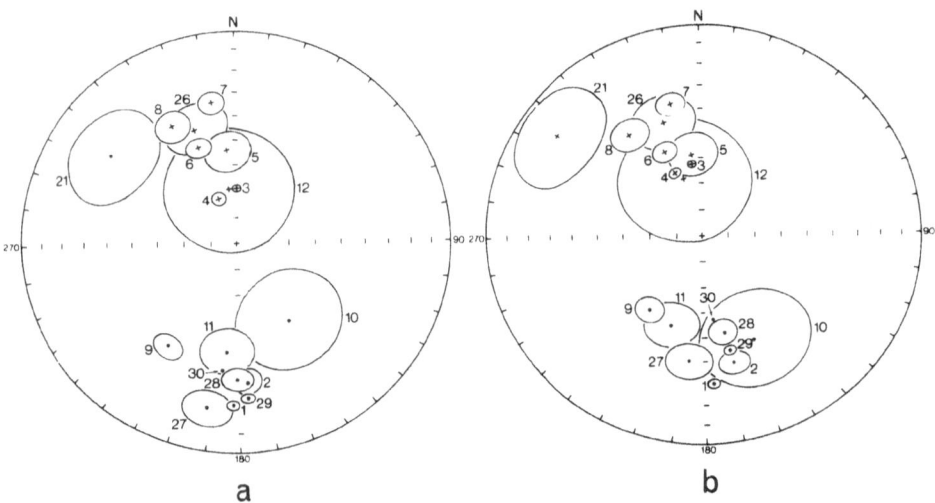

a   b

Figure 5. Site mean directions of magnetisation from lavas of the Osborne Canyon Formation (a) before and (b) after tilt adjustment with 95 percent confidence circles (except for site 30). Dots (crosses) are on upper (lower) hemisphere. Equal Area projections.

3.2 THE CORE COMPLEXES

Of the six sites (13-15, 31-33) sampled in metamorphic rocks of the Buckskin Range (Figure 3), only a few cores possessed detectable components defining vectors passing acceptability criteria (Table II). Most of the metamorphic rocks sampled in this range are psammitic schists showing variable, and often strong, degrees of retrogression with the development of epidote and chlorite (13, 14, 31-33) and copper mineralisation (14). It seems probable that pervasive hydrothermal alteration has largely deleted the remanence carriers in these rocks. The three identified components

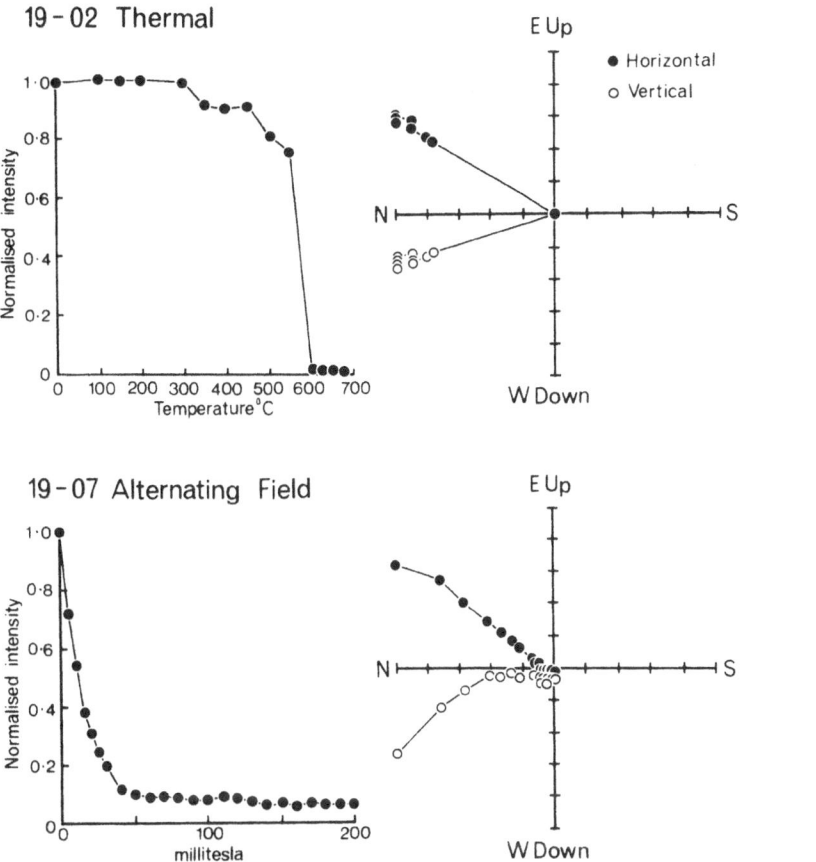

Figure 6. Representative a.f. and thermal demagnetisation results from metamorphic rocks of the Whipple Core complex. Symbols are as for Figure 4.

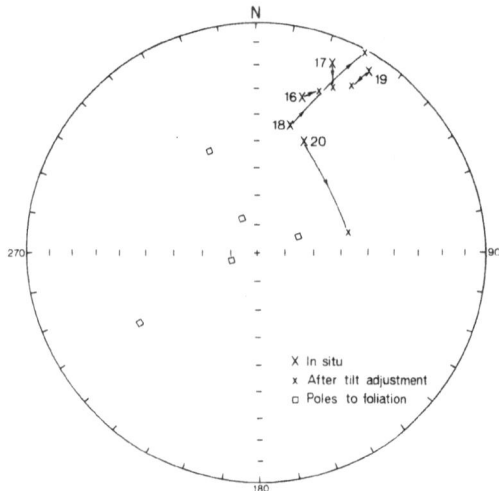

Figure 7. Equal area projection showing site mean directions of magnetisation derived from metamorphic rocks of the Whipple Core complex before and after adjustment for tilt of the foliation.

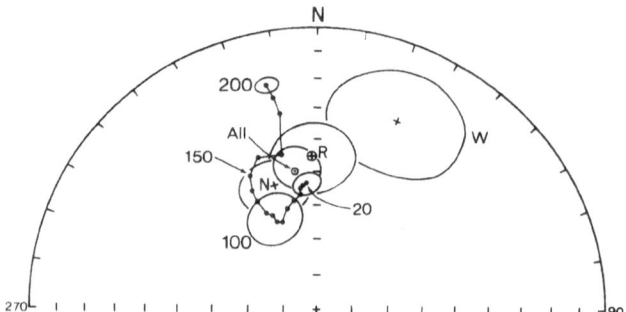

Figure 8. Group mean directions (reversed directions inverted) compared with directions calculated from pole positions for Stable N. America (Harrison and Lindh 1982). *, Direction expected for centred axial dipole.

are scattered (Table II) and uninterpretable without further data.They do however, imply that no pervasive Cenozoic remagnetisation has taken place during emplacement of the Buckskin Range; a similar result was found in the Kern Mountains core complex in Nevada by Hudson and Geissman (1982).

TABLE II : Site and group mean palaeomagnetic results from Core Complexes, Parker District

Site No.	N/n	R	95	In situ D	I	Tilt Adjusted D	I
(i) Buckskin Core complex:							
14	4/2	1.98	39.5	187	71	194	77
15	4/4	3.76	26.8	329	39	-	-
32	4/4	3.98	8.2	193	-39	185	-31
(ii) Whipple Core complex (South Lake Havasu sector):							
16	4/3	3.7	48.1	17	30	21	26
17	2/2	1.99	20.4	22	13	25	22
18	3/2	1.99	7.9	15	43	29	2
19	4/4	3.99	3.3	32	9	30	18
20	3/3	2.72	50.0	23	47	78	57

Group mean Calculations, Whipple Complex						Palaeomagnetic Pole		
(In situ)	5 sites	4.80	17.5	22.3	28.6	13.9E	62.4N	10.5/19.2
(Adjusted for tilt of foliation)	5 sites	4.63	24.3	33.5	27.2	0.9E	53.9N	14.4/26.4

In contrast, the acid to basic gneisses within this sector of the Whipple Complex produced a range of stable components of magnetisation isolated by both thermal and a.f. demagnetisation (Figure 6). Most of these sites show secondary hematite development, especially on joint planes, and although they may well have been profoundly influenced by hydrothermal alteration, the fluids seem to have possessed high oxygen fugacities and their effect has been to enhance or replace the remanence. The directions are uniformly NNE and shallow to intermediate positive.The gneisses exhibit variable degrees of metamorphic foliation and these foliations have directions which are not simply related to the directions of magnetisation (Figure 7). Hence, although an anisotropy effect may be present here, we do not suspect that it has strongly influenced the remanence. When the remanence directions are adjusted to bring the foliation into the horizontal plane the overall grouping deteriorates (with the precision falling from 20.1 to 10.1), although the mean direction is not appreciably affected (D/I=22/29 to 33/27).We therefore infer that this remanence was probably acquired after the broad scale deformation of the inherited metamorphic foliation which presumably accompanied the uplift and emplacement of the core complex.This conclusion is supported by the absence of a correlation between the mean characteristic remanence direction and younger palaeofield directions at the possible ages of formation and metamorphism of the basement rocks in this region (section 4).

4. Discussion

The scale of extension across the Basin and Range Province, as derived from fault geometry, has been estimated to be of the order of 40 per cent (Wernicke and Burchfiel, 1982, Le Pichon and Sibuet, 1981). Coney and Harms (1984) also concluded that a value of 40-60 per cent was reasonable. Bogen and Schweicherf (1985) tested this premise using palaeomagnetic evidence: they noted that since the width of the Basin and Range expands to the north, large extensions in the southern sector of the province would require substantial clockwise rotation of the Sierra Nevada range. By comparing Jurassic data from the Sierra Nevada with results from cratonic North America, they concluded that the limit of this rotation was 4 ± 10 degree (P=0.05) yielding a maximum extension of about 50 per cent and a most probable value of 39 ± 12 per cent which is closely comparable to the values deduced by Wernicke, Coney and others. This analysis defines the net rotation between the "stable" margins of the Basin and Range and does not constrain the possible movements on the rifted and detached terranes in between. The extension probably began in the Oligocene (ca 30 Ma, Frei, Magill and Cox, 1984); reaching a maximum in the Miocene (ca 15 Ma, Proffett 1972, 1977) and has continued to the present. The amount of extension varies along the length of the province and may have been as much as 245 km directed WSW-ENE around latitude 35 degree N, (Frei, Magill and Cox 1984), the region of the present study.

The reference migration path of the palaeofield vector is best computed by calculating mean palaeomagnetic directions or pole positions from the body of data from "stable" cratonic North America lying within overlapping time intervals. Mean poles calculating within a migrating time window (20 or 30Ma) are given by Irving and Irving (1982), Harrison and Lindh (1982) and Piper (1987). In Figure 8 we have used these published poles to compute the expected palaeomagnetic field direction in the Parker District since the beginning of Jurassic times. The cover sequences studied have well constrained stratigraphic ages and merit a direct comparison with this path although the sites from the (Oligocene-Miocene) Gene Canyon and Copper Basin Formations are too few in number to permit analysis. The more numerous results from the lavas of the Osborne Wash Formation (Miocene-Pliocene) have been compared to the predicted direction for 20 Ma (Harrison and Lindh, 1982) and to that for a centred axial dipole. The average direction for the normally magnetised group corrected for observed tilt is shallower and to the west of the predicted

TABLE III : Mean directions and rotation statistics

	D (deg)	I (deg)	A (deg)	R (deg)	F (deg)	d (deg)	Reference direction
Osborne Canyon							
N	341.8	52.0	9.4	-13 ± 12.5	1.1 ± 7.8	1.05 ± 7.8	20 Ma
				-18 ± 12.5	1.6 ± 7.8	1.5 ± 7.8	CAD
R (inverted)	358.1	45.0	10.9	2.9 ± 12.7	8.1 ± 10.1	7.1 ± 9.0	20 Ma
				-2.0 ± 12.7	8.6 ± 10.1	7.5 ± 9.0	CAD
				16.0 ± 17	7 ± 11.3	6 ± 11.3	Normal group
N & R	351	48.6	7.3	-4.1 ± 12.0	4.1 ± 8.2	3.8 ± 8.2	
Whipple							
N	22.3	28.6	17.6	27.2 ± 20.5	24.5 ± 18.3	3.8 ± 8.2	20 Ma

- D = declination, east of true north
- I = inclination, positive downwards
- A = alpha, 95% confidence limit
- Units with A > 20° have been rejected
- R (rotation) and F (flattening) (Beck 1980) and d (Poleward movement) (Frei et al. 1984) with 95% confidence limits calculated according to Demarest (1983)

- 20 Ma reference direction calculated from Harrison and Lindh (1982) directions; the inverted reversed group mean is also shallower but the declination is close to that expected.

Calculation of rotation, R, flattening, F, (Beck 1980, Demarest,1983) and poleward movement P, (Frei et al. 1984) however show that the chance that the observed and predicted directions are different is small (Table III) and the normal group does not differ from the reversed.

Calderone and Butler (1984) sampled lavas in the Buckskin Range and the Plomosa Range, a little to the south, together with three sites in southern Arizona. The 'cooling unit' mean directions for these five areas show a similar range of declination and inclination to that found for the Osborne Canyon normal group reported here. Using a "Range" mean these authors deduced a similar rotation but a greater flattening than reported here for the normal units. They argue that the counter-clockwise rotation is significant. This, of course, would be a net rotation. If the normal and reversed data from the present study are combined, as was done by Calderone and Butler, the rotations are not significant.

Calderone and Butler (op. cit.) noted that the reversed units did not give directions antipodal to the normal units and they attributed this to incomplete sampling of secular variation although they had combined data from five ranges with distinct ages between 25 Ma and 14 Ma. For their cooling units they found that the inverted reversed directions have declinations more westerly than the normal directions; Calderone and Butler (op.cit., Figure 2) found the opposite for their cooling units. A more detailed analysis of the separate age and polarity groups may therefore be warranted.

Although it is convenient to calculate R, the apparent rotation about a vertical axis, F or d, the apparent rotation about a horizontal east-west axis any difference between vectors can be achieved in a variety of ways. In the context of the models for the formation of metamorphic core complexes it may be more appropriate to deduce the apparent rotation about a single horizontal axis which would define the axis of doming of the models. Relative to the 20 Ma predicted direction, the Osborne Canyon normal group suggests a tilt of approximately 10 ± 10 degrees down to the east about an axis 355-175 degrees east, the reversed group a tilt of about 8 ± 11 degrees upwards to the east about an axis 282-102 degrees east. The difference between the normal group and the reversed group implies a 15 ± 14.5 degree rotation about 324-144 degrees east; downward to the east if the normal group is the older. These axes are consistent with the NNW-SSE structural axis (Frei, Magill and Cox, 1984) and a tilt down toward the east consistent with an upper plate fault block on the NE side of the rift. However the amount of tilt cannot be considered significant.

The results from the core complexes currently present a more difficult problem of interpretation. There is no palaeohorizontal reference in these rocks, and whilst the metamorphic foliation probably developed at a low angle, this is not an unambiguous interpretation. Furthermore, in the absence of detailed radiometric coverage, it is necessary to consider the possibility that the remanence recovered from these rocks has survived the middle to late Tertiary thermotectonic event and date from the time of Palaeozoic orogeny or from the mid-Proterozoic age of the basement (Damon 1968). This southern margin of the Laurentian Shield experienced a long and episodic history of incremental growth linked to subduction-related processes in Middle and Upper Proterozoic times (eg. Condie 1983); in this sector it includes the Matatzal belt dated ca. 1500 Ma. The palaeomagnetic pole from the Whipple core (Table II) however, shows no specific correlation with the APW path for these times and indeed correlates with no known Proterozoic pole position after 2100 Ma. The APW path for these times shows a preference for the central and western equatorial Pacific area with a range of excursions executing APW loops (Piper 1987). The closest proximity is achieved at ca. 950 Ma when the extremity of the Grenville Loop moves towards this pole position although it would appear that the extremity of the loop was still ca. 30° from this part of the APWP (McWilliams and Dunlop 1978). A Palaeozoic age for this remanence also seems to be excluded by the absence of a clear correlation with the APWP of the age.

Hence we consider that there are two possible explanations for this remanence: either (i) it is *in situ* and was imparted to the basement during a thermo-tectonic event which has gone unrecorded elsewhere in North America, or more likely (ii) the remanence has been imparted during a time interval represented elsewhere but has since been rotated by tectonic movements associated with the development of the Basin and Range Province.

In the context of the rise of this basement by the isostatic response to crustal attenuation and unloading (section 1), a probable explanation for this remanence must seek an origin at depth and probably during the early-mid Tertiary peak of thermal activity here (section 2).The prehnite-pumpellyite grade of retrogressive metamorphism in the Buckskin rocks, which we consider to be responsible for the paucity of a remanence record here, is likely to record temperatures of 200-300°C at depths of ca. 5-10 km. Constraints are more difficult to place on the depth of origin of the stable Whipple core remanence: although hematite development is superficially widespread and may represent the effects of circulation of high Eh fluids at shallow crustal levels, the effect has probably merely operated to enhance the stability of the magnetic carriers because this characteristic remanence is essentially magnetite-held (section 3 and Figure 6) and has apparently replaced most, or all, of any pre-existing remanence in these rocks. We therefore suspect that it was acquired at levels deeper than the low grade metamorphism prevalent in the Buckskin core complex.

Structural trends in the Whipple core change westwards from ENE to near E-W along the trend of the range with trends of E-W to WNW-ESE in the SE sector of the Whipple Range. The detachments appear to have broadly conformable trends and are aligned approximately N310W in the zone represented by the bulk of the sample sites. A simple rotation of the Whipple remanence about a horizontal axis with this trend by ca. 30° moves this vector towards the predicted Lower-Middle Tertiary field vector (Figure 8). There remains a small clockwise difference (ca. 15) in declination which may not be significant in the context of the error circle on this result. However, this could also record the effect of a clockwise rotation about a near-vertical axis of this crustal segment between detachment zones; the sense of this inferred motion is consistent with the rotational motions of crustal blocks along this cratonic margin of North America (eg. Beck 1980, Jones et al. 1982). If this is the correct explanation for the residual declination difference, then the zone below the detachment in this sector of Whipple core records a rotation which is not evident in the hanging wall rocks; this is a signature of the simple shear model described in section 1.

To date most tectonic analyses of terranes at orogenic margins have sought to explain remanence vectors in terms of rotations about near-vertical axes and declination contrasts are the resultant effect of a range of folding effects linked to strike-slip motions (Macdonald 1980). In this instance such an explanation is not adequate because the difference is largely one of inclination which can only be accommodated by motion about a near-horizontal axis. Hence a simplistic tectonic explanation for the remanence in the Whipple Core (which appears to be required by the absence of any straightforward accordance of this remanence with the known APWP) involves acquisition at a location beneath the detachment zone at a stage during uplift to present levels (which may well date part, or all, of the folding of the foliation (see Figure 7) resulting in a shallowing of the remanence vector acquired in a normal polarity mid-Tertiary (?) field during isostatic adjustment of the crustal section (Figure 9).

Perhaps the main implication of this aspect of the study is to highlight the ambiguities of palaeomagnetic interpretation in the absence of good radiometric control. A full tectonic interpretation of palaeomagnetic data from the metamorphic core complexes requires that the remanence here is closely linked to the cooling history of the rocks as defined by isotopic closure in the metamorphic minerals and particularly, by [39]Ar / [40]Ar studies. We believe that further progress in these studies will require this integration.

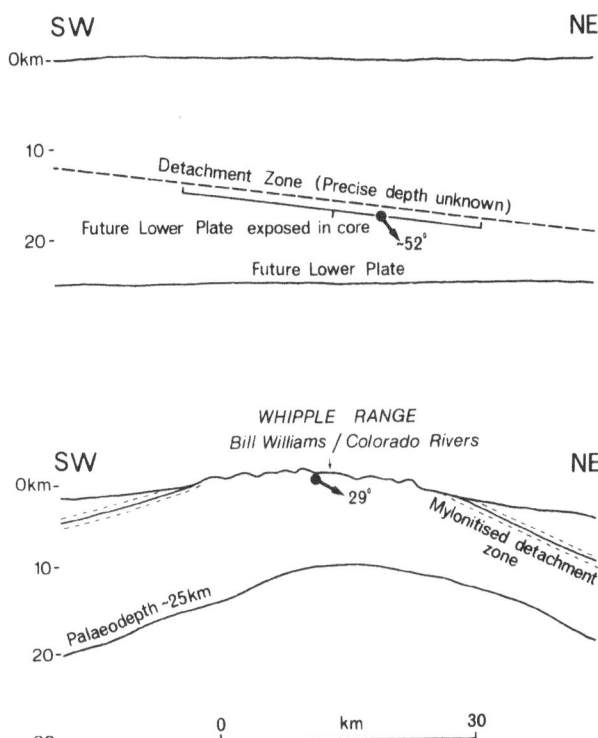

Figure 9. Schematic section through the Whipple Core complex at two stages during its evolution to illustrate how a remanence vector acquired at depth during mid-lower Tertiary times could undergo a decrease in inclination as the crustal section is uplifted and exposed at the surface. No vertical exaggeration.

References

Beck, M.E. Jnr., 1980,. Palaeomagnetic record of the plate-margin tectonic processes along the western edge of North America. *J.Geophys.Res.*, **85**, 7115-7131.

Bogen, N.J., and Schweicherf, R.A., 1985. Magnitude of crustal extension across the northern Basin and Range province: constraints from palaeomagnetism, *Earth Planet.Sci.Lett.*, **75**, 93-100.

Bosworth, W., 1985. Geometry of propagating continental rifts. *Nature*, **316**, 625-627.

Calderone, G., and Butler, R.F., 1984. Palaeomagnetism of Miocene volcanic rocks from south-western Arizona: Tectonic implications. *Geology*, **12**, 627-630.

Condie, 1983. Plate Tectonics model for Proterozoic continental accretion in the south-western United States. *Geology*, **10**, 37-42.

Coney, P.J., 1980. Cordilleran metamorphic core complexes: An overview. *Geol. Soc. Amer. Mem.*, **153**, 7-31.

Coney, P.J., and Harms, T.A ., 1984. Cordilleran metamorphic core complexes: Cenozoic extensional relics of Mesozoic Compression. *Geology*, **12**, 550-554.

Dagley, P. and Ade-Hall, J.M., 1970. Cretaceous, Tertiary and Quaternary Palaeomagnetic Results from Hungary. *Geophys. J. R. Astr.Soc.*, **20**, 65-87.

Damon, P.E., 1968. Application of the K-Ar method to the dating of igneous and metamorphic rocks within the Basin and Ranges of the southwest, In: *Southern Arizona Guidebook III*, (ed. S.R. Titley), Arizona Geol.Soc., 7-20.

Davis, G.A., Anderson, J.L., Martin, D.L., Krummenacher, D., Frost, E.G. and Armstrong, R.L., 1982. Geologic and geochronologic relations in the lower plate of the Whipple-Buckskin-Rawhide mountains, SE California - a progress report. *In: Mesozoic-Cenozoic tectonic evolution of the Colorado River Trough region, California, Arizona and Nevada.* (eds. E.G. Frost and D.L. Martin), Cordilleran Publ., San Diego, California, 408-432.

Davis, G.H., 1980. Structural characteristics of metamorphic core complexes, southern Arizona In: Cordilleran Metamorphic core complexes (ed. M.D. Crittenden, P.J. Coney and G.H. Davis) *Geol.Soc.Amer. Mem.*, **153**, 35-77.

Davis, G.A., Anderson, J.L., Frost, E.G. and Shackelford, T.J., 1980. Mylonitisation and detachment faulting in the Whipple-Buckskin-Rawhide Mountains terrane, SE California and western Arizona, in: Cordilleran Metamorphic core complexes (ed. M.D. Crittenden, P.J. Coney and G.H. Davis), *Geol.Soc.Amer. Mem.* **153**, 79-129.

Demarest, H.H.Jnr., 1983. "Error analysis for the determination of tectonic rotation from palaeomagnetic data".*J.Geophys.Res.*, **88**, 4321-4328.

Frei, L.S., Magill, J.R. and Cox, A., 1984. Palaeomagnetic results from the central Sierra Nevada: Constraints on reconstructions of the western United States. *Tectonics*, **3**, 157-177.

Harrison, C.G.A. and Lindh, T., 1982. A polar wandering curve for North America during the Mesozoic and Cenozoic. *J.Geophys.Res.*, **87**, 1903-1920.

Irving, E., and Irving, G.A., 1982. Apparent polar wander paths, Carboniferous through Cenozoic and the assembly of Gondwana. *Geophys.Surv.*, **5**, 141-188.

Jones, D.L., Cox, A., Coney, P. and Beck, M.E., 1982. The growth of western North America. *Sci. Amer.*, **247**, 70-84.

Kusznir, N.J. and Park, R.G., 1982. Intraplate lithosphere strength and heat flow. *Nature*, **299**, 540-542.

Kusznir, N.J. and Park, R.G., 1987. The extensional strength of the continental lithosphere: its dependence on geothermal gradient and crustal composition and thickness, In: Continental Extensional Tectonics (eds. M.P. Coward, J.F. Deway and P.L. Hancock), *Geol. Soc. Lond. Spec.Publ.*, **28**, 35-52.

Le Pichon, X. and Sibuet, J.C., 1981. Passive margins: A model of formation, *J. Geophys. Res.*, **86**, 3708-3720.

Mabey, D.R., Zietz, I., Eaton, G.P. and Kleinkopf, M.D., 1981. Regional magnetic patterns in part of the Cordillera in the Western United States. *Geol. Soc. Amer. Mem.*, **152**, 93-106.

McDonald, W.D., 1980. Net tectonic rotation, apparent tectonic rotation and the structural tilt correction in palaeomagnetic studies. *J. Geophys. Res.*, **85**, 3659-3669.

McKenzie, D.P., 1978. Some remarks on the development of sedimentary basins. *Earth Planet.Sci.Lett.*, **40**, 25-32.

McWilliams, M.O. and Dunlop, D.J., 1978. Grenville palaeomagnetism and tectonics. *Can. J. Earth. Sci.*, **15**, 687-695.

Piper, J.D.A., 1987. *Palaeomagnetism and the Continental Crust*, (Open University Press/Halstead Press), 434pp.

Proffett, J.M. Jnr.,1972. Nature, age and origin of Cenozoic faulting and vulcanism in the Basin and Range Province (with special reference to the Yerington district, Nevada). *Ph.D. dissertation* Univ. California, Berkeley, 77p.

Proffett, J.M. Jnr., 1977. Cenozoic geology of the Yerington district, Nevada, and implications for the nature and origin of Basin and Range faulting. *Geol. Soc. Amer. Bull.*, **88**, 247-266.

Rehrigg, W.A. and Reynolds, S.J., 1980. Geologic and geochronologic reconnaisance of a northwest-trending zone of metamorphic core complexes in southern and western Arizona. in: Cordilleran metamorphic Core Complexes (eds. Crittenden, M.D., Coney, P.J. and Davis, G.H.) Geol. *Soc.Amer.Mem.*, **153**, 131-157.

Reynolds, S.J. and Spencer, J.E., 1985. Evidence for large-scale transport on the Bullard detachment fault, west-central Arizona. *Geology*, **13**, 353-356.

Shafiquallah, M., Damon, P.E., Lynch, D.J., Reynolds, S.J., Rehrig, W.A. and Raymond, R.H., 1980. K-Ar geochronology and geologic history of south-western Arizona and adjacent areas. *Arizona Geol. Soc. Digest.*, **12**, 201-260.

Shakelford, T.J., 1980. Tertiary tectonic denudation of a Mesozoic-early Tertiary (?) gneiss complex Rawhide Mountains, western Arizona. *Geology*, **8**, 190-194.

Shuey, R.T., Schellinger, D.K., Johnson, E.H. and Alley, L.B., 1973. Aeromagnetics and the transition between the Colorado Plateau and Basin and Range Provinces. *Geology*, **1**, 107-110.

Stewart, J.H., 1978. Basin-range structure in western North America: a review. *Geol. Soc. Amer. Mem.*, **152**, 1-31.

Thompson, G.A. and Burke, D.B., 1973. Rate and direction of spreading in Dixie Valley, Basin and Range Province, Nevada. *Geol. Soc. Amer. Bull.*, **84**, 627-632.

Wernicke, B., 1981. Low angle faults in the Basin and Range Province: Nappe tectonics in an extending orogen. *Nature*, **291**, 645-648.

Wernicke, B., 1985. Uniform-sense normal simple shear of the continental lithosphere. *Can. J. Earth. Sci.*, **22**, 108-125.

Wernicke, B. and Burchfield, B.C., 1982.`Modes of extensional tectonics. *J. Structural Geology*, **4**, 105-115.

Wodzicki, A., Krason, J. and Craver, S.K., 1982. Geology, energy and mineral resources assessment of the Bill Williams area, Arizona. *Geo-explorers International Inc.*, Denver, Colarado.

MECHANISMS OF CENOZOIC TECTONIC ROTATION, PACIFIC NORTHWEST CONVERGENT MARGIN, U.S.A.

R.E. WELLS
U.S. Geological Survey
345 Middlefield Rd., MS 975
Menlo Park, CA 94025
U.S.A.

ABSTRACT. Large clockwise rotations (15-80°) are characteristic of Cenozoic volcanic and sedimentary rocks along the convergent margin of the northwestern United States. Abundant paleomagnetic data from 62-12 m.y. old rocks in forearc, arc, and backarc regions show that rotation increases with age and with proximity to the coast. Paleomagnetic and structural studies both support dextral shear as a significant contributor to tectonic rotation in the Pacific Northwest, with an added contribution from Basin-Range extension. Paleomagnetism of individual Miocene (15-12 Ma) flows of the Columbia River Basalt Group show a well-defined, progressive increase in rotation across the forearc region toward the trench which is most readily explained by shear. This progressive increase in rotation toward the coast is also seen in Oligocene and Eocene rocks, although with steeper gradients, indicating that shear rotation has been important throughout Cenozoic time. The dextral shear may be accommodated by abundant, small-scale strike-slip faults which are acting as Riedel shears. Coast Range faults, usually with well-developed subhorizontal slickensides, occur throughout the stratigraphic section and are most abundant in the oldest rocks. Displacements are small to moderate (10^2-10^3 m), and both dextral and sinistral faults are common. Fault trends are variable, with N20°-45°W faults showing dextral slip and N60°W-S70°W faults showing sinistral slip. The driving force for dextral shear presumably results from coupling of the overlying plate with the subducting Farallon plate, which has been moving northeast throughout most of Cenozoic time. The increase in rotation from north to south along the coast probably represents the contribution of extension in the Basin-Range region to the rotations in Oregon.

1. Introduction

The Pacific Northwest region of the United States (Washington, Oregon, and N. California) has been the locus of convergence between North America and oceanic plates to the west throughout most of Cenozoic time (Atwater, 1970; Engebretson and others, 1985). The margin is characterized by a well-developed forearc high in the Coast Range, a volcanic arc forming the Cascade Range, and backarc volcanism and extension in the Columbia Plateau and Basin and Range (Figure 1a). Subduction of the Juan de Fuca plate is presently oblique and largely reflects right-lateral motion of the Pacific Plate relative to North America, as shown by the San Andreas transform to the south.

Large clockwise rotations (15-80°) are typical of widespread Cenozoic volcanic and sedimentary rocks along the convergent margin, and abundant, high quality paleomagnetic data from 62-12 m.y. (million year) old rocks from the forearc, arc, and backarc regions show that rotation increases with age and with proximity to the coast (Figure 1b). Several plate tectonic explanations for the rotations have been proposed, including: 1) Eocene rotation of an oceanic Coast Range micoplate during its oblique collision with North America (Simpson and Cox, 1977; Magill and others, 1981); 2) Dextral shear rotation of the continental margin throughout

313

C. Kissel and C. Laj (eds.), Paleomagnetic Rotations and Continental Deformation, 313–325.
© *1989 by Kluwer Academic Publishers.*

314

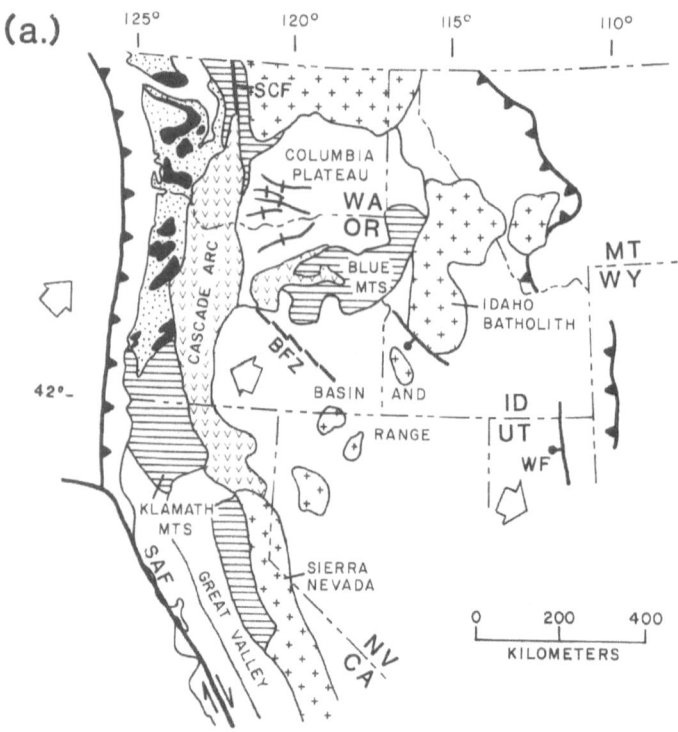

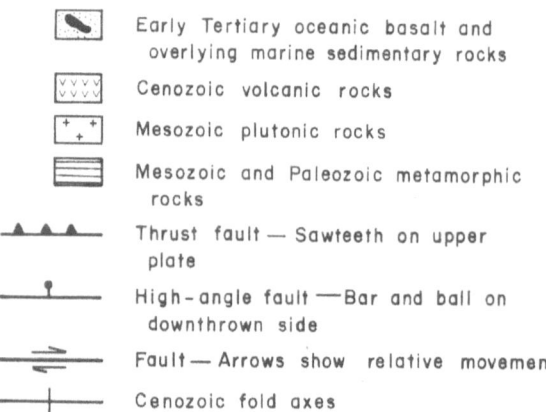

EXPLANATION

⬤ Early Tertiary oceanic basalt and
overlying marine sedimentary rocks

∨∨∨ Cenozoic volcanic rocks

+ + Mesozoic plutonic rocks

▤ Mesozoic and Paleozoic metamorphic
rocks

▲▲▲ Thrust fault — Sawteeth on upper
plate

High-angle fault —Bar and ball on
downthrown side

Fault — Arrows show relative movement

Cenozoic fold axes

315

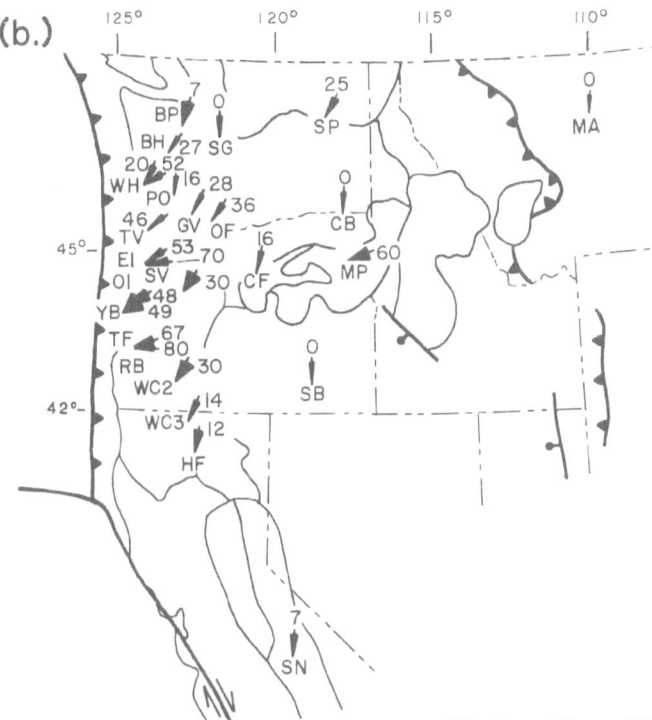

Figure 1. a) Geologic provinces of the Pacific Northwest modified from King and Beikman (1974); SAF, SCF, BFZ, WF represent San Andreas fault, Straight Creek fault, Brothers fault zone, and Wasatch fault, respectively; large arrows show directions of Neogene convergence along margin and extension in Basin and Range. b) Tectonic rotations of rock units, in degrees, slightly modified from compilations by Grommé and others (1986), Magill and others (1981), Bates and others (1981), and original sources; shaded sectors indicate 95% confidence limits on rotations. WASHINGTON: BP - Eocene volcanic rocks at Bremerton-Port Ludlow (Beck and Engebretson, 1982); BH - Eocene Crescent Formation, Black Hills (Globerman and others, 1982); WH - Eocene Crescent Formation, Willapa Hills (Wells and Coe, 1985); GV - upper Eocene Goble Volcanics (Beck and Burr, 1979; Wells and Coe, 1985); PO - Miocene Pomona Member, Saddle Mountains Basalt (Magill and others, 1982) OF - upper Eocene and Oligocene Ohanapecosh Formation of Bates and others (1981); SG - Miocene Snoqualmie and Grotto batholiths (Beske and others, 1973); SP - Eocene Sanpoil Volcanics (Fox and Beck, 1985). OREGON: TV - Eocene Tillamook Volcanics (Magill and others, 1981); EI - Eocene intrusions (Beck and Plumley, 1980); OI - Oligocene intrusions (Beck and Plumley, 1980); SV - Eocene Siletz River Volcanics (Simpson and Cox, 1977); TF - Eocene Tyee Formation (Simpson and Cox, 1977); RB - Paleocene basalt at Roseburg (Wells and others, 1985); YB - upper Eocene Yachats Basalt (Simpson and Cox, 1977); WC1, WC2, WC3 - Oligocene and Miocene volcanic rocks of the western Cascade Range (Magill and Cox, 1980; Beck and others, 1986); CF - Eocene and Oligocene Clarno Formation (Grommé and others, 1986); CB - Miocene Columbia River Basalt Group and Steens Basalt (SB) (Mankinen and others, 1987); MP - Late Jurassic or Early Cretaceous plutons (Wilson and Cox, 1980) CALIFORNIA: HF - Upper Cretaceous Hornbrook Formation (Mankinen and Irwin, 1982); SN - Late Cretaceous Sierra Nevada batholith (Frei and others, 1984). MONTANA: MA - Paleocene and Eocene Montana alkalic province (Diehl and others, 1983). Figure from Wells and Heller (1988).

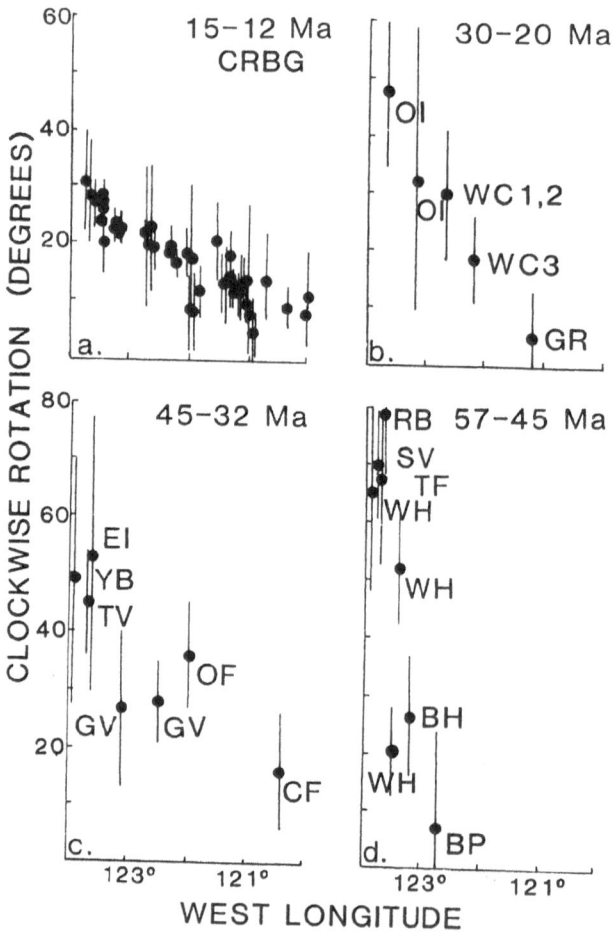

Figure 2. Clockwise tectonic rotation inboard from convergent margin plotted on east-west profile. Progessive increase in rotation toward coast supports distributed dextral shear as rotation mechanism. Unit symbols from Figure 1; data published in Grommé and others (1986) except for middle Miocene Columbia River Basalt Group (CRBG) which is from Wells and Heller (1988).

Cenozoic time, driven by northward-moving oceanic plates to the west (Beck, 1980; Wells and others, 1984; Engebretson and others, 1985); and 3) Late Cenozoic microplate rotation of the arc and forearc in front of intracontinental extension in the Basin-Range region (Magill and Cox, 1980; Magill and others, 1981).

Wells and Heller (1988) have summarized arguments regarding the relative contribution of accretion, shear, and extension to tectonic rotation in the Pacific Northwest, and they concluded that dextral shear and extension were the most important contributors to tectonic rotation. They showed that most of the rotation post dates the Eocene onlap of North American continental shelf strata onto the oceanic basalt basement of the Coast Range and thus limits rotation during oblique collision of an oceanic microplate to less than 10°. The best evidence for dextral shear comes from paleomagnetic studies of certain flood basalt units of the Miocene Columbia River Basalt Group. These flows erupted from sources in the backarc region near Idaho about 15 m.y. ago and flowed westward through a gap in the arc and into the Pacific Ocean, a distance greater than 500 km. These gently tilted flows are ideal rotational strain markers, and paleomagnetic data show a progressive westward increase in clockwise rotation from the arc toward the coast. Wells and Heller (1988) argued that dextral shear rotation was characteristic of continental margin deformation throughout Tertiary time, based on the likelihood of long-term northeast oblique subduction (Engebretson and others, 1985) and the inability of rigid plate models to account for large coastal rotations purely by backarc extension. They also suggested that dextral shear along the coast was responsible for at least 40% of the total rotation and backarc extension in the Basin-Range region was responsible for the remainder. They noted that it was difficult to isolate the component of rotation generated by Basin-Range extension from the dextral shear component in the Miocene paleomagnetic data, but the geology seems to require a substantial rigid plate component, because facies relationships in Eocene coastal sedimentary basins are not disturbed by the rotations, which approach 70°.

In this paper, I want to examine more closely the regional and temporal trends in the paleomagnetic data which tend to support pre-Miocene dextral shear along the convergent margin; and secondly, may indicate the amount of rotation produced by Basin-Range extension. I also examine the relationship between paleomagnetic rotations and local structure in the forearc which supports a model of small block rotations in a predominantly dextral transpressive tectonic regime.

2. Long-term dextral shear along the margin

2.1. PALEOMAGNETIC EVIDENCE

Paleomagnetic results from closely spaced sites in individual flows of the Miocene Columbia River Basalt Group show a well-defined, progressive increase in tectonic rotation from the Cascade arc to the coast, a distance of 200 km (Figure 2a). Sheriff (1984), Wells and others (1983b), and Wells and Heller (1988) interpreted this data as strong circumstantial evidence for post-15 Ma dextral shear, distributed through the arc and forearc. The rotating blocks must be larger than a paleomagnetic site (10^2 m), but substantially smaller than the width of the forearc region (10^2 km), because significant differential rotations between sites are observed over distances of less than 10 km.

A plot of rotation versus distance from the coast for older forearc and arc volcanic rocks shows a similar pattern (Figure 2b-d), although the resolving power is not as good as the single

318

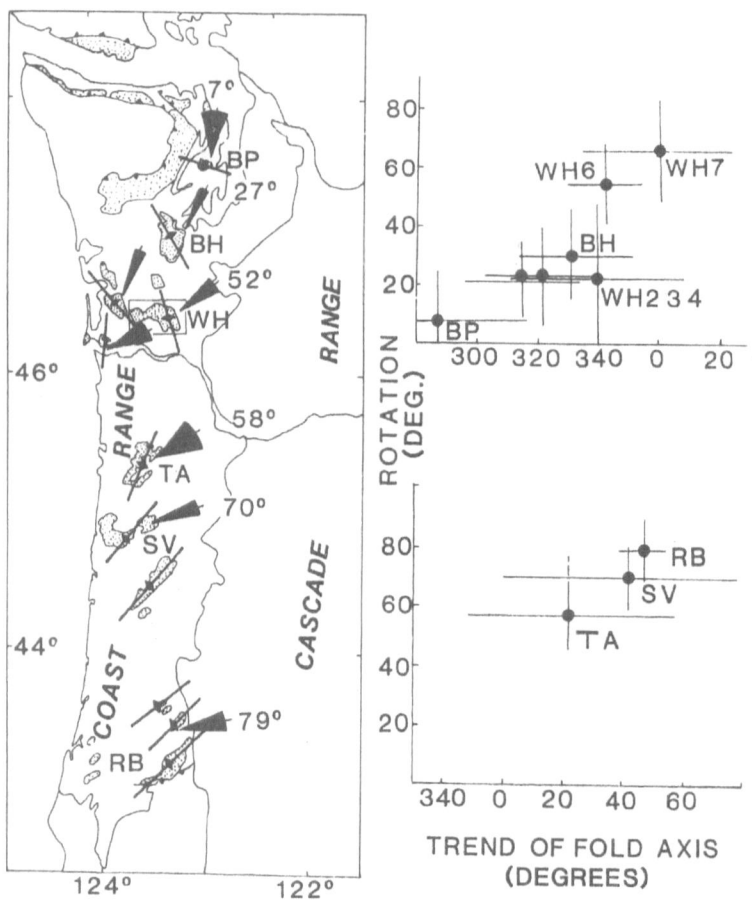

Figure 3. Rotation versus trend of Eocene fold axes in oceanic basement of the Coast Range. T positive correlation indicates differential block rotation after Eocene compression against th continental margin. Map shows outcrop area of Paleocene-Eocene oceanic pillow basalt upli (stipple pattern); shaded sectors with tick marks indicate tectonic rotation in degrees and 95 confidence limits; fold axis symbol indicates trend of best-fit fold axis. Units as in Figure 1 exce TA, which is average of 2 sites from lower Tillamook Volcanics of Magill and others (1981) and W 2, 3, 4, 6, 7, which are structural domains of Wells and Coe (1985). Box shows area of Figure 4.

flow method used in the Columbia River Basalt Group. In most cases, the rotation values are averages of 20 to 50 paleomagnetic sites collected throughout an entire volcanic sequence or formation.

Results from both Oregon and Washington are plotted on the same cross section because data from any one latitudinal band is scarce. In general, Oregon rotations are larger than rotations in Washington, a point I will discuss later. Some uncertainty is also introduced by combining units erupted over a 10-15 m.y. time span, when rotation was occurring. The average rotation rate is about 1.3°/m.y. Given the uncertainties, it is remarkable that rotations in the older units show the same well-defined pattern of progressive increase in rotation toward the coast that is observed in the Miocene data, suggesting that dextral shear has been an important rotation mechanism throughout Tertiary time.

2.2 STRUCTURAL EVIDENCE

2.2.1. Rotation and Fold Trends in the Forearc Basement. Structural domains in the folded, oceanic basalt basement of the Oregon-Washington Coast Range correlate well with observed paleomagnetic rotations on a regional scale (Figure 3). The amount of rotation increases southward, as does the eastward deflection of best-fit fold axes in the Paleocene to middle Eocene pillow basalt basement. Where the age of the folds is constrained by onlapping sedimentary rocks, they are of middle Eocene age and appear to be related to compression and thrusting of the oceanic basalt against the continental margin (Baldwin, 1974; Wells and Coe, 1985). In a detailed study of southwest Washington, Wells and Coe (1985) found that basement structural domains were bounded by faults and had undergone differential clockwise

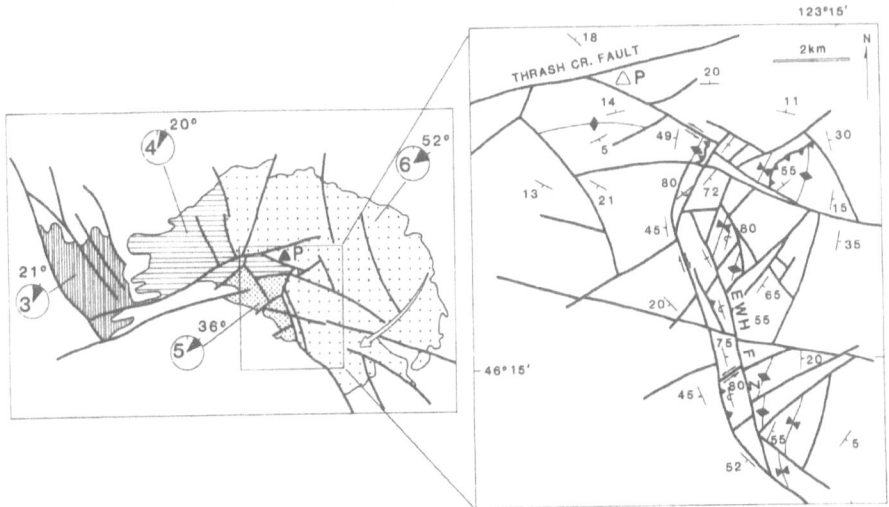

Figure 4. Rotation of oceanic basement structure domains, southwest Washington. Fine lines are fold axes, heavy lines are faults. Most faults are high angle, arrows show sense of motion where known. Thrust faults shown by teeth on upper plate. Sector diagrams show rotation in degrees with 95% confidence limits. Figure modified from Wells and Coe (1985).

320

rotations of 20° to 65° prior to the onlap of middle Eocene sediments (Figures 3 and 4). They showed a correlation between the trend of early-formed folds and tectonic rotation for five basement domains ranging in width from three to thirty kilometers.

There is also an intriguing regional correlation between the amount of basement tectonic rotation and the trend of the steep gravity gradient along the east side of the Coast Range (Figure 5). The large gravity high along the Coast Range correlates directly with basement uplifts of Paleocene-Eocene oceanic basalt. The zig-zag eastern margin correlates with a series of sedimentary basins along the Puget-Willamette lowland between the Coast Range and the Cascade arc. The trend of the gradient strongly suggests that it is structurally controlled. In Washington, the gradient locally parallels the trend of early fold axes in the basement and suggests that the gradient represents the continental buttress against which folding occurred. The present zig-zag pattern may reflect disruption and subsequent block rotation of the Coast Range terrane and its tectonic boundary with the continent.

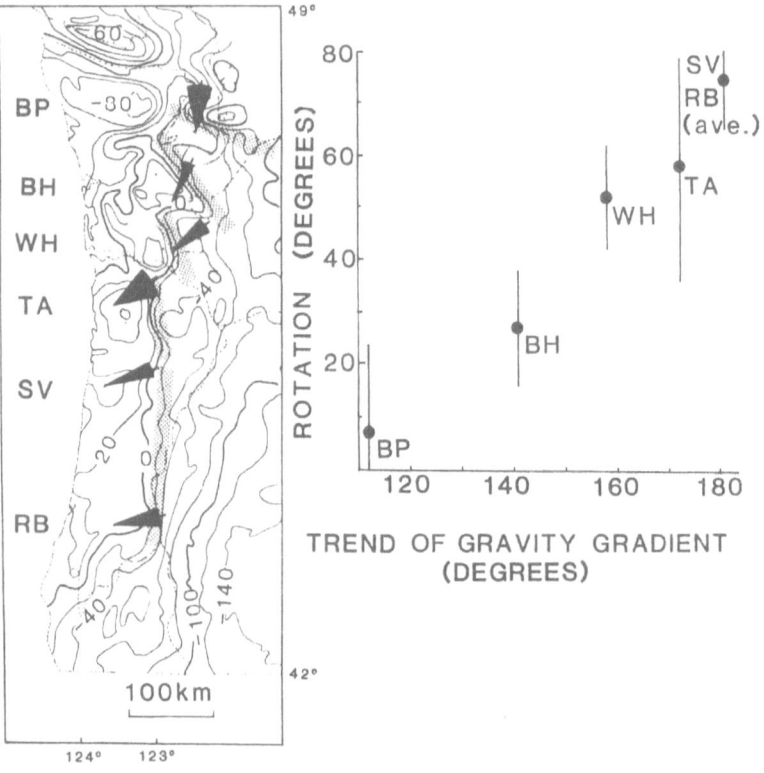

Fig. 5: Complete Bouguer gravity map of the Oregon-Washington Coast Range (modified Magill and others, 1981), showing possible correlation between trend of eastern gradient and tectonic rotation. Zig-zag gradient may indicate outline of large, rotating blocks. Contour interval 20 mGal; units as in Figure 1).

2.2.2. *Rotation and the Trend of Coast Range Faults.* Fault patterns in the Coast Range are complex and reflect a long history of Tertiary deformation. The largest structures are east-dipping thrust or reverse faults that form the western boundary of anticlinal basement uplifts in the Coast Range. In Oregon the uplifts and their boundary faults trend northeast across the axis of the broad Coast Range uplift, whereas in Washington the basement uplifts trend northwest (Figure 3). Local isoclinal or overturned folding is common in some basement uplifts adjacent to the faulted margins, although subhorizontal slickensides also occur along the sheared margins of basement blocks. Where examined in detail, the uplifts are internally deformed by abundant high-angle faults, which offset the fold axes, and by local thrust faults (Figure 4). Most of the high-angle faults have horizontal slickensides and grooves, with N20°-45°W faults showing dextral slip and N60°W-S70°W faults showing sinistral slip. The compressive deformation of the block margins, when combined with north-west-trending dextral slip zones subparallel to the block margins, is consistent with dextral transpression in the forearc driven by oblique subduction of the Farallon plate to the west.

In the southern Oregon Coast Range, N40°E faults are most common and they extend into the western Cascade Range. In contrast, northern Oregon Coast Range faults trend N40°-70°W, with subordinate northeast-trending faults (Wells and others, 1983). The N40°W faults commonly exhibit dextral slip, and the northeast and some of the N60°-70°W faults exhibit sinistral slip. There has been no systematic kinematic analysis of faulting in the Oregon Coast Range, but it seems likely that observed local rotations are being accommodated by the faults.

In southwest Washington, Wells and Coe (1985) made a preliminary interpretation of the fault pattern in terms of a Riedel shear rotation mechanism driven by the northward component of Farallon plate motion along the margin (Figure 6). They suggested that left separation on secondary faults was consistent with the slip required to accommodate observed (~25°) paleomagnetic rotations in late Eocene and Miocene lavas which unconformably overly the previously deformed oceanic basement. These secondary faults form distinct domains in the fault population, with N60°-70°W sinistral faults correlated with regions underlain by strongly rotated oceanic basement (50-65° clockwise) and N45°-80°E sinistral faults typical of regions underlain by less rotated basement (20° clockwise) (Figure 6). Both fault sets offset young stratigraphic units, including the Pomona Member of the Columbia River Basalt Group (12 Ma), which is rotated about 16° clockwise in both domains. The angular difference between the sinistral fault sets (about 50°) is close to the differential rotation observed in the basement (about 40°) and suggests the faults may be inherited structures that were originally subparallel prior to early differential rotation of the basement. The sinistral faults are presently interpreted as Riedel shears that have accommodated post-Eocene clockwise rotation.

3. Basin-Range extension

Previous workers have argued on regional tectonic grounds that Cenozoic extension in the Basin-Range region (Figure 1) has been a major contributor to clockwise rotation of coastal regions in the Pacific Northwest (Simpson and Cox, 1977; Magill and others, 1981; 1982). Magill and others (1982) suggested that the thermally elevated Basin-Range would produce a "ridge-push" force against the southern part of the Cascade arc and forearc in Oregon, thus causing it to rotate clockwise about a fixed point north of the expanding region. They believed that the paleomagnetic data collected along an east-west traverse across the forearc and backarc regions supported a stepwise decrease in rotation within the arc, which they interpreted to be

322

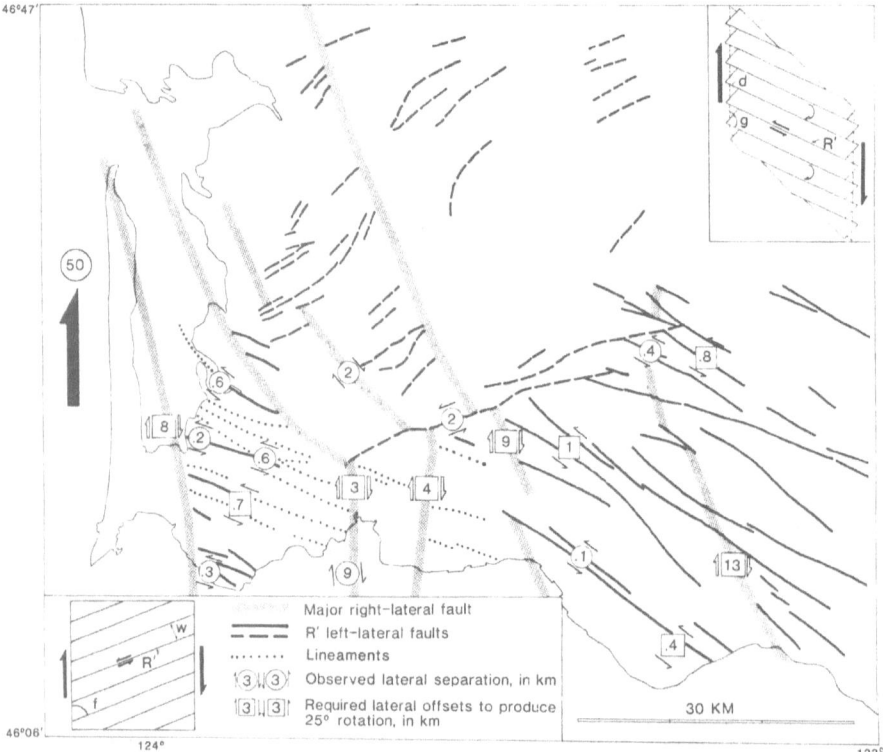

Figure 6. Strike-slip interpretation of post-late Eocene faulting in southwest Washington and comparison with simple shear deformation model of Freund (1974). Rotation of elongate crustal slices is accommodated by sinistral R' shears. Relationship between fault displacement (d), width of fault block (w), and rotation (g-f), is given by cot f - cot g = d/w (Freund, 1974). Large arrow shows northward component of Farallon plate convergence during middle Tertiary time, in kilometers per megaannum, from Engebretson and others, (1985). Figure from Wells and Coe (1985).

the eastern boundary of the rotated block. Using a larger paleomagnetic data set, Wells and Heller (1988) showed that the rotation was distributed across the forearc and into the arc and that this dextral shear overprint effectively obscures a stepwise change in rotation expected from Basin-Range extension. They argued, based on the amount of differential rotation across the forearc region, that dextral shear was responsible for at least 40% of the total rotation, with Basin-Range extension responsible for the remainder.

Although the stepwise increase in rotation due to Basin-Range extension is not resolvable in east-west profiles, the southward increase in rotation along the coast may still be an indicator of a Basin-Range component of rotation. Because the Basin-Range region is inboard of the Oregon Coast Range, extension should have affected Oregon more than Washington. As expected, there is an increase in rotation from north to south along the coast (Figure 1b). The average difference in rotation between Oregon and less rotated Washington, when grouped according to age, is 44° for early and middle Eocene units and 20° for late Eocene and Oligocene units. This compares favorably to the extensional component of rotation for the Oregon Coast Range calculated by Wells and Heller (1988) for post-50 Ma time (43°) and post 37 Ma time (27°). The implication of this observation is that dextral shear and "backarc" extension are subequal contributors to tectonic rotation along the Oregon Coast part of the Pacific Northwest convergent margin.

4. Discussion

The overall pattern of rotations in rocks of Miocene, Oligocene, and Eocene age supports a substantial dextral shear contribution to convergent margin rotations throughout Cenozoic time. The direct correlation between details of forearc structural geology and domains of tectonic rotation is strong evidence for accommodation of the shear rotation by local structure. The fault-block mechanism accommodating the rotations along the convergent margin is similar to that described by Freund (1974), Ron and others (1984), and Garfunkel and Ron (1985), who described a model in which closely-spaced strike-slip faults allow rotation of elongate crustal blocks commensurate with the orientation of the stress field (see Figure 6). Both clockwise and counterclockwise rotations are possible in these models, and in a forearc region undergoing pure shortening, some areas with appropriately oriented faults should rotate counterclockwise. But only clockwise rotations are observed along the coast, and it suggests that simple dextral shear driven by oblique subduction predominates over compression as the driving force for the rotations.

Fault domains in the forearc also have some kinematic similarities to the model of McKenzie and Jackson (1983) in which elongate fault blocks span (in this case) a transpressive zone between two major plates. The blocks can be modeled as floating above a viscous substrate in which the relative motion between the major plates is distributed linearly across the zone of deformation. Given relative motions between the Farallon and North American plates we can calculate an expected rotation rate from the model of 3.8°/m.y. This is substantially higher than the paleomagnetically determined rotation rate of 0.7-1.3°/m.y. (depending on the contribution of backarc extension). This may reflect structural complexities in the upper plate; the width of forearc fault domains is much narrower than the total width of the forearc deformation zone. Alternatively, the slower rotation rate may be telling us something about the efficiency of coupling between the forearc and the downgoing slab. It implies that 20 to 35% of the relative dextral slip component between the plates is converted to rotational strain in the upper plate. The efficiency of coupling appears greatest near the coast and decreases inland where the upper plate is thicker and hotter. The decrease in rotation inboard from the coast is also similar to that predicted by Sonder and others (1986) for continental deformation along strike-slip zones. Their continuum calculations for southern California, which they modelled as a thin viscous sheet with a power law rheology, successfully accounted for the eastward decrease in paleomagnetically determined finite rotations. A similar approach may prove useful for analyzing rotations along the convergent margin of the Pacific Northwest.

Acknowledgements. I am thankful to R.W. Simpson (U.S.G.S.) for continued discussion of these problems and to Simpson and D.M. Miller for comments on the manuscript.

References

Atwater, T., 1970, Implications of plate tectonics for the Cenozoic evolution of western North America: *Geological Society of America Bulletin*, v. **81**, p. 3513-3536.

Baldwin, E.M., 1974, Eocene stratigraphy of southwestern Oregon: *Oregon Department of Geology and Mineral Industries Bulletin*, **83**, 40 p.

Bates, R.B., Beck, M.E., Jr., and Burmester, R.F., 1981, Tectonic rotations in the Cascade range of southern Washington: *Geology*, v. **9**, p. 184-189.

Beck, M.E., Jr., 1980, Paleomagnetic record of plate-margin tectonic processes along the western edge of North America: *Journal of Geophysical Research*, v. **85**, p. 7115-7131.

Beck, M.E., Jr., and Burr, C.D., 1979, Paleomagnetism and tectonic significance of the Goble Volcanic Series, southwestern Washington: *Geology*, v. **7**, p. 175-179.

Beck, M.E., Jr., and Engebretson, D.C., 1982, Paleomagnetism of small basalt exposures in the west Puget Sound area, Washington, and speculations on the accretionary origin of the Olympic Mountains: *Journal of Geophysical Research*, v. **87**, p. 3755-3760.

Beck, M.E., Jr., and Plumley, P.W., 1980, Paleomagnetism of intrusive rocks in the Coast Range of Oregon: Microplate rotation in middle Tertiary time: *Geology*, v. **8**, p. 573-577.

Beck, M.E., Jr., Burmester, R.F., Craig, D.E., Grommé, C.S., and Wells, R.E., 1986, Paleomagnetism of middle Tertiary volcanic rocks from the Western Cascade series, Northern California: Timing and scale of rotation in the southern Cascades and Klamath Mountains: *Journal of Geophysical Research*, v. **91**, p. 8219-8230.

Beske, S.J., Beck, M.E., Jr., and Noson, L., 1973, Paleomagnetism of the Miocene Grotto and Snoqualmie Batholiths, central Cascades, Washington: *Journal of Geophysical Research*, v. **78**, p. 2601-2608.

Diehl, J., Beck, M.E., Jr., Beske-Diehl, S., Jacobson, D., and Hearn, C.B., Jr., 1983, Paleomagnetism of the late Cretaceous-early Tertiary north central Montana alkalic province: *Journal of Geophysical Research*, v. **88**, p. 10,593-10,610.

Engebretson, D.C., Cox, A., and Gordon, R.G., 1985, Relative motions between oceanic and continental plates in the Pacific Basin: Geological Soc. of America Special Paper, **206**, 59p.

Fox, K.F., and Beck, M.E., Jr., 1985, Paleomagnetic results for Eocene volcanic rocks from northeastern Washington and the Tertiary tectonics of the Pacific Northwest: *Tectonics*, v. **4**, p. 323-341.

Frei, L.S., Magill, J.R., and Cox, A., 1984, Paleomagnetic results from the central Sierra Nevada: Constraints on reconstructions of the western United States: *Tectonics*, **3**, 157-178.

Freund, R., 1974, Kinematics of transform and transcurrent faults: *Tectonophysics*, v. **21**, p. 93-134.

Garfunkel, Z., and Ron, H., 1985, Block rotation and deformation by strike-slip faults 2. The properties of a type of macroscopic discontinuous deformation: *Journal of Geophysical Research*, v. **90**, p. 8589-8602.

Globerman, B.R., Beck, M.E., Jr., and Duncan, R.A., 1982, Paleomagnetism and tectonic significance of Eocene basalts from the Black Hills, Washington Coast Range: *Geological Society of America Bulletin*, v. **93**, p. 1151-1159.

Grommé, C.S., Beck, M.E., Wells, R.E., and Engebretson, D.C., 1986, Paleomagnetism of the Tertiary Clarno Formation of central Oregon and its significance for the tectonic history of the Pacific Northwest: *Journal of Geophysical Research*, v. **91**, p. 14089-14103.

King, P.B., and Beikman, H.M., 1974, Geologic map of the United States, scale 1:2,500,000: U.S. Geological Survey, Washington, D.C.

Magill, J.R., and Cox, A., 1980, Tectonic rotation of the Oregon western Cascades: Oregon *Department of Geology and Mineral Industries Special Paper* **10**, 67 p.

Magill, J.R., Cox, A., and Duncan, R., 1981, Tillamook Volcanic Series: Further evidence for tectonic rotation of the Oregon Coast Range: *J. of Geophysical Res.*, v.**86**, 2953-2970.

Magill, J.R., Wells, R.E., Simpson, R.W., and Cox, A.V., 1982, Post 12 m.y. rotation of southwest Washington: *Journal of Geophysical Research*, v. **87**, p. 3761-3776.

Mankinen, E.A., and Irwin, W.P., 1982, Paleomagnetic study of some Cretaceous and Tertiary sedimentary rocks of the Klamath Mountains province, Califonia: *Geology*, v. **10**, p. 82-87.

Mankinen, E.A., Larson, E.E., Grommé, C.S., Prevot, M., and Coe, R.S., 1987, The Steens Mountain (Oregon) geomagnetic polarity transition 3. Its regional significance: *Journal of Geophysical Research*, v. **92**, p. 8057-8076.

McKenzie, D., and Jackson, J., 1983, The relationship between strain rates, crustal thickening, palaeomagnetism, finite strain and fault movements within a deforming zone: *Earth and Planetary Science Letters*, v. **65**, p. 182-202.

Ron, H., Freund, R., Garfunkel, Z., and Nur, A., 1984, Block rotation by strike slip faulting: structural and paleomagnetic evidence: *Journal of Geophysical Research*, v. **89**, p. 6256-6270.

Sheriff, S.D., 1984, Paleomagnetic evidence for spatially distributed post-Miocene rotation of western Washington and Oregon: *Tectonics*, v. **3**, p. 397-408.

Simpson, R.W., and Cox, A., 1977, Paleomagnetic evidence for tectonic rotation of the Oregon Coast Range: *Geology*, v. **5**, p. 585-589.

Sonder, L.J., England, P.C., and Houseman, G.A., 1986, Continuum calculations of continental deformation in transcurrent environments: *Journal of Geophysical Research*, v. **91**, p. 4797-4810.

Wells, R.E., and Coe, R.S., 1985, Paleomagnetism and geology of Eocene volcanic rocks of southwest Washington, Implications for mechanisms of rotation: *Journal of Geophysical Research*, v. **90**, p. 1925-1947.

Wells, R.E., and Heller, P.L., 1988, The relative contribution of accretion, shear, and extension to Cenozoic tectonic rotation in the Pacific Northwest: *Geological Society of America Bulletin*, v. **100**, 325-338.

Wells, R.E., Niem, A.R., MacLeod, N.S., Snavely, P.D. Jr., and Niem, W.A., 1983a, Preliminary geologic map of the west half of the Vancouver (Wa-Ore) 1°x2° sheet, Oregon: U.S. Geological Survey Open-File Report 83-591, scale 1:250,000

Wells, R.E., Simpson, R.W., Kelly, M.M., Beeson, M.H., and Bentley, R.D., 1983b, Columbia River Basalt stratigraphy and rotation in southwest Washington: *EOS, Transactions, American Geophysical Union*, v. **64**, p. 687.

Wells, R.E., Engebretson, D.C., Snavely, P.D., Jr., and Coe, R.S., 1984, Cenozoic plate motions and the volcano-tectonic evolution of western Oregon and Washington: *Tectonics*, v. **3**, no. 2, p. 275-294.

Wells, R.E., Kelly, M.M., Levi, S., Schultz, K., and McElwee, K., 1985, Folding and rotation of Paleocene pillow basalt near Roseburg, Oregon: *EOS, Transactions, American Geophysical Union*, v. **66**, p. 863.

Wilson, D., and Cox, A., 1980, Paleomagnetic evidence for tectonic rotation of Jurassic plutons in the Blue Mountains, eastern Oregon: *Journal of Geophysical Research*, v. **85**, p. 3681-3689.

ROTATION OF CENTRAL AND SOUTHERN ALASKA IN THE EARLY TERTIARY: OROCLINAL BENDING BY MEGAKINKING?

R. S. COE[1], B. R. GLOBERMAN[2], and G. A. THRUPP[3]
Earth Sciences,
University of California
Santa Cruz, California 95064,
USA

[1] *Presently on leave at Centre Géologique et Géophysique,*
Université des Sciences, 34060 Montpellier Cedex, France
[2] *Now at Pacific Geoscience Centre,*
Geological Survey of Canada, Sidney, British Columbia,
Canada
[3] *Now at Department of Earth Sciences,*
Macquarie University, North Ryde, NSW,
2109 Australia

ABSTRACT. Systematic counterclockwise rotations of paleomagnetic declinations relative to those expected from North American cratonic reference poles are exhibited by latest Cretaceous and early Tertiary volcanic rocks from the western two-thirds of central and southern Alaska. These data suggest that the entire region underwent a tectonic rotation of 44 ± 11° counterclockwise. The most probable rotation mechanism is one that resembles kink folding about a vertical axis, but on a gigantic scale, with the axial plane vertical and nearly coincident with the 148th meridian. The flexural slip required by such megakinking agrees with geological estimates of displacements on known faults. Moreover, this mechanism has the virtue of straightening the sharply bent part of the structural grain of Alaska without opening a huge sphenochasm. A plausible cause of the rotation is convergence between Eurasia and North America, which is predicted to have occurred at the correct time by several analyses of the marine magnetic anomalies in the North Atlantic and Arctic Oceans.

1. Introduction

The idea that a large part of Alaska may have undergone significant tectonic rotation goes back at least as far as S. W. Carey's orocline hypothesis (1955). Carey examined the geotectonic consequences of straightening various curved fold belts around the world, including those of Alaska. He concluded that the western two-thirds of Alaska had rotated 28° counterclockwise (CCW) with respect to the eastern part since the Triassic, about a pivot point in the Alaska Range near the conspicuous bend in the Denali fault (Fig. 1). Associated with this rotation, Carey believed, was the opening of the Arctic Basin "sphenochasm," which in turn he concluded was driven by a post-Triassic expansion of the earth's radius by twenty percent (Carey, 1958). Although his mechanism involving such a large and recent expansion of the earth appears highly unlikely from our present perspective, Carey's geometrical and tectonic insights have inspired many paleomagnetic investigations around the globe to test for the presence or absence of predicted azimuthal rotations.

C. Kissel and C. Laj (eds.), Paleomagnetic Rotations and Continental Deformation, 327–342.

Such a test for the Alaska orocline was first attempted by Packer and Stone (1972). They came to a tentative negative conclusion on the basis of a paleomagnetic pole that they obtained from Jurassic sedimentary rocks on the Alaska Peninsula, which suggested clockwise (CW) rotation when compared to the Jurassic North American reference pole. This early work, however, was carried out before the extent of pervasive regional overprinting was generally recognized, when more rudimentary magnetic cleaning was considered sufficient to isolate the primary component. Only blanket rather than progressive alternating field demagnetization was used, and neither thermal demagnetization nor analysis of magnetic components was undertaken. In light of CCW rotations reported just to the north in more recent and detailed studies of latest Cretaceous and early Tertiary rocks, Packer and Stone's early conclusion needs to be reconsidered. In this paper we summarize those results, analyze their implications, and discuss mechanisms that could have produced CCW rotation.

Two subsequent models for the formation of the orocline were proposed somewhat after Carey's. In one, northern Alaska rotated 70° CCW away from the Canadian Arctic Islands about a more northerly pivot point sometime between the Late Jurassic and Late Cretaceous, opening the Canada Basin in its wake and producing intense folding and thrusting in the Brooks Range (Tailleur, 1969 and 1973; Rickwood, 1970). The paleomagnetic data bearing on this model are discussed in a companion paper in this volume by D. B. Stone. In the other model, which figures prominently in the present paper, the western two-thirds of central and southern Alaska rotated 40 to 55° CCW during latest Cretaceous and Early Tertiary time (Grantz, 1966). In contrast to the earlier ideas of Carey (1955), rotation is accommodated not simply by the pivoting of one large block, but by sympathetic movement of several narrower blocks separated by major strike-slip faults, much like one limb of a huge kink fold with its vertical axial plane coinciding approximately with the 148th meridian (Fig. 1). Note that neither of these subsequent models excludes the other, because the rotations that are invoked by each occur at different times and involve different parts of Alaska.

2. Paleomagnetic Evidence

The region of direct concern for this paper lies south of the Kaltag fault and west of the 148th meridian, an area of more than 400,000 square kilometers in central and southern Alaska (Fig. 1). The paleomagnetic data available are almost entirely derived from lava flows of Latest Cretaceous and Tertiary age. Lava flows are very reliable recorders of the paleomagnetic field, and their primary high-temperature thermoremanent magnetization is often easier to distinguish and separate from secondary overprints than is the case for the remanent magnetization of sedimentary rocks. However, it is usually more difficult to determine the structural correction necessary to restore lava flows to their pre-folding orientation and to sample a sufficient span of time well enough to average out secular variation of the paleomagnetic field. This is important because even relatively small errors in paleomagnetic direction translate into significantly large errors in paleomagnetically inferred tectonic rotation at high latitudes, where the inclination of the field is steep. A more detailed discussion of the relative merits of lava flows and sediments for tectonic studies and of strategies to minimize the above difficulties can be found in Coe et al. (1985) and Thrupp and Coe (1986).

The locations of all but one of the paleomagnetic studies and the tectonic rotations inferred from them are shown in Figure 1. The reference poles required to calculate these rotations relative to North America were interpolated from the lists of Irving and Irving (1982) and Diehl et al. (1983) and are given in Coe et al. (1985). In all studies stepwise thermal and alternating-field demagnetization were performed, and the stable components that were isolated by the two methods yielded mean directions that are comparable and, with one exception, have 95 percent confidence limits between 4 and 8°.

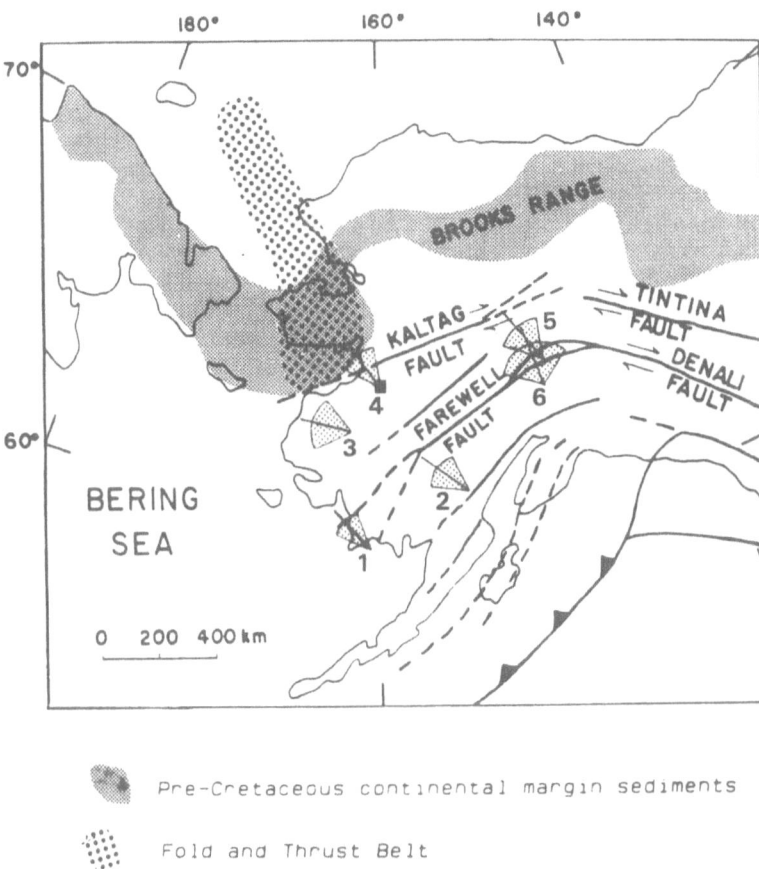

Figure 1. Simplified tectonic map of Alaska (after Patton and Tailleur, 1977) showing conspicuous bends in great faults that mark the "orocline" in central and southern Alaska, evidence for east-west compression in northwest Alaska, and sites of Late Cretaceous and early Tertiary paleomagnetic studies. Paleomagnetically inferred tectonic rotation relative to North America is indicated at each site, with 95 percent confidence limit of the rotation depicted by the half-angle of each stippled fan: (1) Bristol Bay flows, 39±23° CCW; (2) Lake Clark lava flows, 55±28° CCW; (3) Yukon River tuffaceous sediments, 77±37° CCW; (4) Blackburn Hills lava flows, 38±23° CCW; (5) Teklanika lava flows, 29±99° CCW; (6) Talkeetna lava flows, 37±48° CCW.

Progressing generally from west to east and south to north in Figure 1, we now describe briefly each of the paleomagnetic studies.

2.1. BRISTOL BAY REGION, SOUTHWESTERN ALASKA.

A CCW rotation of 39±23° was found in a study of 84 latest Cretaceous lava flows from the Bristol Bay region in southwestern Alaska (Globerman and Coe, 1983a; Globerman, 1985). Three of these flows have been radiometrically dated by the K-Ar method, giving a mean age of 68±3 Ma. Two well-exposed sections comprising about 1800 and 2700 meters of stratigraphic thickness were sampled on three islands along the northern coast of Bristol Bay. The flows are mostly subaerial, and selected trace-element ratios and Rare Earth Element patterns are consistent with eruption in an island arc setting. Bedding attitudes were determined unusually well from both sedimentary interbeds and clearly defined flow boundaries. Each section dips homoclinally, but the variation in attitudes between them is 70°. The regional tilt test (McFadden and Jones, 1981) indicates a pre-folding age of magnetization at greater than 99 percent confidence, supporting our belief that the characteristic component is primary. However, all of the flows have normal polarity, raising the question of whether they might have been erupted in a time too short to average out secular variation adequately. Other evidence suggests that this is not so. From latest Cretaceous to early Paleocene time there are several periods of normal polarity lasting as long as 0.5 to 1.0 Ma within which the entire sequence could have been erupted. The large number of flows, thickness of section, and numerous sedimentary interbeds suggests to us that the sequence spanned at least several tens of thousands of years. Moreover, the angular dispersion is consistent with a full sampling of secular variation, as is the presence of three geomagnetic excursions recorded in the sequence. The 10 deviant flow mean directions that record these excursions were excluded from the final formation mean.

2.2. LAKE CLARK REGION, SOUTHWESTERN ALASKA.

A CCW rotation of 55±28° was found 350 km to the northeast of (1) by Thrupp and Coe (1986) in a study of 30 Paleocene(?) basaltic lava flows just north of Lake Clark. Several scattered K-Ar dates for these flows indicate an Eocene or Paleocene age. Ten better K-Ar determinations on more acidic plutons and extrusive rocks in the same general area are consistent with a Paleocene age. The range of possible ages does not significantly affect the inferred CCW rotation because the reference pole did not move very much during that interval. The flows are contained in two shallowly dipping sections about 4 km apart, and collectively comprise a sequence of at least 200 m that spans several reversal boundaries and includes many soil and ash layers. Thus it is likely that the samples average at least several hundred thousand years of secular variation. Reasonably reliable mean bedding attitudes were constructed for each section from apparent dips shot from a distance, but they were too similar to allow a conclusive fold test. Closely antipodal normal and reversed directions, however, afford a positive reversal test, adding support to our contention that the remanence is primary and that secular variation has been sufficiently averaged to obtain a good approximation of the axial dipole field.

2.3. LOWER YUKON RIVER, WESTERN ALASKA.

A much less well constrained CCW rotation of 77±37° was found 400 km to the northwest of (2) by Globerman et al. (1983b) in a study of Lower Cretaceous tuffaceous sediment exposed along the Yukon River. Twenty-seven sites spanning 1500 meters of stratigraphic thickness all yielded normal polarity. At three of them an individual sample revealed a reversed characteristic direction after removal of a large normal overprint. A weak fold test was inconclusive, but did demonstrate a small increase in scatter after tectonic correction. For all these reasons the characteristic remanence is probably a secondary normal overprint, but its direction is entirely distinct from that of the recent field. Assuming that it was acquired at 70 Ma, the time of peak magmatic activity in the region, and that no tilting occurred after that time, the paleomagnetically inferred rotation is 77° CCW. Exploring the consequences of other plausible assumptions, such as increasing the age of the overprint by 20 to 50 Ma or allowing up to 30 percent of the folding to be post-remagnetization, the amount of rotation can be cut in half but always remains significantly CCW.

2.4. BLACKBURN HILLS, WESTERN ALASKA

A CCW rotation of 38±23° was found 200 km to the north of (3) by Thrupp and Coe (manuscript in preparation) in a study of 42 Paleocene basaltic-andesite flows from the Blackburn Hills, a little south of the Kaltag fault and west of the Yukon River. Published K-Ar ages from this igneous complex range from 56-65 Ma (Moll and Patton, 1982), and later unpublished determinations also point to an age of about 60 Ma. Three sections, each 200 to 400 m thick, were sampled on both limbs of a large syncline. We took local bedding attitudes wherever possible in the field, but the attitudes that we employed for the structural corrections were obtained by least-square fitting planes to points obtained by a technician of the Topographic Maps Division of the U. S. Geological Survey, who used a PG-2 stereo plotter on traces of individual lava flows that are visible on the aerial photos. The fold test resulting from these highly precise and unbiased structural corrections is positive at a very high confidence level. In addition, a positive reversal test is also obtained, strongly suggesting that the characteristic remanence is primary and that secular variation has been adequately averaged. Qualitative support for this CCW rotation is provided by a rotation of 60±40° reported by Harris et al. (1987) for several lava and pyroclastic flows that are exposed nearby along the Yukon River and dated at about 55 Ma.

2.5. CANTWELL BASIN, CENTRAL ALASKA.

A CCW rotation of 29±99° can be estimated from the results of the study by Hillhouse and Grommé (1982) on 18 flows of the Teklanika Volcanics associated with the Paleocene Cantwell Formation, located 550 km to the east of (4) between the two major strands of the Denali fault. The 95 percent confidence limit on the rotation is very large because the alpha-95 circle around the paleomagnetic pole contains the sampling site as well. K-Ar dates indicate an age of 60 Ma or perhaps slightly older. The generalized fold test is strongly positive. No reversal test is possible because all the flows have reversed polarity. The main argument that sufficient time had probably been sampled to average out the effects of secular variation was that the flows were sampled over a large geographical area. Additional sampling of 16 Teklanika lavas has been carried out by Panuska and Macicak (1986). Depending on the method they use for combining and averaging flows in the two data sets, CCW rotations ranging from 12 to 32° are obtained (B. Panuska, personal communication, 1987).

2.6. TALKEETNA MOUNTAIN REGION, SOUTHERN ALASKA.

A CCW rotation of 37±48° can be calculated from the results of Hillhouse et al.'s (1984) study of 26 Eocene lava flows in the Talkeetna Mountains, 75 km to the south of (5). K-Ar ages from these and associated intrusive rocks give an age of about 52 Ma. The sampling localities were spread over a large area, and both fold and reversal tests were positive. Additional sampling of 24 of these Eocene flows further to the south by Panuska and Stone (1985) yielded a mean direction only 3° different from that of the former study. A detailed account of this latter study is nearing publication (D. B. Stone, personal communication, 1988), and it is possible that combination of the two sets of data will reduce the error to the point that the rotation is significant at 95 percent confidence.

2.7. McGRATH REGION, WEST-CENTRAL ALASKA.

A CW rotation of 2° was obtained by Plumley (1984) in a study of seven lava flows of the Paleocene Nowitna Volcanics, 350 km to the northwest of (6) in the McGrath region (not shown on Fig. 1). The small number of flows leads to a much larger uncertainty in the mean direction than for the other studies, and thus, for the same reason as in (5) above, a huge and poorly defined uncertainty in the paleomagnetically inferred rotation. Moreover, exposures were poor, so that attitudes had to be estimated from sloping benches that formed on the hillsides above some of the flows. Even though the flows have only moderate dip, the structural correction is probably also the most uncertain of those used in this paper. After removal of large, scattered overprints due to lightning, reversed directions were found for six of the flows and a stable, intermediate direction for the seventh (excluded from the final mean). Although no fold or reversal test is available, we have no reason not to suppose that the characteristic remanence is primary. Moreover, the lack of serial correlation in the directions and the presence of an intermediate (excursion?) direction suggest that the flows span at least a couple of thousand years. Although that is shorter than one would like for averaging out the effects of secular variation, the time-average of the geomagnetic pole during the past 2,000 years does approximate the geographic axis fairly well (Champion, 1980). Thus we are reluctant to exclude this result, but its extremely large uncertainty must be kept in mind.

3. Discussion

3.1. THE CASE FOR REGIONAL ROTATION

The azimuthal rotations inferred from the above paleomagnetic studies show a strong CCW bias. Although the validity of even the best determined of these rotations taken alone might be doubted, owing to the sensitivity of the paleomagnetic declination to systematic errors because the inclination is steep, the consistency of the sense of rotation from study to study is difficult to dismiss. Only one of the seven studies (7) gives a small CW rotation, and it is by far the most uncertain of all the results. Weighting all seven equally without regard to confidence limits gives a mean CCW rotation of 39±19, where the plus or minus value is the 95 percent confidence interval of the mean.

Restricting the average to the three results (2.1, 2.2 and 2.4) that are each clearly significant at the 95 percent confidence level, the mean rotation is 44±11° CCW.

It is important to note in passing that little or no latitudinal movement relative to North America is implied by the paleomagnetic results. The values of northward displacement derived from the seven studies range between -8 and +9° and have a mean value of 2±4°. Only one of them is significant at 95 percent confidence. Recalling that 9° of northward displacement at these high latitudes corresponds to only 5° shallowing of inclination, it appears unwarranted to make any tectonic interpretation based on the geographic distribution of the apparent northward and southward displacements.

3.2. ROTATION MECHANISMS

The pronounced CCW bias of the azimuthal rotations strongly suggests the operation of some kind of regional rotation mechanism. A number of possibilities come to mind:

3.2.1.*Simple Oroclinal Bending.* Carey's (1955) classic orocline hypothesis, adapted to explain the post-latest Cretaceous rotations presented in this paper, implies either great extension or great shortening of the crust, neither of which fits the geological record. The rotational pivot point favored by Carey himself would require the opening of a huge sphenochasm in Alaska that is simply not observed (Fig. 2). A pivot point much farther north in the Mackenzie River delta has the virtue of explaining the opening of the Canada Basin, but then requires great east-west compression

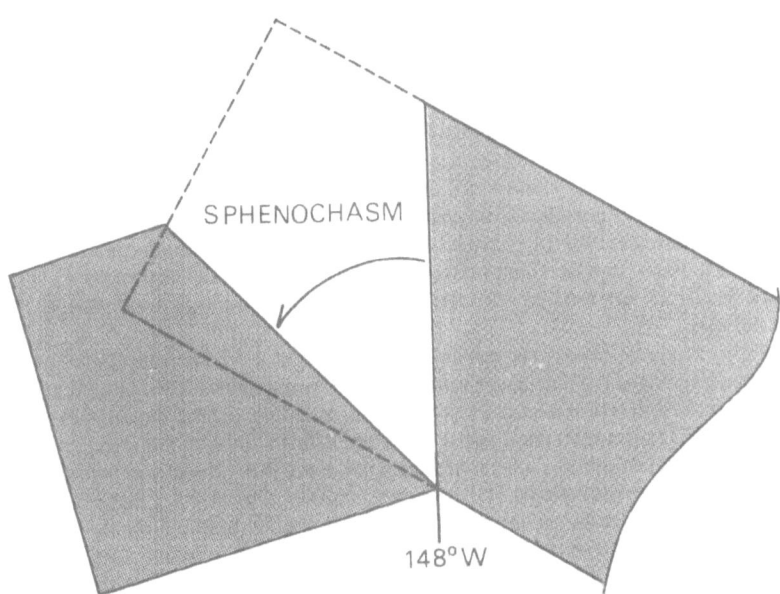

Figure 2. Simple oroclinal bending model of Alaska. Regional CCW rotation of the amount implied by the paleomagnetic results would open a huge sphenochasm.

to the south that is also not observed. Moreover, the Canada Basin opened before the rotation of central and southern Alaska that is of concern here (Sweeney, 1985).

3.2.2. *Local Rotations In Transcurrent Shear Zones.*

In two seminal papers, Beck (1976 and 1980) demonstrated that the systematic CW declination anomalies found in numerous paleomagnetic studies of rocks along the western margin of North America occurred as block rotations in the right-lateral megashear zone associated with the transform and oblique subduction boundaries that existed from late Mesozoic to the present time between the North America and the Pacific, Farallon, and Kula plates. This would appear at first glance to be a promising explanation for systematic rotations in Alaska. The tectonic map of Alaska is dominated by a series of long, parallel strike-slip faults that pass through the region of interest, and several of the study sites are close to one or the other of them (Fig. 1). The geologically inferred sense of offset on all of these faults, however, is right-lateral (Grantz, 1966; Patton, 1973; Lanphere, 1978), the opposite of what is needed to explain the observed CCW rotations by Beck's hypothesis.

Even earlier, Freund (1970 and 1974) examined the kinematics of block rotations in transcurrent shear zones accommodated by strike-slip displacement on bounding cross-faults. Many of his ideas are supported by results of later studies combining paleomagnetic and structural information (Luyendyk et al., 1980; Ron et al., 1984; Wells and Coe, 1985; Hornafius, 1985; Wells and Heller, 1988), and they have been given a firm mechanical basis by Nur et al. (1986). One geometrical case considered by Freund (1974) predicts CCW rotation in a right-lateral shear zone, and thus is potentially capable of explaining the 44±11° CCW paleomagnetic rotation of southern and central Alaska. To do so, however, would require that the entire region was subdivided into elongate blocks by numerous east-west cross-faults between the southwesterly trending master faults (Fig. 1), evidence for which is presently lacking. Moreover, the observed rotation of 44° lies at the upper limit of what can be expected mechanically with only one such set of cross-faults (Nur et al., 1986).

3.2.3. *Rotations of Blocks Pushed Around Sharply Bent Faults.*

Stout and Chase (1980) pointed out that some or all of the curvature of structural grain in Alaska that has been termed the Alaska orocline might in fact be original. A crucial feature for their model is that the sharply curved part of the Denali fault can be closely approximated by a small circle. They applied this idea to that segment of the fault by assuming locally rigid plates, and were able to explain with considerable success the orientation and sense of displacement of secondary faults in the area. Their model also provides a plausible mechanism for significant azimuthal rotation of terranes south of and close to the big bend in the Denali fault.

Recently this model has been adapted on a much larger scale by Panuska (1987) to the Tintina-Kaltag fault system (Fig. 1) in an attempt to account for paleomagnetically rotated blocks far beyond the sharp bend. There are at least two major difficulties to this explanation. First, fault displacements much larger than geologically inferred offsets would be required for the terranes of westernmost Alaska to have passed around the bend. For instance, to account for the paleomagnetically inferred rotation of the Blackburn Hills by this model, a displacement of at least 750 km on the Kaltag fault since the Early Paleocene would be needed, more than three and a half times the well documented offset since Late Cretaceous time (Grantz, 1966; Patton, 1973). Moreover, Kula plate motion relative to North America throughout the early Tertiary was not in a direction that could have pushed a block along the Kaltag or Farewell faults in their present orientations (Engebretson, et al., 1985). Second, both the Denali-Farewell and the Tintina-Kaltag fault systems depart markedly from small-circle geometry on this scale (Fig.3), so that pushing terranes into, around, and out of the sharp bends in either of these fault systems would be mechanically difficult. It would necessitate significant compression followed by compensatory extension to satisfy the spatial constraints imposed by the radically varying curvature of the backstop, and would leave a complex structural imprint that is not apparent in the rotated rocks of western Alaska.

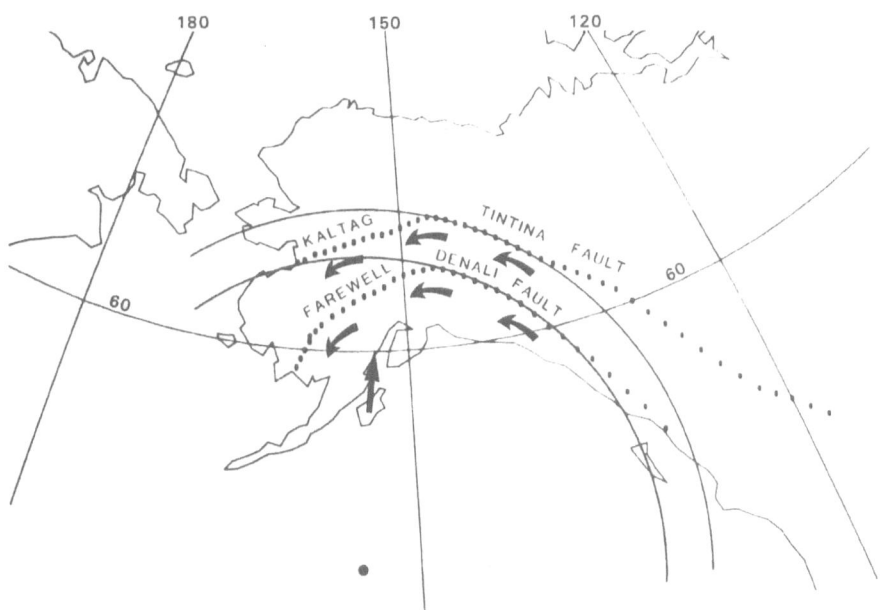

Figure 3. "Railroad car" model--rotation produced by pushing blocks around curved faults, as suggested by curved arrows. Small circles drawn about solid dot illustrate that, on the large scale, the fit of the Denali-Farewell and Tintina-Kaltag fault systems to small circles is not very good. Straight arrow shows average direction of motion of Kula plate in Alaska region (fixed North America reference frame) for the interval 74 to 43 Ma, not a favorable orientation for pushing blocks in the direction supposed by this model. Moreover, the geologically inferred offsets on the Kaltag and Farewell faults are much too small to produce the observed paleomagnetic rotations (Figure 1).

The poleward displacement of the most outboard terranes of Alaska and their collision with the Alaska backstop during late Cretaceous and early Tertiary time that is documented by paleomagnetic (e. g., Coe et al., 1985) and geological (e. g., Csejtey et al., 1982) evidence prompts an analogy to the indentation and extrusion tectonics used to describe the India-Asia collision (e. g., Molnar and Tapponnier, 1975; Tapponnier et al., 1983). The significantly greater curvature of the tectonic grain in southern Alaska compared to that farther north (Fig. 1) does suggest intensification by indentation from the south, and recent structural data around the Gulf of Alaska support this idea (Bol et al., 1987). For the same reasons given in the paragraph above, however, extrusion around a sharply curved backstop is not a likely explanation for the paleomagnetic rotations observed many hundreds of kilometers to the west.

3.2.4. *Megakinking.* The explanation that best fits the available paleomagnetic, geological, and plate tectonic evidence is the "megakink" version of the oroclinal hypothesis, first suggested by Grantz in 1966. As mentioned in the Introduction and illustrated in Figure 4, this model supposes that bending of Alaska occurred in response to northwest-southeast compression by a mechanism analogous to kink folding about a vertical axis. The zones of weakness that accommodated the right-lateral flexural slip required by such folding were the pre-existing Kaltag, Iditarod, Farewell, and Castle Mountain-Lake Clark faults that now divide central and southern Alaska conspicuously into northeast-southwest trending strips (Fig. 1). The axial plane of the kink runs more or less along the 148th meridian, and the eastern limb is postulated not to have rotated, a reasonable constraint because it is backed up by all of North America.

Applying the geometrical relationships for such folding, Grantz (1966) showed that the best estimates for displacement on these faults since Late Cretaceous time predict CCW azimuthal rotations of 30 to 55° of central and southern Alaska. This range is in excellent agreement with the paleomagnetically inferred rotations, and is also the right order to straighten the structural grain of Alaska in the sharply bent region. Moreover, in contrast with the prediction of simple oroclinal bending, the megakink mechanism does not require the opening of a giant sphenochasm (c. f. Fig.

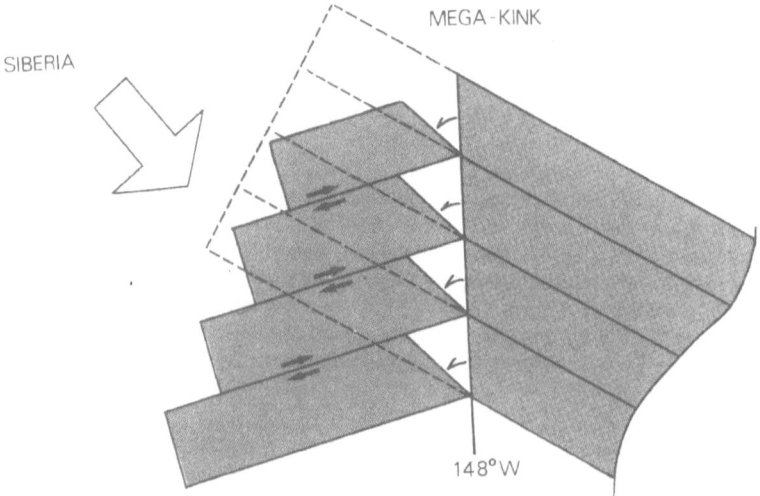

Figure 4. Megakink model of rotation of Alaska. Simultaneous rotation of fault-bounded blocks by the paleomagnetically inferred amount, possibly in response to Eurasia-North America convergence, produces the correct sense and approximately correct displacement on faults, and also does not open a giant sphenochasm.

2 and 4). Finally, another attraction is that the existence of a more gently curved ancient continental margin than at present would have facilitated the large northward displacements of southern Alaskan terranes during Cretaceous time that are indicated by other paleomagnetic results (Hillhouse, 1977; Coe et al., 1985; Thrupp and Coe, 1986).

The key element of Grantz's (1966) model is block rotation accommodated by strike-slip displacement on bounding faults, and thus is similar to the later, more detailed kinematic models of Freund (1974) and Garfunkle (1974). The enormous scale of the rotated blocks, however, is unprecedented. We use the term "kink" in this paper because Grantz originally stressed the idea of flexural slip folding about a vertical axis, and "kink" conveys the idea that distributed deformation is localized around the hinge region. As is evident from Figure 1, the hinge boundary of the Alaska "megakink" is not mathematically sharp. Figure 4 neglects the complex deformation around the hinge as well as problems of extension or compression normal to the strike-slip faults, and should be regarded as an illustrative cartoon rather than an accurate kinematic representation. Moreover, it does not show how rotation of such a large part of Alaska could be accommodated at its other boundaries. We will return to this latter topic shortly.

In summary, we should emphasize that the paleomagnetic data themselves do not *prove* the rigid or even quasi-rigid behavior of the regions between the major strike-slip faults in central and southern Alaska. The density of determinations is low and the uncertainties in declination, and thus in paleomagnetic rotation, are necessarily large because of the steep inclination of the field. Moreover, even perfect paleomagnetic data cannot delineate the block size in a region undergoing uniform rotation. What can be concluded is that, considering both the paleomagnetic and dating errors, the data *allow* the hypothesis of uniform rotation of this part of Alaska by 35 to 55° CCW. In combination with the regional geological evidence, the case for block rotation on this large scale becomes more convincing. The first order discontinuities apparent in the field and on aerial photos are indeed long, straight faults, and the geologically inferred strike-slip offsets across them are consistent in both sense and magnitude with those required to produce the mean paleomagnetic rotation by rigid rotation of the intervening blocks. To test further the integrity of these blocks and the degree of accommodation of the regional CCW rotation by right-lateral displacement on these prominent faults, however, will require extensive field and structural studies to refine estimates of timing and amount of offset, to search for less obvious sets of faults that might have played an important role, and to evaluate the contribution of smaller scale, more continuous deformation in both the limb and the hinge zone of the presumed "megakink."

3.3. CAUSE OF ROTATION

Thirty years ago Carey (1958) had already linked his idea of the Alaskan orocline with the opening of the Arctic Ocean, and, indeed, of the entire Atlantic Ocean as well. He drew attention to the similarity between the rotation angle of 28° that he estimated for the Arctic Ocean sphenochasm and the 30° angle between the Cretaceous-Tertiary portions of the North American and European apparent polar wander paths (Irving, 1958). However, opening of the relevant part of the Arctic Ocean (the Canada Basin) is now dated at 140 to 80 Ma (Sweeney, 1985), too early to drive the rotation of central and southern Alaska that we are concerned with here. Moreover, Carey's geometrical argument for linking opening with bending was based on different, more gently curved structural trends that run westward through northern Alaska.

Grantz (1966) called upon Eurasia-North America convergence to produce the Late Cretaceous-Early Tertiary rotation of central and southern Alaska. He also referred to paleomagnetic evidence for the opening of the Atlantic Ocean during that time, but emphasized the concomitant eastward movement of Siberia toward North America that such opening implied. Patton and Tailleur (1977) extended the picture to the north, suggesting that northwestern Alaska and eastern Siberia buckled and shortened under the east-west compression (Fig. 1). They further supposed that this compressive deformation was largely decoupled by the Kaltag fault from the block rotation that was occurring simultaneously to the south.

Comparison of the recent paleomagnetic data defining the rotation with plate-motion models provides a further test of Grantz's model. The paleomagnetic evidence cited above demonstrates a CCW rotation of 44±11° since about 65 Ma. Thirty million years or less later most of the rotation was complete, as indicated by other paleomagnetic results of Thrupp (1986), derived from more than 70 flows from 12 localities widely distributed on the Alaska Peninsula with average age 35 Ma, that yield a small and statistically insignificant CCW rotation of 11±26°. Pitman and Talwani (1972), in the first detailed plate-motion analysis of the opening of the North Atlantic using marine magnetic anomalies, noted that their results implied Eurasia-North America convergence during Late Cretaceous and early Tertiary time. Later analyses have upheld their conclusion. Thus the timing of convergence fits the paleomagnetic observations well.

In principle, the plate-motion analyses also allow a quantitative test of the model. Using the more detailed analysis of Srivastava (1978), which is similar to that of Pitman and Talwani, Coe et al. (1985) calculated that the amount of convergence between Siberia and Alaska was sufficient to have caused a maximum of 43° of CCW rotation of southern and central Alaska between 67 and 53 Ma. This startlingly close agreement between the land-based paleomagnetic and the marine magnetic observations, however, appears to have been merely a coincidence in light of more recent analyses of the sea-floor anomalies, including one by Srivastava and Tapscott (1986) in which the convergent displacement during the time of interest is reduced to one-third of the previous value. As noted by Harbert et al. (1987), the main problem is that the stage poles defining the ocean opening lie too close to the Arctic region, so that the uncertainty in their positions radically affects the magnitude and direction of the relative motions. Nonetheless, from a qualitative standpoint the plate-motion models are in reasonable agreement with the hypothesis that rotation was driven by Eurasia-North America convergence.

3.4. BOUNDARIES OF THE ROTATED REGION

An important question that arises when we try to fit the rotation scenario into the tectonic history of the rest of Alaska and the Arctic is the location and nature of the boundaries of the CCW rotated region. Where and how was the rotation accommodated spatially? The eastern boundary is the hinge region of the presumed megakink, which, as we have already noted, lies in the vicinity of the 148th meridian. The northwestern boundary is covered in part by the Bering Sea, but is probably a broad and complex zone of distributed deformation recording Siberia-Alaska convergence (Fig. 1). The southern (southwestern and southeastern) boundary was a subbduction zone where, during the interval of interest, the Kula plate, and, at the very end, the Pacific plate dived beneath Alaska (Engebretson et al., 1985). In the southwest it was probably at the Bering Sea shelf until, in early Tertiary time, it stepped back and formed the present Aleutian Arc (Scholl et al., 1975). Farther east in southern Alaska its precise location as a function of time is complicated by questions concerning the timing of arrival of exotic terranes. The important point for our purposes, however, is not exactly where it was, but that it was a subduction boundary, because subduction provides a natural means of accommodating the hundreds of kilometers of southeasterly displacement of western Alaska relative to North America that are required by the megakinking model.

An unresolved and critical problem, however, is the northern extent of the rotated region. As mentioned above, the Kaltag fault is a probable boundary (see Fig. 1 and also Patton and Tailleur, 1977). Also consistent with this limit is the paleomagnetic result of Harris et al. (1987) on overprinted rocks just north of the Kaltag fault. Using their estimated age of 50 Ma for the overprinting and supposing, on the basis of the strongly negative fold test, that little or no post-remagnetization deformation occurred, we calculate only 2±33° of CCW rotation since 50 Ma. If the boundary of the rotated region is the Kaltag fault, however, then substantial extension must have occurred in latest Cretaceous and early Tertiary time to the north of it. In fact, there is a huge triangular basin, the Yukon-Koyukuk basin, south of the Brooks Range and mainly north of the Kaltag fault that, in terms of its size, shape, and position, could account very well for much of the required extension. The problem is that it is filled with Cretaceous sediments that are apparently older than the various volcanic units that record the rotation (Patton, 1973). Could a major part of the extension in this basin have occurred substantially later than is presently believed? If not, where else could the extension be taken up? In normal faulting distributed throughout northwestern Alaska? Offshore even farther to the north? Further paleomagnetic, structural, and seismic data are needed to answer this question, as well as other questions concerning the position and the nature of the lower boundary of the rotating blocks.

4. Conclusions

Systematically CCW azimuthal rotations relative to North America are demonstrated by six of seven paleomagnetic studies of latest Cretaceous and early Tertiary volcanic rocks in the western two-thirds of southern and central Alaska. The CCW bias is highly significant statistically: the mean CCW rotation for all seven results is 39±19°, whereas for the three best results it is 44±11°. These data appear to record a regional rotation of an area greater than 400,000 square kilometers in southern and central Alaska that lies west of the 148th meridian and at least as far north as the Kaltag fault. Additional paleomagnetic results suggest that most or all of the rotation was completed by 35 Ma.

The most probable rotation mechanism is oroclinal bending by megakinking; that is, simultaneous rotation of at least four elongate blocks in a manner analogous to the rotation of one limb of a huge kink fold about a vertical fold axis. Convergence at the appropriate time between Siberia and North America, predicted by most analyses of marine magnetic anomalies in the North Atlantic and Arctic Oceans, provides a plausible motive force. The flexural slip required by the megakinking mechanism to accomplish the paleomagnetically observed rotation agrees with geological observations of offsets, which have both the expected sense and approximately correct amount of displacement, on prominent strike-slip faults bounding the blocks. This mechanism has the advantage of straightening the structural grain of Alaska without opening an enormous sphenochasm where one is known not to exist. Moreover, the more gently curved continental margin would conveniently facilitate the northward translation of the most outboard terranes of southern Alaska during Cretaceous and early Tertiary time that is demanded by other paleomagnetic data. Unresolved questions remain, however, concerning the northern extent of the rotated region and the manner in which the extension that is expected to the north of it was accommodated.

340

Acknowledgments

This research was supported by National Science Foundation Grants EAR-8417369 and EAR-8609784, and by grants and logistical support from the U. S. Geological Survey, Arco Alaska, Arco Oil and Gas, Amoco Production Company, Mobil Exploration and Producing Services, and Exxon Production and Research Company.

References

Beck, M. E., Jr., 1976, Discordant paleomagnetic pole positions as evidence of regional shear in the western Cordillera of North America: *American Journal of Science,* v. **276**, p. 694-712.
Beck, M. E., Jr., 1980, Paleomagnetic record of plate-margin tectonic processes along the western edge of North America: *Journal of Geophysical Research,* v. **85**, p. 7115-7131.
Bol, A. J., and Coe, R. S., 1987, Terrane accretion in an active oroclinal bend: Comparison of structure and kinematics along the Contact Fault in E. and W. Prince William Sound, Alaska: *Geological Society of America Abstracts with Program,* v. **19**, p. 360.
Carey, S. W., 1955, The orocline hypothesis in geotectonics: *Proceedings of the Royal Society of Tasmania,* v. **89**, p. 255-288.
Carey, S. W., 1958, A tectonic approach to continental drift, *in* Carey, S.W., ed., Continental Drift- A Symposium: *University of Tasmania,* p. 177-355.
Champion, D. E., 1980, Holocene geomagnetic secular variation in the western United States: implications for the global geomagnetic field: *U. S. Geological Survey Open-File Report* **80-824**, 326 p.
Coe, R. S., Globerman, B. R., Plumley, P. W., and Thrupp, G. A., 1985, Paleomagnetic results from Alaska and their tectonic implications, *in* Howell, D. G., ed., Tectonostratigraphic terranes of the Circum-Pacific region: *American Association of Petroleum Geologists Circum Pacific Earth Sciences Series,* no. **1**, p. 85-108.
Csejtey, B., Jr., Cox, D. P., Evarts, R. C. , Stricker, G. D., and Foster, H., 1982, The Cenozoic Denali fault system and the Cretaceous accretionary development of southern Alaska: *Journal of Geophysical Research,* v. **87**, p. 3741-3754.
Diehl, J. F., Beck, M. E., Jr., Beske-Diehl, S., Jacobson, D., and Hearn, B. C., Jr., 1983, Paleomagnetism of the Late Cretaceous-Early Tertiary Central Montana alkalic province: *Journal of Geophysical Research,* v. **88**, p. 10593-10609.
Engebretson, D. C., Cox, A., and Gordon, R. G., 1985, Relative motions between oceanic and continental plates in the Pacific Basin: *Geological Society of America Special Paper* **206**, 59 p.
Freund, R., 1970, Rotation of strike-slip faults in Sistan, southeastern Iran: *Journal of Geology,* v. **78**, p.188-200.
Freund, R., 1974, Kinematics of transform and transcurrent faults: *Tectonophysics,* v. **21**, p. 93-134.
Garfunkel, Z., 1974, Model for the late Cenozoic tectonic history of the Mojave Desert, California: *Geological Society of America Bulletin,* v. **85**, p. 1931-1944.
Globerman, B. R., 1985, A paleomagnetic and geochemical study of Upper Cretaceous to Lower Tertiary volcanic rocks from the Bristol Bay region, southwestern Alaska: *Ph.D. Dissertation,* University of California, Santa Cruz, 292 p.
Globerman, B. R., and Coe, R. S., 1983, Paleomagnetic results from Upper Cretaceous volcanic rocks in northern Bristol Bay, SW Alaska, and tectonic implications, *in* Howell D. G., Jones,

D. L., Cox, A., and Nur, A., eds., Proceedings of the Circum-Pacific Terrane conference: *Stanford University Publications in the Geological Sciences,* v. **18,** p. 98-102.

Globerman, B. R., Coe, R. S., Hoare, J. M., and Decker J., 1983, Paleomagnetism of Lower Cretaceous tuffs from Yukon-Kuskokwim delta region, western Alaska: *Nature,* v. **305,** p. 516-520.

Grantz, A., 1966, Strike-slip faults in Alaska: *U.S. Geological Survey, Open- File Report,* **267,** 82 p.

Harbert, W., Frei, L. S., Cox, A., and Engebretson, D. C., 1987, Relative motions between Eurasia and North America in the Bering Sea region: *Tectonophysics,* v. **134,** p. 239-261.

Harris, R. A., Stone, D. B., and Turner, D. L., 1987, Tectonic implications of paleomagnetic and geochronologic data from the Yukon-Koyukuk province, Alaska: *Geological Society of America Bulletin,* v. **99,** p. 362-375.

Hillhouse, J. W., 1977, Paleomagnetism of the Triassic Nikolai greenstone, McCarthy quadrangle: *Canadian Journal of Earth Sciences,* v. **14,** p. 2578- 2592.

Hillhouse, J. W., and Grommé, C. S., 1982, Limits to northward drift of the Paleocene Cantwell Formation, central Alaska: *Geology,* v. **10,** p. 552-556.

Hillhouse, J. W., Grommé, C. S., and Csejtey, B., Jr., 1984, Paleomagnetism of Early Tertiary volcanic rocks in the northern Talkeetna Mountains, *in* Howell, D. G., Jones, D. L., Cox, A., and Nur, A., eds, Proceedings of the Circum-Pacific terrane conference: *Stanford University Publications in the Geological Sciences,* v. **18,** p. 111-114.

Hornafius, J. S., 1985, Neogene tectonic rotation of the Santa Ynez Range, western Transverse Ranges, California, suggested by paleomagnetic investigations of the Monterey Formation: *Journal of Geophysical Research,* v. **90,** p. 12503-12522.

Irving, E., 1958, Rock magnetism: a new approach to the problems of polar wandering and continental drift, *in* Carey, S. W., ed., Continental drift-- a symposium: *University of Tasmania.*

Irving, E., and Irving, G. A., 1982, Apparent polar wander paths Carboniferous through Cenozoic and the assembly of Gondwana: *Geophysical Surveys,* v. **5,** p. 141-188.

Lanphere, M. A., 1978, Displacement history of the Denali fault system, Alaska and Canada: *Canadian Journal of Earth Sciences,* v. **15,** 817-822.

Luyendyk, B. P., Kamerling, M. J., and Terres, R., 1980, Geometric model for Neogene crustal rotations in southern California: *Geological Society of America Bulletin,* v. **91,** p. 211-217.

McFadden, P. L., and Jones, D. L., 1981, The fold test in paleomagnetism: *Geophysical Journal of the Royal Astronomical Society,* v. **67,** p. 53-58.

Moll, E. J., and Patton, W. W., Jr., 1982, Preliminary report on the Late Cretaceous and early Tertiary volcanic and related plutonic rocks in western Alaska: *U. S. Geological Survey Circular* **844,** p. 73-76.

Nur, A., Ron, H., and Scotti, O., 1986, Fault mechanics and the kinematics of block rotations: *Geology,* v. **14,** p. 746-749.

Packer, D. R., and Stone, D. B., 1972, An Alaskan Jurassic paleomagnetic pole and the Alaskan orocline: *Nature Physical Science,* v. **237,** p. 25-26.

Panuska, B., 1987, Is there an orocline in south-central Alaska?: *Geological Society of America Abstracts with Program,* v. **19,** p. 409.

Panuska, B., and Macicak, M., 1986, A revised paleolatitude for the Paleocene Teklanika Volcanics, Central Alaska Range, Alaska: *EOS,* v. **67,** p. 921.

Panuska, B., and Stone, D. B., 1985, Confirmation of the pre-50 my accretion age of the southern Alaska superterrane: *EOS,* v. **66,** p. 863.

Patton, W. W., Jr., 1973, Reconnaissance geology of the northern Yukon-Koyukuk province, Alaska: *U. S. Geological Professional Paper* **774-A**, 17 p.

Patton, W. W., Jr., and Tailleur, I. L., 1977, Evidence in the Bering Strait region for differential movement between North America and Eurasia: *Geological Society of America Bulletin,* v. **88**, p. 1298-1304.

Pitman, W. C., III, and Talwani, M., 1972, Sea-floor spreading in the North Atlantic: *Geological Society of America Bulletin,* v. **83**, p. 619-646.

Plumley, P. W., 1984, A paleomagnetic study of the Prince William terrane and Nixon Fork terrane, Alaska: *PhD dissertation,* University of California, Santa Cruz, 190 p.

Rickwood, F. K., 1970, The Prudhoe Bay Field, *in* Adkison, W. L., and Brosgé, M. M., eds., Proceedings of the Geological Seminar on the North Slope of Alaska: *Pacific Section American Association of Petroleum Geologists,* p.   L1-L11.

Ron, A., Freund, R., Garfunkel, Z., and Nur, A., 1984, Block rotation by strike-slip faulting: Structural and paleomagnetic evidence: *Journal of Geophysical Research,* v. **89**, p. 6256-6270.

Scholl, D. W., Buffington, E. C., and Marlow, M. S., 1975, Plate tectonics and the structural evolution of the Aleutian-Bering Sea region, *in* Forbes, R. B., ed., Contributions to the geology of the Bering Sea basin and adjacent regions: *Geological Society of America Special Paper,* **151**, p. 1-32.

Srivastava, S. P., 1978, Evolution of the Labrador Sea and its bearing on the early evolution of North America: *Geophysical Journal of the Royal Astronomical Society,* v. **52**, p. 313-357.

Srivastava, S. P., and Tapscott, C. R., 1986, Plate kinematics of the North Atlantic, *in* Vogt, P. R., and Tucholka, B. E., eds., The Geology of North America, Volume M, The Western North Atlantic Region: *Geological Society of America,* p. 379-404.

Stout, J. H., and Chase, C. G., 1980, Plate kinematics of the Denali fault system: *Canadian Journal of Earth Sciences,* v. **17**, p. 1527-1537.

Sweeney, J. F., 1985, Comments about the age of the Canada Basin: *Tectonophysics,* v. **114**, p. 1-10.

Tailleur, I. L., 1969, Rifting speculation on the geology of Alaska's North Slope: *Oil and Gas Journal,* v. **67**, p. 128-130.

Tailleur, I. L., 1973, Probable rift origin of Canada basin, *in* Pitcher, M. G., ed., Proceedings of the Second International Symposium on Arctic Geology, San Francisco: *American Association of Petroleum Geologists Memoir* **19**, p. 526- 535.

Thrupp, G. A., and Coe, R. S., 1986, Early Tertiary paleomagnetic evidence and the displacement of southern Alaska: *Geology,* v.**14**, p.213-217.

Wells, R. E., and Coe, R. S., 1985, Paleomagnetism and geology of Eocene volcanic rocks of southwest Washington: Implication for mechanisms of tectonic rotation: *Journal of Geophysical Research,* v. **90**, p. 9986-10010.

Wells, R. E., and Heller, P. L., 1988, The relative contribution of accretion, shear, and extension to Cenozoic tectonic rotation in the Pacific Northwest: *Geological Society of America Bulletin,* v. **100**, p. 325-338.

PALEOGEOGRAPHY AND ROTATIONS OF ARCTIC ALASKA - AN UNRESOLVED PROBLEM.

D. B. STONE
Geophysical Institute,
University of Alaska,
Fairbanks, AK., 99775,
U.S.A.

ABSTRACT. Although most of the paleomagnetic data available for arctic Alaska show evidence of a magnetic overprint, which commonly obscures the original magnetization, there are a few localities which appear to have recorded an original magnetic field. However, any conclusions drawn on the basis of these apparently reliable paleomagnetic data have to be considered tentative, since they have not all had rigorous stability tests applied. With this caveat in mind, it is found that the northernmost localities indicate a post Early Cretaceous counterclockwise rotation commensurate with a rotation of arctic Alaska away from the region of the Canadian arctic islands, followed by a smaller clockwise rotation recorded in sites further to the west, but still on the Arctic Slope. South of the Brooks Range, sediments of Cretaceous age from the Yukon-Koyukuk Province on-lap the Brooks Range and are thus considered part of Arctic Alaska. These sediments give paleomagnetic directions that indicate variable amounts of clockwise rotation. This rotation is most easily interpreted in terms of ball-bearing style motions due to dextral strike-slip motion along the southern margin of Arctic Alaska. These clockwise rotations are in contrast to the counterclockwise motion seen throughout southwest Alaska.

1. Introduction

The tectonic history and setting of arctic Alaska is very poorly understood, and has generated a wide range of models. The existing paleomagnetic data for the area suggest that at least two scales of rotation have taken place. One involves a counterclockwise rotation of a very large block, Arctic Alaska, which may have extended from the Canadian border/McKenzie Delta region in the east and included parts of Chukotka in the west (Figures 1,2,3). This proposed rotation was about 70 degrees about a pole near the McKenzie Delta (Halgedahl and Jarrard, 1987) and must have involved the opening of the Canada Basin of the Arctic Ocean behind it. The other involves clockwise rotations of apparently relatively small areas on the southern margins of Arctic Alaska. These latter rotations are not well defined, but are a persistent feature of the paleomagnetic data (Harris, 1985; Harris et al., 1987) and may represent ball-bearing type motion along a dextral slip margin (Beck, 1980). This clockwise motion along the southern margin is in direct contrast to the counterclockwise rotations seen just a little further south, south of the Kaltag Fault (Coe, this volume). A third rotation may also be implied by limited paleomagnetic data from the western North Slope of Alaska (Witte et al., 1987). These latter data are most easily explained in terms of an overshoot of the original large-scale counterclockwise rotation producing a subsequent clockwise

C. Kissel and C. Laj (eds.), Paleomagnetic Rotations and Continental Deformation, 343–364.
© *1989 by Kluwer Academic Publishers.*

rebound.This model has circumstantial support from the available paleolatitude data (Stone and Witte, 1983).

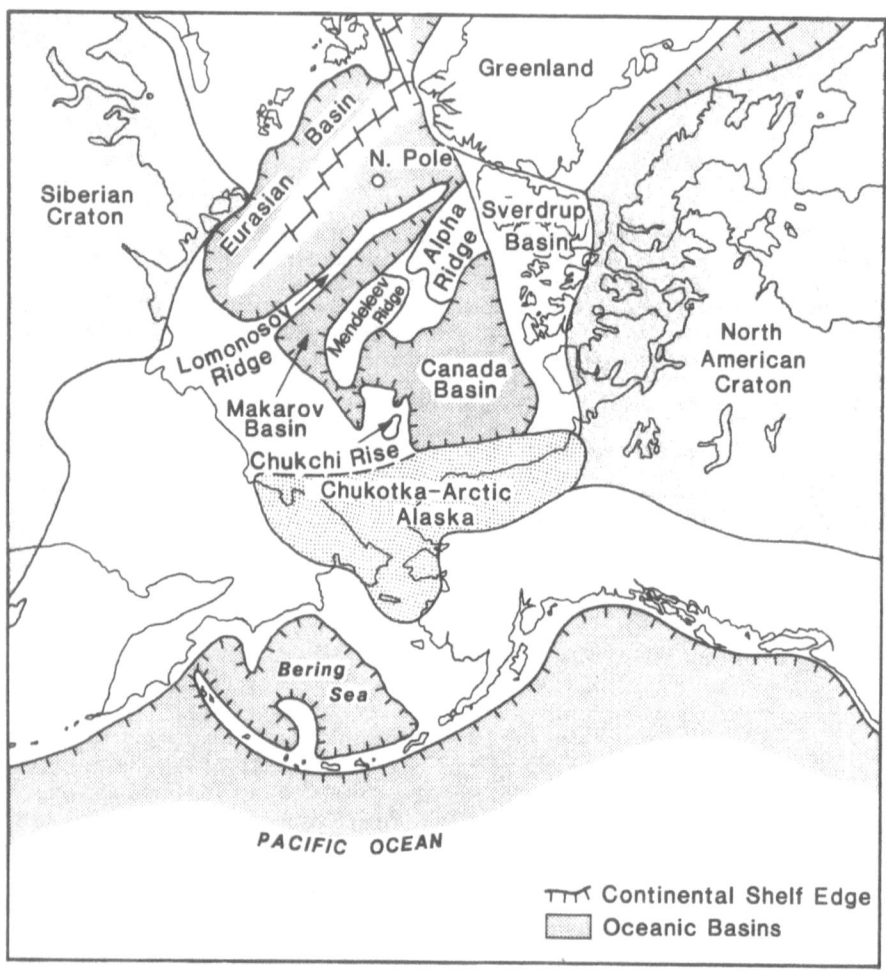

Fig. 1. A schematic map based on Churkin and Trexler, 1981, showing the main geologic elements that have to be considered in any reconstructions of the Arctic in general, and Arctic Alaska in particular. The large areas to the southeast, south and southwest of Chukotka-Alaska are generaly considered to consist of accreted and displaced terranes and associated fold and thrust belts.

As has been pointed out by a number of authors, including Halgedahl and Jarrard (1987), the use of the words rotation and translation can be misleading since any movement on a sphere can be considered a rotation about an appropriate Euler pole. The relative motions discussed in this paper divide into those with Euler poles close to the areas in question, labelled rotations,and those with distant Euler poles, labelled translations. This is based on common useage and on the appearance of the motions as rotations and translations on map representations.

Any attempts to unravel the geologic, tectonic and rotational history of arctic Alaska are intimately related to the paleogeography of the overall Arctic region. It is now generally accepted that much of southern Alaska achieved its present geography through the accretion of tectonostratigraphic (or lithotectonic) terranes (see for instance Howell, 1985 and Jones et al., 1987). Interior Alaska may also consist of terranes, though in general these are not considered to be so far-travelled as the southern terranes, and there is little or no paleomagnetic data available for most of them (Plumley, 1984).

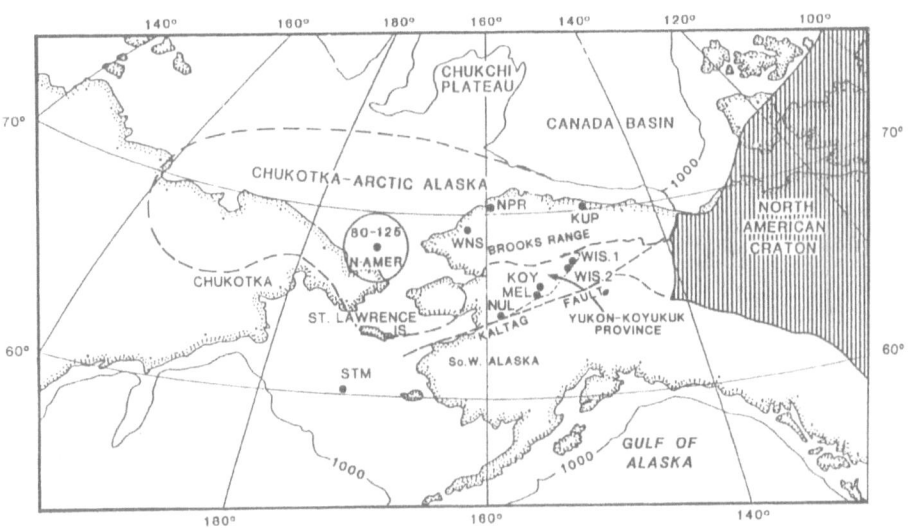

Fig. 2. The sampling sites within Chukotka-Arctic Alaska (WNS,NPR,KUP), together with those in the adjacent Yukon-Koyukuk Province (NUL,MEL,KOY,WIS) and St Matthew Island (STM) are shown, together with the 80-125 Ma paleomagnetic pole for cratonic North America (Irving and Irving, 1982). The boundary for Chukotka-Arctic Alaska is modified from Churkin and Trexler, 1981).

The term Arctic Alaska has been used in different ways by different authors. When considering possible large-scale motions, such as the proposed counterclockwise rotation away from the Canadian Arctic Islands, the Arctic Alaska block is generally considered to include at least the Brooks Range and North Slope of Alaska, usually the Seward Peninsular, parts of St. Lawrence Island and a large part of Chukotka, areas south of the Brooks Range and variable amounts of the extensive continental shelf to the north (Figure 2.) (see for instance Tailleur, 1973, Newman et al., 1977, Churkin and Trexler, 1981). Occasionally parts of the Yukon-Koyukuk Province and some of the other terranes adjacent to the Brooks Range are also included (Figure 3). The detailed boundaries used to define Arctic Alaska are commonly dictated by geometric constraints imposed by the paleogeography being discussed, such as whether or not there is room to move Chukotka-Alaska as a unit , or whether it should be subdivided. The models proposed for the paleogeographic development of Arctic Alaska range from no motion at all with respect to cratonic North America, to travel histories involving origins in all parts of the Arctic Ocean and the North Pacific. Most of the

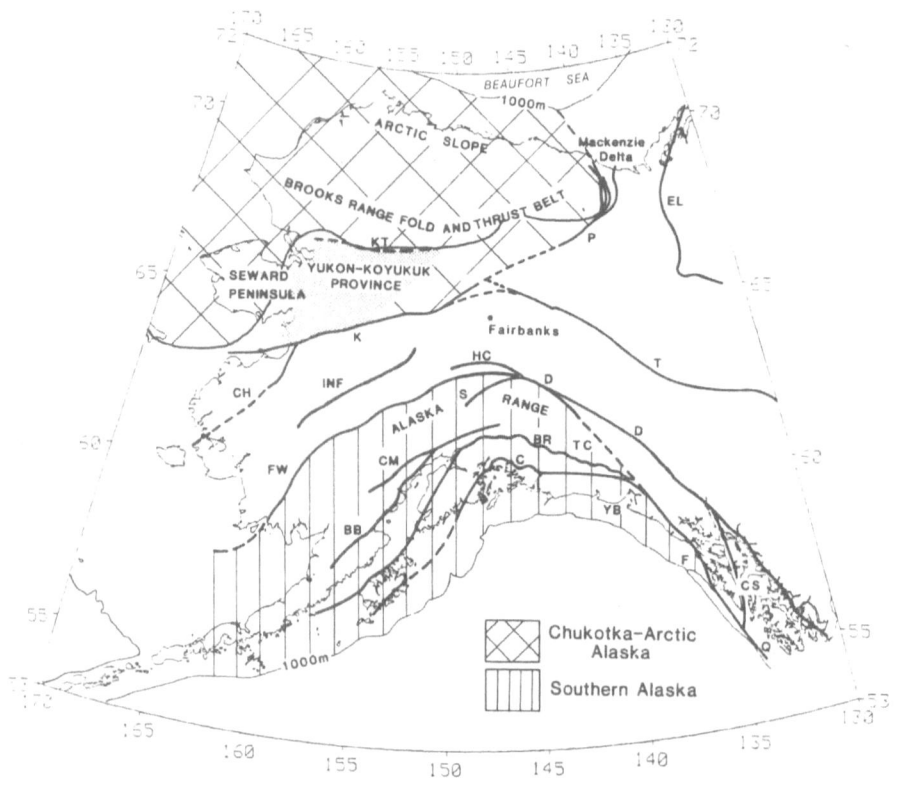

Fig. 3. Alaska can be subdivided into three related areas: arctic Alaska, the southern boundary of which is here considered to be the Kaltag-Porcupine fault system, and thus includes the shaded portions of the Yukon-Koyukuk Province; interior Alaska, lying between the Kaltag-Tintina and the Denali-Farewell fault systems; and southern Alaska between the Pacific and interior Alaska. The major faults shown are: BB, Bruin Bay; BR, Border Ranges; C, Contact; CH, Chiroskey; CM, Castle Mountain; CS, Chatham Straight; D, Denali; EL, Eskimo Lakes; F, Fairweather; FW, Farewell; HC, Hines Creek; INF, Innoko-Nixon Fork; K, Kaltag; KT, Kobuk Trench; P, Porcupine lineament; Q, Queen Charlotte; S, Susitna; T, Tintina; TC, Totschunda-Connecting.

models that have been suggested require sea-floor spreading in the Canada Basin of the Arctic Ocean, thus the geologic history of the Arctic Ocean region is also critical to the history of Alaska.

The most persistent model for the tectonic history of Arctic Alaska is commonly known as the "rotation" model. This model considers that Arctic Alaska (with or without Chukotka) was adjacent to the Canadian arctic islands, and rotated counterclockwise about a point in the vicinity of the

Fig. 4. Almost all possibilities for the tectonic origins of Arctic Alaska have been proposed. Of the four main ones, three are shown here in cartoon form against an Early Cretaceous reconstruction of the major landmasses. The coastlines shown are present-day for easy reference. The fourth model, which keeps Arctic Alaska more or less in place with respect to cratonic North America is not shown.

Mackenzie River delta in latest Jurassic or Early Cretaceous time (Figure 4). At face value, paleomagnetic techniques would seem to be ideal for discriminating between this and the various other models that have been presented in the literature (the other two common models are also shown in Figure 4.) however, the results obtained to date are ambiguous. The available data are discussed in more detail below, but the key points are that the majority of the rocks sampled have been magnetically overprinted, and those that give apparently reliable paleomagnetic directions, allow for either complicated (involving multiple rotations) or ambiguous interpretations. One of the problems with existing, apparently reliable data sets is that they are all of Cretaceous age. Although this is a time when a lot of relative motion is thought to have occured in the Arctic, the mean paleomagnetic poles are located too close to the sites sampled to allow easy determination of declination anomalies, or to allow easy discrimination between different directions and styles of

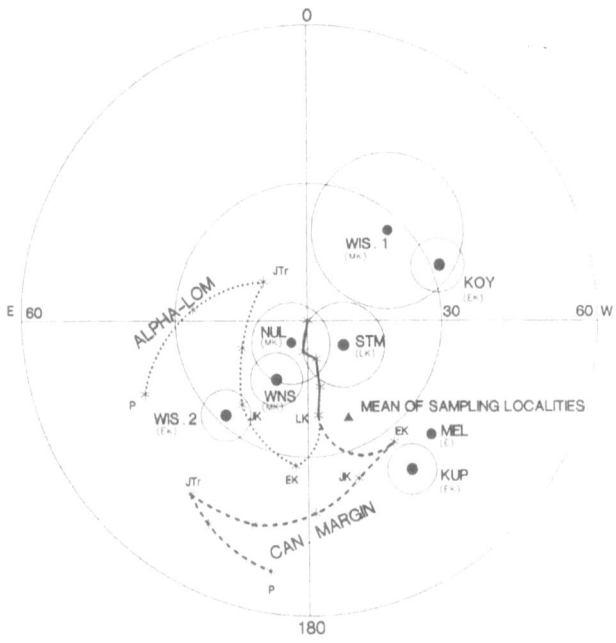

Fig. 5. A polar stereographic projection showing the Apparent Polar Wander (APW) paths expected for arctic Alaska (heavy lines) if it was attached to the North American craton from Permian (P) to Early Cretaceous (EK) time at the location of and adjacent to: 1) the Alpha-Lomonosov ridges (ALPHA-LOM), then translated across the Arctic Ocean during the Cretaceous normal polarity interval: 2) the Canadian arctic continental shelf (CAN MARGIN), then rotated about a pole near the McKenzie River Delta, also during the Cretaceous normal interval. The two paths join in Late Cretaceous (LK) time, and have cusps during Early Cretaceous and near the Jurassic-Triassic (JTr) boundary. The dots represent the paleomagnetic data listed in Table II. The circles of confidence at the 95% level are shown except for MEL, the circle of confidence for which is large and confuses the diagram. The triangle represents the mean location of the localities sampled.

motion. In contrast, Triassic and Jurassic aged rocks have a far greater spread of reconstructed pole positions which are also farther away from the sampling sites (Figure 5). With careful selection of sampling sites, and very careful processing of the paleomagnetic signals, it may be possible in the future to identify pole positions for these ages, and thus put some clearer constraints on possible paleogeographic models.

Correlations of the geology across the reconstructed boundaries formed by the continental margins should also allow discrimination between models. However, the geology of the Arctic margins is seldom known in sufficient detail for unambiguous correlations to be made. For most of the margins relevant to the reconstructions, the geologic settings seem to have been very similar, thus producing similar rock types and sequences.

2. Background

Better knowledge of the geologic history of the Arctic Ocean would considerably restrict possible models for the geologic development of Arctic Alaska.Unfortunately, the Arctic Ocean ranks among the least known areas of the world, particularly in terms of its age and tectonic development. The general morphology and bathymetry are known (see for instance Perry and Fleming, 1986), but for most of the observed features, the age, geology, structure and origin can only be guessed at, with very few positive constraints on the guesses. The exception to this is the Eurasian Basin (Figure 1) which is a relatively young basin formed by the northward extension of the Atlantic spreading center. Based on the well-defined magnetic stripes and on the active seismicity, the spreading in the Eurasian basin began about 60 ma and is continuing today (Husebye et al., 1985). Separating the Eurasian basin from the other basins of the Arctic Ocean is the Lomonosov Ridge.The morphology of this ridge, combined with limited geophysical data across it, indicate that it is a slice of the Barents Sea continental shelf, sliced off by the northward migration of the Atlantic spreading center (see for instance the collection of papers in Johnson and Kaminuma, 1986, and Husebye et al., 1985). The Amerasian Basin, bounded by the Lomonosov ridge, the Siberian, Alaskan and Canadian continental shelves is more complicated and more enigmatic. Various authors subdivide it in a number of ways, but the main features are the Alpha Ridge or Cordillera, the Mendeleev Ridge which may or may not be an extension of the Alpha Ridge, the Makarov basin between the Lomonosov and Alpha-Mendeleev ridges and the Canada Basin adjacent to the Alaskan and Canadian margins (Figure 1). In addition, there is the Chuchki Plateau which extends from the Alaska-Siberian shelf into the Canada Basin. This very poorly understood feature is located in a position that greatly complicates the geometry of paleogeographic reconstructions of the Arctic region, particularly with regard to the rotation model which juxtaposes the Alaska-Siberia continental shelve against the Canadian shelf. One possibility is that the Chukchi Plateau post-dates the proposed paleogeographic reconstructions in either time of origin or achieving its present location.

Among the first workers to attempt a reconstruction of Arctic paleogeography was Professor Carey from the University of Tasmania. The Alaska orocline and the Arctic sphenochasm (both of these terms were first defined in this paper) were a consequence of his reconstructions of the opening of the Atlantic (Carey 1958). The discovery of magnetic stripes and the mechanics of sea-floor spreading show that his model is no longer plausible as proposed, however, some of the broad principles may still hold. Some early re-assemblies of the continents to form Gondwanaland, such as those of Dietz and Holden (1970) show the Arctic Ocean, which the authors termed Oceanus Borealis, as a more or less permanent feature, at least as far back as Paleozoic time.

Although more recent reconstructions of Gondwanaland are better constrained, there is still enough variability in the reconstructions to allow the Arctic Ocean to be a long-lived feature.

When Dietz and Holden (1970) assembled their reconstructions, western north America in general, and Alaska in particular were considered to be part of the stable craton. Since that time, the relative mobility of quite large areas surrounding the cratons has become apparent (see for instance, Howell, 1985; Stone and McWilliams, 1988).

Among the early arguments in favor of allowing relative motion between Arctic Alaska and the rest of North America was the apparent need for a northern source for many of the rocks of Paleozoic age. The limited width of the continental shelf and the need for a large sediment source area, led to the idea that either a land mass had rifted away from the northern Alaskan continental margin, or that Alaska had rifted away from some other margin. Based in part on the curvature of the Alaskan oroclines, and in part on the broad brush geologic correlations, several models rotating arctic Alaska away from the Canadian continental shelf were proposed, usually involving a rotation pole in the vicinity of the McKenzie Delta (see for instance Tailleur, 1969, 1973; Grantz et al., 1981). A variation on this was proposed by Freeland and Dietz (1973) whose model involved counterclockwise rotations of most of Alaska through fan-like motions, with relative motion taking place along the major curved faults that had already been mapped. Other evidence used to support models involving rotations about a pole located near the McKenzie Delta came from claims that the amount of shortening in the western end of the Brooks Range fold and thrust belt was considerably greater than that in the east. However, the low estimates of shortening in the eastern Brooks Range have since been challenged (Oldow et al., 1987).

As more sophisticated models became available for plate tectonics in general, and the opening of the Atlantic in particular, a wider range of models were proposed to explain the present lack of a sediment source area for the older rocks of arctic Alaska (see Nilsen, 1981 for a review). Several of the proposed models involve the creation of ocean floor in the Canada Basin of the Arctic Ocean, including the more fully developed rotation models such as that of Grantz et al., (1981), the models of Herron et al., (1974), who rifted Alaska away from the Siberian margin, and those of Dutro (1981), Smith (1987) and Hubbard et al., (1987a, 1987b) who proposed rifting Alaska away from the Barents shelf region (from the outboard edge of the reconstructed Lomonosov Ridge). Since these models all involve creating new sea floor, they constrain, or are constained by the age of the Canada Basin. Other models, including those of Jones (1980, 1982), who proposed translating Arctic Alaska from the north Pacific by transform motion along the western margin of North America, and those that involve little or no relative motion between Alaska and North America (for example, Churkin and Trexler, 1980, 1981), imply that older ocean floor has been trapped to form the present Canada Basin. Unfortunately, there is no definitive evidence available at present as to the age of the oceanic basement north of Alaska.

Since they were first proposed, many of these models have been modified and improved, but none have been unambiguosly rejected. The currently popular models are those that involve translation of Arctic Alaska across the Arctic Ocean from the reconstructed continental margin of the Barents Shelf (Smith et al., 1987, Hubbard et al., 1987a,b, Crane, 1987) and variations on the original theme of rotating Alaska away from the Canadian Arctic Islands (Grantz et al., 1981). A modification of the rotation model has been proposed by Zonenshain (1988), who postulates that an arctic continental mass, Arctida, once occupied most of what is now the Arctic Basin, and that this continent broke up through multiple rifting, spreading its component parts around the present Arctic Ocean margins. One part, arctic Alaska, was rotated and translated into its present position.

The paleogeography of the areas south of Arctic Alaska are equally critical to any model of the overall paleogeographic development. As was mentioned briefly above, it is now commonly accepted that southern Alaska (south of the Denali and associated faults) is made up of accreted terranes, many of them far-travelled. What is more debateable is the time of arrival in their present locations (Stone and Wallace, 1987). There is considerable paleomagnetic evidence that everything was more or less in-place by Eocene time, and that the major terranes were probably there by late Cretaceous or soon after (see for instance Hillhouse and Gromme, 1984, Panuska, 1985). Interior Alaska, loosely defined as those areas lying between the Denali Fault system and Arctic Alaska, is more problematic (Churkin et al., 1982). There are few paleomagnetic data for pre-Eocene times (Plumley, 1984), and their interpretation is clouded by evidence of remagnetization effects. With the available paleomagnetic data it would be hard to detect significant longitudonal motion produced by strike-slip motion along the Tintina-Rocky Mountain Fault system. From the fragmentary evidence available, it seems probable that interior Alaska has been within about 10 to 15 degrees of its present latitude since at least mid-Cretaceous time, and perhaps since the Paleozoic. As a result, interior Alaska may have acted as a backstop for the northward migrating southern terranes, and probably for the southward moving arctic terranes too.

Summarizing, in order to reach an understanding of the paleogeography of Arctic Alaska, which is a prerequisite to determining and understanding the various rotations that appear to have taken place, it is neccessary to distinguish between the dramatically different models for the evolution of the arctic region. The principal constraints that have to be satisfied (not neccessarily exclusively by arctic Alaska) include:

1. The need for a northerly or northeasterly sediment source for the rocks underlying the North Slope and the Brooks Range from about Middle Paleozoic through about Middle Mesozoic time.

2. The need for a north-south (present coordinates) compressional setting in latest Jurassic and Early Cretaceous time capable of producing the many-fold shortening across the Brooks range which produced the fold and thrust belt seen today (Mayfield et al., 1983).

3. The on-lap relationships of the Cretaceous sediments from the Yukon-Koyukuk Province and of the Kuskokwim Group further south, require that Arctic Alaska be within about 10 to 15 degrees of its present location by at least latest Cretaceous time.

4. The apparent similarity, and by implication, continuity of the geology of the Brooks Range with the cordillera of western North America, noting that arguments can also be made for continuity through to Chukotka and the Canadian arctic.

5. The requirement that paleogeographic development is consistent with the formation of, or trapping of the oceanic Canada Basin of the Arctic Ocean. Unfortunately, the age of the basin has only been determined using inspired guesses based on cooling rates and depth (Lawver and Baggeroer, 1983) sediment thicknesses and estimated sedimentation rates and the lack of well defined magnetic anomalies which may indicate formation during the reversal free period in the Cretaceous (approx. 120-85 Ma).

It is apparent that if unambiguous geologic ties could be established between the various parts of the Arctic Ocean rim, this would significantly reduce the number of possible models. Similarily, if the age of the ocean floor of the Canada Basin could be determined, this would at least allow the rejection or acceptance of the trapped old ocean floor models.

3. Paleomagnetism

An ongoing approach to the large-scale problem is paleomagnetic. As mentioned above, the paleomagnetic data are still somewhat ambiguous, but give some hope of being able to resolve at least part of the problem.

Before exploring the paleomagnetic data available for Alaska, it is neccessary to establish the various assumptions used in interpreting the data.

The general paleomagnetic techniques of collecting oriented samples from in situ rock outcrops or boreholes and determining the "characteristic" magnetic vector directions are well established.In general, the characterisitic magnetization direction is the most stable one recorded by the rock, but this does not imply that it is the magnetization acquired at the time of formation of the rock sampled. The most stable magnetization, especially in the case of arctic Alaska, is often that of an overprint of unknown origin. If, through the use of established stability tests such as the fold or reversal test, the magnetization can be demonstrated to date from a time close to the time of origin of the rock, then there are a number of ways in which these data can be interpreted.

The vector direction of the characteristic magnetization can be converted to an equivalent virtual geomagnetic pole (VGP) by first converting from the vector direction with respect to present horizontal (geographic reference frame) to the vector with respect to ancient horizontal as represented by bedding planes or similar indicators (stratigraphic reference frame). It should be recognized that estimating the attitudes of the ancient horizontal can often be a siginificant source of error, and one that can be particularly important at high paleomagnetic latitudes. The inclination of the stratigraphic vector gives a direct measure of the paleomagnetic latitude, and is inherently more accurate than the declination or the derived VGP. This is because the ancient declination of the vector is determined using a single rotation about the strike direction to correct for the bedding attitude at the outcrop sampled. However, if there have been multiple episodes of tilting to achieve the observed attitude, rotation about a single strike can produce significant errors (up to 180 in extreme cases). Obviously, if area-wide rotations are being sought in the data, great care has to be taken to either unravel or avoid any structural complexities. Unfortunately, in Alaska, both of these are problematic. In one case because of a lack of detailed mapping, and in the other because available logistic support commonly dictates the sites to be sampled.

On the assumption that the geomagnetic field, if time-averaged over a sufficiently long period (ten thousand to a million years depending on the geomagnetic secular variation model used) represents a geocentric axial dipole, then the paleomagnetic paleolatitude gives the ancient geographic latitude of the site. This latitude can then be compared with the latitude expected for the site if it was fixed with with respect to some other locality for which the paleolatitude is also known.

The VGPs for a given locality can be compared with the equivalent VGPs for other localities or with those available for the craton itself. If sections of an apparent polar wander path are available for both localities, these can be superimposed to get the relative longitudes as well as the relative latitudes.

In specific studies problems often arise as to which VGPs or polar wander paths should be used for comparisons with the data obtained. For arctic Alaska it has been common practice to make comparisons with the poles for the North American craton. However, when the circum-Arctic and associated cratons are reconstructed based on all available data, including data from the oceanic magnetic stripes, transform faults et cetera, it is found that the North American paleomagnetic poles for the Cretaceous are displaced from the combined Cretaceous poles for all the reassembled cratons.

Since much of the available data for arctic Alaska comes from rocks of Cretaceous age, obtaining a reliable reference pole is very important. The Cretaceous VGPs for the North American craton alone are very well defined, but are commonly displaced Pacificwards from the combined poles for the reconstructed cratons, the exact amount depending on the reconstruction made (see for instance Harrison and Lindh, 1982; Andrews, 1985; Smith et al, 1981). The discepancies between the North American and the combined poles are usually of the order of 10 degrees, which at the high latitudes of arctic Alaska are quite significant. This is especially true when considering rotations, since the sampling site, the derived VGP and the reference pole are commonly all very close together.

Within North America there is general agreement and internal consistency with regard to the locations of the Cretaceous VGPs, however, further back in time the poles and APW path are less clearly defined, and depend on the data selected as representing the poles, and on the techniques employed to combine and weight the data (see for instance Irving and Irving 1982, Gordon et al., 1984, May and Butler 1986).In this study, a composite polar wander path was used based on the compiled pole positions of Irving and Irving, (1982), Coe et al., (1985) and Gordon et al., (1984), who used the paleomagnetic Euler pole technique (Table I).

Table I. Pole positions for stable (cratonic) North America compiled from Coe et al., 1985 (A); Irving and Irving, 1982 (B); Mankinen, 1978 (C); Gordon et al., 1984, (deduced using paleomagnetic Euler poles) (D). The radius of the circles of confidence (alpha 95) are for the mean values or the larger of the two estimated confidence limits for Gordon et al., 1984.

Nominal	Age (ma)	Lat.N	Long.E	α_{95}	Ref.
10	(0 - 20)	88	101	3	A
15	(10 - 20)	89	156	5	B
30	(20 - 40)	85	134	4	B
40	(25 - 55)	84	168	4	A
45	(40 - 44)	83	169	4	A
50	(44 - 54)	83	170	3	A
55	(50 - 60)	82	182	3	A
60	(55 - 67)	81	193	3	A
65	(60 - 70)	78	191	7	A
70	(55 - 85)	74	190	7	A
75	(70 - 80)	71	191	7	A
80	(65 - 95)	68	191	7	A
80	-125	68	186	2	C
130		70	182	5	D
140		69	162	5	D
150		68	144	5	D
160		67	126	5	D
170		65	110	5	D
180		64	95	5	D

4. Paleomagnetic data for Alaska

There is a considerable amount of paleomagnetic data available in the literature for southern Alaska, much of which indicates a generaly northward motion of the terranes sampled (see for instance the compilation of Stone and McWilliams, 1988). Although there is still much debate as to when the terranes arrived, where they came from, and where they made landfall against the craton, there is general acceptance of the concept that southern Alaska is made up of accreted terranes. There is less data available for interior Alaska, and that which has been obtained from older rock units are generally overprinted, making interpretations more speculative. However, Plumley (1984) concludes that the Paleozoic carbonates of interior Alaska cannot have moved far with respect to North America.

For reasons cited earlier, paleolatitude (inclination) data for Alaska are inherently more reliable than rotation (declination) data and relative paleolongitudes are notoriously hard to establish. Thus, even though significant latitude changes may not be evident, it does not prohibit more or less east-west motion. This means that there is considerable uncertainty as to when the terranes south of Arctic Alaska achieved their present relative locations in an east-west sense, and thus they only poorly constrain the relative motions of the Arctic Alaska block itself. That relative east-west motion has occured between arctic and interior Alaska can be inferred from the apparent clockwise rotation of blocks along the boundary between them. However, it should be noted that following careful selection of the available data for southwestern Alaska, Coe (this volume) has found consistent declination anomalies indicating a significant counterclockwise rotation since Eocene time. The implication of these data is that although Alaska was probably more or less assembled by Cretaceous time, considerable jostling, reorganisation and perhaps large-scale deformation continued in post-Eocene time.

In contrast to these uncertainties, data from Eocene aged rocks throughout Alaska show a paleolatitude consistent with their present locations with respect to the North American craton.

For arctic Alaska, despite many attempts, there are very few data sets that appear to have recorded the magnetic field at the time of origin of the rocks, and those that exist are all confined to rocks of Cretaceous age or younger. As mentioned earlier, there is a problem in terms of defining Arctic Alaska and thus thus there is a problem in terms of selecting which sampling areas can be considered as representative of it. Early terrane maps have subdivided arctic Alaska into many individual terranes (Jones et al., 1981), while later ones have recombined many of them (Jones et al 1987). There seems to be general agreement that the Brooks Range and the North Slope of Alaska have behaved as a more or less coherent block (exclusive of the deformation recorded in the fold and thrust sequences), and that the two terranes along the south flank of the Brooks Range, the Ruby and Angayucham terranes have also been directly involved with Arctic Alaska, at least since Jurassic time. This is basically the Arctic Alaska block defined earlier. Further south, and apparently onlapping onto the Angayucham terrane, and thus onto Arctic Alaska, is the Yukon-Koyukuk province. The Yukon-Koyukuk Province consists of an Early Cretaceous island arc sequence known as the Koyukuk Terrane, and an extensive, thick sequence of mid to upper Cretaceous deep water marine turbidites and shallow and brackish water deposits including coal units. These Upper Cretaceous sediments are commonly considered to be coeval with, and directly related to the sediments of the Kuskokwim group to the south (Patton, 1973, Wallace and Engebretson, 1986). They may also be related to the Cretaceous sediments that underlie the Bering Sea shelf to the west.

In discussing the so-called reliable paleomagnetic data sets of Cretaceous age from Arctic Alaska (those that have passed at least a fold, reversal, or conglomerate test) three are from localities north of the Brooks Range, and thus within the well-defined Arctic Alaska block, four others are from the apparently onlapping sediments and volcanics of the Yukon-Koyukuk Province, and one is from St. Matthew Island further to the west in the Bering Sea (and thus has a somewhat tenuous link to Arctic Alaska)(Figure 2).

In the interest of being perhaps overcautious, and keeping in mind the fact that all but three of the localities within Arctic Alaska show clear evidence of remagnetization, it should be noted that the standard stability tests can be somewhat misleading. With the exception of St. Matthew Island, all the Cretaceous data used in this study come from sediments, this means that there is a possibility that the apparently positive reversal and conglomerate tests could be due to differential fluid flow causing remagnetizations at different times in bedding units with slightly different mechanical properties. Similarly, fold tests only show stability post-folding, the timing of which is commonly uncertain. The whole data set should be viewed with this caveat in mind.

The most extensive and best constrained data set for Arctic Alaska is that of Halgedahl and Jarrard (1987) for rocks of Early Cretaceous age from oriented bore-hole cores from the Kuparuk River Formation on the Arctic Slope. These data show steeply dipping vectors of both polarities, thus pass the reversal test. The tectonic tilt is too small to apply a fold test. The age control is based on both paleontology and on correlations of the magnetic reversals with the marine magnetic anomalies known as the M-series. The mean paleomagnetic pole obtained for the Kuparuk Formation (KUP, Figure 2) is displaced in a counterclockwise sense from the equivalent pole for cratonic North America (Table II, Figures 5,6) but the two poles coincide if Arctic Alaska is rotated back into a position adjacent to the Canadian Arctic Islands, as is shown schematically in figure 4. This reconstruction gives the best fit of the Kuparuk River pole to the equivalent pole for North America, the other paleogeographic models all requiring significant local rotations of the sampling localities about a vertical axis. These data clearly imply that a counterclockwise rotation of Arctic Alaska has occured since Early Cretaceous time.

Table II . Basic paleomagnetic pole positions and derived rotations and displacements of the sampling sites with respect to cratonic North America as determined using the North American paleomagnetic poles listed in Table I. Columns: 1. 3-letter codes for localities (Figures 2,6); 2. rounded north latitudes/east longitudes; 3. approximate ages (Ma); 4. north latitude/east longitude of mean virtual geomagnetic pole for the locality; 5, Alpha 95 for the VGP; 6. number of values used; 7, min/mean/max paleomagnetic paleolatitudes; 8, expected paleolatitude if the locality remained fixed with respect to cratonic North America; 9,10, min/mean/max rotations and displacements of the localities with respect to the equivalent North American paleopoles.

Site Name	Lat/Long N/E	Age Ma	VGP N/E	Alpha 95	N	Paleo-Lat estim.	exp.	Rotation (wrt N.Amer.Paleopole)	Displacement
MEL	66/205	45	54/225	22	8	55/74/90	71	69/153/240	16/-3/-19
STM	60/187	80	80/235	10	38	59/67/76	79	-70/10/88	3/12/20
NUL	65/202	100	84/141	9	116	59/67/76	83	6/43/78	4/15/25
WNS	69/197	100	74/150	6	8	67/74/82	86	0/56/120	4/12/19
WIS.1	67/209	100	64/316	17	18	38/51/69	81	76/114/154	10/30/45
NPR	70/200	100		3	20	66/70/75	85		10/15/19
WIS.2	67/209	120	61/138	6	25	55/60/65	81	-13/21/47	16/21/26
KUP	70/210	130	49/214	5	89	64/69/73	81	-62-105-154	8/12/17
KOY	66/205	140	59/293	6	39	44/50/56	74	84/116/150	18/24/30

In contrast to the data for the Kuparuk Formation, paleomagnetic data obtained from surface exposures of the Nanushuk Formation at localities at the western end of the North Slope of Alaska

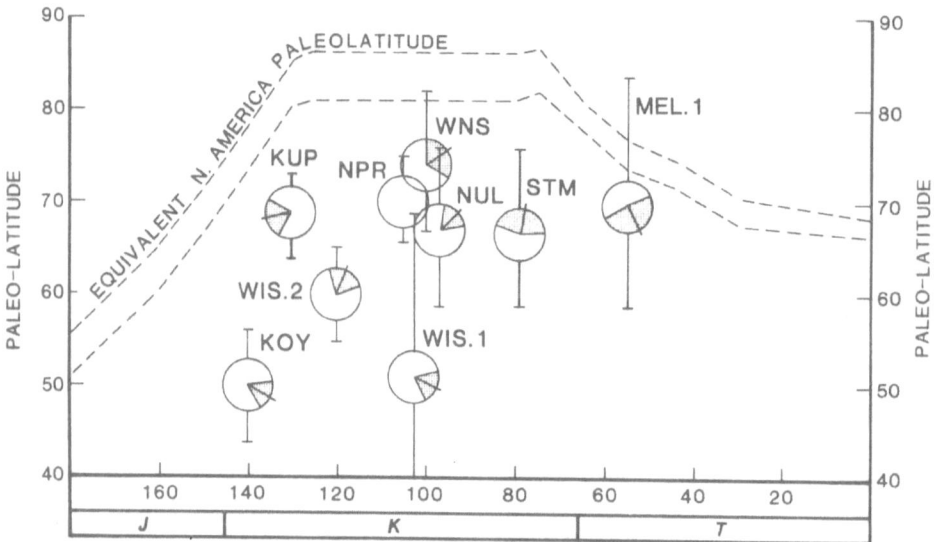

Fig. 6. Paleolatitude and rotations versus time. The centers of the circles represent the paleomagnetic latitudes for the localities, and the vertical bars represent the 95% confidence limits. The circles show the apparent rotations and their 95% confidence limits with respect to the equivalent North American poles as listed in Table I. The "Equivalent N. America Paleolatitude" curve shows the latitudes to be expected if arctic Alaska remained fixed with respect to North America. Note that all localities show southerly paleolatitudes, and with the exception of KUP, show clockwise rotations.

(WNS, Figure 2) indicate clockwise rotations (Table II, Figures 5,6) (Witte, 1982, Witte et al., 1987). The Nanushuk Formation is of Late Cretaceous age based on paleontologic evidence and on one unpublished K-Ar age of about 100ma from a volcanic ash deposit within a coal unit (Triplehorn, pers. comm. 1982). The paleomagnetic data are less well constrained than those from the Kuparuk Formation. They pass a fold test, but are all of normal polarity as would be expected if they were all deposited within the Cretaceous normal interval. These data give a far-sided paleomagnetic pole rotated in a clockwise sense with respect to the equivalent North American pole.The far-sidedness implies a northward motion of Arctic Alaska with respect to North America of about 12 degrees, and the rotation is in the opposite sense to that implied by the Kuparuk River Formation data, noting that the implied rotation and latitudonal motions are both small (Table II, Figures 5,6).

Paleomagnetic data were also obtained from bore-hole cores from the National Petroleum Reserve in western Arctic Alaska. These cores are from the Upper Cretaceous Nanushuk and Torok Formations (NPR, Figure 2) (Stone and Witte, 1983), and were not oriented with respect to azimuth, thus they carry no information regarding rotation. These data pass a reversal test based on a single reversed horizon, but there was not sufficient tectonic tilt to apply a fold test. The inclination data give similar paleolatitudes to those obtained from surface outcrops at WNS (Table II, Figure 6) and are thus considered circumstantial evidence supporting the stability of the samples from WNS (Witte et al., 1987) . This leaves us in the situation where data from the only two apparently reliable sites definitively within Arctic Alaska, give opposite senses of rotation, but a more or less consistent paleolatitude with respect to cratonic North America. The older (Early Cretaceous) data set implies about 12 degrees of poleward motion and about 100 degrees of counterclockwise rotation (about a vertical axis through the site. The younger data set, from rocks of lowermost Upper Cretaceous age, gives a similar poleward motion (12 degrees) but a clockwise rotation (56 degrees).

Since the Upper Cretaceous sediments of the Yukon-Koyukuk Province apparently onlap the southern margin of Arctic Alaska (Patton, 1973; Dillon,1983) the paleomagnetic record from the province can be used to infer the motions of the northern block. However, it should be noted that several major faults that trend more or less parallel to the southern margin of Arctic Alaska cross the Yukon-Koyukuk Province. These faults raise the question as to where to draw the boundary between those parts of the Yukon-Koyukuk Province that reflect motions related to Arctic Alaska and those parts which relate more closely to southwest and interior Alaska. The Kaltag Fault (Figures 2,3) is currently used as a dividing line based in part on the somewhat circular argument that the sites on either side of it appear to have different rotational histories. Sites to the south of the fault behave as parts of southwestern Alaska as described by Coe (this volume) and show counterclockwise rotations, while sites to the north show clockwise rotations.

Samples were collected in sediments of Upper Cretaceous age along the Yukon River, south of the junction with the Koyukuk River, and north of the Kaltag Fault. The samples were taken from a wide range of localities, and from a number of different lithologies within the fluvial-deltaic sequence (NUL, Figure 2). These samples gave both overprinted and apparently reliable data. The apparently reliable data are well constrained with respect to their paleontologic age, and pass a generalised (area wide) paleomagnetic fold test. A successful reversal test was also performed, however, there is some question regarding the attitudes of the reversed units. The combined data give a VGP (NUL, Table II, Figures 5,6) which is far-sided and rotated in a clockwise sense with respect to the equivalent North American pole (Harris et al 1987).

Three localities were sampled at the northeastern end of the Yukon-Koyukuk Province (WIS, Figure 2) of which two are associated with conglomeratic deposits of early Late Cretaceous age (WIS.1) and the other is in the late Early Cretaceous turbidites (WIS.2) dated on the basis of plant fossils (Patton, 1973). The paleomagnetic data were tested for stability by checking that the clasts were randomly magnetized, thus indicating that there had been no pervasive remagnetization event (Harris, 1985, Harris et al., 1987). Because of the small number of localities involved, the precision of these data are not high, but both sets show clockwise rotations and southerly paleolatitudes (Table II, Figures 5,6).

Another study in the Yukon-Koyukuk Province involved volcanic rocks of Tertiary age (MEL, Figure 2). The paleolatitudes determined from this data set indicates that the province was in-place with respect to North America by this time, but show a large clockwise rotation of about 150 degrees (MEL, Table II, Figures 5,6) (Harris et al 1987).

Within the Yukon-Koyukuk Province and north of the Kaltag Fault is an Early Cretaceous island arc complex, the Koyukuk Arc (KOY, Figure 2). The island arc is thought to have been north-facing, and thus moving with respect to the Arctic Alaska block, hence the localities sampled are not directly related to it. However, they can constrain in a general way any large-scale relative motions because of the inter-relationship between the arc, the slightly younger sediments of the Yukon-Koyukuk Province and the Angayucham terrane to the north. The VGPs obtained for the Koyukuk Arc (Harris et al., 1987; Hillhouse and Gromme 1985) are far-sided (displaced about 24 degrees) and rotated clockwise (116 degrees) with respect to the equivalent North American pole

(Table II, Figures 5,6). This result is echoed by the sediments of Early Cretaceous age (WIS.2) sampled to the east (Table II, Figures 5,6) however, these sediments give less well defined VGPs (Harris et al. 1987).

Data collected from St Matthew Island in the Bering Sea (STM, Figure 2) have also been included in the Arctic Alaska compilation (Table II, Figures 5,6) these localities can only be loosely tied to the Arctic Alaska block. The tie is a somewhat long extrapolation from the onshore units in the Yukon-Koyukuk Province, through the subsurface units seen in seismic refraction profiles across the Bering Sea to the volcanic rocks exposed on St Matthew Island. These volcanic rocks are well dated by the K-Ar technique at 78Ma.The paleomagnetic data pass a reversal test, the reversals correlating with marine anomaly 33 (Wittbrodt, 1985, Wittbrodt et al., 1988). At face value the data also pass a fold test, but due to the similarity of the tectonic corrections that had to be applied, the fold test was not statisticaly strong. The overall paleomagnetic directions for St Matthew Island give a slightly far-sided VGP with respect to the equivalent North American pole, but the poles are too close to it to be able to determine a rotation.

In addition to the studies described above, there have been many other attempts to obtain paleomagnetic directions for Arctic Alaska (see Hillhouse and Gromme 1980, 1983; Stone, 1985; Van Alstine, 1986; Plumley and Tailleur, 1987; Newman et al., 1977). Many of these attempts sampled Paleozoic and Early Mesozoic sequences in and around the Brooks Range (Figure 7). As can be seen from Figure 8, most of them showed a very consistent, very steep direction of magnetic overprinting, and most of them are normally magnetized. This normal polarity bias, combined with the steep directions, and a known 80 to 100 Ma overprint of the K-Ar ages for much of Arctic Alaska (Turner et al., 1979, Turner, 1984) has led to the speculation that the remagnetization took

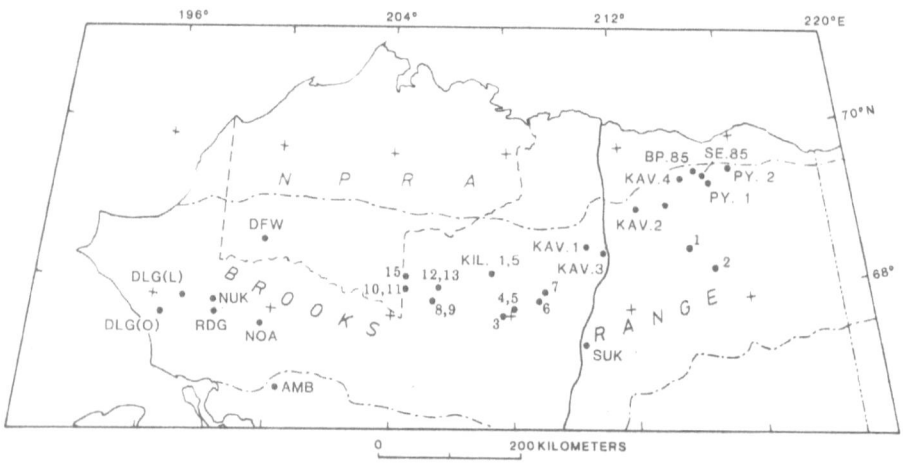

Fig. 7. An outline map of the Brooks Range and Arctic Slope of Alaska showing the localities from which the samples display clear signs of remagnetization. Vector directions from localities with letter codes from the western end, and those with number codes from the central parts of the Brooks Range are shown in Figure 8. The heavy north-south line represents the location of the Trans-Alaska pipeline.

place during the Cretaceous long normal interval. During this time the North American pole was near the present location of northwestern Alaska. Unpublished data (Van Alstine, 1986) from the more eastern parts of Arctic Alaska show an increase in the number of still steeply dipping but reversely magnetized vectors. This reversely magnetised overprint would be consistent with an eastward migrating (possibly chemical?) remagnetization front, thus placing the timing of the eastern end remagnetization at the end of the Cretaceous long normal chron (approx. 85 Ma). If it is assumed that there has been no significant tectonic tilting or rotation of Arctic Alaska since the overprinting, then the mean VGP derived from the overprinted rocks indicates a slightly near-sided and counterclockwise- rotated pole. The assumption that there has been little or no tectonic tilting since Late Cretaceous time is hard to justify. However, the internal consistency of the overprinted vector directions from end to end of Arctic Alaska is circumstantial evidence that there have not been many large-scale relative rotations within the block. The implicit assumption in the above discussion is that the remagnetization is of a chemical origin. This is based on the lack of evidence for significant post-depositional heating of the majority of the Brooks Range. Evidence from Conodont color alteration indices (Harris et al., 1983) implies that temperatures of about 200 degrees Celsius were reached in the Brooks Range, which Halgedahl and Jarrard (1987) argue is a sufficiently high temperature to cause a thermal remagnetization if the elevated temperatures were maintained for a sufficiently long period of time. However, there is growing evidence that the

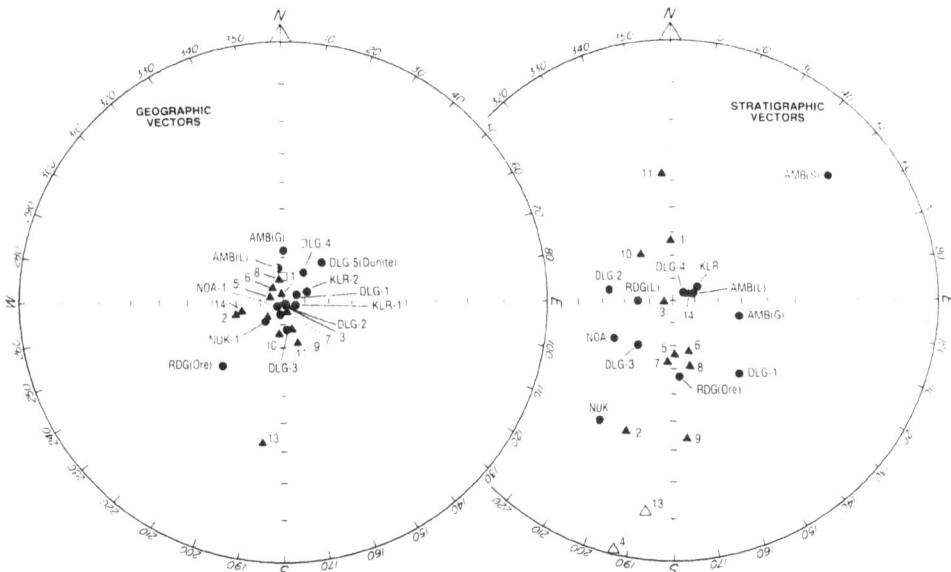

Fig. 8. A composite of the characteristic paleomagnetic vector directions (after demagnetization) in both the geographic (present north and horizontal) and stratigraphic (present north and ancient horizontal) reference frames for the western (letter codes) and central (number codes) Brooks Range, from Stone (1985) and Hillhouse and Gromme (1983). The low dispersion prior to correcting for tectonic tilt indcates a pervasive, very steeply directed remagnetization. The open box represents the field due to an axial dipole.

fluids migrating in front of fold and thrust belts are capable of causing low temperature chemical remagnetizations (McCabe et al., 1983) as has been recognised in the Appalachians and elsewhere (Scotese et al., 1983) a hypothesis which is currently being investigated in Arctic Alaska.

5. Discussion

The immediate problem with trying to interpet the paleomagnetic available data for Arctic Alaska is reconciling the fact that the apparently well constrained Early Cretaceous data from the Kuparuk River Formation indicate a counterclockwise rotation of the Arctic Alaska block and a slightly southerly paleolatitude with respect to cratonic North America. This rotation is in direct contrast to the data from mainly Late Cretaceous aged rocks from both within the Arctic Alaska block, and from rocks associated with it directly to the south (in the Yukon-Koyukuk Province). The paleomagnetic data from these latter localities indicate a similar southerly paleolatitude and clockwise rotation with respect to North America. South of this again, in southwestern Alaska, the data from sites of mid-Tertiary and older ages indicate counterclockwise rotations (Coe, this volume).

Explanations of these various declination anomalies by calling on local rotations are possible, however, there is little or no evidence for small scale block rotation in the local geology of the Kuparuk River area, including the evidence derived from seismic reflection lines. Block rotations on a larger scale (but still small compared with the rotation needed to open the Arctic Ocean) are possible, but the known structure beneath the North Slope of Alaska is difficult to incorporate in models that would allow the 100 degrees of rotation that the Kuparuk River Formation data indicate. If the rotation implied by the Early Cretaceous Arctic Slope data is correct, then an explanation is needed for the small clockwise rotation seen in the Upper Cretaceous Arctic Slope sites (Witte et al., 1987) which also come from an area in which local rotations are unlikely. An explanation in terms of an overshoot of a counterclockwise rotation of Arctic Alaska away from the Canadian Arctic islands, followed by a recovery of about 10 degrees of clockwise rotation can explain the western Arctic Slope data, but does not explain the clockwise rotation seen in the Cretaceous localities from the Yukon-Koyukuk Province. In this case, local block rotations can be invoked, perhaps driven by Beck-style (Beck 1980) "ball-bearing" motions adjacent to the right-lateral strike-slip boundary between Arctic and interior Alaska.

A possible scenario thus involves the rotation of Arctic Alaska away from the Canadian Arctic Islands in post Early Cretaceous time, and at roughly the same time that the terranes of southern Alaska were migrating northwards. The eventual subduction of the material between Arctic and southern Alaska, or the arrival of the southern terranes by transform motion along the edge of western North America, reverses the sense of motion of the Arctic Alaska block, giving the approximately 10 degrees of poleward motion seen in the data of Witte et al., (1987). The only Early Cretaceous rocks from terranes south of, and associated with Arctic Alaska are from the island arc complex known as the Koyukuk Terrane (Jones et al., 1987), however, these rocks pre-date the onlap sequence, thus could have undergone significant relative motion prior to being "locked onto" Arctic Alaska. The younger on-lapped sequences show variable amounts of clockwise rotation which is most easily interpreted as the result of deformation associated with right-lateral strike-slip motion along the southern margin of Arctic Alaska. Further south again, the terranes of southwest Alaska show an apparent clockwise rotation (Coe, this volume). It has been postulated that this could be due to a rigid indentor impinging on south central Alaska, in a similar fashion to that described for the Indian sub-continent by Molnar and Tapponier (1975). A probable effect of such an indentor would be the westward expulsion of southwest Alaska resulting in the dextral strike-slip motion observed on many of the major faults and a clockwise rotation as the various blocks followed the curvature produced by the Arctic Alaska "backstop".

Acknowledgements

I would like to thank John Roe for all the effort put into compiling the various data sets, including the many that were unuseable, Doug Christensen for reviewing the manuscript, and Carlo Laj for setting up the NATO meeting which stimulated me to re-look at the Arctic data.

References

Andrews, J.A., 1985. True polar wander: an analysis of Cenozoic and Mesozoic paleomagnetic poles. *J. Geophys. Res.*, **90**, 7737-7750.

Beck, M.E., 1980. Paleomagnetic record of plate margin tectonic processes along the western edge of North America. *J. Geophys. Res.*, **85**, p7115-7131.

Carey, S.W., 1958. A tectonic approach to continental drift, *in: Carey, S.W., ed., Continental Drift*. A Symposium. Pub. by University of Tasmania, Hobart, 177-355.

Churkin, M, Foster, H.L., Chapman, R.M., Weber, F.R., 1982. Terranes and suture zones in East Central Alaska. *J. Geophys. Res.*, **87**, 3718-3730.

Churkin M., Trexler J.H., 1980. Circum-Arctic plate accretion - isolating part of a Pacific Plate to form the nucleus of the Arctic Basin. *Earth Plan. Sci. Lett.*, **48**, 356-362.

Churkin M., Trexler J.H., 1981. Continental plates and accreted oceanic terranes in the Arctic, *in:* Nairn A.E.M., Churkin M., Stehli, F.G.,*The Ocean Basins and Margins*, **5**, Plenum Press, New York, p1-20.

Coe, R.S., 1989. Large-scale rotation in west-central Alaska. *This volume.*

Crane, R.C., 1987. Arctic reconstruction from an Alaskan point of view. *In:* Tailleur I.L., Weimer P., eds., *"Alaska North Slope Geology"*, Vol.2, 769-784.

Dietz, R.S., Holden, J.C., 1970. Reconstruction of Pangea: Breakup and dispersion of continents, Permian to present. *J. Geophys. Res.*, **75.**,4939-4956.

Dillon, J.T., 1983. Informal presentation, *University of Alaska.*

Dutro, J.T., 1981. Geology of Alaska bordering the Arctic Ocean. *In:* eds., Nairn A.E.M., Churkin M., Stehli F.G., *"The Ocean Basins and Margins"*, Vol.5, p21-36.

Freeland, G.L., Dietz, R.S., 1973. Rotation history of Alaskan tectonic blocks. *Tectonophysics*, **18**, 379-389.

Gordon, R.G., Cox, A., O'Hare, S., 1984. Paleomagnetic Euler poles and the apparent polar wander and absolute motion of North America since the Carboniferous. *Tectonics*, **3.**, 499-537.

Grantz, A., Eittreim, S., Dinter, D.A., 1981. Geology and tectonic development of the continental margin north of Alaska. *In:* Nairn A.E.M., Churkin M., Stehli, F.G., *"The Ocean Basins and Margins"*, Vol **5**, Plenum Press, New York, p439-492.

Halgedahl, S.L., Jarrard, R.D., 1987. Paleomagnetism of the Kuparuk River Formation from oriented drill-core: Evidence for rotation of the Arctic Alaska plate. *In:* Tailleur I.L., Weimer P., eds., *"Alaska North Slope Geology"*, Vol.2, p581-617.

Harris, A.G., Elleersieck, I.F., Mayfield, C.F., Tailleur, I.L., 1983. Thermal maturation values (Conodont color alteration indices) for Paleozoic and Triassic rocks, Chandler Lake, DeLong

Mountains, Howard Pass, Killik River, Misheguk Mountain, and Point Hope Quadrangles, northwest Alaska, and subsurface NPRA: U.S.Geological Survey Open-File Report 83-505. 15p.

Harris, R.A., 1985. Paleomagnetism, geochronology and paleotemperature of the Yukon-Koyukuk Province, Alaska. *M.S. thesis,* University of Alaska, 143p.

Harris, R.A., Stone, D.B., Turner, D.L., 1987. Tectonic implications of paleomagnetic and geochronologic data from the Yukon-Koyukuk province, Alaska. *Geol. Soc. Amer. Bull.,* **99**, 362-375.

Harrison, G.C.A., Lindh, T., 1982. A polar wandering curve for North America during the Mesozoic and Cenozoic. *J. Geophys. Res.,* **87**, 1903-1920.

Herron, E.M., Dewey, J.F.,Pitman, W.C.III., 1974. Plate tectonics model for the evolution of the Arctic. *Geology,* **2**, 377-380.

Hillhouse, J.W., Gromme, C.S., 1980. Cretaceous overprint revealed by paleomagnetic study in the northern Brooks Range. *U.S.Geological Survey Circular* **844**, 43-46.

Hillhouse, J.W., Gromme, C.S., 1985. Early Cretaceous paleolatitude of the Yukon-Koyukuk basin. *EOS* , **66**, 1103.

Hillhouse, J.W., Gromme, C.S., 1983. Paleomagnetic studies and the hypothetical rotation of Arctic Alaska. *J.. Alaska Geol. Soc.,* **2**, 27-39.

Hillhouse, J.W., Gromme, C.S., 1984. Northward displacement and accretion of Wrangellia: New paleomagnetic evidence from Alaska. *J. Geophys. Res.,* **89**, 4461-4477.

Howell, D.G., editor, 1985. Tectonostratigraphic Terranes of the Circum-Pacific Region. Circum-Pacific Council for Energy and Mineral Resources, *Earth Sci. Ser.* No.**1**, Houston, Texas, 581p.

Hubbard, R.J., Edrich, S.P., Rattey, R.P., 1987a. Geologic evolution and hydrocarbon habitat of the "Arctic Alaska Microplate". *In:* Tailleur I.L., Weimer P., eds., *"Alaska North Slope Geology",* Vol.2, 797-830.

Hubbard, R.J., Edrich, S.P., Rattey, R.P., 1987b. Geologic evolution and hydrocarbon habitat of the "Arctic Alaska Microplate". Marine and Petroleum Geology, Vol.4, p1-34.

Husebye, E.S., Johnson, G.L., Kristofferson, Y., 1985. eds., "Geophysics of the Polar Regions", special issue of *Tectonophysics,* **1**-4, 470.

Irving E., Irving G.A., 1982. Apparent polar wander paths Carboniferous through Cenozoic and the assembly of Gondwana. *Geophys. Surveys,* **5**, 141-188.

Johnson, G.L., Kaminuma, K., eds., 1986. Special issue on Polar Geophysics, *J. Geodynamics,* **6**, 406.

Jones, D.L., Silberling, N.J., Berg, H.C., Plafker, G., 1981. Map showing tectonostratigraphic terranes of Alaska, columnar sections, and summary descriptions of terranes. *U.S.Geol. Surv.* Open-file Report **81-792**, 20 p.

Jones, D.L., Silberling, N.J., Coney, P.J., Plafker, G., 1987. Lithotectonic Terrane Map of Alaska (west of the 141st meridian). *U.S.Geol. Surv.* map 1:2,500,000, MAP MF-1874-A.

Jones, P.B., 1980. Evidence from Canada and Alaska on plate tectonic evolution of the Arctic Ocean Basin. *Nature,* **285**, 215-217.

Jones, P.B., 1982. Mesozoic rifting in the western Arctic Ocean Basin and its relationship to Pacific sea-floor spreading. *In:* Embry A.F., Balkwill H.R., eds., "Arctic Geology and Geophysics", *Canada Soc. Petroleum Geol. Memoir* , **8**, p23-50.

Lawver, L.A., Baggeroer, A., 1983. A note on the age of the Canada Basin. *J. Alaska Geol. Soc.,* **2**, 57-66.

May, S.R., Butler, R.F., 1986. North American Jurassic polar wander: Implications for plate motion, paleogeography, and Cordilleran tectonics. *J. Geophys. Res.,* **91**, 11519-11544.

Mankinen, E.A., 1978. Paleomagnetic evidence for a Late Cretaceous deformation of the Great Valley Sequence, Sacramento Valley, California, *J. Res. U.S.Geol. Surv.*, **6**, 383-390.

Mayfield, C.F., Tailleur I.L., Ellersieck, I., 1983. Stratigraphy, structure and palinspastic synthesis of the western Brooks Range, Northwestern Alaska. *U.S.Geological Survey Open-File Report* **83-779**, 57p.

McCabe C., Van der Voo R., Peacor D.R., Scotese C.R., Freeman R., 1983. Diagenetic magnetite carries ancient yet secondary remanence in some Paleozoic sedimentary carbonates. *Geology*, **11**, 221-223.

Molnar, P., Tapponier, P., 1975. Cenozoic tectonics of Asia: effects of a continental collision. *Science*, **189**, 419-426.

Newman, G.W., Mull, C.G., Watkins, N.D., 1977. Northern Alaskan paleomagnetism, plate rotation, and tectonics, *in:* Sisson, A., ed., "Relationship of Plate Tectonics to Alaskan Geology and Resources", *Alaska Geol. Soc. meeting,* Anchorage, pC1-C7.

Nilsen, T.H., 1981. Upper Devonian and Lower Mississipian redbeds, Brooks Range, Alaska. *In:* Miall A.D., ed., "Sedimentation and Tectonics in Alluvial Basins" *Geol. Assoc. Canada, Special Paper* , **23**, 187-219.

Oldow, J.S., Ave Lallament, H.G., Julian, F.E., Seidensticker, C.M., 1987. Ellsmerian(?) and Brookian deformation in the Franklin Mountains, northeastern Brooks Range, Alaska, and its bearing on the origin of the Canada Basin. *Geology*, **15**, 37-43.

Panuska, B.C.,1985. Paleomagnetic evidence for a post-Cretaceous accretion of Wrangellia. *Geology*, **13**, 880-883.

Patton, W.W., 1973. Reconnaissance geology of the northern Yukon-Koyukuk province, Alaska. *U.S.Geol. Surv. Prof. Paper* , **774A**, 16p.

Perry, R.K., Fleming, H.S., 1986. Bathymetry of the Arctic Ocean, Geol. Soc. Amer. *Map and Chart Series* MC-56, 1:4,704,075.

Plumley W.P., 1984. A paleomagnetic study of the Prince William Terrane and Nixon Fork Terrane, Alaska. University of California Santa Cruz, *PhD thesis*, 190p.

Plumley, P.W., Tailleur, I.L., 1987. Paleomagnetic results from the Sadlerochit and Shublik mountains, eastern North Slope, Alaska. *In:* Tailleur I.L., Weimer P., eds., *"Alaska North Slope Geology"*, Vol.2, 580.

Scotese C.R., Van der Voo R., McCabe C., 1983. Paleomagnetism of the Upper Silurian and Lower Devonian carbonates of New York State: Evidence for secondary magnetizations residing in magnetite. *Phys. Earth and Plan. Inter.*, **30**, 385-395.

Smith, A.G., Hurley, A.M., Briden, J.C., 1981. *"Phanerozoic paleocontinental world maps".* Cambridge Earth Science Series, eds., Harland W.B., Cook A.H., Hughes N.F., Sclater J.G., Richardson S.W. Cambridge Univ. Press, Cambridge, 102p.

Smith, D.G., 1987. Late Paleozoic to Cenozoic reconstructions of the Arctic. *In:* Tailleur I.L., Weimer P., eds., *"Alaskan North Slope Geology"*, Vol.2, 785-796.

Stone D.B., McWilliams, M.,1988. Paleomagnetic evidence for relative terrane motion in western North America. *In:* Ben-Avraham Z., ed., *"Evolution of the Pacific Ocean Margins"*, Oxford Univ. Press, in press.

Stone, D.B., Wallace, W.K., 1987. *A geological framework of Alaska. Episodes*, Vol.10, 283-289.

Stone, D.B., 1985. Paleomagnetism, paleolatitudes and magnetic overprinting on the North Slope, Alaska. *Amer. Assoc. Petroleum Geol.*, **69**, 680.

Stone, D.B., Witte, W.K., 1983. Paleomagnetism of National Petroleum Reserve bore-hole cores. *Report to ARCO-Alaska* (available from the authors), 20p.

364

Tailleur, I.L., 1973. Probable rift origin of the Canada Basin, Arctic Ocean. *In:* Pitcher, M.G., ed., Proceedings of the Second International Symposium on Arctic Geology. *Amer. Assoc. Petrol. Geol., Mem.* **19**, 526-535.

Tailleur, I.L., 1969. Rifting speculation on the geology of the Alaska's North Slope. *Oil and Gas J.,* **67**, 128-130.

Turner, D.L., 1984. Tectonic implications of widespread Cretaceous overprinting of K-Ar ages in Alaskan metamorphic terranes. *Geol. Soc. Amer. Abs.* with Prog., **16,** p338.

Turner, D.L., Forbes, R.B., Dillon, J.T.,1979. K-Ar geochronology of the southwestern Brooks Range, Alaska. *Can. Jour. Earth Sci.,***16,** 1789-1804.

Van Alstine, D.R., 1986 and 1987. Normal-and reversed-polarity synfolding CDM along the Brooks Range mountain front. *In:* Tailleur I.L., Weimer P., eds., *"Alaska North Slope Geology",* Vol.**2**, p580, 1987, and *EOS,* Vol.**67**, p269, 1986.

Wallace, W.K., 1984. Engebretson, D.C., Relationships between plate motions and Late Cretaceous to Paleogene magmatism in southwestern Alaska. *Tectonics,* **3.**, 293-315.

Wittbrodt P.R., 1985. Paleomagnetism and petrology of St. Matthew Island, Alaska. Univ. of Alaska *MS thesis,* 210p.

Wittbrodt, P.R., Stone, D.B., Turner, D.L.,1988. Paleomagnetism and geochronology of St. Matthew Island, Bering Sea. *Submitted to Can. Jour. Earth Sci.*

Witte, W.K., Stone, D.B., Mull, C.G., 1987. Paleomagnetism, paleobotany and paleogeography of the Arctic Slope, Alaska. *In:* Tailleur I.L., Weimer P., eds., *"Alaska North Slope Geology",* Vol.**2**, 571-579.

Zonenshain, L.P., Natapov, L.M., 1988. Tectonic history in Arctic region from Ordovician to Cretaceous. *In preparation.*

PALAEOMAGNETIC ESTIMATES OF ROTATIONS IN COMPRESSIONAL REGIMES AND POTENTIAL DISCRIMINATION BETWEEN THIN-SKINNED AND DEEP CRUSTAL DEFORMATION.

E. McCLELLAND[1] and A. M. McCAIG[2]
[1]Department of Earth Sciences,
Oxford University,
Oxford OX1 3PR,
U.K.

[2]Department of Earth Sciences,
Leeds University,
Leeds LS2 9JT,
U.K

ABSTRACT. Palaeomagnetism is a powerful tool for identifying rotations about any axis. Palaeomagnetic study has shown that a thrust sheet in ORS red marls from SW Dyfed, Wales has rotated during emplacement about an inclined axis plunging 55° to southwest in response to regional transpression across the Hercynian orogenic front and local climb of the thrust sheet over a lateral ramp. A palaeomagnetic study in basement thrust sheets in the southern Axial zone, Pyrenees, has demonstrated that 3-D rotations can be identified in basement terrain where no other palaeohorizontal markers exist. Three different movement histories are described within a stack of six thrust sheets. The lowermost unit has experienced doming and tilting about a horizontal axis at a different time to clockwise rotation about a vertical axis of 25°. The middle four sheets have rotated clockwise by 35 to 60° and have been tilted by 40 to 65° to the north. The uppermost sheet has experienced no rotation and tilting, indicating that the thrust (or reverse fault) had an original attitude similar to its present steep northward dip. This study illustrates the potential of the palaeomagnetic method for discriminating between models of the evolution of the Pyrenees which invoke steep structures in the central region to have had an originally steep attitude, either rooted into a zone of inhomogeneous shortening in the lower crust or decollen on a shallow fault, and those which invoke originally shallow structures which have been subsequently back-steepened.

1. Introduction

Crust and lithosphere respond in a variety of ways to tectonic stress. At deeper levels, quasiplastic flow is common and on a large scale deformation can be treated as continuous, even if in practice it may be concentrated within shear zones. At higher levels, crustal rocks are more rigid and deformation may be more or less restricted to brittle fault zones. If faults are sufficiently closely spaced, deformation may still be approximately continuous, but it is clear that the difficulty of deforming the material between the faults must have an important influence on the geometry of active fault networks in the upper crust (cf. Jackson and McKenzie 1983). Any general deformation is likely to lead to rotations of both faults and the intervening fault blocks; palaeomagnetic study of such rotations can place important constraints on fault geometry and sequence of activity and on

365

C. Kissel and C. Laj (eds.), Paleomagnetic Rotations and Continental Deformation, 365–379.

the relationship between high level, fault-dominated deformation and more continuous strain at depth.

In many foreland thrust belts deformed sedimentary rocks are separated from a more or less undeformed middle and lower crust by a shallow decollement. Great strides in the understanding of such belts have been made by constructing balanced cross sections (Dahlstrom 1969; Boyer and Elliott 1982). Such sections are only valid if no material has moved in or out of the plane of section during deformation (plane strain). If rotations about axes which are not perpendicular to the section line have occurred, sections should not be expected to balance. Despite the fact that such rotations are generally acknowledged to be common due to complexities such as lateral and oblique ramps, and differential movement on faults (eg. Coward and Potts 1983; Boyer and Elliott 1982), most published balanced sections take no account of such rotations. The ideal of 3-D section balancing requires the characterization of 3-D rotations; this cannot be achieved using field data alone since rotations about axes at high angles to bedding cannot be constrained. Palaeomagnetic data provide an additional constraint on 3-D rotations and we believe it to be an essential test before any 2-D balanced section can be considered admissible. In addition, information may be extracted concerning the evolution of fold and thrust structures, if remanence components acquired during deformation can be identified (McClelland Brown 1983).

A central assumption of section balancing is that sedimentary bedding can be restored to an initially horizontal attitude. This is used to constrain the original orientation of tilted thrusts and the shape of thrust slices at depth. A number of authors (eg. Butler 1986; Butler and Coward, 1984) have suggested that similar geometries including staircase thrust trajectories and shallow decollements can be assumed in basement terrains, where no sedimentary bedding is present with which to define the palaeohorizontal or the extent of tilting. McCaig (1986) has used a number of indirect arguments to distinguish originally steep from originally shallow, backsteepened faults in the Pyrenees. However, in many basement terrains palaeomagnetism may provide the only constraint other than surface mapping with which to constrain the geometry and kinematic evolution of fault systems.

In this paper we will first discuss a palaeomagnetic study of the Hercynian deformation front in SW Dyfed, Wales, where Siluro-Devonian redbeds are folded and thrust involving a considerable clockwise rotation and downtilting to the west. In such terrain, a combined structural and palaeomagnetic study can constrain the total rigid rotation if the palaeomagnetic vector is significantly different from the pole to bedding. Secondly, we describe a preliminary structural and palaeomagnetic study in the Axial zone of the Pyrenees, where Alpine rotations during thrusting and folding of Hercynian basement rocks can be constrained using palaeomagnetic measurements on late Hercynian dykes.

2. Rotations in thin-skinned thrusting; an example from SW Dyfed, Wales.

In SW Dyfed (Pembrokeshire), Cambro-Ordovician to Carboniferous rocks are involved in a thin-skinned northward verging thrust belt of Hercynian age (Fig.1). Boyer and Elliott (1982) have described the range of structures produced in thrust belts, allowing prediction of the types of rotation to be expected. Movement of thrust sheets over ramps will generally lead to folding about sub-horizontal axes, although ramps and hence related fold axes can have any orientation with respect to thrust transport direction. Similarly, when duplexes or imbricate fans form by footwall collapse (Boyer and Elliott 1982), rotations about horizontal axes will occur, leading to backsteepening of thrust sheets or folding into antiformal stacks. Rotations about steep axes will occur if lateral pinning of thrust sheets occurs, and when thrust sheets move over combinations of frontal and lateral ramps. Complex rotations may also occur when shortening above a decollement leads to folding, as appears to have occurred in the southern part of Figure 1. Such folds are often periclinal in form, suggesting lateral propagation of the fold axis as its amplitude increases. As the fold grows, rotations about axes other than the fold axis are likely, as reflected in variations in fold plunge.

Two palaeomagnetic studies (McClelland Brown 1983 and McClelland 1987) on red marls from the Milford Haven group in the Old Red Sandstone of SW Dyfed, have indicated significant clockwise rotations and westward tilting of at least the western part of the area. These rotations occurred during thrusting.

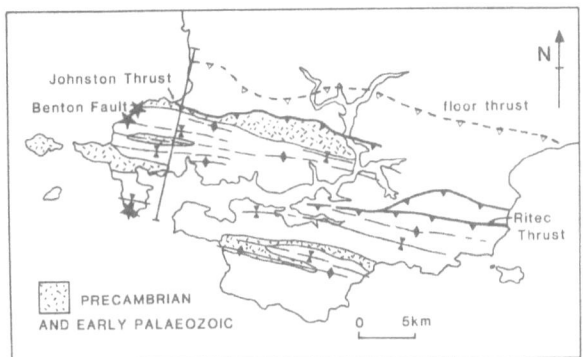

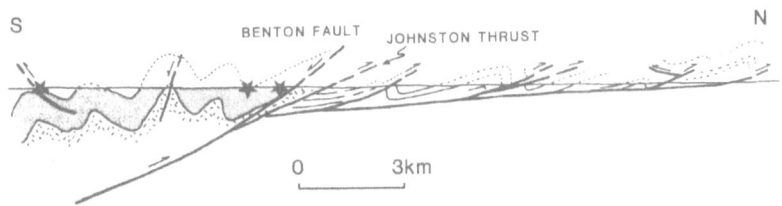

Fig. 1. Simplified tectonic map of SW Dyfed, Wales (after Coward and Smallwood 1984). Main fold axes and thrusts are illustrated. South of the Johnston thrust, red sandstones of Siluro-Devonian age are folded as shown by the shaded ornament in the cross section. Stars indicate the fold structures described.

The magnetization in these redbeds is particularly suited to such a study because the remanence consists of two discrete remagnetizations which were acquired during deformation. This enables us to form a picture of the 3-D path the rotating thrust sheets took. Undeformed ORS in the Anglo-Welsh cuvette (Chamalaun 1964) has a two-component remanence, a low blocking temperature (T_B) remanence acquired during the Permo-Carboniferous which has the expected direction of ~190°/-20°, and a high T_B remanence directed upwards to the northeast in a very similar direction to the expected Siluro-Devonian directions from the rest of Britain. McClelland Brown (1983) showed that these two directions were obtained from sites within the Freshwater anticline, a fold with an amplitude of several kilometers, after structural corrections had been applied to restore the tilted bedding to horizontal. However, both directions were rotated 30° clockwise from their expected positions. This region of SW Dyfed (structural zone 1a of Hancock et al (1983)) is characterized by a few large amplitude folds with very few parasitic mesofolds. In contrast, Hancock et al.'s zone 1b has large scale folds of smaller amplitude and abundant parasitic folds with amplitudes on the scale of tens of metres. Palaeomagnetic study of fold structures in this zone (indicated by stars in Figure 1) has shown no evidence for the preservation of a primary Siluro-Devonian remanence, but two components of remanence can be identified which we believe were acquired during deformation in the Permo-Carboniferous.

Palaeomagnetic samples were collected from around three fold and fold-and-thrust structures (Figure 1), the fold axes strike between 275° and 295° and plunge from 22° to 2°W although many folds in this area have easterly plunge. Detailed thermal demagnetization reveals two components of remanence in each sample which have distinctively different blocking temperatures, these are referred to hereafter as low- and high-T_B components. Each component from each sample would originally have had the same orientation irrespective of the sample's present position in the fold structure, but now these insitu remanence directions have been distributed along small circles about the fold axis by the folding process. We can determine the shape of the fold at the time that each component was acquired, by incrementally rotating the bedding (and its associated remanence directions) back towards the horizontal at each sample location. When the directions of a component

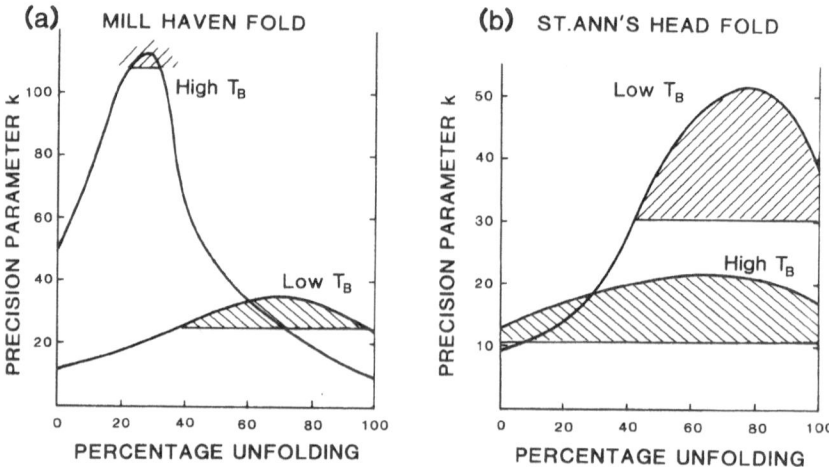

Fig. 2. Incremental fold tests to determine the age of remagnetization relative to folding. Precision parameter, k varies during incremental unfolding. Hatched area shows the region (a) in which partial unfolding achieves a statistically significant improvement in grouping over the insitu grouping (low TB remanence was acquired at 27+5% unfolding, and high TB remanence was acquired at 70+28% unfolding) or (b) below which grouping is significantly worse than for the percentage unfolding which achieves a maximum value of k. Low TB remanence was acquired at 75+33% unfolding, but there is no statistically significant improvement in grouping of the high TB remanence directions from this structure at any stage of structural correction, however, by comparison of magnetic properties with those from (a), we believe that the high TB remanence is earlier.

from around the structure all point in the same direction then we have identified the percentage of unfolding necessary to restore the fold to the shape it had when the remagnetization event occurred. This procedure has been described in detail by McClelland Brown (1983), together with the necessary statistical tests. Stepwise unfolding of the data indicates that the high-T_B component is earlier and was acquired between a quarter folded and prefolding, and the low-T_B component was acquired a quarter to three-quarters through folding (McClelland 1987 and Figure 2). We have assumed that the bedding was originally horizontal at the start of folding and we have unplunged the bedding before incrementally unfolding. The order in which unplunging and unfolding are applied can be important since the operations are non-commutative; however, in this case varying the order makes a difference of at most a few degrees since the plunges are small. We calculate the mean direction of remanence at optimum unfolding for each component. Since we have collected samples spread symmetrically about the fold axes, the direction of this mean is hardly affected by the chosen unfolding level (see for example Table 2, McClelland Brown 1983); choosing an optimum unfolding level provides relative dating of the components and gives an assessment of the true error on the mean.

Mean directions of remanence at the optimum structural correction are shown in Figure 3. These two components of remanence were acquired during folding in the Hercynian, and therefore were originally parallel to the Permo-Carboniferous field direction (~190°/-20°). They have been rotated away from this reference direction by rotations other than those due to the local folding (on the scale of tens of metres). It is possible to fit the low and high T_B directions on a great circle path which passes through the reference Permo-Carboniferous direction. This model assumes a single rotation of 45° about an inclined axis plunging approximately 55° to 120° (Figure 3). The high T_B component would have been acquired first as folding began and was parallel to the reference Permo-Carbonife-

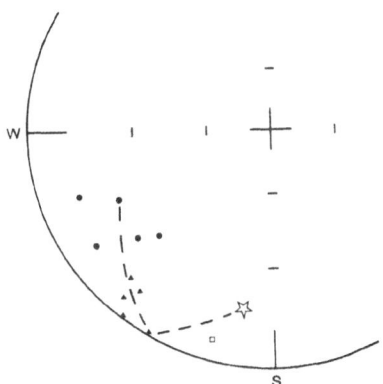

Fig. 3. Mean remanence directions calculated at optimum unfolding level. Circles represent high blocking temperature remanence, triangles represent low blocking temperature remanence from Mill Haven, St.Ann's Head and St.Brides' Haven. Square represents post-folding remanence from Llandstadwell formation from Stearns and Van der Voo (1987). Star is reference direction for Permo-Carboniferous. Closed symbols are projected on the lower hemisphere, open symbols on the upper hemisphere.

-rous direction. After ~20° of rotation had occurred the low T_B component was acquired, and then a further ~25° of rotation occurred until the present orientation of vectors was achieved. This model implies a tilting of bedding of 25° to the south since the acquisition of the high-T_B component. Scatter in the data would also fit a model of partitioning of rotation into a clockwise rotation about a vertical axis and a separate rotation about another axis (which could be horizontal or inclined), causing distribution of the vectors along a small circle. Clockwise rotation of about 35° about a vertical axis and separate rotation about a horizontal axis striking to 240° would also fit the data. However, this would require 130° of tilting of bedding down to the south (i.e. overturning by 40°) since the acquisition of the high-T_B component and 50° since the acquisition of the low-T_B component. This does not fit with the known structure (Figure 1). At present, we favour the model of a single inclined rotation axis, since this minimizes the necessary tilting.

Stearns and Van der Voo (1987) have shown that Old Red sandstones in a small fold 20km east of the sites discussed here, and lying between the Ritec and Johnston thrusts, was remagnetized before folding but in a direction close to the reference Permo-Carboniferous direction. This structure is probably a late stage fold, formed after the main rotation of SW Dyfed, and their observation of this direction confirms that it is valid to compare the Permo-Carboniferous reference direction with our rotated syn-deformational components of remanence.

This palaeomagnetic study has identified a clockwise rotation of the Johnston thrust sheet, probably about an inclined axis at 120°/55°. Clockwise rotation occurred in response to the oblique angle between the main Hercynian transport direction (towards the northwest) and the edge of the pre-Hercynian basin margin in southern Britain (Coward and Smallwood 1984). The Johnston thrust increases in displacement to the west, dying out to the east, giving further evidence of clockwise rotation. The probable inclined axis of rotation indicates that the thrust sheet climbed a lateral ramp tilting it westwards while undergoing clockwise rotation due to more regional forces. We believe that both local and regional rotation mechanisms operated at the same time, because two remagnetization events occurred at different times during deformation and these give us a picture of the 3-D path the rotating thrust sheets took.

3. Rotations in Basement thrust sheets; an example from the Axial zone, Spanish Pyrenees

3.1 GEOLOGICAL BACKGROUND

The geometry of structures in the internal parts of orogenic belts can be difficult to determine since the palaeohorizontal within igneous or polydeformed basement rocks is not known. This problem is of great significance in the Pyrenees where a number of conflicting and fundamentally different models of the deep structure have been proposed (Figure 4). The surface expression of Tertiary thrust faults in the Pyrenees shows a fanning geometry which has been described by many authors. Williams and Fischer (1984) and Parish (1984) favour a thin skinned model in which all thrusts are assumed to have originated at shallow angles and to have been subsequently steepened by backthrusting or by piling of lower imbricates into antiformal stacks. Seguret and Daignieres (1986), Seguret (1972), and Choukroune and Seguret (1973) prefer an interpretation in which the thrusts steepen downwards into a zone of inhomogeneous shortening in the lower crust. Deramond et al (1985) propose a thick-skinned model in which thrusts root down to join a basal detachment at the Moho. McCaig (1986) attempted to identify originally steep and shallow structures on the basis of their relationship to Hercynian isograds. He suggested some steep structures may have developed only in the hangingwall of a shallow-dipping decollement. The ECORS deep reflection seismic survey across the Pyrenees (ECORS Pyrenees team 1988) has constrained the shallow level thrusts on the north and south margins to have a similar geometry to that predicted in the thin-skinned models, and has identified lower Iberian crust 'subducted' under the Pyrenees. Unfortunately, since the data from the centre of the profile at the French-Spanish border of poor quality, many problems remain regarding the continuation of thrust faults to depth in the Pyrenees.

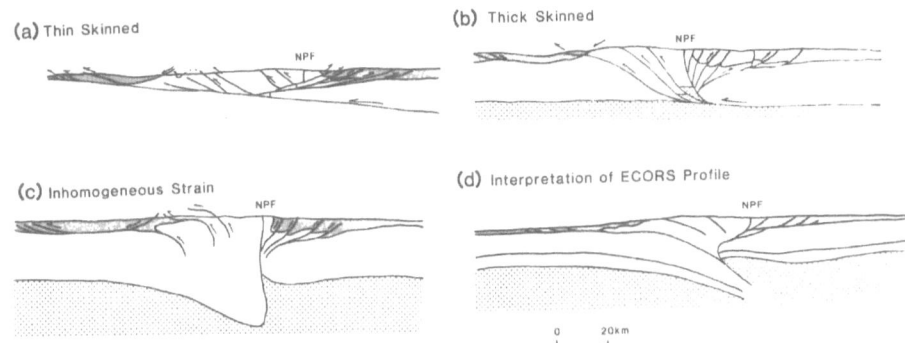

Fig. 4. Possible models for the deep structure of the Pyrenees. (a) Williams and Fischer (1984) (b) Deramond et al (1985) (c) Seguret and Daignieres (1986) (d) ECORS Pyrenees team (1988). Dotted ornament represents mantle material, shaded ornament represents cover rocks. NPF refers to surface outcrop of North Pyrenean Fault.

Palaeomagnetism has the potential to determine the palaeohorizontal within basement thrust sheets at the end of the Hercynian, testing some of the predictions of the models shown in Figure 4. We have therefore initiated a study of late Hercynian dykes south of the Maladeta granodiorite, where the presence of Triassic rocks unconformably overlying and folded with the basement provides an independent constraint on rotations.

The geology of this area, north of Pont de Suert, has been described by Mey (1969). Figure 5 shows a much simplified sketch map of the area with sampling localities. Hercynian deformation has affected sediments of Cambro-Ordovician to lower Carboniferous age. These are cut by intrusives of the late Carboniferous Bono complex, which is probably associated with the Stephanian volcanics found at Erill Castell. Unconformably overlying the Hercynian deformed basement are upper Carboniferous volcanics and coals, then Permian and Triassic conglomerates and red sandstones. The southward directed Alpine thrusting has imbricated sheets of basement and occasional Triassic redbeds into an antiformal stack. Six separate thrust sheets can be identified. The lowermost sheet (sheet 1) is folded into an E-W trending anticline defined by the orientation of Triassic redbeds exposed on both the northern and southern margins of the sheet (Figs. 8 and 9). Thrust sheets 2 to 6 have northward-dipping thrust contacts. The thrusts below sheets 2, 3 and 4 die out west of Figure 5 into a zone of upright folds of the Triassic unconformity. Immediately south of the study area, Triassic redbeds form downward facing hangingwall anticlines of the Nogueras zone at the front of the antiformal stack (Bates 1987). The Bono complex consists of a central intrusion of diorite to granodiorite composition from which a suite of quartz and feldspar phyric intermediate to felsic dykes radiate. These dykes outcrop in all thrust sheets in the study area. Here, we concentrate on a palaeomagnetic investigation of the remanence carried by these dykes.

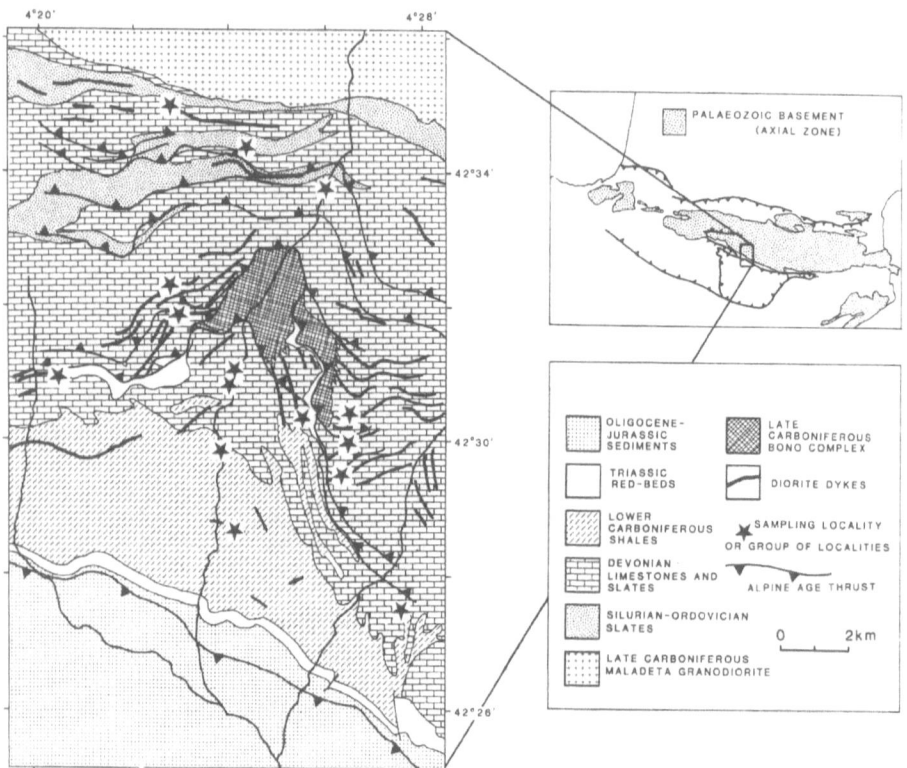

Fig. 5. Simplified geological sketch map of southern Axial zone of the Pyrenees, south of the Maladeta granodiorite. Thrust sheets are numbered in Fig. 8.

3.2 PALAEOMAGNETIC RESULTS

Palaeomagnetic samples have been collected from 30 dykes in 5 different thrust sheets. An average of eight cores were drilled from each dyke and where the host rock was sampleable, a series of cores were drilled across the contact zone. At this stage in our study we have thermally demagnetized 108 samples from dykes and contacts. The remanence structure has been analysed by using the Linefind algorithm (Kent,Briden and Mardia 1983) to rigorously determine the direction of magnetization components.

A total of 80 samples from 23 of the dykes and contacts, have a well defined characteristic remanence determined by detailed thermal demagnetization. In some samples the remanence is single component, and in others a component in the present earth's field direction is demagnetized at low treatment temperatures before the characteristic direction is revealed. In some dykes, the remanence is carried exclusively by magnetite, while other dykes have a single component remanence carried by both magnetite and hematite. The primary nature of the characteristic dyke remanence directions has been demonstrated by four full and two partial positive contact tests. In a full contact test, the host rock close to the dyke margin has a single component remanence statistically indistinguishable from that of the dyke, indicating that the host rock was heated above the Curie point of its remanence carriers during the intrusion of the dyke. Further from the dyke, where the host rock was heated to a temperature less than the Curie point, the remanence has two components, the lower blocking temperature component having been acquired during the contact heating, and the higher blocking temperature component predating the heating. This indicates that no remagnetization event has occurred since the intrusion. In a partial contact test, the host rock close to the dyke margin has the same remanence direction as the dyke, but was not sampled further away. Figure 6 shows a full

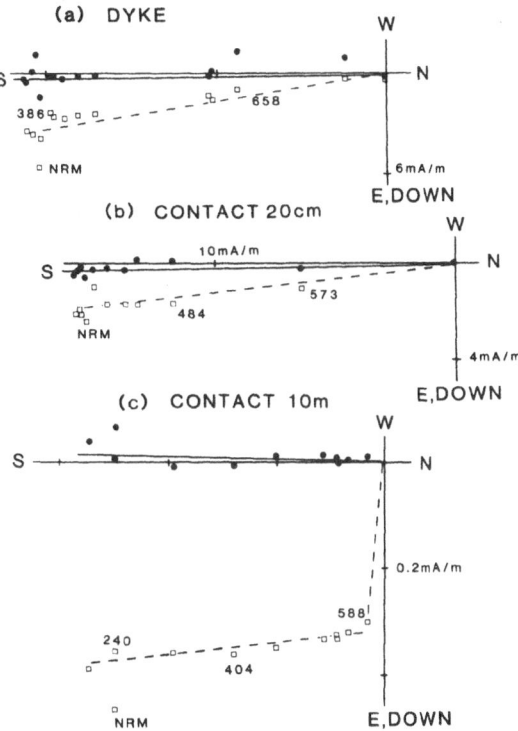

Fig. 6. Contact test from dyke 87p16 in thrust sheet 1: vector plots of thermal demagnetization of remanence from dyke and contact rocks. Filled circles represent projection of vector in horizontal (N,S,E and W) plane, open squares represent projection of vector in vertical (Up,N,Down and S) plane (see Dunlop 1979 for explanation). Numbers next to points indicate temperature of thermal treatment.

374

contact test from site 16, where the dyke (Figure 6(a)) and close contact rock (Figure 6(b)) have the same single component remanence, and the host rock 10m from the 5m wide dyke (Figure 6(c)) has this same component with blocking temperatures below 580°C. The direction of the high temperature component is not properly isolated because above 580°C thermal alteration begins, and the magnetization direction becomes erratic. However, the high temperature direction is significantly different from that in the dyke.

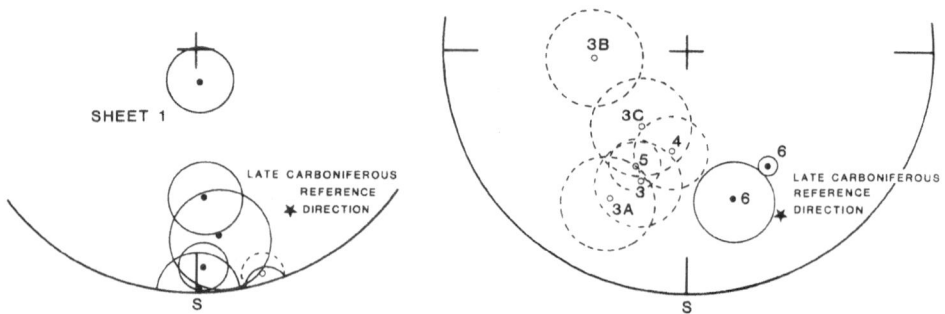

Fig. 7. Segment of equal angle projection of mean remanence directions. Filled symbols indicate a downward pointing, lower hemisphere vector, open symbols indicate upward pointing, upper hemisphere vectors. Dotted circles for thrust sheets 3 to 6 indicate error circles projected onto the upper hemisphere.

Mean Characteristic dyke directions from groups of dykes close to each other and in the same thrust sheet,have been calculated. Figure 7 shows the distribution of mean directions. Directions from locations within sheet 1 are distributed along a north/south and vertical great circle, (ie their declinations are close to 180°) and the inclination of remanence is related to the geographical position of the site. The most southerly site has the steepest inclination and the most northerly site has the shallowest inclination. The circles indicate the cone of 95% confidence in the mean direction (α_{95}). All directions from sheets 3 to 5 are upward directed to the south-west, whereas the directions from sheet 6 are directed downwards to the south-east. Figure 8 shows the mean remanence directions in their geographical locations, represented as an arrow in the direction of declination and a number indicating the inclination, superimposed on a map of the thrust sheets. The errors on palaeomagnetic estimates in this preliminary study are approximately +15°.

3.3 INTERPRETATION

The remanence in the dykes has been shown to date from the time of intrusion in the late Carboniferous. The expected geomagnetic field direction for stable Iberia at this time has been proposed by Vandenberg and Zijderveld (1982), to be Dec=150°, Inc=+15°, this direction takes account of the 30o counterclockwise rotation of Iberia. This is, therefore, presumed to be the original orientation of the magnetization in the dykes. Any significant directional difference between the present direction of dyke magnetization and this reference field direction indicates a block rotation of the dykes after intrusion and most probably during the evolution of Alpine age thrust sheets. The amount of 3-D rotation of the thrust sheets can be estimated in this way. The orientation of the

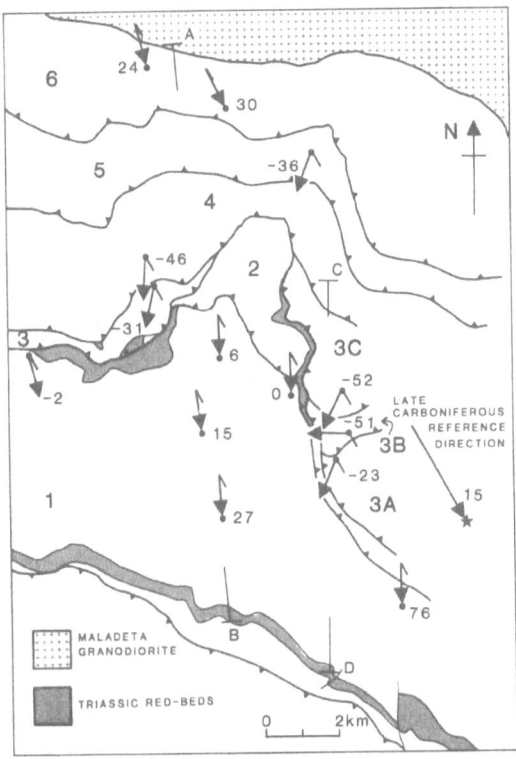

Fig. 8. Mean remanence directions plotted at geographic location. Direction of arrow indicates actual declination of magnetization vector, number by arrow indicates inclination. Filled circle indicates sampling site or group of sites, arrow points away from site for negative inclinations, towards site for positive inclinations. Small line at end of arrow indicates late Carboniferous reference direction, the angle between this line and the arrow is the estimate of clockwise rotation about a vertical axis. Thrust sheets are numbered 1 to 6. Cross-sections A-B and C-D are shown in Fig. 9. Errors in declination are approximately +15°.

Carboniferous field is ideal for constraining rotations about the generally E-W trend of fold axes in the Pyrenees, as well as about any steeply dipping axis. However, we cannot constrain any rotations which may have occurred about shallow southerly axes similar to the remanence direction.

Figure 9 shows two schematic N-S cross-sections through the area, with the palaeomagnetic estimates of tilt about a horizontal axis. The arrows are oriented at the angle of tilt estimated from the mean remanence direction from the group of dykes at that locality. The doming of the rocks within thrust sheet 1 is clearly shown in this diagram.

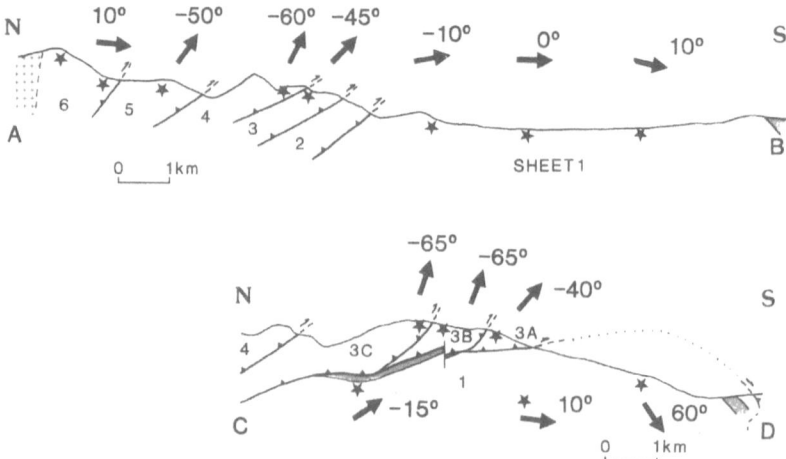

Fig. 9.   Schematic N-S cross-sections showing palaeomagnetic estimates of tilt about a horizontal axis. The arrows are oriented at the estimated angle of tilt obtained from the group of dykes below the arrow and the number next to each arrow is the tilt estimate. Errors in tilt are approximately +15°.

Poles to bedding in Triassic redbeds from thrust sheet 1 are distributed on a north-south girdle indicating an easterly, sub-horizontal mean fold axis. The distribution of remanence directions from dykes in this thrust sheet (Figure 7) is consistent with the dykes and their remanent magnetism having undergone two independent components of rotation since their intrusion. All declinations are about 25° west of the expected late Carboniferous declination of 150°, i.e. they have been rotated 25° clockwise about a vertical axis. The remanence directions also appear to have been rotated about the sub-horizontal fold axis defined by poles to Triassic bedding, as passive markers during folding of the thrust sheet. For example, the southernmost group of dykes has a remanence which has been tilted downwards to the south by 60°. This locality is the closest to the southern unconformity with overlying Trias where the Triassic bedding dips south at about 60°. Similarly, the two northernmost groups of dykes which lie just under the northern unconformity where the Triassic bedding dips 10 to 20° north, have remanences which have been tilted downwards to the north (or upwards to the south) by 10 to 15°, away from the reference direction. Hence the palaeomagnetic estimates of tilt about a horizontal axis agree well with the amount of tilting of the Triassic bedding, and therefore can be used as an independent tilt estimate in higher thrust sheets, where Triassic sediments are not preserved.

Remanence directions from thrust sheets 3 to 5 indicate that more rotation has occurred than in sheet 1. All remanences are upward directed in the southwest quadrant (Figure 7) indicating tilting upwards to the south by 40 to 65° and clockwise rotation about a vertical axis by 35 to 60° (Figs. 8 and 9). With the present data set, it is not possible to distinguish between remanence directions from sheets 3 to 5. No information can be gained about the possible partitioning of rotation in these thrust sheets from this grouping of data. However, it is very likely that this thrust stack has experienced the same 25° clockwise rotation about a vertical axis as the underlying sheet 1, and that this probably occurred at a different (later?) time to the rest of the clockwise rotation and tilting. Remanence directions from thrust sheet 6, which is in contact with the Maladeta granodiorite, are, however, significantly different from those in underlying sheets. They lie within the experimental error of 15° of the expected late Carboniferous field direction, implying no significant rotation or tilt has occurred. Rotations about a hypothetical vertical axis for all localities are illustrated in Figure 8.

These palaeomagnetic estimates enable us to make a preliminary interpretation of the sequence of Alpine-age thrusting in this area. Thrust sheets 3 to 5 moved first and, being pinned to the west of the area, this movement caused clockwise rotation and some back-tilting. Movement must then have occurred on a deeper thrust coming up under sheet 1, which caused the doming and clockwise rotation of sheet 1 and the further back-steepening and rotation of sheets 3 to 5. Lastly, thrust sheet 6 may have moved on a breaching reverse fault at a steep angle similar to its present attitude causing no rotation or tilting, and which postdates the underlying antiformal stack. This would place the main root zone for the Nogueras zone thrusts north of the Maladeta granodiorite, in contrast to the interpretation of Williams and Fischer (1984), whose section goes through the area we have studied. We intend to test this hypothesis by further sampling and fieldwork.

3.4. DISCUSSION

Palaeomagnetic work by Bates (1987) in Triassic redbeds immediately south of our area, shows that the downward-facing folded and thrust sheets of the Nogueras zone have been rotated clockwise by an average of 45° during the late stages of their development. We identify similar rotations in thrust sheets 3 to 5 which therefore are likely to be part of the same antiformal stack. The above interpretation has assumed that all rotations have occurred during Alpine age thrusting. All rocks with Iberian affinity have undergone anticlockwise rotation due to the opening of the Bay of Biscay, but this rotation is implicitly assumed by using the Iberian reference field for comparison. Episodic rifting after the Hercynian orogeny formed basins in which Permian and Triassic redbeds were deposited (Puigdefabrigas and Souquet 1986). The unconformity between Permian and Triassic is tilted in this area, and Triassic sediments lie in half-grabens. This indicates that our study area will have experienced some rotations between Hercynian and Alpine orogenies. The 25° clockwise rotation about a vertical axis defined in sheet 1 may be a regional rotation which affected the whole of the southern Pyrenees, and the age of this rotation is not constrained to be Alpine. However, we believe that, although the absolute rotations defined by comparison with the Iberian reference pole may be somewhat in error, relative rotations between thrust sheets are valid.

The success of this preliminary palaeomagnetic study of 3-D rotations in basement rocks of the Axial zone of the Pyrenees is striking. The method obviously has great potential for addressing the fundamental problem of whether the vertical structures in the central part of the Axial zone were originally steep and root down into a zone of inhomogeneous shortening in the lower crust, or whether they had originally shallow attitudes and have been back steepened by tectonic processes.

4. Conclusions

Palaeomagnetism is a powerful tool for identifying rotations about any axis. In folded and thrust cover rocks in external parts of orogenic belts, palaeomagnetic and structural study can totally constrain three-dimensional rotations. It is necessary to use two preferably orthogonal direction markers since rotations about the direction of a single marker cannot be identified. Balanced cross sections constructed using structural data alone are likely to be in error, as are estimates of shortening, the errors may be great in regions where large rotations have occurred. Both examples discussed here show rotations of 60° or more out of section.

Palaeomagnetism has even more exciting potential for constraining geometries and rotations in internal parts of orogenic belts, where section balancing techniques cannot be applied directly to igneous or polydeformed basement rocks since no palaeohorizontal orientation is known.

Large variations in amounts of rotation have been observed over a small section only 15km in length, in our study in the Pyrenees (e.g. Figure 9). These rotations are clearly related to Alpine thrusting, and would be very difficult to interpret if the details of the structure were not known. This highlights the necessity for doing palaeomagnetic studies in great detail to define the scale on which rotations occur (at least over part of a study area), and demonstrates the essential link with structural analysis. Palaeomagnetic studies on fold and thrust belts also have a scale problem, generally small-scale folds with amplitudes of tens of metres are sampled (this scale is ideal for incremental fold tests as samples can be taken from a single bed around the fold structure so all samples should have the same magnetic properties). However, these may well be parasitic folds on limbs of much larger amplitude folds as is the case in our study in SW Dyfed. Restoring the bedding from the small folds back to horizontal during the incremental fold test may not be correct, since the parasitic folds may have formed on an already tilted limb of a larger-scale fold. A much more detailed study in this area is underway at the moment, which will hopefully better define the necessary structural corrections. The scale on which rotations about a vertical axis occur in strike-slip zones is still uncertain, and again integration of detailed structural and palaeomagnetic analysis will be necessary to solve this problem.

Acknowledgments

E.M and A.M.M. gratefully acknowledge a Research Grant from the Natural Environment Research Council for field expenses, E.M. is supported by a Royal Society Research Fellowship. Thanks to Thomas Jenkin, who did almost all the experiments on rocks from the Pyrenees, for his humour and all his hard work, and to our field assistant Mark Bennett for bravery beyond the call of duty.

References

Bates, M.P., 1987. 'Palaeomagnetic studies of fold and thrust geometry in the Southern Pyrenees' Unpubl. PhD Thesis, Univ. of Leeds, 474pp.

Boyer, S.E. and Elliot, D., 1982. 'Thrust systems' *AAPG Bull.*, **66**, 1196-1230.

Butler, R.W.H., 1986. 'Thrust tectonics, deep structure and crustal subduction in the Alps and Himalayas' *J. Geol. Soc.*, **143**, 857-873.

Butler, R.W.H., and Coward, M.P., 1984. 'Geological constraints, structural evolution, and deep geology of the Northwestern Scottish Caledonides' *Tectonics*, **3**, 347-365.

Chamalaun, F.H., 1964. 'Origin of the secondary magnetization of the Old Red Sandstones of the Anglo-Welsh Cuvette' *J. Geophys. Res.*, **69**, 4327-4337.

Choukroune, P. and Seguret, M., 1973. 'Tectonics of the Pyrenees; role of compression and gravity'. In: K. de Jong and R. Scholten (Eds), Gravity and Tectonics. Wiley, New York, pp. 141-156.

Coward, M.P. and Potts, G.J., 1983. 'Complex strain patterns developed at the frontal and lateral tips to shear zones and thrust zones' *J. Struct. Geol.*, **5**, 383-399.

Coward, M.P. and Smallwood, S., 1984. 'An interpretation of the Variscan tectonics of SW Britain' In: D.H.W. Hutton and D.J.Sanderson (Eds.), Variscan Tectonics of the N.Atlantic Region. *Geol .Soc. London, Spec. Publ.*, **14**, 89-102.

Dahlstrom, C.D.A., 1969.'Balanced cross sections' *Can. J. Earth Sci.,*6, 743-758.

Deramond, J., Graham, R.H., Hossack, J.R., Baby, P. and Crouzet, G.,1985. 'Nouveau modele de la chaine des Pyrenees' *C.R.Acad.Sci. Paris,* Ser. 2, 1205-1210.

Dunlop, D.J., 1979. 'On the use of Zijderveld vector diagrams in multicomponent palaeomagnetic studies' *Phys. Earth Planet. Inter.,* 20, 12-24.

ECORS Pyrenees team, 1988. 'The ECORS deep reflection seismic survey across the Pyrenees' *Nature,* 331, 508-511.

Hancock, P.L., Dunne, W.M. and Tringham, M.E., 1983. 'Variscan Deformation in Southwest Wales' In: The Variscan fold belt in the British Isles, Hilger, Bristol. pp. 47-73.

Jackson, J. and McKenzie, D., 1983. 'The geometrical evolution of normal fault systems' *J. Struct. Geol.,* 5, 471-482.

Kent, J.T ,Briden, J.C. and Mardia, K.V., 1983. 'Linear and planar structure in ordered multivariate data as applied to progressive demagnetization of palaeomagnetic remanence' *Geophys. J.R. astr. Soc.,* 62, 699-718.

Mey, P.H.W., 1969. 'Geology of the upper Ribagorzana and Tor valleys, Central Pyrenees, *Spain' Leidse Geol. Meded.,* 41, 229-292.

McCaig, A.M., 1986. 'Thick- and thin-skinned tectonics in the Pyrenees' *Tectonophys.,* 129, 319-342.

McClelland Brown, E., 1983. 'Palaeomagnetic studies of fold development and propagation in the Pembrokeshire Old Red Sandstone' *Tectonophys.,* 98, 131-149.

McClelland, E., 1987. 'Palaeomagnetic results from the Lower Devonian Llandstadwell Formation, Dyfed, Wales- Discussion' *Tectonophys.,* 143, 335-336.

Parish, M., 1984. 'A structural interpretation of a section of the Gavarnie nappe and its implications for Pyrenean geology' *J.Struct. Geol.,* 6, 247-255.

Puigdefabrigas, C. and Souquet, R., 1986. 'Tectono-sedimentary cycles and depositional sequences of the Mesozoic and Tertiary from the Pyrenees' *Tectonophys.,* 129, 173-204.

Seguret, M. and Daignieres, M., 1986. 'Crustal scale balanced cross-section of the Pyrenees; discussion' *Tectonophys.,* 129, 303-318.

Seguret, M., 1972. 'Etude tectonique du versant sud des Pyrenees centrales: nappes et series decolles, role du serrage et de gravity' Presses Univ.Sci.Tech.Languedoc, Montpellier.

Stearns, C. and Van der Voo, R., 1987. 'Palaeomagnetic results from the Lower Devonian Llandstadwell formation, Dyfed, Wales' *Tectonophys.* 143, 329-334.

VandenBerg, J. and Zijderveld, J.D.A., 1982. 'Palaeomagnetism in the Mediterranean area'. In: Berekhemer, H and Hsu, K. (Eds), Alpine-Mediterranean Geodynamics. Geodynamics Series, 7, Am. Geophys Union, 216pp.

Williams, G.D. and Fischer, M.W., 1984. 'A balanced section across the Pyrenean orogenic belt' *Tectonics,* 3, 773-780.

PALAEOMAGNETIC EVIDENCE FOR BLOCK ROTATIONS AND DISTRIBUTED DEFORMATION OF THE IBERIAN-AFRICAN PLATE BOUNDARY

M.L. OSETE[1], R. FREEMAN[2], R. VEGAS[3]
[1]*Departamento de Física de la Tierra, Astronomía y Astrofísica*
Universidad Complutense
Madrid 28040, Spain.

[2]*Institut für Geophysik*
ETH-Hönggerberg
CH - 8093 Zürich, Switzerland.

[3]*Departamento de Geodinamica*
Universidad Complutense
Madrid 28040, Spain.

ABSTRACT. Palaeomagnetic evidence from Jurassic limestones and basic submarine volcanic rocks from the Subbetic Zone of the Betic Cordillera indicate that block rotations have played a significant role in the tectonic evolution of this area. Two models can account for the observed clockwise rotations: 1) a shear zone generated during the Miocene thrusting and accretion of the Alboran microplate onto the southern margin of Iberia, or 2) an intracontinental domain of distributed deformation produced during the welding of Africa and Iberia since the Miocene thrusting.

1. Introduction

The southern border of the Iberian Peninsula, the site of the Betic Orogeny, has been affected by transcurrent movements since the openning of the South Atlantic in the Mesozoic and initiation of the Atlantic-Azores transform fault along which differential movement continues today (Paquet, 1972, Araña and Vegas, 1974; Hermes, 1978; Bousquet, 1979; van der Fliert et al., 1980; Vegas et al., 1980; De Smet, 1984; Leblanc and Olivier, 1984; Udias, 1986; Dercourt et al., 1986). However, the role, magnitude, and sense of movement of crustal blocks in this region are poorly defined.

Palaeomagnetic declinations are sensitive indicators of rotations about vertical axes, movements which could be expected within a zone experiencing transcurrent faulting. Nevertheless, until 1985 very few palaeomagnetic investigations were carried out in the Betics and none for the purpose of identifying possible block rotations related to the known transcurrent tectonic style. The few published investigations (Vandenberg, 1980; Ogg et al., 1984, and Mäkel et al., 1984) were directed either towards large-scale comparative studies or magnetostratigraphic purposes. The paradigm - still strong today - that large-scale N-S thrusting dominated the tectonics in the Betics may have played a role in the delayed awareness of the possible role of block rotations in this zone.

C. Kissel and C. Laj (eds.), Paleomagnetic Rotations and Continental Deformation, 381–391.
© *1989 by Kluwer Academic Publishers.*

382

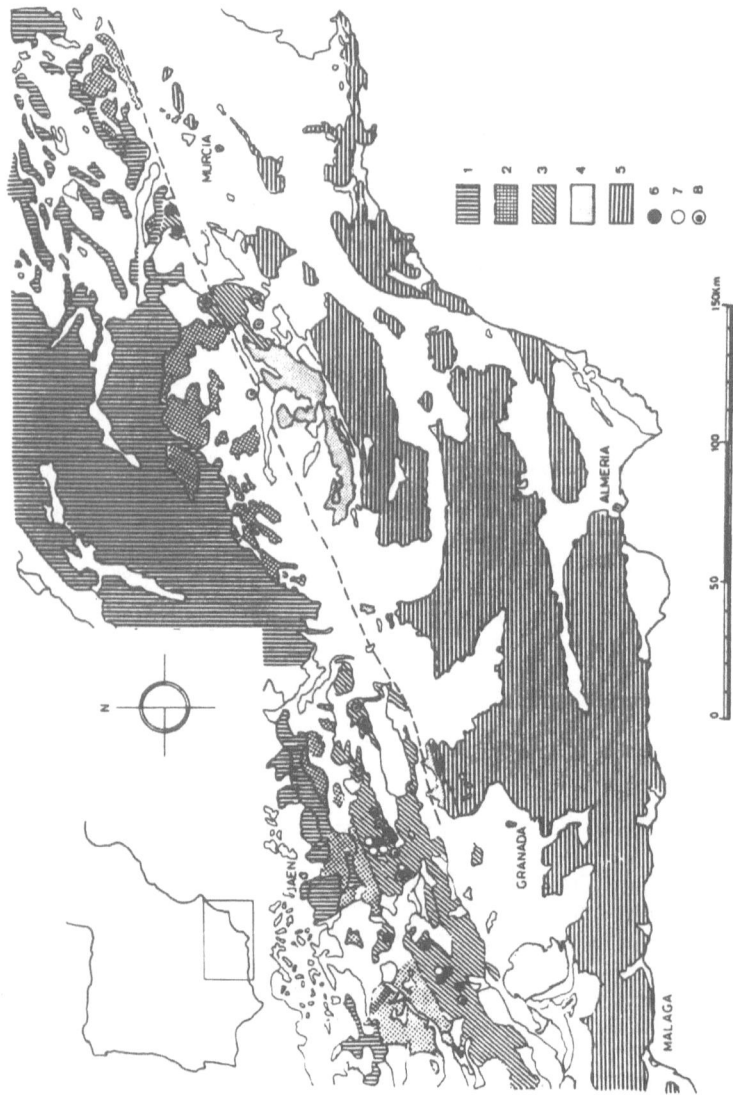

Fig. 1. Geological map of the distribution at the External and Internal Zones of the Betic Cordilleras (simplified from Azema et. al., 1979) and location of the sampled sites. External Zones (1-4): 1. Prebetic, 2. External Subbetic, 3. Middle Subbetic, 4. Internal Subbetic, 5. Internal Zones including the "Dorsale Calcaire", 6. palaeomagnetic sites in volcanic rocks, 7. palaeomagnetic sites in limestones, 8. Jurassic volcanism. The dashed line is the main trace of the Crevillente Fault; we have not extended this line west of Granada since the exact path is uncertain, although the fault itself certainly continues westwards.

In 1985 we initiated a palaeomagnetic study in the Betic Cordillera investigating both limestones and volcanics rocks from the Subbetic Zone. First results of this survey are discussed in Osete et al. (1988). Several recent magnetostratigraphic sudies in the Subbetics (Mazaud et al., 1986; Steiner et al., 1987; Ogg et al., 1988) have provided additional palaeomagnetic information which has been included in the present review.

The Subbetic Zone belongs to the Betic Cordillera which is divided into the Internal and External Zones. The latter comprises rocks from the Mesozoic margin of the Iberian plate and is itself subdivided into the Prebetic and Subbetic Zones (Figure 1). The Subbetic Zone is made up of hemipelagic and pelagic sediments, and submarine volcanics extruded onto the southern margin of the Iberian Plate during the Mesozoic breakup of Pangea. Vegas and Muñoz (1984) have ascribed the Jurassic sedimentation of the Subbetic Zone to horst-and-graben tectonics related to oblique extension between Africa and Iberia. According to their model, the widespread Jurassic submarine volcanic activity of the Subbetic Zone can be related to a broad system of volcanic outpourings that occurred during the peak of the Jurassic transtensional event along the entire transform zone between the Iberian Chain of eastern Spain (Vegas, 1985) and the Atlas System in southern Morocco (Jenny, 1984).

2. Palaeomagnetic Results

Table 1 gives the palaeomagnetic results obtained in Jurassic-Lower Cretaceous rocks from the Subbetic Zone. Data from an additional five sites is not included in this list because we have not been able to define satisfactory tectonic corrections.

We briefly comment on the treatment and results obtained for each site. More details can be found in the references. Samples from sites AC, AB, BB, GNV, GNC, GO and SB were demagnetized either by stepwise thermal heating, alternating field treatment, or a combination of both, to isolate the characteristic magnetization for each site. Bulk susceptibility was monitered to detect formation of new ferromagnetic minerals during thermal demagnetization. In addition, thermomagnetic (Curie) curves were obtained on samples from the volcanic sites (AB, BB, GNV, and GO). Acquisition of isothermal remanent magnetization (IRM) and subsequent stepwise thermal demagnetization of two perpendicular IRM components were carried out on selected limestone samples from sites AC, GNC, and SB. In the studies carried out by other workers, thermal demagnetization was the common procedure. In nearly all cases magnetite (or titano-magnetite) carries the characteristic magnetization (Steiner et al., 1986; Osete et al., 1988). Ogg et al., (1988) report differences of up to 20° of palaeomagnetic declination between normal and reversed site means as a result of incomplete cleaning. Nevertheless, this difference is still less than the observed declination deviations from the expected values for stable Iberia. There are at present not enough data to quantify strain rates and other tectonic parameters (some of our long-term goals) and we restrict ourselves in this brief review to a qualitative discussion.

All the sites of pillow-lavas yielded anomalous inclinations: they were consistently shallower than inclinations measured in nearby limestones and in most cases shallower than expected for a reasonable estimate of the Jurassic palaeolatitude. This observation has been reported by workers from other areas as well and Osete et al., (1988) review several explanations for the anomalous inclinations. The ultimate cause is still open and we therefore do not draw any interpretations on the palaeolatitude from these volcanics.

384

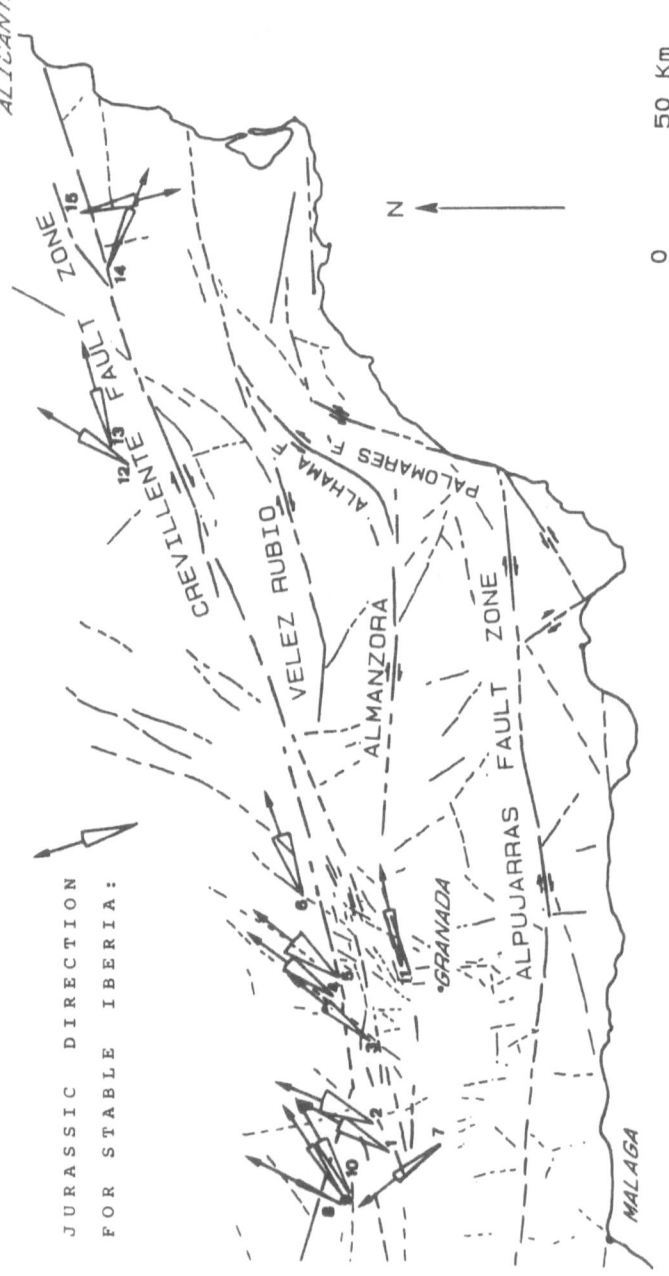

Fig. 2. Map of the southern border of Iberia showing the sites of the units sampled and the palaeomagnetic directions for each site with the a95 (radius of the 95% circle of confidence) indicated on each arrow. Numbers identify the sites: 1- AC, 2-AB (100 m from site 1, separated here for clarity), 3-BB, 4-GNV, 5-GNC, (100 m from site 4, separated here for clarity), 6-GO, 7-SG, 8-C, 9-BJ, 10-BC, 11-BH, 12-VCY, 13-VCZ, 14-SL, 15-SB. Data from Mazaud et al, 1986; Ogg et al., 1984: Ogg et al., 1988; Steiner et al., 1987). Insert top left: average direction for stable Iberia (data from Vandenberg 1980 and Schott et al., 1981). Lineations from Landsat images and field observations (including unpublished data from R. Vegas and L.M. Barranco).

Table 1. Palaeomagnetic data

Site	LAT(°N)	LON(°W)	Rock type	N	Dec	Inc	k	α95	Age	Reference
1. AC	37.34	4.13	marly limestone	14	25.8	43.0	28.7	7.6	Aalenian	1)
2. AB	37.34	4.13	pillow-lava	6	22.1	17.4	42.2	10.4	Aalen.-Bajocian	1)
3. BB	37.40	3.72	basalt	11	35.9	32.4	71.7	5.4	Middle Jurassic	1)
4. GNV	37.50	3.65	basalt	10	32.7	44.7	56.7	6.5		1)
5. GNC	37.50	3.65	marly limestone	10	36.3	38.9	15.9	12.5		1)
6. G0	37.58	3.27	pillow-lava	13	68.4	11.8	32.8	7.3	1. Tithonian -1. Berriasian	1)
7. SG	37.2	4.1	limestone	97	323.2	30.0	34.7	2.5	Tithonian - Kimmeridgian	2)
8. C	37.5	4.3	limestone	105	26.3	38.3	66.4	1.7		2)
9. BJ	37.5	4.3	limestone	196	53.8	47.0	13.6	2.8	Bathonian - Bajocian	3)
10. BC	37.5	4.3	limestone	112	59.4	42.9	12.1	4.0		3)
11. BH	37.3	3.5	limestone	107	75.8	46.6	8.0	5.2		3)
12. VCY	38.07	1.81	limestone	21	29.4	47.6	52.0	4.4	Berriasian- -Valanginian	4)
13. VCZ	38.07	1.81	limestone	36	72.4	47.7	39.0	3.8		4)
14. SL	38.13	1.11	limestone	169	111.7	46.6	12.7	3.0	Kimmeridgian -Berriasian	5)
15. SB	38.31	0.86	limestone	12	167.8	47.4	35.8	7.4	Neocomian	6)

1) Osete et al., 1988; 2) Ogg et al., 1984; 3) Steiner et al., 1987; 4) Ogg et al., 1988; 5) Mazaud et al., 1986; 6) R. Freeman (unpubl.)

3. Discussion

The magnetic declinations of the sites summarized in Table 1 are illustrated in Fig. 2. In addition, the Jurassic direction for stable Iberia (data from Vandenberg, 1980 and Schott et al., 1981) is given for comparison. The major tectonic alignments in the studied region drawn up from Landsat images (modified from Barranco, 1986) and field mapping are also depicted in Fig. 2. Solid lines indicate major faults seen both on the Landsat images and mapped in the field; broken lines identify clearly defined linear features seen only in the Landsat images, some of them being partly covered by Neogene sediments. Several major lineations in the Betics correspond to large right-lateral, E-W-trending strike-slip fault zones: the Crevillente, Velez-Rubbio, and Alpujarras faults. A series of mostly left-lateral, NE-SW strike-slip faults to the east of the region represented in Fig. 2

(Carboneras, Palomares, and Alhama de Murcia faults) have also been active in the Neogene (Bousquet, 1979).

The palaeomagnetic results presented in this review reflect the total sum of all tectonic movements since the Jurassic-lower Cretaceous up to the present. Thus, the strikingly consistent clockwise trend of the declination deviations may be attributed to rotations accumulated *during* transtension in the Jurassic *and* transpression since the upper Cretaceous. However, considering the intensity of the Alpine orogeny in this region, we tend to favour Tertiary tectonic events as more likely to have caused the bulk of the movements represented by the palaeomagnetic data.

We consider two dynamic models to explain the palaeomagnetic results. The first envisages clockwise rotation of blocks related to the last collisional event that induced dextral shear in the Iberian margin. This collisional event could be due to the Eocene-Oligocene impingement of the Alboran block (now comprising the extra-Iberian internal units of the Betics) (Fig. 3). The Alboran block was accreted to Iberia and included in a broad orogenic zone astride Africa and Iberia. If we assume this scenario for the rotations in the Subbetic Zone, a substantial amount of crustal shortening during the collision was accommodated in an intracontinental dextral shear zone generated at the Iberian margin. Within this shear zone the fractured cover reacted with discrete eastward rotations. These rotations must be added to any acquired during the thrust and nappe formation that lasted until middle Miocene. Hence the rotations of cover blocks may be coeval with tectonic transport.

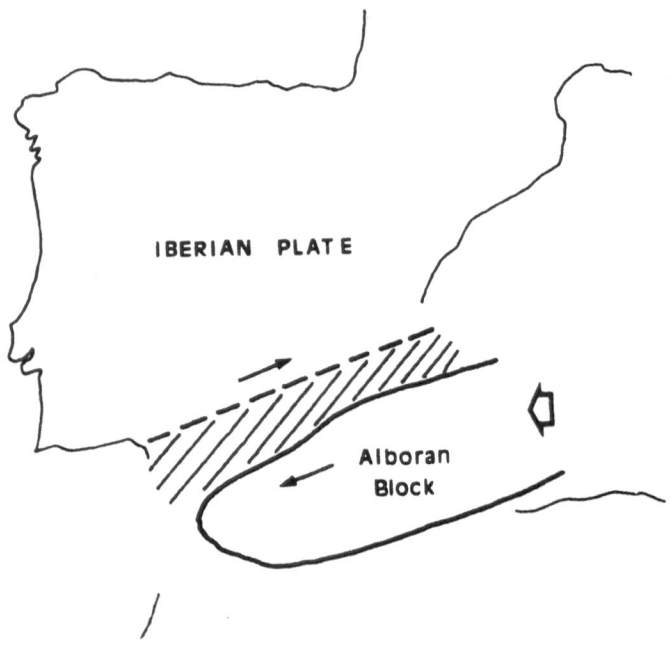

Fig. 3 Geodynamic setting of the dextral shear zone induced in the Iberian margin. In the Eocene-Oligocene time span the Alboran block was accreted to the Iberian plate.

387

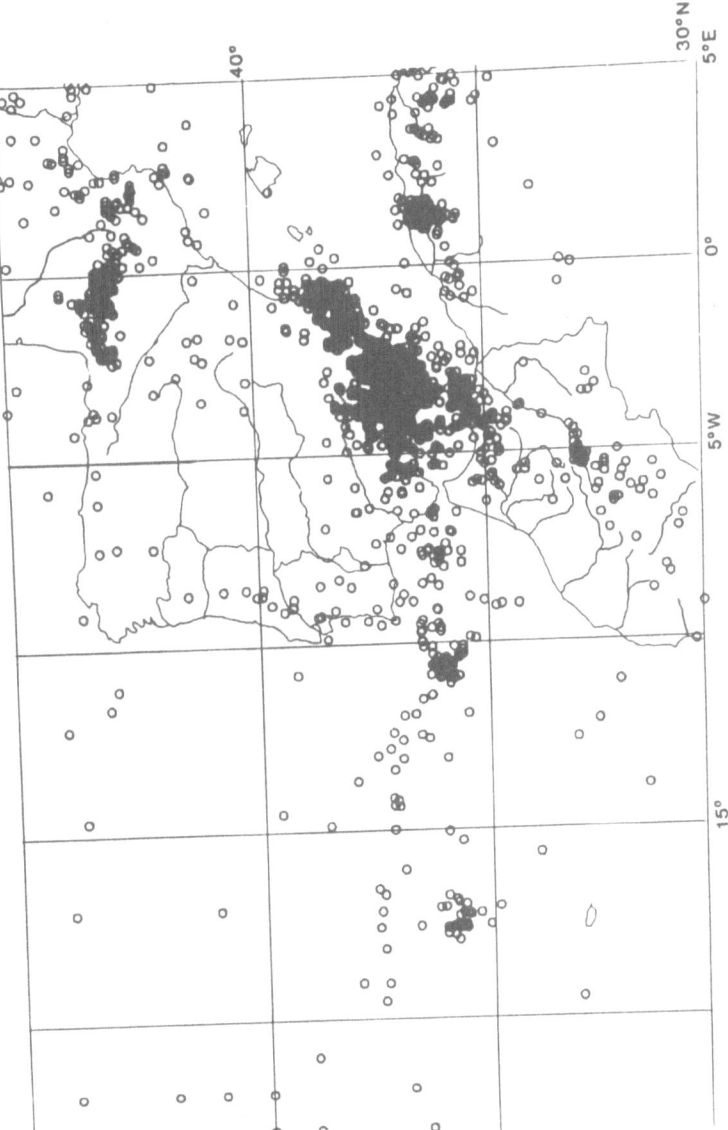

Fig. 4 Seismicity of the Ibero-Africa plate boundary for earthquake epicenters located historically and instrumentally from 880 B.C. through 1959 by the National Geographic Institute of Spain, and from 1960 through 1985 by the U.S. Geological Survey. This data base represents the location of earthquakes for which there is an assigned magnitude and for which a depth of focus is available (Udias et al., 1986).

The second model ascribes the rotations to a much larger deforming zone between Africa and Iberia. This zone can be delimited by the diffuse seismicity (Fig. 4) in the western end of the Eurasian-African plate boundary (Vegas, 1985). As mentioned above, this broad deforming zone resulted in the welding of Africa and Iberia and subsequent locking of these plates in the westernmost Mediterranean. The beginning of this tectonic scenario can be fixed at the end of the nappe-and-thrust tectonics in close relation to the phase of extension described by Platt (1986) and Aldaya et al. (1984). This broad zone of deformation has been described tentatively as an intracontinental domain of distributed deformation (Fig. 5). Within this zone plate convergence is accommodated by rotations of blocks (Vegas, 1988). Actually, the present Africa-Eurasia pole of rotation implies a compressive behaviour in the continental interface of Africa and Iberia. A model of block rotations in the zone can offer a clue to understanding the complex seismicity of this area.

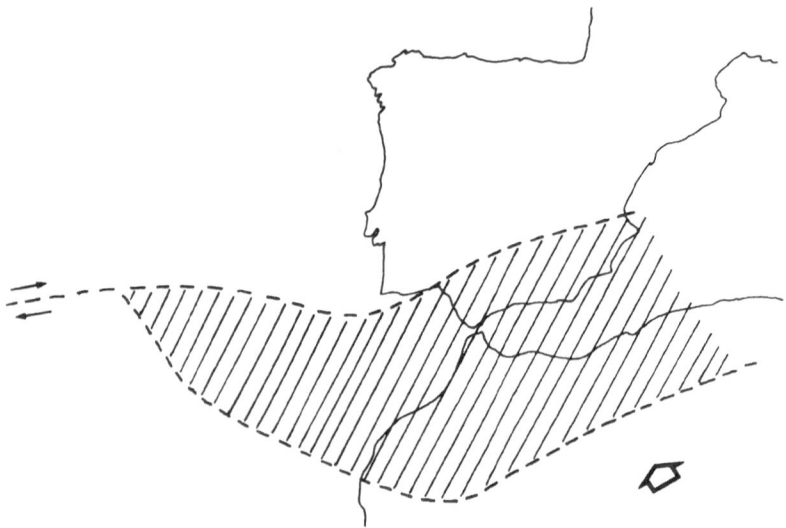

Fig. 5. Geodynamic setting for the middle Tortonian-Present of the broad zone of intermediate deformation between Iberia and Africa. Open arrow: sense of convergence between Iberia and Africa.

One site (Nr. 15 in Fig. 2 and Table 1) is located just north of the Crevillente fault zone This roughly 2 x 1 km block of Mesozoic limestone has evidently rotated (it is ambiguous whether clockwise or counterclockwise) at least 160°, considerably more than theoretically possible for a mosaic of fault-bounded blocks. We conjecture that it acted as a relatively small block swimming within a matrix of Triassic evaporites and Quartenary fault gouge in a megashear zone (De Smet, 1984).

Another site encountered in our review (located in the Sierra Gorda and studied by Ogg et al., 1984) exhibits westerly declinations which contrasts with the consistent easterly declinations from all other sites. Recent palaeomagnetic work in the area east of Granada (Platzmann, pers. commun, 1988) confirms that the Sierra Gorda block is rotated counterclockwise.The same investigation shows conclusively that this Sierra Gorda block is an exception: all the other measured sites exhibit clockwise declinations. At present we have no certain explanation for this counterclockwise-rotated block in a predominately dextral shear zone. Theoretically, some of the more complex models of block rotations discussed by Garfunkel and Ron (1985) show simultaneous but opposite rotations of adjacent blocks. According to these authors, the sense of movement depends on the original orientation of the faults bounding the blocks relative to the overall stress regime. In fact, Kamerling and Luyendyck (1985) found a few cases of counterclockwise-rotated blocks among predominantly clockwise-rotated blocks from another, well-known dextral strike-slip zone - the San Andreas Fault Zone in California.

In the Subbetic Zone we need more data to define the geometry, sense, and timing of the movements of the blocks in this region. In order to answer some of these questions several palaeomagnetic studies are being carried out, one on the Tertiary volcanics from the southeastern Iberia (Calvo et al., 1987). We hope that future work integrating palaeomagnetics, seismicity and structural geology will help contribute to a more detailed understanding of the region.

Aknowledgements

This work has been supported by the Direction General de Investigacíon Cientifica y Tecnologica, project PB 86.0431 C05.1. Publications nrs 307 (Catedra de Geofisica) and 573 (Institut für Geophysik).

References

Aldaya, F., Campos, J., Garcia-Dueñas, V., Gonzales Lodeiro, F. and Orozco, M. (1984) El contacto Alpujarrides/Nevado Filábrides en el vertiente meridional de Sierra Nevada: implicaciones tectónicas. In: El borde medtierraneo éspañol: Evolución del orógeno bético y geodinamica de las depresiones neógenas. Dept. Inv. Geol. C.S.I.C. Univ. Granada, Vol. 1, 18-22

Araña, V. and Vegas, R., 1974. Plate tectonics and volcanism in the Gibraltar Arc. *Tectonophysics*, **24**: 197-212.

Barranco, L.-M., 1986. Analisis de la fracturation en el sureste de España, implicaciones geodinamicas. Tesis de Licenciatura, Universidad Complutense de Madrid, pp. 190 (unpublished).

Bousquet, J.C., 1979. Quaternary strike-slip faults in southeastern Spain. *Tectonophysics* **52**: 277-286.

Calvo, M., Osete, M.L., Vegas, R. and Barranco, L.-M., 1987. Preliminary results of palaeomagnetic investigations in Tertiary volcanic rocks from the south-eastern part of the Iberian Peninsula. Abstracts. Servei Geologic de Catalunya, Barcelona, p. 3.

Dercourt, J., Zonenshain, L.P., Ricou, L.-E., Kazmin, V.G., Le Pichon, X., Knipper, A.L., Grandjacquet, C., Sbortshikov, I.M., Geyssant, J., Lepvrier, C., Pechersky, D.H., Boulin, J., Sibouet, J.-C., Savostin, L.A., Sorokhtin, O., Westphal, M., Bazhenov, M.L., Lauer, J.P. and Biju-Duval, B., 1986. Geological evolution of the Tethys belt from the Atlantic to the Pamirs since the Lias. In: Aubouin, J., Le Pichon, X. and Monin, A.S. (Editors) Evolution of the Tethys. *Tectonophysics*, **123**: 241-315.

De Smet, M.E.M., 1984. Wrenching in the external zone of the Betic Cordilleras, southern Spain, *Tectonophysics*, **107**: 57-79.

Garfunkel, Z. and Ron, H.,1985. Block rotation and deformation by strike-slip faults. 2. The properties of a type of macroscopic discontinuous deformation. *J. Geophys. Res.*, **90 (B10)**: 8589-8602.

Hermes, J. J., 1978. The stratigraphy of the subbetic and southern prebetic of the Velez Rubio-Caravaca area and its bearing on transcurrent faulting in the Betic Cordilleras of southern Spain. *Proc. K. Ned. Akad. Wet.*, **81(1)**: 41-72.

Jenny, J., 1984. Dynamique de la phase tectonique synsédimentaire du Jurassique moyen dans le Haut Atlas central (Maroc). *Eclogae geol. Helv.*, **77 (1)**: 143-152.

Kamerling, M.J. and Luyendyk, B.P., 1985. Paleomagnetism and Neogene tectonics of the Northern Channel Islands, California. *J. Geophys. Res*, **90 (B14)**: 12485-12502.

Mazaud, A., Galbrun, B., Azema, J., Enay, R., Fourcade, E. and Resplus, L., 1986. Données magnétostratigraphiques sur la Jurassique Supérieur et la Berriasien du NE des Cordilléres Bétiques. *C. R. Acad. Sc. Paris*, t. 302, Série II, **18**: 1165-1170.

Mäkel, G.H., Rondeel, H.E. and Vandenberg, J., 1984. Triassic palaeomagnetic data from the Subbetic and the Malaguide Complex of the Betic Cordilleras (*Southeast Spain*). *Tectonophysics*, **101**: 131-141.

Ogg, J. G., Steiner, M. B., Oloriz, F. and Tavera, J.M., 1984. Jurassic magnetostratigraphy, 1. Kimmeridgian-Tithonian of Sierra Gorda and Carcabuey, southern Spain, *Earth Planet. Sci. Lett.*, **71**: 147-162.

Ogg, J. G., Steiner, M. B., Company, M. and Tavera, J.M., 1988. Magnetostratigraphy across the Berriasian-Valanginian stage boundary (Early Cretaceous) at Cehegin (Murcia PRovince, southern Spain). *Earth Planet. Sci. Lett.*, **87**: 205-215.

Osete, M.L., Freeman, R. and Vegas, R., 1988. Preliminary palaeomagnetic results from the Subbetic Zone (Betic Cordillery, southern Spain): kinematic and structural implications. *Phys. of the Earth and Planet. Inter.* (in press).

Paquet, J., 1972. Charriages et coulissements dans l'Est des Cordillères Bétiques (Espagne), *Int. Geol. Congr.*, 24th, Montreal, **3**: 395-404.

Platt, J.P., 1986. Dynamics of orogenic wedges and the uplift of high-presssure metamorphic rocks. *Geol. Soc. Am. Bull.*, **97**: 1037-1053.

Schott, J.J., Montigny, R. and Thuizat, R., 1981. Paleomagnetism and potassium-argon age of the Messejana Dike (Portugal and Spain): angular limitation to the rotation of the Iberian Peninsula since the Middle Jurassic. *Earth Planet. Sci. Lett.*, **53**: 457-470.

Steiner, M.B., Ogg, J. and Sandoval, J. 1987. Jurassic magnetostratigraphy, 3. Bathonian-Bajocian of Carcabuey, Sierra Haran and Campillo de Arenas (Subbetic Cordillera, southern Spain). *Earth Planet. Sci. Lett.*, **82**: 357-372.

Udias, A. Espinosa, A.F. Mezcua, J. Buforn, E., Vegas, R., Nishenko, S.P., Martinez-Solares, J.M. and López-Arroyo, A., 1986. Seismicity and tectonics of the North African-Eurasian plate boundary (Azores-Iberia-Tunisia). *U.S. Geol. Surv. Open File Rep.* 86/626.

Vandenberg, J., 1980. New palaeomagnetic data from the Iberian peninsula, *Geol. en Mijnb.*, **59(1)**: 49-60.

Van de Fliert, J. R., Graven, H., Hermes, J. J. and De Smet M. E.M., 1980. On stratigraphic anomalies associated with major transcurrent faulting, *Eclogae geol. Helv.*, **73(1)**: 223-237.

Vegas, R., 1985. Tectónica del área Ibero-Mogrebf. In: Udías, A., Muñoz, D. and Buforn, E. (Editors) Mecanismo de los terremotos y tectónica. Publ. Univ. Complutense Madrid: 197-215.

Vegas, R., 1985. Tectonic model for the seismicity of the Ibero-Mahgrebian region, western end of the Africa-Eurasia plate boundary, (Abstract). *Annales Geophysicae Special Issue* EGS XII General Assembly, Bologna, p. 14.

Vegas, R. and Muñoz, M., 1984. Sobre la evolucion geodinámica del borde meridional de la Placa Ibérica. I Congr. Español de Geol. 3: 105-118.

Vegas, R., Fontboté,J.M. and Banda, E. 1980. Widespread Neogene rifting superimposed on alpine regions of the Iberian peninsula, Proc. EGS symposium "Evolution and Tectonics of the western Mediterranean and surrounding areas". *Inst. Geog. Nac. Madrid, Spec. Publ.,* **201**: 109-128.

FAULT BLOCK ROTATIONS IN OPHIOLITES: RESULTS OF PALAEOMAGNETIC STUDIES IN THE TROODOS COMPLEX, CYPRUS.

S. ALLERTON,
Centre des Faibles Radioactivités,
Laboratoire mixte C.N.R.S./C.E.A.,
Avenue de la Terrasse,
91198 Gif-sur-Yvette Cedex,
France.

ABSTRACT. Results of palaeomagnetic surveys in the Troodos ophiolite indicate the existence of major fault block rotations, about both sub-horizontal and steeply inclined net tectonic rotation axes, apparently related to deformation at a ridge and at a transform fault.
The palaeomagnetic results from dykes in the sheeted complex and the extrusives are analysed by a technique that restores the dykes to vertical, and the measured stable magnetisation to a previously deduced mean Troodos magnetisation directon (dec.= 276°; inc.=32°). The method produces two solutions for the initial dyke strike that are symmetric about the mean Troodos magnetisation direction, and two corresponding solutions for the net tectonic rotation that effected the site. The appropriate solution has to be selected from external criteria. It is assumed that dykes from a given area will have the same initial strike.
The Solea graben is a possible fossil axial valley that lies directly to the north of the Mount Olympos plutonic centre. The results of a palaeomagnetic survey across the structure give a best solution giving dykes a north-westerly initial strike and net tectonic rotation axes parallel to the initial dyke strike, and sub-horizontal. They are also parallel to the dominant fault trend in the area. These results are consistent with rotational normal faulting.
Palaeomagnetic and structural studies on the western flank of the Larnaca graben, at the eastern edge of the ophiolite, give similar results to those obtained in the Solea graben. The north-westerly initial dyke solution,with sub-horizontal rotation axes is preferred. The maximum recorded rotation is 115°, from flat-lying dykes. The sense of rotation is variable; both clockwise and anticlockwise about north-westely directed axes. Four east-west elongate domains with similar senses and degrees of dip can be identified, each bounded to the north and south by transfer faults. The main normal faults in this area trend north-west, parallel to the initial dyke strike, suggesting that the block rotations occurred by movement on a set of ridge-parallel normal faults.
The Lefkara area, in the eastern part of the ophiolite, immediatly to the north of the east-west trending Arakapas fossil transform fault, has dykes in both the sheeted complex, and the extrusives, with a north-easterly present-day strike. Palaeomagnetic evidence suggests that these had an initial north-westerly strike, and were rotated some 110° clockwise about steeply inclined axes. Cross-cutting dyke relationships suggest that dykes were intruded during this rotation. At least two generations of dominantly strike-slip faults can be identified. The rotations were apparently accommodated by movement on both macrofaults (spacing 1-2km) and mesofaults (spacing 1-10m), associated with dextral slip on the Arakapas transform fault.

1. Introduction

Studies of the Troodos ophiolite, Cyprus, suggest that much of the internal structure of the complex is primary, related to its formation and subsequent deformation at a spreading centre

C. Kissel and C. Laj (eds.), Paleomagnetic Rotations and Continental Deformation, 393–410.
© *1989 by Kluwer Academic Publishers.*

adjacent to a transform fault (Moores and Vine, 1971, Varga and Moores, 1985, Murton, 1986). It thus provides an ideal opportunity to investigate the nature of faulting and block rotation within such an environment.

Many of the structural features that are believed to exist in the oceans have also been recorded in ophiolites. The Josephine ophiolite in California has an oceanic sequence that is tilted with respect to its overlying (oceanic) sediments, and serpentinised faults that cut the crustal and mantle sequences (Harper, 1982). In the Troodos ophiolite, Cyprus, Verosub and Moores (1981) and Varga and Moores (1985) have identified extensional features including tilted fault blocks and detachment faults. These features all predate the emplacement of the ophiolite, and are apparently related to the spreading process. In the southern part of this ophiolite extreme extension has occurred, with tilted pillow lavas directly overlying tectonised harzburgites (MacLeod, in prep.) on a sub-horizontal serpentinised decollement.

Many authors have commented on the differences between ophiolites and "normal" oceanic crust (e.g., Miyashiro, 1973). Certainly the Troodos ophiolite exhibits geochemical features that seem to preclude a mid-oceanic setting for its formation, consistent with palaeogeographic reconstructions of Western Tethys (Robertson and Woodcock, 1980). A distinctive "subduction component" (Miyashiro. 1973, Pearce and Cann, 1973, Robinson, et. al., 1983) is apparently present in the geochemistry, although no island arc can be identified. Whilst the actual plate tectonic setting of the complex remains contentious, the internal structure is generally consistent with a spreading mode of formation (Moores and Vine, 1971, Cann, 1974).

In this paper the results of a combined palaeomagnetic and structural survey of the Troodos ophiolite are presented. I will restrict myself to a discussion of palaeomagnetic and structural evidence of the nature of faulting and block rotation at ridges and transform faults. The implications for the spreading structure of the Troodos ophiolite are not discussed in detail, as they are given elsewhere (Allerton, and Vine, 1987, Allerton, 1988, Allerton, in prep.a).

2. Crustal Structure of the Troodos ophiolite

In the Troodos complex a complete ophiolite (Penrose conference definition, Anon. 1972) stratigraphy is preserved. The gross structure of the complex is relatively simple (Gass, 1980, Figure 1), attesting to its gentle emplacement. The upper crustal lithologies have a metamorphic stratigraphy (Gass and Smewing, 1973) similar to that recorded in oceanic samples, and in contrast to many other (Alpine) ophiolites. Early after the formation of the complex, in the upper Cretaceous (mid-late Turonian, Blome and Irwin, 1985) the ophiolite was rotated as a coherent unit by 90° anticlockwise about a vertical axis (Vine and Moores, 1969, Clube and Robertson, 1986). In this paper the present day orientations of features are described.

The east trending Arakapas fault belt was first proposed to be a fossil transform fault by Moores and Vine (1971), and confirmed as such by Simonian and Gass (1978). Simonian and Gass (1978) also document the deviation of northerly striking sheeted dykes through northeast to east as the transform is approached from the north. They suggest that this feature may be formed by either intrusion in a sigmoidal stress field, or by some fault drag mechanism, both at a.dextral-slipping transform fault. Recently, comparison with present oceanic transforms has led Murton (1986), and Varga and Moores (1985), to propose that the feature was generated by dyke intrusion in a reorientated stress field at a sinistral-slipping transform. To the south of this transform fault is the Limassol Forest Complex, which is believed to have largely been formed within the transform zone (MacLeod, in prep., Murton, 1986).

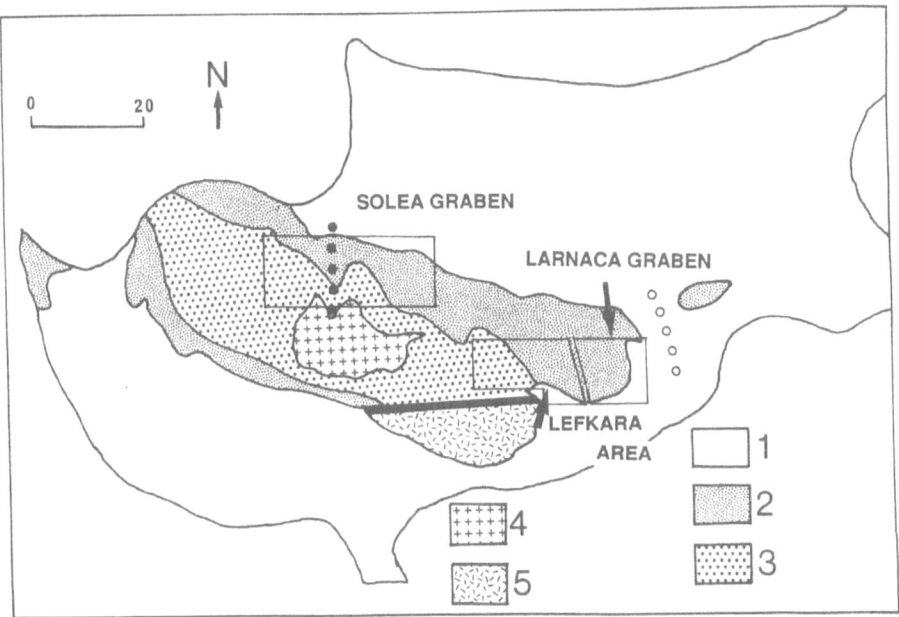

Figure 1. Troodos ophiolite, Cyprus: General structure and location of study areas. Key; 1= Sediments, 2=Pillow lavas, 3=Sheeted complex, 4=Plutonics, 5=Limassol Forest Complex. Heavy black line = exposed extent of Arakapas transform fault. Black dots = Solea graben axis, white dots = Larnaca graben axis.

3. Palaeomagnetic units

Gass and Smewing (1973) have shown that the boundary between the zeolite and greenschist facies corresponds approximately to the lithological contact between the extrusives and the sheeted units of the ophiolite. This metamorphic contact is also reflected in the magnetic mineralogy (Banerjee, 1980, Hall et al.,1987). Wang et al. 1984, have shown in their study of the CY-1 borehole, in the zeolite facies units, that the grains are generally clean, skeletal, titanomagnetites or titanomaghemites with a magnetic mineralogy consistent with an oceanic origin. They conclude that "the upper part of the Troodos ophiolite is a reliable guide to ocean crust magnetization." Demagnetization (both thermal and A.F.) of samples from the zeolite facies units generally gives extremely clean, simple results (Figure 2.). A viscous remanent component is occasionally present, although this is generally removed at a field of 10mT, or at a temperature of 300 °C.

The greenschist facies rocks frequently record a far more complex alteration history than their zeolite facies counterparts. Hall, et al. (1987) describe an alteration sequence from almost ubiquitous early deuteric alteration of titanomagnetites by exsolution to magnetite and ilmenite, followed by hydrothermal oxidation of the ilmenite to sphene. In extreme cases the grains exhibit complete recrystallisation. These grains are typically large ($\approx 50\mu m$), although the development of the lamelli seems to considerably influence the magnetic behaviour, giving pseudo-single domain

396

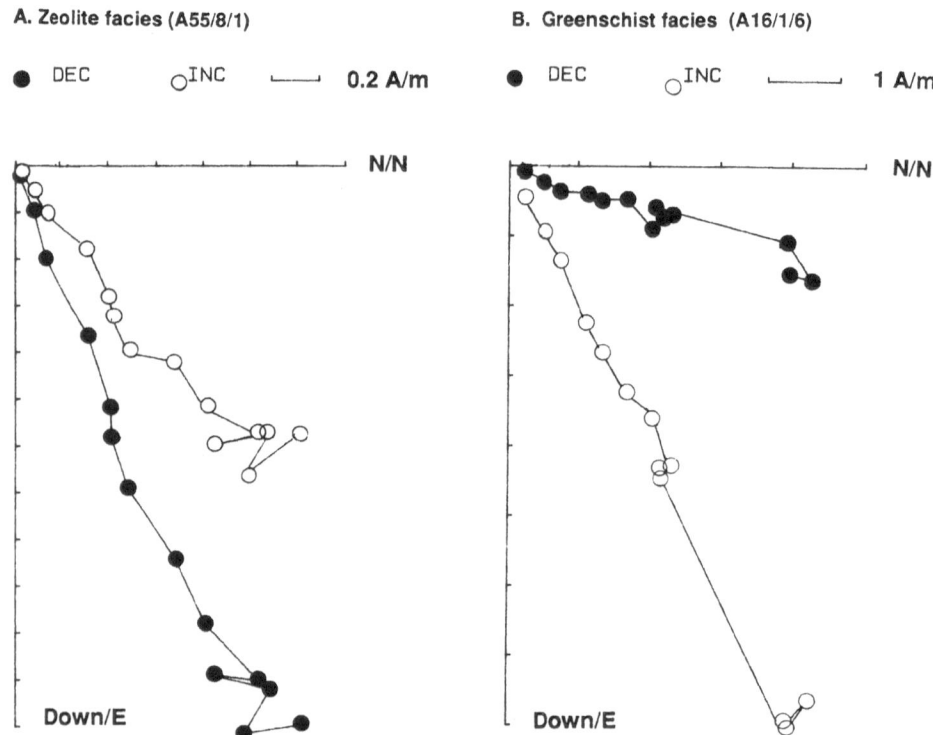

Figure 2. Zijderveld plots of thermal demagnetisation of samples from dykes in A) Zeolite facies and B) Greenschist facies.

characteristics. Whilst the magnetisation recorded within the greenschist facies thus probably contains subsequent C.R.M.s associated with hydrothermal alteration as well as N.R.M.s acquired on cooling, these were all probably acquired at the spreading axis, before fault blocks were rotated (Allerton and Vine, 1987, Allerton, 1988). Both thermal and A.F. demagnetisation give stable components after 15mT or 300°C (see Figure 2.). Bonhommet et al. (1988) also gain stable components from thermal demagnetisation.

4. Method of analysis

The method of analysis of the stable magnetisation vectors and the dyke orientation is that initially described by Allerton and Vine (1987), and more completely in Allerton (1988) and Allerton (in prep.b). This method assumes 1) that the primary magnetisation direction is known, 2) that the magnetisation predates the structure, 3) that dykes are intruded vertically, and 4) that deformation is essentially brittle, so that the angle between the dyke and the magnetisation direction is invariant throughout deformation.

The mean magnetisation vector of the complex is well known for the complex (Vine and Moores, in prep., dec. =276°, inc.=32°, Clube and Robertson, 1986, dec. =274°, inc.=36°.). A value: dec. =276°, inc.=32° is taken as the mean Troodos magnetisation vector for this survey. The above assumptions are justified in detail in Allerton and Vine (1987) and Allerton (1988).

This technique allows the initial dyke strike to be derived, giving two possible solutions, symmetric about the mean Troodos magnetisation direction. For each of these solutions a (typically obliquely inclined) net rotation axis and angle can be calculated. The correct solution generally has to be separated by external criteria, such as the local structure. Where pillow lavas exist at a site, they can be restored to their presumed original orientation for both models. Frequently this gives one reasonable (initial dip < 45°) and one unreasonable (initial dip > 60° or overturned) solution.

Where cross-cutting dykes exist at a site the palaeomagetic results can be separated into groups of similar dyke orientation and magnetisation direction. At these sites, where the intrusion order is known, the later rotation can be removed from the early rotation, to give the net tectonic rotation that occurred between the intrusion of the two dyke sets.

Use of this analytical technique facilitates a completely different sampling strategy to that normally employed. It gives best results for large rotations, as with small rotations errors in the data are magnified in the estimate of the rotation. In regional palaeomagnetic studies, sites are generally sampled from structurally simple locations, whilst this technique allows highly complicated folded and faulted terrains to be analysed.

5. Extensional deformation

5.1. SOLEA GRABEN

A palaeomagnetic survey of the Solea graben (Figure 1, Allerton and Vine,1987, Allerton, 1988) was carried out to investigate the assertion of Varga and Moores (1985), that the structure represented a fossil axial valley preserved by a ridge jump. This graben has an axis that trends due north from the plutonic complex centered on Mt. Olympus, parallel to the strike of the dykes of the sheeted complex. The brittle deformation of the upper crustal units is partitioned from the lower crust (isotropic gabbros and cumulates) by a major decollement, the Kakopetria detachment, at the top of the isotropic gabbros. The nature of this fault zone is disputable. Varga and Moores (1985), recognise mylonitic fabrics, whilst Hurst, pers. comm., considers the structure to be essentially brittle.

The sites were sampled in an approxiimately axis-perperndicular profile within the sheetred complex. The orientations of the mean site magnetisation vectors and poles to dykes is shown inset in Figure 3 and in Table 1. Two sites have magnetisations that may have been reset in the Troodos magnetisation direction after rotation, and are not considered further.

The data was analysed by the technique alluded to above. The results are sorted by similar initial strike, which seems appropriate for dykes formed at a single spreading axis. The solution with an originally northwesterly dyke strike is preferred, (Figure 3, Table 2.) as it generally gives lower angles of rotation than the model with a northeasterly original strike, and because the rotation axes cluster well, parallel to the dominant fault strike in the area.

The preferred model, with dykes that had an initial northwesterly strike, rotated about a dyke-parallel sub-horizontal axis is consistent with rotation by normal faults parallel to the ridge axis, indicated by the original dyke strike.

SOLEA GRABEN

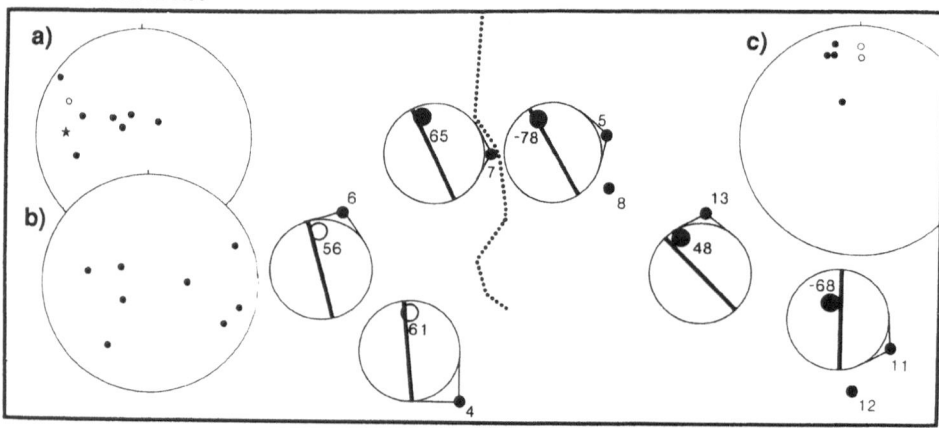

Figure 3. Results of palaeomagnetic survey in the Solea graben: Initial dyke orientation (lines in nets); mean site rotation vectors (open circles = upper hemisphere; solid circles = lower hemisphere, equal area projection) and angles of rotation (positive angle corresponds to anticlockwise rotation by faulting) for the solutions with a northwest-southeast original dyke strike (the preferred solution). Insets, A) mean stable site magnetisation vectors (star = mean Troodos magnetisation direction), B) present orientation of poles to dykes, and C) mean site rotation vectors (equal area projection).

Table 1. Cleaned (20 mT AF demagnetisation) mean site magnetisation vectors and mean dyke orientations.

Site	mean site magnetisation				mean dyke orientation			
	dec.	inc.	α_{95}	N	strike	dip	α_{95}	N
Solea Graben:								
S4	48.4	73.4	13.4	6	029.5	24.5 E	6.6	6
S5	294.8	-25.3	15.5	5	359.5	29.7 W	21.0	5
S6	334.0	71.5	18.5	5	011.5	48.3 E	4.9	5
S7	295.4	74.0	14.1	10	327.8	24.1 E	15.8	7
S8	288.8	41.8	23.6	4	016.1	75.4 W	8.8	4
S11	304.9	6.3	25.9	2	029.3	68.4 W	34.3	2
S11 (Sill)	318.9	26.2	18.2	2	336.0	56.0 E		1
S12	253.5	36.7	17.9	5	337.6	75.0 W	12.9	5
S13	302.8	64.2	13.0	6	304.7	58.5 E	10.4	6

Table 2. Axes and angles of rotation for both solutions.

| Site | NW solution | | | | NE solution | | | |
	dyke strike	dec.	rotation inc.	angle	dyke strike	dec.	rotation inc.	angle
Solea graben:								
S4	357.5	1.1	-15.8	70.1	014.5	350.9	8.9	73.8
S5	329.1	335.7	14.9	-73.6	042.9	249.0	-7.5	86.1
S6	347.8	357.7	-23.9	53.2	024.2	328.6	26.5	76.8
S7	001.5	343.0	21.0	65.1	019.9	334.1	35.6	85.8
S8	325.8	273.0	43.5	-74.3	046.3	295.2	23.8	45.0
S12	322.6	259.1	11.4	-44.8	049.5	279.0	45.5	82.7
S13	310.7	343.5	-0.8	38.3	061.3	290.8	46.8	153.7

5.2. LARNACA GRABEN

This north-west trending structure has an axis that lies in a sediment filled trough lying between the main part of the Troodos Complex, and the Trouli inlier to the east. The western flank of the graben is exposed on the far eastern edge of the main massif (Figure 1.). This area is approximately flat lying, presenting a plan view of the oceanic structure, which is unconformably overlain by untilted Cretaceous sediments to the east. Units from both the pillow lavas and the sheeted complex are exposed, and were sampled for palaeomagnetic study. 15 palaeomagnetic sites were sampled in both greenschist and zeolite facies. Samples were cleaned by AF treatment at 15mT. Dykes generally have an approximately northwesterly strike and a variable sense and degree of dip. They are occasionally approximately flat lying.

Table 3. Cleaned (20 mT AF demagnetisation) mean site magnetisation vectors and mean dyke orientations. (D = discarded)

| Site | mean site magnetisation | | | | mean dyke orientation | | | | |
	dec.	inc.	α_{95}	N	strike	dip		α_{95}	N
Larnaca graben: Zeolite facies.									
A3	263.9	-8.5	8.1	6	144.5	38.5	S	16.9	3
A20	270.9	6.0	22.2	3	333.4	68.3	W	12.4	3
A32	317.3	46.9	12.6	4	301.5	70.1	N	15.5	4
A49	271.3	-5.5	8.9	8	320.4	49.9	W	27.8	2
A54	323.8	57.9	5.4	9	330.0	45.0	E		1
A55	351.0	66.8	8.7	8	019.6	60.8	E	14.1	4
A57	342.9	23.9	11.4	9	328.0	31.8	S	31.9	3
Larnaca graben: Greenschist facies.									
A4	318.1	-32.4	8.4	7	331.3	7.3	N	6.9	7
A5	254.9	15.1	26.7	3	344.3	59.2	W	16.1	3
A6	43.4	62.8	12.4	8	277.4	21.1	S	13.9	8
A20	270.9	6.0	22.2	3	333.4	68.3	W	12.4	3
A24	312.9	-27.7	20.4	6	325.6	16.5	N	17.4	6
A48 Gp1	272.1	13.4	44.6	4	070.0	10.0	S	45.0	2 D
Gp2	289.1	10.7	18.2	3	330.0	50.0	W		1
A56	310.9	61.9	9.7	6	352.4	68.1	E	30.5	2

Stable remanent magnetisations (Table 3.) plot in both upper and lower hemispheres, dominantly in the northwest quadrant (Figure 4, inset a) of an equal area projection. The derived net tectonic rotations and original dyke strikes are shown in Table 4. and Figure 4. The solution for dykes with an initial northwest strike is preferred on the basis of the original pillow orientation. Of the 10 sites that have 2 solutions 8 give an initial pillow dip < 45° and 10 sites < 55° for the northwest dyke strike solution whilst the northeasterly solution gives 2 sites with a dip < 55° and the rest > 80° or overturned.

The northwest dyke solution is illustrated in Figure 4. The rotation axes are approximately sub-horizontal, with northwesterly declinations, sub-parallel to the initial dyke strike. Rotations of up to 115° are recorded. These results, similar to those obtained on the Solea graben, are interpreted as being due to rotations on normal faults that have the same trend as the dykes, and thus the spreading axis.

Table 4. a. Axes and angles of rotation for both solutions (Larnaca graben).

| Site | | NW solution | | | | NE solution | | | |
		dyke strike	rotation dec.	inc.	angle	dyke strike	dec.	rotation inc.	angle
A3		326.0	139.1	4.5	52.0	046.0	247.6	16.9	92.1
A4		312.8	331.2	17.6	-103.4	059.2	074.3	21.2	-93.9
A5	Gp1	328.4	102.9	-2.6	52.3	063.8	247.8	38.0	66.4
	Gp2	354.9	108.4	-60.5	98.8	017.1	256.1	74.3	-65.9
A20		343.4	194.6	14.3	27.6	028.6	257.8	21.0	86.5
A22		316.8	313.2	6.2	65.6	055.2	300.1	43.0	-179.3
A24		302.1	318.7	14.6	-114.5	069.9	70.5	19.5	-85.3
A32		281.0	150.7	41.8	-34.6	271.1	294.8	40.8	175.3
A48	Gp1	287.2	263.4	24.3	89.8	084.8	281.7	21.3	-105.2
	Gp2	302.2	313.6	32.8	-48.1	069.8	275.1	17.2	110.9
A49		324.9	158.1	0.4	41.5	047.1	257.8	14.6	105.4
A54		315.1	150.8	7.2	-45.4	046.9	301.1	41.3	139.5
A55		322.4	207.3	58.3	-71.4	049.6	323.8	29.5	89.6
A56		322.5	201.7	47.7	-44.8	049.5	305.1	37.8	101.2

Table 4 b. Axes and angle of rotation for groups with 1 solution only (Larnaca graben).

Site	Dyke strike and dip		Rotation dec.	inc.	angle
A6	006.0	73.4	E 170.4	7.6	-77.3
A57	006.0	73.1	E 307.0	6.5	103.5

Within the Larnaca graben the analysis gives a fairly consistent northwesterly, (dyke-strike parallel) sub-horizontal rotation axis for all the sites except A55, which has a steeper inclination. The angle of rotation is variable, with a maximum rotation of 115° being recorded at site A24. The senses of rotation are reversed for some of the sites. Sites with similar senses of rotation can be separated into "domains", analogous to those defined by Verosub and Moores (1981), although on a smaller scale. These domains are east-west elongate strips, and are separated by major east-west trending transfer faults (Gibbs,1984, Lister et.al. 1986).

401

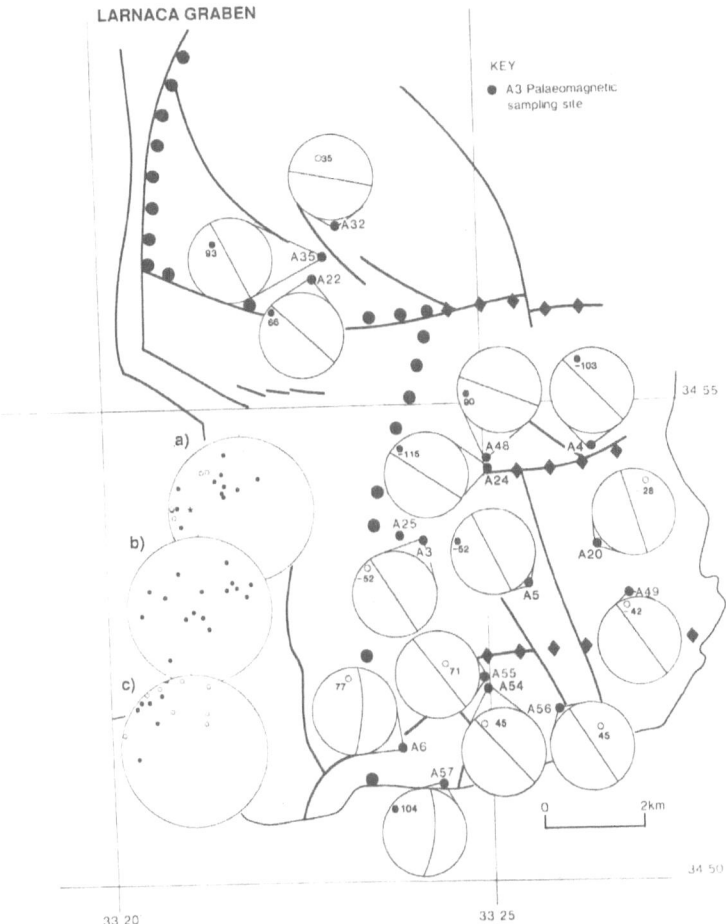

Figure 4. Results of palaeomagnetic survey in the Larnaca graben: Initial dyke orientation (lines in nets); mean site rotation vectors (open circles = upper hemisphere; solid circles = lower hemisphere, equal area projection) and angles of rotation (positive angle corresponds to anticlockwise rotation by faulting) for the solutions with a northwest -southeast original dyke strike (the preferred solution). Insets; A) mean stable site magnetisation vectors (star = mean Troodos magnetisation direction), B) present orientation of poles to dykes, and C) mean site rotation vectors (equal area projection).

5.3. STYLES OF EXTENSIONAL DEFORMATION

The locally extreme extensions of these grabens is in marked contrast to large areas of the complex, between the grabens, which are structurally extremely simple, having little deviation of dyke-dips from vertical, and regular dyke parallel features with throws of typically 100m (Boyle and Robertson, 1984). This suggests that the extension by faulting, rather than by dyke intrusion, is ephemeral in nature, rather than part of a steady state accretionary process.

Extensional domains can be identified from the palaeomagnetic sites and regional mapping. These domains are east-west elongate, with similar senses and degrees of rotation within them, and are bounded by east-west striking transfer faults. The widths of these domains is approximately 2-5km. Such features have frequently been identified within continental extension regimes (Gibbs, 1984, Lister, et al. 1987), although normally on a larger scale. The existence of these domains with opposite senses of rotations, and similar internal rotations also suggest that the extensional event was relatively rapid, effecting the whole of the graben area simultaneously. The major faults in the area can be identified where they correspond to a lithological contrast, although additional major faults may not have been observed. Thus the estimate of the typical spacing of the normal faults, between 1 and 2km, can be considered as a maximum. Transfer faults have also been identified within the extensional regime of the Eastern Limassol forest, where they have a spacing of some 5km (MacLeod, in prep.).

Within the Larnaca graben the basal decollement, if it exists, is not observed. The Kakopetria detachment, at the top of the isotropic gabbro is believed to act as a basal fault to the Solea graben (Varga and Moores, 1985), whilst in the Eastern Limassol Forest area a major detachment within the mantle has been recognised. Thus the depth to the decollement horizon is at about 3km for the Solea graben, and at about 5-7.5 km for the Eastern Limassol Forest.

The across strike extent of the Solea graben is ≈20km. The eastward extent of the extensional zone of the Larnaca graben is obscured by sediment cover, although the axis of the graben lies in the sediment filled trough to the east, giving it a "half width" of ≈10km. The width of the Eastern Limassol Forest extensional region is probably greater than these other two examples.

Continental extensional areas are typically of the order of 100-500km across (McKenzie, 1978). In comparison, oceanic and ophiolitic grabens are between 20-40km (Tapponier and Francheteau, 1978, Harper, 1985, Karson, in press, b, Varga and Moores, 1985), consistent with what might be expected from the thermal structure and the lithospheric stength. Whilst the graben widths differ by an order of magnitude the depths to decollement surfaces are relatively similar; 3-12km for ocean lithosphere (Harper, 1985) and ophiolites, and 10-20km for continental crust (Kusznir and Park, 1987). Continental graben are thus relatively "thin-skinned". Fault spacing, typically about 20km for continental crust (Barr, 1987), and 1-2km for oceanic/ophiolitic crust gives a similar ratio of fault spacing to graben width (approx. 1 to 10) for both.

Deformation is confined to the boundaries of oceanic plates because the lithospheric strength of oceanic crust increases dramatically with age, and therefore distance from the boundary. For example, 10 My old oceanic lithosphere has an extensional strength similar to that of stable continental cratons, whilst continental areas of active extensional tectonics have strengths comparable to oceanic lithosphere less than 3 My old (Kusznir and Park, 1987, Lynch and Morgan, 1987). Thus at a ridge with a slow spreading rate ($\approx 10^{-2}$m.y^{-1}) extensional tectonics may occur up to about 15km on each side of the ridge. In comparison, heat flow and lithospheric strength are broadly distributed in continental lithosphere.

6. Strike-slip deformation

6.1. LEFKARA AREA

This area, in the south-east part of the ophiolite, immediately to the north of the Arakapas fossil transform fault, includes the change in dyke strike adjacent to the transform fault.

Within this area 35 palaeomagnetic sites with 45 groups were sampled, in both greenschist and zeolite facies lithologies (Figure 5, Table 5). The palaeomagnetic results, after cleaning by AF demagnetisation at 15mT, are quite scattered, as are the present dyke orientations. Magnetisations close to the mean Troodos directon generally correspond to steeply inclined dykes with a northwesterly to northerly present strike, and are believed not to have been significantly rotated.

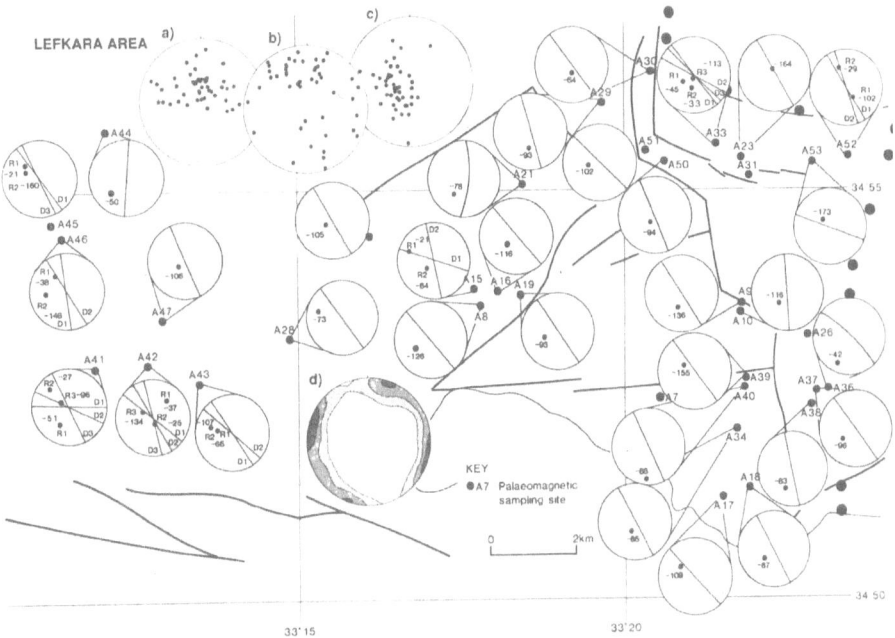

Figure 5. Results of palaeomagnetic survey in the Lefkara area: Initial dyke orientation (lines in nets); mean site rotation vectors (open circles = upper hemisphere; solid circles = lower hemisphere, equal area projection) and angles of rotation (positive angle corresponds to anticlockwise rotation by faulting) for the solutions with a northwest -southeast original dyke strike (the preferred solution). Where two groups that have been rotated exist for a site both intermediate rotations and both original dyke orientations are shown. The earliest dyke and rotation are labelled R_1 and D_1. Insets; A) mean stable site magnetisation vectors (star = mean Troodos magnetisation direction), B) present orientation of poles to dykes, C) mean site rotation vectors (equal area projection) and D) contoured equal area projection of all faults measured in the western area. Contour intervals are at 1%,2%,3% and 4% per 1% area. N=365.

The preferred solution for the analysis of the palaeomagnetic data from this area gives dykes with a northwesterly initial strike (Figure 5, Table 6). For the 10 sites with pillow lavas in this area, 9 give initial pillow orientations of less than 45° for this solution. The solution with a northeasterly initial strike the original pillow dips would all be greater than 70° or overturned. One site gives only one solution, with dykes with an initial 78° E dip, and a pillow tilt of 55°.

Table 5. Cleaned (20 mT AF demagnetisation) mean site magnetisation vectors and mean dyke orientations. (D = discarded)

Site	mean site magnetisation				mean dyke orientation				
	dec.	inc.	α_{95}	N	strike	dip		α_{95}	N
Lefkara area, early dykes: Zeolite facies.									
A10	61.1	70.3	14.2	6	294.7	46.9	S	11.9	6
A21	344.7	76.6	18.7	5	073.1	30.8	S	16.7	5
A23	70.2	33.6	13.0	4	314.7	85.6	N	10.6	3
A29 Gp1	53.1	57.0	11.8	4	075.1	37.1	S	20.1	4
Gp2	293.7	64.3		1	070.0	35.0	S		1 D
A30	348.9	45.8	8.9	5	022.9	70.6	E	14.2	5
A33 Gp1	29.8	30.3	12.8	2	070.0	90.0			1
Gp2	63.9	43.5	6.1	2	080.0	80.0	S		1
Gp3	102.9	49.4	13.6	5	325.0	77.4	S	24.1	2
A37 Gp3	345.1	74.8	9.6	4	040.0	65.0	S		1
A38	41.6	71.4	6.5	8	052.5	26.9	S	33.1	4
A39 Gp1	75.4	0.9	34.5	2	165.0	35.0	W		1 D
Gp2	13.3	68.8	11.2	7	288.8	71.4	S	34.5	2 D
Gp3	40.4	53.1	16.9	2	170.0	45.0	W	pillow	1 D
A40 Gp1	84.5	59.5	8.9	3	105.0	25.0	N	intrusive	1
A50	7.2	60.3	6.2	10	053.4	65.6	S	7.3	3
A51	350.6	59.6	14.7	6	048.5	85.0	S	15.3	2
A52 Gp1	46.4	49.8	16.4	4	085.0	65.0	S		1
A53	61.5	53.2	12.2	4	282.4	72.5	N	35.0	2
Lefkara area, early dykes: Greenschist facies.									
A7 Gp1	346.8	61.4	7.7	3	327.5	43.7	W	74.7	3 D
Gp2	26.8	83.4	28.5	7	023.5	23.8	W	15.3	7 D
A8	211.2	86.9	15.0	4	056.0	54.0	S	24.1	4
A9	92.6	41.4	13.2	5	290.6	38.7	S	11.8	5
A15 Gp1	356.0	62.0	28.9	4	005.4	83.7	E	10.2	4
Gp2	8.9	62.2	17.2	2	058.6	53.8	S	45.8	2
A16	29.8	71.4	8.7	5	054.0	63.8	S	8.3	5
A17	339.7	33.4	10.9	5	067.5	67.6	N	10.1	5
A18	10.6	67.4	4.3	3	052.2	55.6	S	22.0	4
A19 Gp1	318.6	60.1	19.0	3	045.0	80.0	S	21.6	2 D
Gp2	300.0	60.3	45.8	5	075.1	34.8	E	20.5	4 D
A27	12.1	57.0	7.5	5	069.8	73.1	S	11.8	5
A28	287.7	33.4	5.2	5	021.4	70.8	W	5.3	5
A34	8.9	59.2	8.1	6	061.7	60.9	S	9.3	6
A41 Gp1	354.0	26.4	15.5	2	065.0	70.0	N		1
Gp2	337.6	50.4	15.2	3	080.2	47.4	N	19.5	2
Gp3	4.9	17.8	16.0	2	050.0	42.5	N	10.9	2
A42 Gp1	346.1	85.4	11.4	3	281.0	53.9	S	49.2	2
Gp2	109.1	56.2	34.9	2	285.0	60.0	S		1
Gp3	343.1	63.8	4.0	4	282.2	57.5	S	58.3	2

Table 5, continued;

Site	mean site magnetisation				mean dyke orientation			
	dec.	inc.	α_{95}	N	strike	dip	α_{95}	N
A43 Gp1	261.7	70.4	32.4	2	280.0	40.0 S		1
Gp2	172.7	80.9	21.7	3	gabbro			
Gp3	303.6	50.8	21.2	3	047.5	80.0 S	10.8	2
A44	337.4	60.7	6.1	6	046.3	59.3 S	4.6	3
A45 Gp1	338.6	63.7	15.1	4	290.0	50.0 S		1
Gp2	359.3	70.5	15.7	5	290.0	69.3 S	6.2	3
A46 Gp1	120.4	66.4	16.0	2	No orientation			
Gp2	297.6	82.5	2.5	2	290.0	35.0 S		1
Gp3	273.5	59.1	5.9	4	043.7	72.5 S	30.4	2
A47	11.1	61.4	6.7	7	075.0	67.5 S	5.4	4
Lefkara area, late dykes.								
A36 Gp1	270.5	38.0	11.9	3	345.5	65.6 W	131.1	2
					(2 dykes 010 70 W and 140 65 W)			
Gp2	293.7	64.3		1	045.0	20.0 N	Sill	1
A37 Gp1	291.7	23.4		1	025.0	80.0 W		1
Gp2	267.3	47.8		1	335.0	80.0 W		1
A40 Gp2	237.3	30.0	15.5	3	025.0	45.0 W	pillow	1
A52 Gp2	292.7	14.7	31.3	2	005.0	70.0 W		1

Table 6. a. Axes and angles of rotation for both solutions.

Site		NW solution				NE solution			
		dyke strike	dec.	rotation inc.	angle	dyke strike	dec.	rotation inc.	angle
Lefkara area:									
A8		318.1	244.1	55.3	-125.9	053.9	2.2	12.6	57.4
A9		327.7	192.2	59.4	-136.1	044.6	4.1	3.8	106.6
A10		359.9	222.3	63.5	-115.9	012.1	208.2	55.8	-100.9
A15 Gp1		287.8	224.7	67.3	-83.8	084.2	318.2	36.1	114.0
Gp2		349.0	194.5	59.3	-75.8	023.0	162.4	12.8	-64.8
A16		319.2	240.9	66.8	-116.4	052.9	347.4	4.7	70.8
A17		317.6	306.7	57.8	-109.1	054.4	128.9	21.7	-60.0
A18		334.8	214.2	62.6	-86.6	037.2	339.7	6.6	70.0
A19 Gp2		325.2	262.8	58.4	-93.0	046.8	326.9	6.9	49.6
A23		330.2	347.8	82.5	-163.9	041.8	173.9	45.7	-115.8
A27		329.9	249.1	73.8	-105.1	042.1	158.7	12.2	-69.4
A28		321.6	280.7	39.7	-72.7	050.4	283.2	22.7	51.6
A29	Gp1	345.4	184.8	48.4	-92.8	026.6	349.4	0.3	87.0
A30		321.6	176.5	71.5	-64.2	050.4	315.6	20.8	103.3
A33	Gp1	317.2	103.6	88.5	-112.8	054.8	152.4	6.0	-103.5
	Gp2	298.1	229.2	83.1	-142.9	074.0	351.0	2.6	103.1
	Gp3	320.3	275.9	81.4	174.1	051.8	195.1	35.2	-112.6
A34		342.3	208.2	67.9	-84.6	029.7	159.1	10.9	-67.0
A35		330.5	308.3	38.9	92.9	041.5	282.9	50.8	-167.9
A37	Gp1	356.1	306.1	49.1	-35.3	015.9	065.0	35.3	-17.2
	Gp2	310.9	248.2	28.7	-42.1	061.1	280.9	42.4	101.6
	Gp3	321.4	237.4	59.7	-95.6	050.6	336.7	19.7	66.5

Table 6. a. continued

Site		NW solution				NE solution			
		dyke	rotation			dyke		rotation	
		strike	dec.	inc.	angle	strike	dec.	inc.	angle
A38		345.8	196.8	47.3	-82.5	026.2	350.3	4.0	72.7
A41	Gp2	271.7	297.1	52.1	-148.1	280.3	140.2	0.3	-61.6
	Gp3	291.8	336.2	61.3	-119.2	080.2	135.0	11.0	-99.5
A42	Gp1	328.4	166.2	60.5	-153.3	043.6	204.6	35.6	-67.6
	Gp2	309.4	237.3	73.5	-170.0	062.7	193.4	10.3	-93.4
	Gp3	348.6	281.3	60.1	-134.4	023.4	235.8	64.9	-83.2
A43	Gp1	326.6	269.7	51.2	-172.0	045.4	231.7	37.6	-68.4
	Gp3	309.8	276.8	49.4	-107.2	062.2	312.3	13.1	45.1
A45	Gp1	355.1	283.6	58.1	-140.0	016.9	263.4	63.0	-105.1
	Gp2	323.6	287.7	60.4	-160.3	048.4	196.9	49.4	-68.1
A46	Gp2	352.3	261.4	57.3	-142.2	019.7	237.7	52.5	-99.6
	Gp3	324.3	262.1	44.3	-112.1	047.7	355.1	14.6	28.2
A47		336.1	247.5	71.1	-106.0	035.9	075.8	22.3	-64.6
A50		336.0	230.2	70.6	-93.9	036.0	159.3	9.7	-66.1
A51		312.5	262.3	67.8	-101.9	059.6	327.6	10.2	71.5
A52	Gp1	330.4	204.8	75.2	-116.1	041.6	166.5	10.2	-88.9
	Gp2	343.1	343.7	45.8	-29.3	028.9	269.8	10.2	62.8
A53		285.1	313.4	72.7	-172.9	086.9	175.1	16.5	-89.5

Table 6 b. Axes and angle of rotation for groups with 1 solution only.

Site		Dyke			Rotation		
		strike	and	dip	dec.	inc.	angle
A21	Gp1	006.0		75.2 E	218.2	53.2	-78.3
A44		006.0		89.9 E	235.1	42.2	-50.0

Results of the analysis indicate about 100° of clockwise rotation about a steeply inclined axis (mean declination = 240°; inclination = 66°; α_{95} = 31°). The large variation in the angle of rotation may be related to either a highly inhomogeneous rotation of the area, or dyke intrusion during the rotation. The later is preferred, as cross-cutting dykes at various sites, i.e. A33, have different magnetisation vectors. The early cross-cutting dykes generally appear in the first 50° of rotation, in contrast to the later set of cross-cutting dykes which have not been rotated. In this case the mean angle of rotation probably represents an underestimate of the total rotation of the area.

The orientation of mesofaults (faults visible at outcrop with a throw of less than 5m, Hancock, 1985), slickensides, mineral growths and other kinematic indicators for sites in the western part of the area are shown inset in Figure 5d. The faults are generally steep, with fault lineations indicating a strike-slip sense of motion. The strikes of the faults are variable over the area as a whole, although at individual sites they are more consistent. The fault sets are largely consistent with rotation about steeply inclined axes as described by the palaeomagnetic results.

Within this area, particularly close to the transform fault, brittle fracturing is so intense that fault rocks (generally breccias) are the dominant lithologies. Larger scale faults also occur, and have been identified by lithological mapping. An early, northeast-southwest striking set, mineralised, and with a normal component of motion is cut by a later, cross-cutting set, striking north to north-north-west. These early faults appear to have a spacing of 1-2km.

6.2. FAULT BLOCK ROTATION

The Lefkara area exhibits significant block rotations about inclined axes for about 10km away from the transform fault. Dykes were intruded during this rotational episode, suggesting that this rotation occured adjacent to the ridge axis. In the Bay of Islands ophiolite (Karson, in press, a) the region of ductile shear associated with the transform fault is between 5 and 10km wide, similar to the dimensions of the brittle deformation of the Troodos ophiolite. The thermal conditions of the ridge transform intersection considerably effect the rheology. If the transform has a moderate (>20km) offset then the southern wall adjacent to the ridge-transform intersection will be relatively cold and rigid, and act as a stable barrier to ductile deformation, which would be confined to the northern plate. As this plate moves along the transform, it would progressively cool, and thicken, giving it a more brittle character.

McKenzie and Jackson, 1983, suggest that in a zone of distributed deformation fault blocks rotate in response to viscous fluid flow at depth, describeable by fluid continuum mechanics. If their model is applied to the situation described above, the amount of slip along the transform fault can be calculated independent of the spreading rate, giving 110° of rotation over 20km of fault movement. Rotated dykes occur all along the transform fault (some 50km), with no observable increase in the amount of rotation, suggesting that this is a relatively steady state phenomena, and not the result of a single rotational episode. Thus blocks were rotated, with contemporaneous dyke intrusion, close to the ridge, where the lithosphere was warm and weak enough to deform. When the lithosphere had cooled sufficiently the slip was accommodated entirely within the transform valley.

The rotations that affected the area were about inclined, and not purely vertical axes. An implication of this is that the slip vector cannot be parallel to the transform fault, as the rotation requires a component of extension or compression. The mean rotation axis derived from the palaeomagnetic results is similar to the mean fault intersections (corresponding to estimates of the σ_2 direction. Hancock, 1985) from the early fault set (dec.=222°; inc. = 78°, and from fault lineations on all the sets (dec. = 225°; inc. = 72°). Such a relationship is reminiscent of simple shear, although it is not clear how such a mechanism could generate the large degrees of rotation observed. The development of at least two fault generations in a region that has been rotated by about 110° is consistent with the domain theory proposed by Nur et al, 1986, which predicts that a single fault set may only accommodate some 45° of block rotation. Where sites which have more than one group of dykes have been sampled, the intermediate rotations between the intrusion of the first and the second dyke (early rotation), and after the intrusion of the last dyke (late rotation) can be calculated. The mean axis of early rotation (dec. = 293°; inc. = 78°; α_{95} = 31°) is similar to the mean late rotation (dec. = 271°; inc. = 65°; α_{95} = 17°), and both similar to the total mean rotation axis. This suggests that the fault blocks rotated about a fairly constant axis.

The nature of the fault block rotations appears paradoxical. The area is cut by major faults which occasionally have considerable offsets, and apparently a long history of movement, and yet it is not clear how these faults can generate the observed rotations. The areas between the major faults are themselves highly fractured, and locally within the sheeted dykes fault breccia becomes the dominant rock type. The rotation axes are approximately parallel to the intersection of the local fault sets, and to the no-movement direction on the fault planes. It is suggested that the rotations were accommodated by both rotation on major oblique/normal slip faults with relatively long active life, and by internal deformation of the fault bounded blocks by minor slip on a pervasive set of mesofractures.

7. Conclusions

Areas of extreme extension by faulting have been observed in the Troodos ophiolite, where large (locally >90°) rotations have been recorded by palaeomagnetics about sub-horizontal axes, approximately parallel to the initial dyke strike. Associated structural features include originally sub-horizontal detachment faults, and transfer faults that separate opposite dipping domains.

Whilst the nature of the extensional deformation of the ophiolite parallels its continental counterparts, the scales are clearly different. There is approximately an order of magnitude difference in the horizontal dimension, but only a factor of about two in the vertical, giving ophiolitic/oceanic graben a relatively thick-skinned geometry when compared to continental crust. This is probably due to the rapidity of lateral changes of geothermal gradient in oceanic/ophiolitic lithosphere during extension by faulting, relative to continental graben.

In the Lefkara area palaeomagnetics indicate major block rotations about steeply inclined axes associated with dextral motion along the east-west trending transform fault. These rotations occur in a band about 10km wide parallel to the transform. If a model of brittle faulting above a ductilely deforming shear zone is invoked, then some 20km of movement parallel to the transform fault is required to generate the observed rotations. This is probably part of a continuous process of rotation at a ridge-transform intersection. After a point on the plate adjacent to the transform has moved 20km away from the ridge the plate cools, and its strength increases such that brittle failure is confined to the transform trace.

A relationship between the local mesofault systems, and the rotation axes is noted, which may indicate that internal deformation of fault blocks is an important mechanism for rotation.

Acknowledgements

The author wishes to express his gratitude to F.J.Vine, whose supervision, encoragement and advice were invaluable. Thanks also to R.Walcott, for useful comments on an early draft of this paper. The work was funded by a NERC grant, held at the School of Environmental Sciences, University of East Anglia, Norwich, England.

References

Allerton, S., 1988, Palaeomagnetic and structural studies of the Troodos ophiolite, Cyprus. *Ph.D. Thesis, University of East Anglia, Norwich, England.*

Allerton, S., *in prep,a,* Paleomagnetic and structural studies in the Troodos Complex. *in Proceedings of the Symposium on Ophiolites and Oceanic Lithosphere-TROODOS 87.*

Allerton, S., *in prep.b.* The derivation of the net tectonic rotation from palaeomagnetic data.

Allerton, S., & Vine, F.J., 1987, Spreading structure of the Troodos ophiolite, Cyprus: Some paleomagnetic constraints. *Geology,* 15, 593-597.

Anon. , 1972, Penrose field conference, Ophiolites; *Geotimes,* 24-25.

Banerjee, S.K., 1980, Magnetisation of the oceanic crust: evidence from ophiolite complexes. *Journal of Geophysical Research*, **85**, 3557-3566.

Barr, D., 1987, Lithospheric stretching, detached normal faulting and footwall uplift. *in Coward, M.P., Dewey, J.F., & Hancock, P.L., Continental extensional Tectonics, Geological Society Special Publication No. 28*, **28**, 53-65.

Blome, C.D., & Irwin, W.P., 1985, Equivalent radiolarian ages from ophiolitic terrains in Cyprus and Oman. *Geology*, **13**, 401-404.

Bonhommet, N., Roperch, P., & Calza, F., 1988, Paleomagnetic arguments for block rotations along the Arakapas fault (Cyprus). *Geology*, **16**, 422-425.

Boyle, J.F., & Robertson, A.H.F., 1984, Evolving metallogenesis at the Troodos spreading axis, *in Gass, I.G., Lippard,S.P., and Shelton, A.W., eds., Ophiolites and oceanic lithosphere: Geological Society of London Special Publication* **13**, 169-181.

Cann, J.R., 1974, A model for oceanic crustal structure developed. *Geophysical journal of the Royal Astronomical Society*, **39**, 169-187.

Clube, T.M.M., & Robertson, A.H.F., 1986, The paleorotation of the Troodos microplate, Cyprus, in the Late Mesozoic-Early Cenozoic plate tectonic framework of the Eastern Mediterranean. *Surveys in Geophysics*, **8**, 375-437.

Gass, I.G., 1980,The Troodos massif: Its role in the unravelling of the ophiolite problem and its significance in the understanding of constructive plate margin processes, *in Panayiotou, A., ed., Ophiolites, Proceedings , International ophiolite symposium, Cyprus, 1979: Nicosia, Cyprus Geological Survey Department*, 23-35.

Gass, I.G., & Smewing, J.D., 1973, Intrusion, extrusion and metamorphism at constructive margins: Evidence from Troodos massif, Cyprus. *Nature*, **242**, 26-29.

Gibbs, A.D., Structural evolution of extensional basin margins. *Journal of the Geological Society of London*, **141**, 609-620.

Hall, J.M., Fisher, B.E., Walls, C.C., Hall, S.L., Johnson, P.H., Bakor, A.R., Agrawal, V., Persaud, M., & Sumaiang, R.M., 1987, Vertical distribution and alteration of dikes in a profile through the Troodos Ophiolite. *Nature* , **326**, 780-782.

Hancock, P.J., 1985, Brittle microtectonics: Principals and practice. *Journal of Structural Geology*, **7**, 437-457.

Harper, G.D., 1982, Evidence for large scale block rotations at spreading centers form the Josephine ophiolite. *Tectonophysics*, **82**, 25-44.

Harper, G.D., 1985, Tectonics of slow-spreading mid-ocean ridges, and consequences of a variable depth to the brittle-ductile transition. *Tectonics*, **4**, 395-409.

Karson, J.A., *in press* (a). Factors controlling the orientation of dykes in ophiolites and oceanic crust. *in. Hall, H.C., & Fahrig, W.F., Mafic Dyke Swarms, Geological Association of Canada, Special paper, 33*.

Karson, J.A., *in press* (b). Seafloor spreading on the Mid-Atlantic Ridge: Implications for the structure of ophiolites and oceanic lithosphere produced in slow-spreading environments. *in Proceedings of the Symposium on Ophiolites and Oceanic Lithosphere-TROODOS 87*.

Kusznir, N.J., & Park, 1987, *in Coward, M.P., Dewey, J.F., & Hancock, P.L., Continental extensional Tectonics, Geological Society Special Publication No. 28*, **28**, 53-65.

Lister, G.S., Etheridge, M.A., & Symonds, P.A., 1986, Detachment faulting and the evolution of passive continental margins. *Geology*, **14**, 246-250.

Lynch, H.D., & Morgan, P., 1987, The tensile strength of the lithosphere and the localisation of extension. *in Coward, M.P., Dewey, J.F., & Hancock, P.L., Continental extensional Tectonics, Geological Society Special Publication No. 28*, **28**, 53-65.

MacLeod, C.J.,1988, Upper Cretaceous evolution of the Eastern Limassol Forest Complex: the role of the Southern Troodos microplate. *in Troodos 87 Symposium; Ophiolites and oceanic lithosphere.* In prep.

McKenzie, D.P., 1978, Active tectonics of the Alpine-Himalayan Belt: the Aegean Sea and surrounding regions. *Geophysical Journal of the Royal Astronomical Society*, **55**, 217-254.

McKenzie, D.P., & Jackson, J.A., 1983. The relatonship between strain rates, crustal thickening, paleomagnetism, finite strain and fault movements within a deforming zone. *Earth & Planetary Science Letters*, **65**, 182-202, and Correction to the above, **ibid**, 1984, **70**, 444.

Miyashiro, A., 1973, The Troodos ophiolite was probably formed in an island arc. *Earth & Planetary Science Letters*, **19**, 218-224.

Moores, E.M., & Vine, F.J., 1971, The Troodos Massif, Cyprus and other ophiolites as oceanic crust: Evaluation and implications. *Philosophical transactions of the Royal Society of London, Series A*, **268**, 443-466.

Murton, B.J., 1986, Anomalous oceanic lithosphere formed at a leaky transform fault: evidence from the Western Limassol Forest Complex, Cyprus. *Journal of the Geological Society of London*, **143**, 845-854.

Nur, A., Ron, H., & Scotti, O., 1986, Fault mechanics and kinematics of block rotation.*Geology*, **14**, 740-749.

Pearce, J.A., & Cann, J.R., 1971, Ophiolite origin investigated by discriminant analysis using Ti, Zr and Y. *Earth & Planetary Science Letters*, **12**, 339-349.

Robertson, A.H.F., & Woodcock, N.H., 1980, Tectonic setting of the Troodos massif in the east Mediterranean. *in Panayiotou, A., ed., Ophiolites, Proceedings , International ophiolite symposium, Cyprus, 1979: Nicosia, Cyprus Geological Survey Department*, 36-49.

Robinson, P.T., Melson, W.G.,O'Hearn, T. & Schminke, H.-U., 1983, Volcanic glass compositions of the Troodos ophiolite, Cyprus. *Geology*, **11**, 400-404.

Simonian, K., & Gass, I.G., 1978, Arakapas fault belt, Cyprus: A fossil transform fault. *Bulletin of the Geological Society of America*, **89**, 1220-1230.

Tapponier, P., & Francheteau, J., 1978, Necking of the lithosphere and the mechanics of slowly accreting plate boundaries. *Journal of Geophysical Research*, **83**, 3955-3970.

Varga, R.J., & Moores, E.M., 1985, Spreading structure of the Troodos ophiolite, Cyprus. *Geology*, **13**, 846-850.

Verosub, K.L., & Moores, E.M., 1981, Tectonic rotations in extensional regimes and their paleomagnetic consequences for oceanic basalts. *Journal of Geophysical Research*, **86**, 6335-6349.

Vine, F.J., & Moores, E.M., 1969, Paleomagnetic results from the Troodos igneous massif. *EOS*, **50**, 131.

Wang, B.-X., Walls, C., & Hall, J.M., 1984, Cyprus drillhole CY-1: Oxide petrography, magnetic properties and alteration in a section through the uppermost half kilometer of Troodos type oceanic crust. *Publication no. 7, Center for Marine Geology, Dalhousie University*.

PALEOMAGNETISM IN SE ASIA: SINISTRAL SHEAR BETWEEN PHILIPPINE SEA PLATE AND ASIA.

M. FULLER, R. HASTON and E. SCHMIDTKE.
Department of Geological Sciences,
Univ. of California,
Santa Barbara, CA., 93106
USA.

ABSTRACT. Paleomagnetic data from the Philippine Sea Plate indicate clockwise rotation since Oligocene time. This has rotated the Central Basin Spreading Center from being parallel to the Asian shoreline to its present NW/SE strike. The initial configuration is consistent with a back-arc origin of the West Philippine Sea Province. The subsequent clockwise rotation has defined a sinistral shear zone between the Philippine Sea Plate and Asia.
The Philippines lie at present in the region of sinistral shear. Within a substantial part of Luzon, paleomagnetic results reveal a Late Miocene/ Pliocene clockwise rotation compatible with Philippine Sea Plate motion. Prior to that, this same region exhibited a period of rapid counterclockwise rotation from about 20 to 10 Ma. which is incompatible with Philippine Sea plate motion. This counterclockwise rotation coincided with the collision of the Philippine archipelago with the Palawan microcontinent, as it was carried south by sea floor spreading in the South China Sea. Between Late Eocene and Early Miocene, additional counterclockwise rotation was accompanied by northward motion of the region. It therefore appears that at least substantial parts of Luzon had an origin on a counterclockwise rotating plate, or microplate to the west of the Philippine Sea Plate.

1. Introduction.

The active tectonics of SE Asia have attracted generations of geologists as a modern analogue for ancient tectonic processes. With the advent of plate tectonics, the region serves as a model for accretionary plate margins with the associated complexity of subduction and collision. Among the most important is the recognition of the role of indenter tectonics (Molnar and Tapponnier, 1975), which has had a profound effect on the understanding of the tectonic features throughout the region. Another key aspect is the interpretation of the sea floor spreading of the South China Sea (Taylor and Hayes, 1983 and Pautot et al., 1986). Our work has been centered around the South China Sea so that eventually the motion of the various units can be tied back through the known history of South China Sea motion to the Asian landmass. In this paper, the principal discussion concerns the Philippines and the Philippine Sea Plate which can be regarded as forming an eastern boundary to SE Asia. We present paleomagnetic results from the Philippines and from islands on the Philippine Sea Plate to establish plate and microplate motions in the region.

2. Philippine Sea Plate.

The history of Philippine Sea Plate motion is poorly understood, despite considerable shipborne geology and geophysics. Even determination of its present motion is difficult because it is

C. Kissel and C. Laj (eds.), Paleomagnetic Rotations and Continental Deformation, 411–430.
© *1989 by Kluwer Academic Publishers.*

412

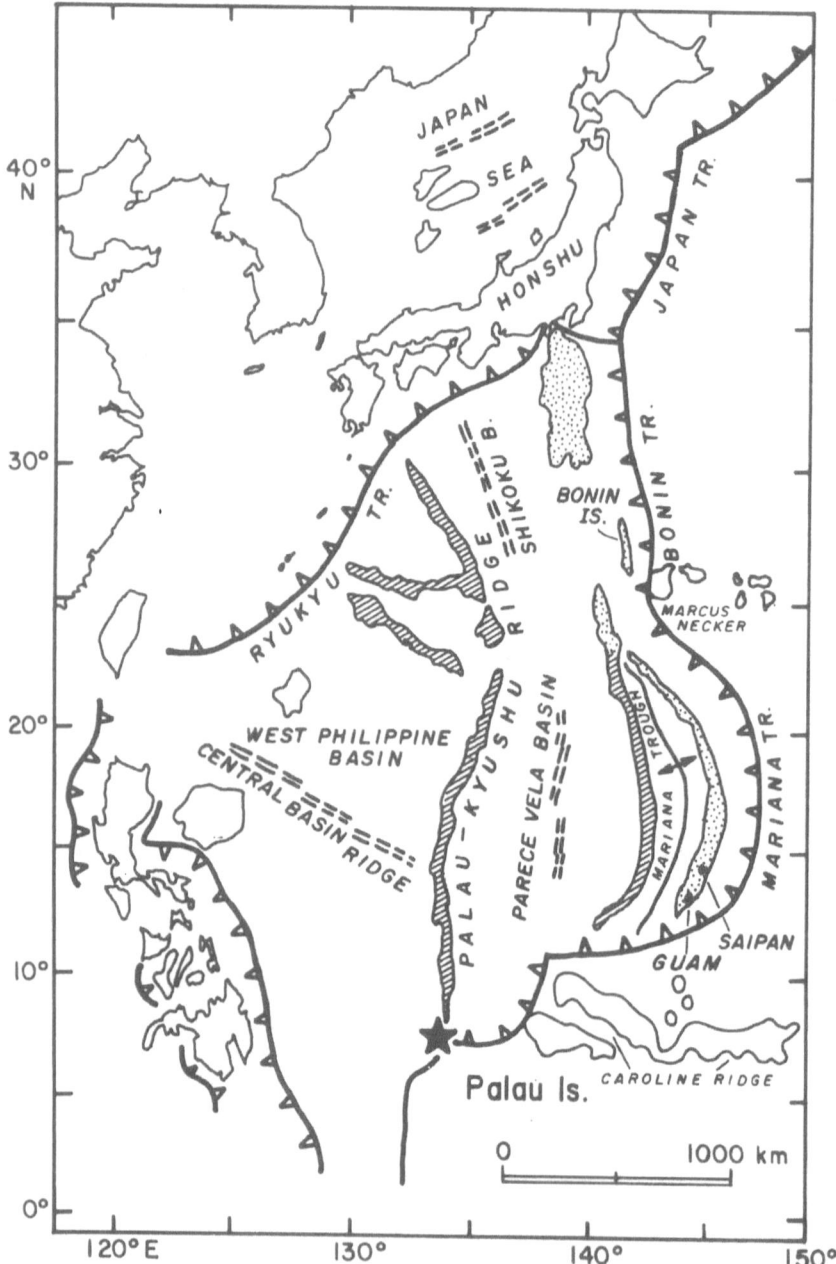

Fig. 1. Major features of the Philippine Sea Plate (adapted from Keating and Helsley, 1985).

surrounded by subduction zones. Nevertheless relative to Asia, a clockwise rotation of 1.5 to 2.0 degrees per million years about a pole in the vicinity of Japan has been proposed by Ranken et al., (1984). The plate consists of two distinct tectonic provinces, the east and west Philippine Sea provinces (Figure 1). In the West Philippine Sea Province (WPSP) the sea floor anomalies are approximately east- west and spreading has been dated between approximately 60 and 35 Ma. To the east in the East Philippine Sea Province (EPSP), anomalies are approximately north - south and episodes of spreading and arc volcanism have continued until the present.

There are two principal models of the origin of the Philippine Sea Plate (Figure 2a). Uyeda and Ben-Avraham (1972), Hilde et al., (1977) and Hilde and Lee (1984) all suggest that the West Philippine Sea Province is trapped Pacific Plate crust, isolated from the Pacific Kula spreading system when the Pacific plate motion changed at about 44 Ma. The trapping was facilitated by the presence of north - south transforms which permitted the initiation of west dipping subduction. In contrast, Lewis ET AL., (1982) suggested that the WPSP originated by back arc spreading behind a northeast dipping subduction zone (Figure 2b). Karig (1971,1975) first suggested that the EPSP is a series of remnant arcs and back arc basins. Since then other authors have proposed additional models, but the concept of arc splitting and back arc spreading is central to most. While the origin of the younger EPSP is therefore commonly agreed upon, the origin of the WPSP has emerged as a central unresolved problem of the Asian accretionary margin. Paleomagnetism affords a possible means of resolving this question and so we have applied it to try to establish the history of Philippine Sea plate motion.

New paleomagnetic results from the Philippine Sea Plate. The Palau Islands are located at the southern end of the Palau-Kyushu ridge. They provide the only geologic record of the WPSP from the Eocene to the present, so that they afford a unique paleomagnetic opportunity. They are composed of island arc sub-aerial and submarine volcanic rocks and reef limestones ranging in age from the Eocene to Holocene. The stratigraphic relations between units was established by Mason et al., (1956) as illustrated in Figure 3. Preliminary age determinations were carried out by Meijer et al., (1983).

A total of 278 samples were collected from 24 sites in lavas, dykes and tuffs, on Babeldaup, Koror, Arakabesen and Malakal islands. Each sample was progressively demagnetized and measured using a Molspin fluxgate spinner magnetometer. For the most part AF demagnetization was used, but thermal demagnetization was carried out in addition. Twelve sites give well defined stable directions, readily selected using equal area and Zijderveld orthogonal plots. As described in more detail elsewhere (Haston et al., 1988), the results, which are primarily from flows and dykes, satisfy a reversal test and a modified attitude test. They yield a particularly clear picture of clockwise rotation of between 60 and 70° since the Oligocene (Figure 4). Despite the low latitude of the site, which could give rise to an equatorial ambiguity, the sense of rotation can be obtained by comparing the Miocene and Oligocene results. It is less clear what happened in Early Oligocene and Late Eocene times. Possibly the islands experienced a period of rapid clockwise rotation in the Early Oligocene associated with initial rifting in the Parece-Vela Basin. The interpretation of the Late Eocene result remains unclear, although we favour remagnetization in Oligocene times. In view of the difficulties in the interpretation of results of units older than Mid-Oligocene, we consider principally, the tectonic implications of results from the younger well established data. The paleomagnetic rotations appear to be accompanied by slight northward motion, although it is presently at the limit of our resolution. These results are consistent with, but extend the earlier results of Aoki (1972).

414

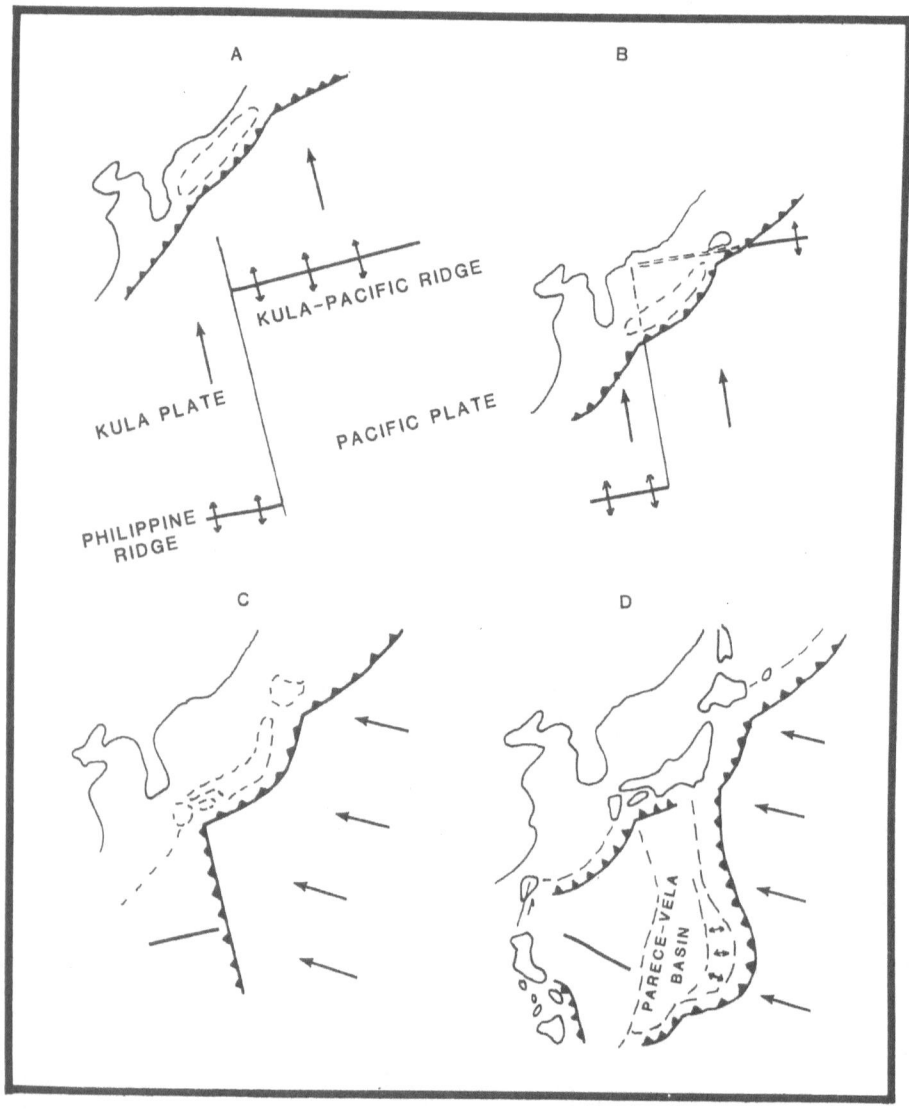

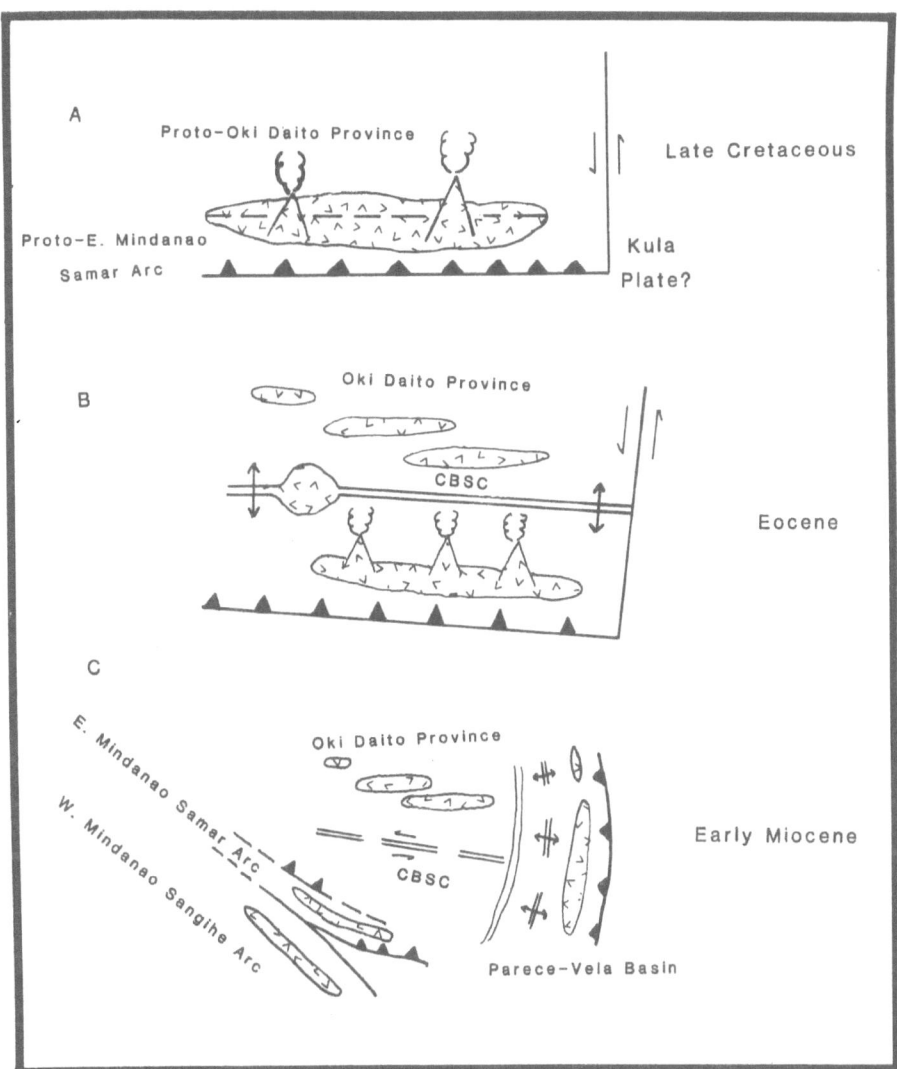

Fig. 2. Models for the origin of the Philippine Sea Plate , (a) Uyeda and Ben-Avraham (1972), (b) Lewis et al., (1982).

416

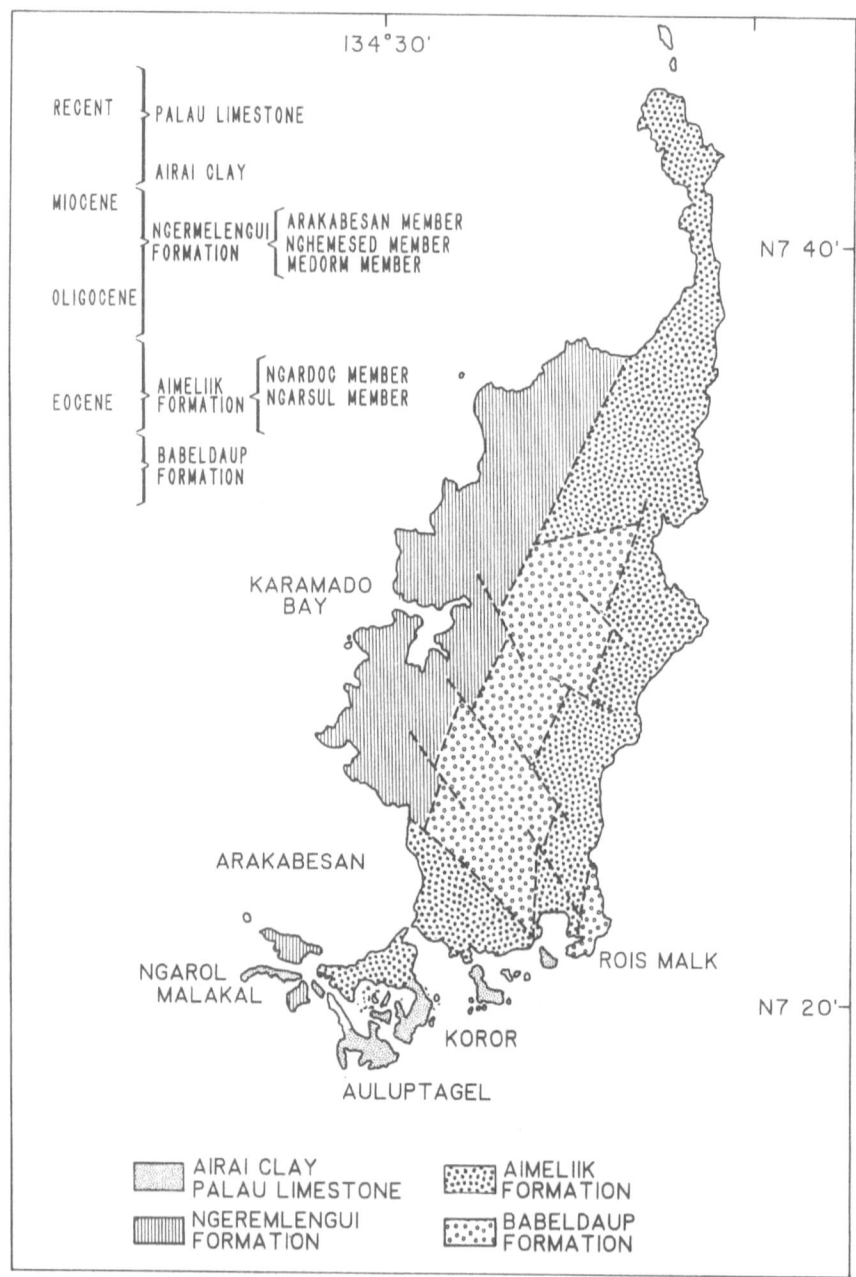

Fig. 3. Generalized geological map of the Palau Islands (after Mason et al., 1956).

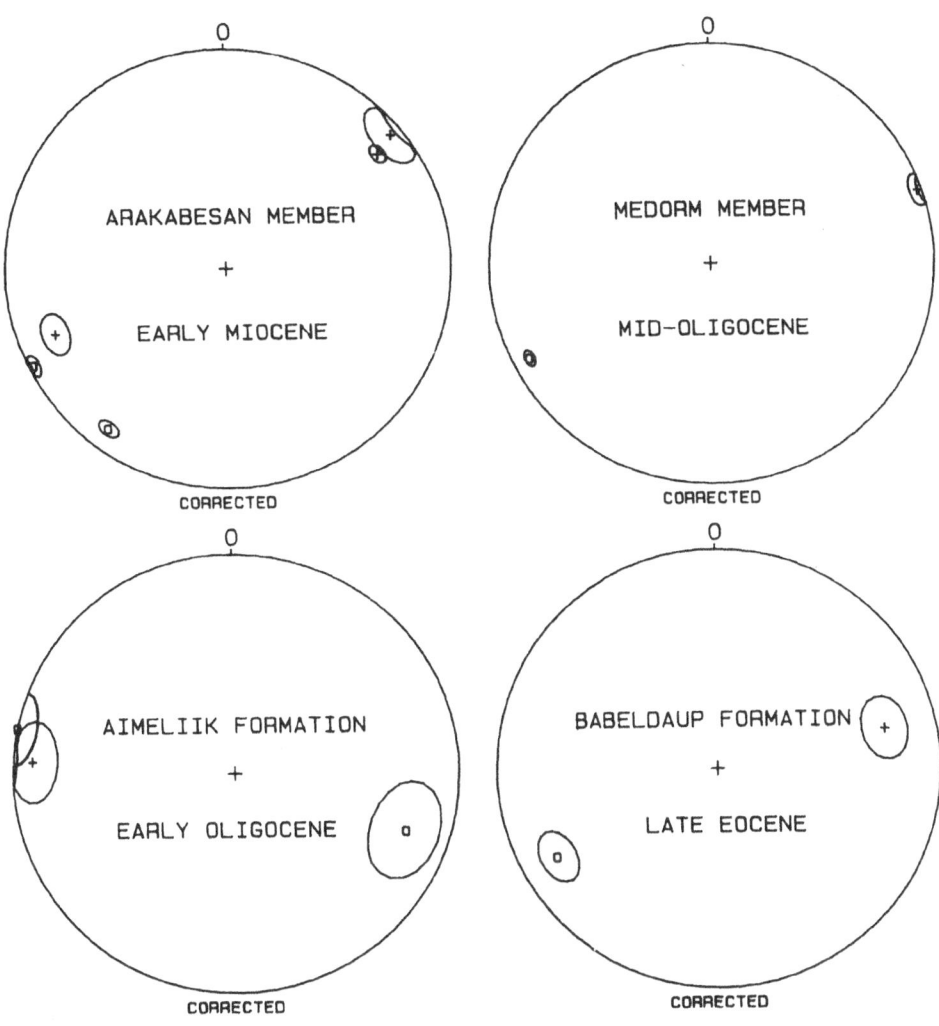

Fig. 4. Paleomagnetic results from the Palau Islands (from Haston et al., 1988). + are lower hemisphere, o are upper hemisphere, alpha 95 confidence intervals are shown.

The paleomagnetic record of the Palau Islands since Mid-Oligocene can be interpreted in the following three ways - (1) local rotations within the Palau Islands, (2) rotation of the WPSP or (3) a combination of local and plate rotations. Because the Palau Islands are located near to a plate boundary, local rotation of the area is a strong possibility. However, in the vicinity of the islands, the Palau-Kyushu ridge does not reflect any such local rotation. There is no evidence in the geology of the islands of tectonic events likely to cause such rotations. The paleomagnetic data are consistent with predicted clockwise rotation of the WPSP found using phase shifted magnetic anomaly data from the WPSP (Louden, 1977, Shih, 1980). However, estimates for the magnitude of the rotation vary as much as 70°. The paleomagnetic data are also consistent with ubiquitous clockwise rotations seen in data from Guam (Larson et al., 1975), Saipan (Fuller et al., 1980) and the Bonin Islands (Kodama et al., 1983, Keating and Helsley, 1985). For these reasons, we favour the interpretation of the paleomagnetic results from the Palau Islands as an expression of Philippine Sea Plate rotation.

The tectonic implications of clockwise rotation of the magnitude proposed and slight northward motion of the Philippine Sea Plate are considerable. First, it suggests that the origin of the WPSP by the trapping of Pacific crust in the manner proposed by Ben-Avraham and Uyeda (1972) is not likely; the critical role of the north-south transforms is precluded by the paleomagnetic rotation. When the WPSP spreading center is returned to its initial orientation, it is seen to have been roughly parallel to the Asian margin, giving credence to the idea of the back arc origin. Second, the results suggest that a region of sinistral shear will be established between the Philippine Sea Plate and Asia. This may provide a driving mechanism for the postulated translation of terranes in the Philippines (e.g. Karig et al., 1986). Third, it suggests that earlier use of paleomagnetic data to support models of deformation of the Mariana arc may may not be appropriate (Larson et al., 1975, McCabe and Uyeda, 1983, McCabe, 1984). For example, if the Philippine Sea Plate has rotated between 60 and 70° since the Oligocene, there is no need to develop special models to explain such rotations in Guam.

3. Luzon, northern Philippines.

The Philippine archipelago is a geologically complex assemblage of ophiolites, arc remnants and at least one continental fragment (e.g. Karig, 1983). From whence these various terranes came and how they were assembled into their present configuration is not clear. Paleomagnetism has proved a useful tool in solving such problems. Thus, we have carried out paleomagnetic studies in Luzon to test the various models of the tectonic history of the region.

Recent models of the tectonic history of the Philippines can be divided into three types. One follows the ideas of Holloway (1982). The Philippine archipelago is considered to have originated as an Early Tertiary E/W arc at equatorial latitudes. With time this arc rotated counterclockwise, moved northwards and collided with the Palawan microcontinent, as it was carried south by South China Sea floor spreading (Figure 5a). This collision may have stopped South China Sea spreading. Such a model requires counterclockwise paleomagnetic rotations and northward motion and was consistent with some of the early results, (e.g. Hsu, 1977). The model ignores interaction with the Philippine Sea Plate which is also moving northward, but rotating clockwise, as noted above.

Another view of Philippine tectonics has been proposed by Karig, (1983), Karig et al., (1986) and by Sarewitz and Karig, (1986). An important aspect of their studies has been the recognition of the importance of large scale strike-slip deformation within the Philippines (Figure 5b). In this view of Philippine tectonic history, the Philippines have had a long history of association with the Philippine Sea Plate. This contrasts with the Holloway model. The strike slip motion is an expression of a left lateral shear zone between the Philippine Sea Plate and Asia, within which the Philippines presently lie.

Still another view has been proposed by McCabe et al., (1986), who place the Philippines on the advancing boundary of the Philippine Sea Plate, thereby accounting for interpreted east dipping subduction in the western Philippines. Eventually, convergence between the Philippine Sea Plate and the Palawan microcontinent brings about the same collision postulated by Holloway (1972).

The different models of the tectonic evolution of the Philippines make some predictions which should be distinguishable paleomagnetically, although all three models have elements in common. Unfortunately, distinction between the Holloway model and those of Karig and Sarewitz cannot be made by paleomagnetism alone. Paleomagnetism cannot distinguish between northward motion and counterclockwise rotation of a rigid beam, as proposed by the former, and northward motion and counterclockwise of units within a sinistral shear zone, as proposed by the latter.

Paleomagnetism of Luzon. The principal area for which results are available is northwestern Luzon which is dominated by the Zambales ophiolite. By determining the latitude at which it was formed and its subsequent history of motion, paleomagnetism may be able to play an important role in unravelling the tectonics of the region. However, paleomagnetism has proved difficult in the Philippines. First, the location of successive arcs in Luzon gives rise to multiple intrusion events with associated hydrothermal systems. These have caused widespread secondary magnetization. Second the tectonic complexity of the region coupled with the limited outcrop has led to uncertainty in tectonic corrections, the absence of simple fold tests, and a general lack of structural control. Third the low latitude of the Philippines makes the demonstration of the sense of rotations more difficult than at intermediate latitudes. In light of this situation, paleomagnetic results from the Philippines are likely to remain controversial until large amounts of good data are assembled.

The pioneering paleomagnetic studies were carried out in the Philippines by Hsu (1971). Later, McCabe et al., (1982) reported data from six Cenozoic sites in Panay which showed easterly declinations. Comparing them with data from Luzon they argued that the easterly declinations was due to the collision of Panay with the Palawan microcontinent. Results from about 50 Cenozoic sites in Luzon were reported by Fuller et al., (1983) and in Fuller (1985). A small clockwise rotation of between 10 and 20° in Late Miocene or Early Pliocene was reported. In contrast, immediately prior to that Early and Mid-Miocene sites exhibited evidence for rapid counterclockwise rotation. Preliminary results from Late Eocene and Oligocene sites in the Zambales ophiolite suggested additional counterclockwise rotation and northward movement. McCabe et al., (1986) presented results from the Central Philippines and challenged the interpretation of counterclockwise rotation in Luzon. The case for counterclockwise rotation has been restated elsewhere (Fuller et al., 1988). The sense of rotation is a key aspect of the results, essentially determining whether Luzon originated on the Philippine Sea Plate, or on a plate to the west which was rotating counterclockwise.

Approximately 50 sites were collected in the ophiolite (Figure 6). For the most part standard paleomagnetic methods were used. The samples were measured on a 2G RF driven SQUID rock magnetometer. Both thermal and AF demagnetization were used. A thermal demagnetization system with which measurements could be made at high temperature was particularly helpful in resolving some of the multi-component magnetization. Another useful technique which helped the separation of primary and secondary magnetizations was the use of NRM : IRM (s) demagnetization plots (Fuller et al., 1988). Finally, it has been our interpretation that in a

420

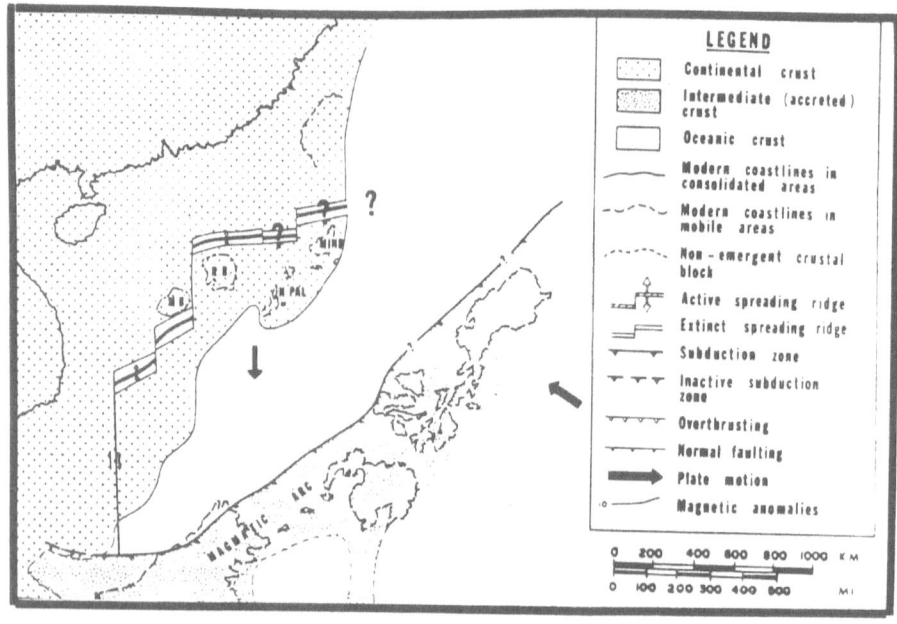

Fig. 5. Tectonic models of the Philippines, (a) Holloway (1982) and (b) Sarewitz and Karig (1985).

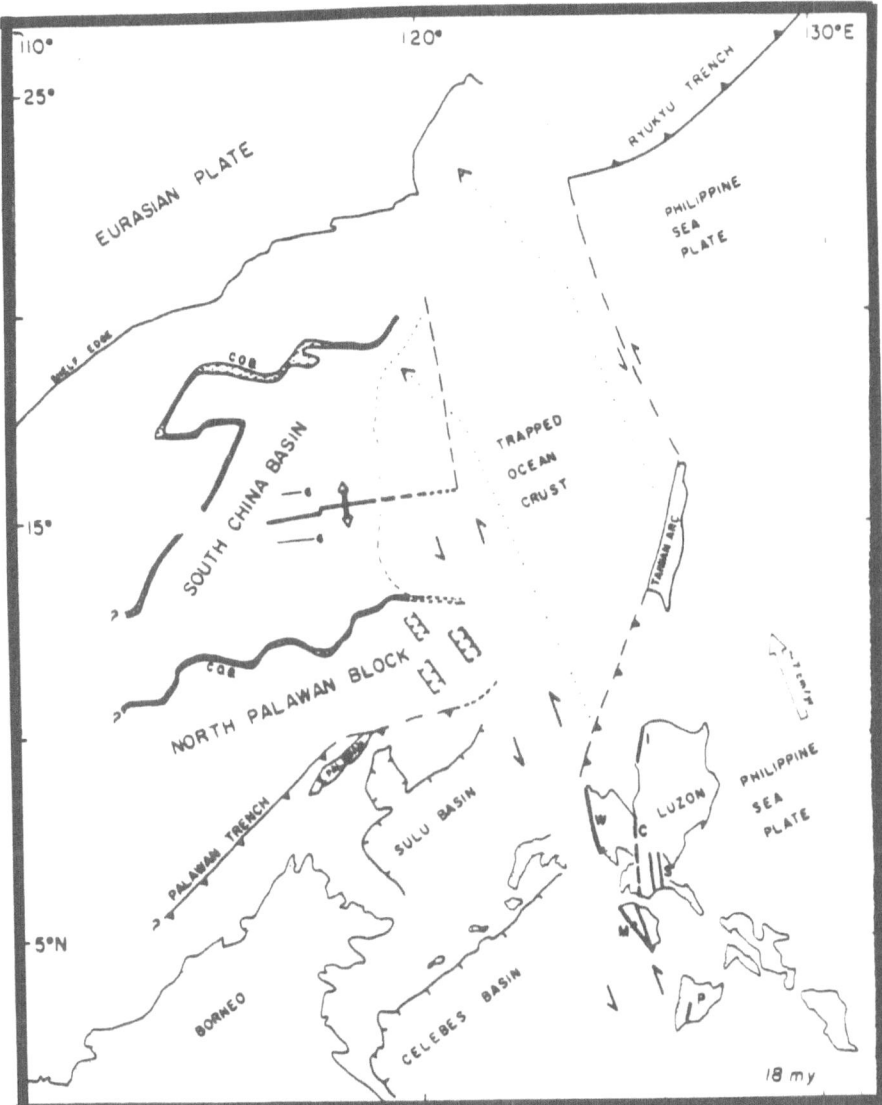

422

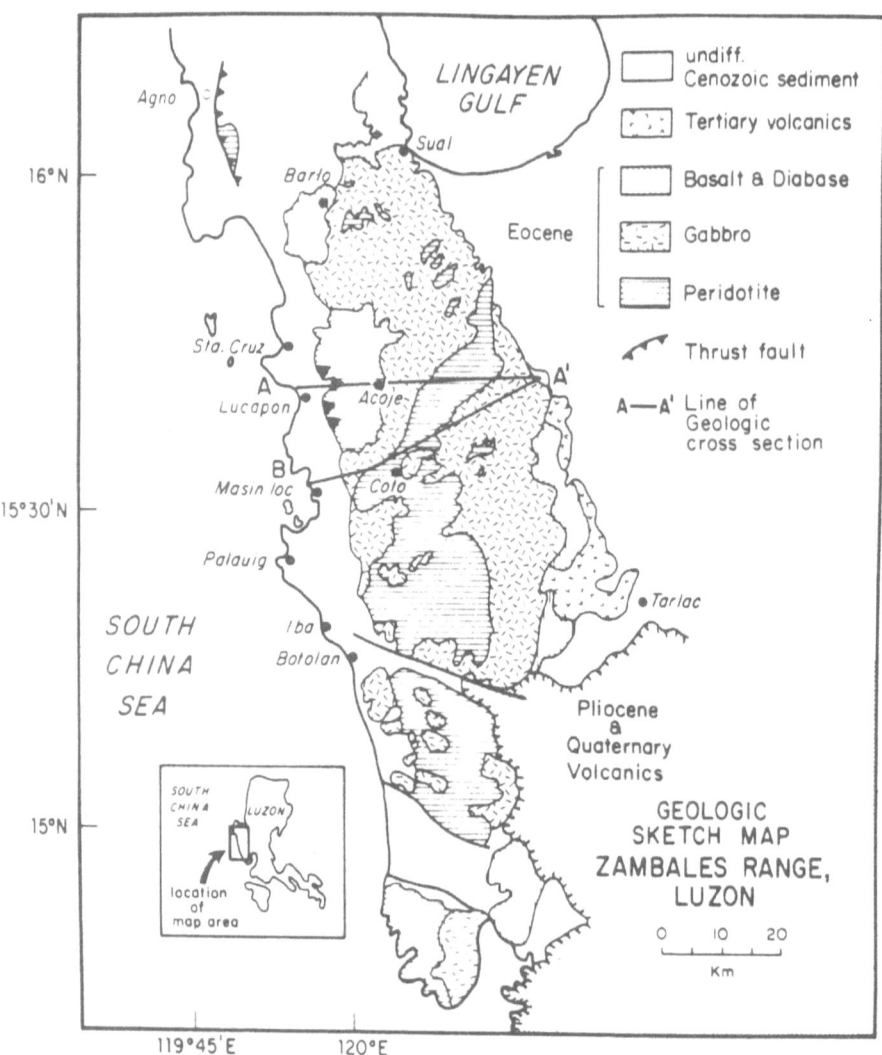

Fig. 6. Geology of the Zambales ophiolite: a) generalized geologic map of the Zambales ophiolite, b) generalized cross sections (Hawkins and Evans, 1983).

Zambales Range - Cross section A-A' Through Acoje Mine Area

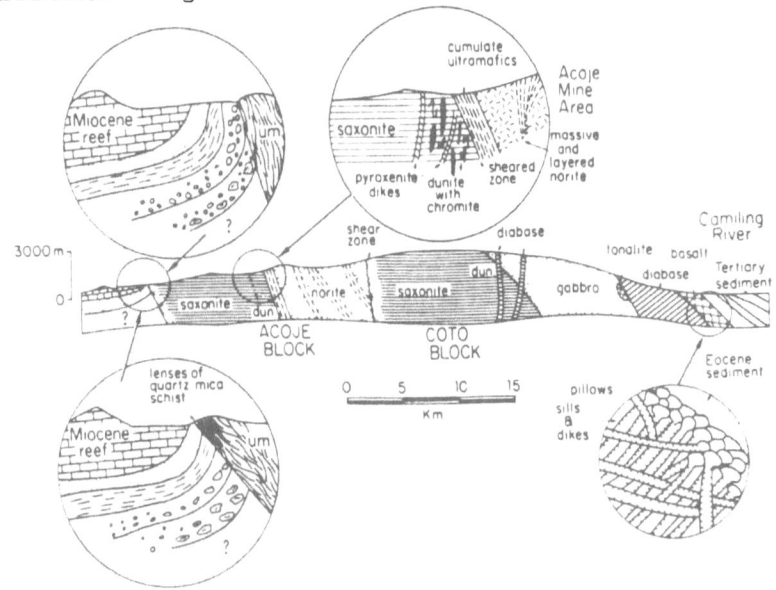

Zambales Range - Cross section B-A' Through Coto Mine Area

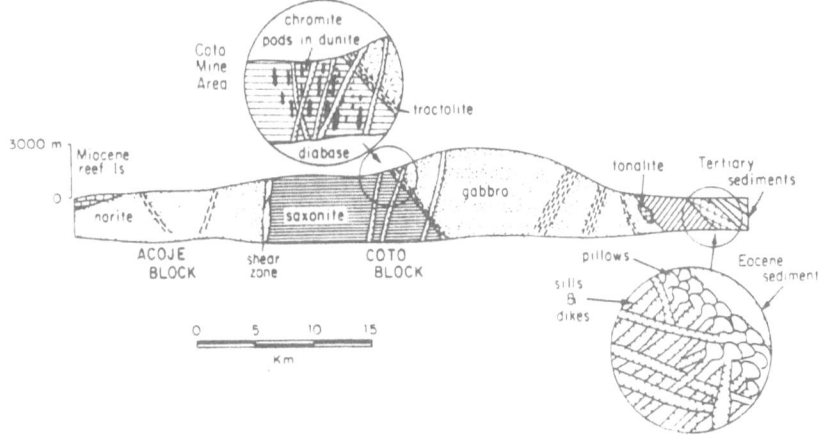

424

tectonically active region such as the Philippines, when younger rocks exhibit a similar direction of magnetization to that seen in older units, the magnetization of the latter is a remagnetization acquired no earlier than the age of the younger unit.

Paleomagnetic results from Zambales. The new results come from the Acoje block of the Zambales ophiolite in the north and from the Coto Block to the south. They are presented in Figure 7, in the form of a summary plot discussed in more detail in Fuller et al., (1988). In this figure, the results are mapped into the interpreted normal polarity in the interest of clarity. The polarity was assigned by tracing the progression of rotations back in geological time. The youngest rocks are Late Miocene or Pliocene intrusions; they show a minor clockwise rotation. This same direction is seen in many other older units, and is interpreted as a secondary magnetization associated with substantial hydrothermal systems generated by these intrusions. The next youngest sites studied are in the Early to Mid-Miocene Zambales Formation in the north and the Moriones Formation in the south. Both carry similar directions of magnetization, although it cannot be demonstrated that either is a primary magnetization. This direction is also seen in older rocks suggesting that it represents a time of significant remagnetization. It is not however a time of intrusion in this region, so a time of collision with associated migration of fluids may be indicated as suggested by Oliver (1986). Eocene-Oligocene sediments were also collected in the north and south. There were plentiful secondary magnetizations, but the directions shown in the figure were not seen in younger units and are tentatively interpreted as the primary Eocene-Oligocene directions. The gabbros, dykes and sill and flows of the ophiolite show consistent directions over distances of about 100 kms. Some of these directions are similar to those seen in the sediments overlying the ophiolite. We interpret these as secondary magnetization acquired as a result of unroofing, or as part of the process of obduction of the ophiolite in the Late Oligocene, or earliest Miocene. There are more

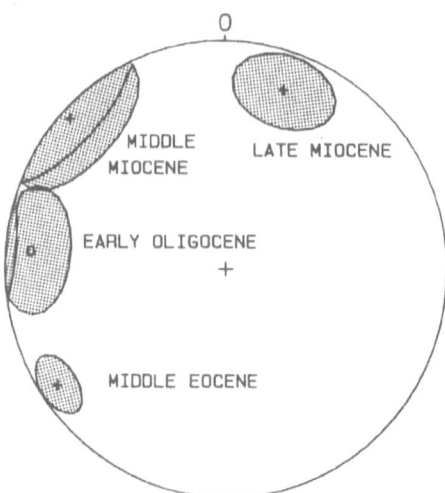

Fig. 7. Paleomagnetic results from Zambales (from Fuller et al., 1988).

strongly rotated directions seen in the 46.6 ±- 5.1 Ma. dykes at the Coto Mine and in a sill of 44.1 ±- 3.0 Ma. age near the Barlo Mine. The AF demagnetization characteristics of these magnetizations and their very high blocking temperature suggest to us that they are the primary direction of the ophiolite.

The paleomagnetism of the Zambales ophiolite reflects a history of active tectonics. Since the Late Miocene a small amount of clockwise rotation has taken place. Prior to that counterclockwise rotation and northward motion had dominated since Early Oligocene times. The most rapid rotation was in the Miocene after the main period of northward motion was completed. Taken literally, the results indicate a period of southerly motion in the Late Eocene, but this is less than the limit of our resolution.

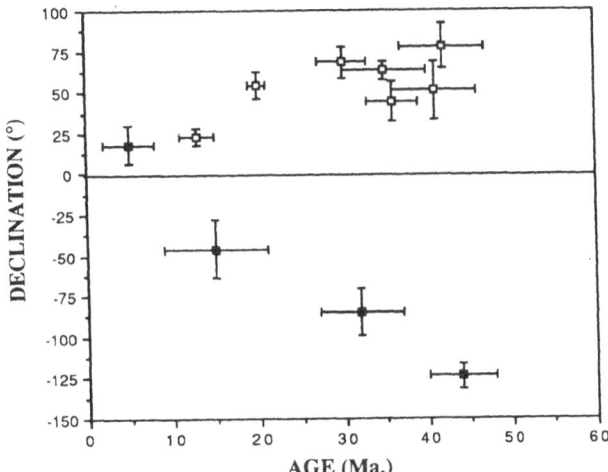

Fig. 8. Comparison of declination values observed in the Philippines and values predicted for Philippine Sea Plate. Open squares are data from the Philippine Sea Plate, solid squares are data from Luzon. Age dates are the assumed age of magnetization.

Considered in conjunction with earlier results from Luzon, the new results define a region including parts of the Cordillera, Zambales and Batangas, all of which show a similar history of Mid - Miocene strong counterclockwise rotations followed, where data are available, by indications of Late Miocene clockwise rotations (Fuller et al., 1988). This has led us to suggest that substantial parts of Luzon suffered a similar tectonic history. Moreover since this region includes a number of shear zones, these features cannot have involved sufficient motion to alter the paleomagnetic vector (i.e. motion of as much as a thousand kms). The results also demonstrate that Zambales and the other parts of this region cannot have had an origin on the Philippine Sea Plate. If for example, we plot the declination values for the Philippine results and compare those with the Philippine Sea Plate results, it is clear that prior to Late Miocene times the results diverge (Figure 8). That other parts of the Philippine archipelago may have been derived from the

426

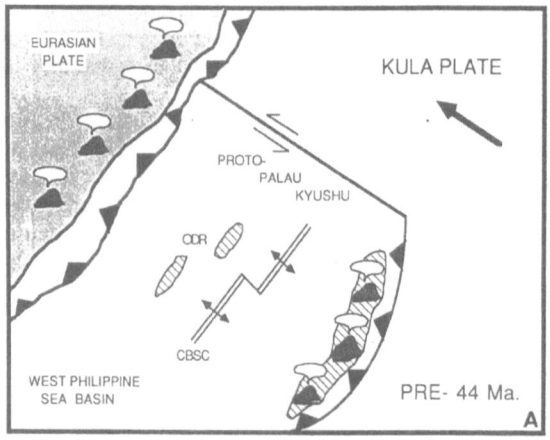

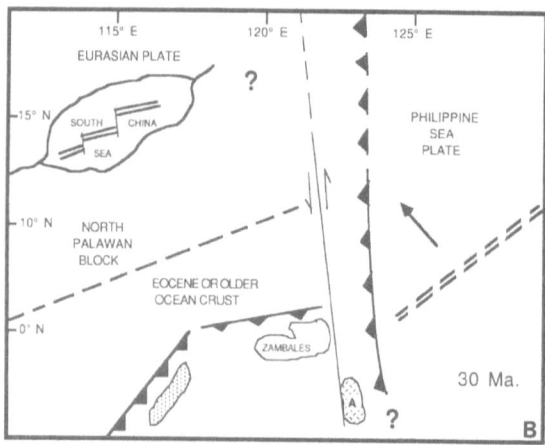

Fig. 9. Tectonic reconstructions; (a) 50 Ma. - Schematic reconstruction of the West Philippine Sea Basin. CBSC - Central Basin Spreading Center, ODR - Oki-Daito Ridge. The CBSC has been unrotated approximately 60-70° counterclockwise. Spreading along the CBSC has rifted off the pieces of remnant arc associated with the Oki-Daito Ridge complex. Andean subduction was occurring along the Eurasian margin. (b) 30 Ma. - Spreading has started in the South China sea, rifting off the North Palawan microcontinent. West dipping oblique subduction along the eastern boundary was consuming Philippine Sea plate. Behind the oblique subduction, left-lateral strike-slip motion was transporting terranes (shown as A) up from the south. Zambales was moving northward and rotating counterclockwise behind south dipping subduction.

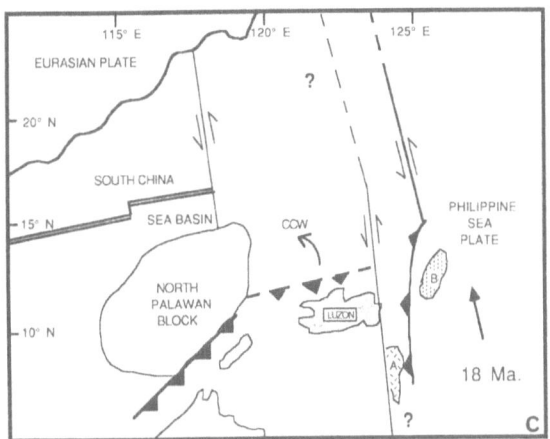

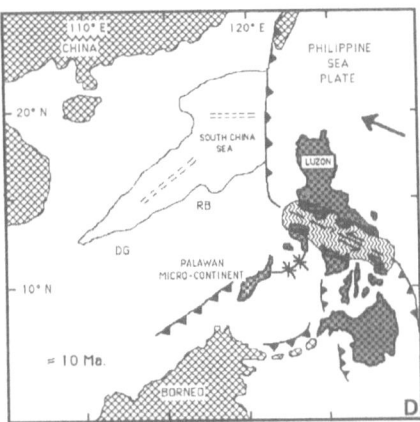

(c) 18 Ma. - Spreading was nearly complete in the South China Sea. Oblique subduction of Philippine Sea plate continues along the eastern boundary, with the continued northward translation of terranes (landmass A) along left-lateral shear zones. The North Palawan Block had collided with the Palawan Trench, shutting off the subduction on the southern end of the trench. At this time, Zambales had been amalgamated to a substantial portion of Luzon, and continued subduction on the northern end of the Palawan trench, caused a rapid counterclockwise rotation around the North Palawan Block. Some landmasses (B) moved with the rotating Philippine Sea plate, outboard of the subduction. The Benham rise may be an example of this. (d) 10 Ma. - Spreading in the South China Sea had ceased, along with subduction on the Palawan trench. The terranes of Luzon had been assembled at this time. East dipping subduction along the Manila trench was active at this time, and was linked by a diffuse zone of deformation through the Philippine archipelago to the west dipping Philippine Trench. Luzon was now associated with the clockwise rotating Philippine Sea plate.

Philippine Sea Plate is very likely, but the region we have studied appears to require an origin on a plate to the west of the Philippine Sea Plate.

4. Tectonics of Philippines and Philippine Sea Plate.

While it is premature to attempt detailed plate tectonic reconstructions of the interaction of the Philippines and the Philippine Sea Plate, the paleomagnetic results do require certain modifications of earlier models. As we noted above, the clockwise rotation of the Philippine Sea Plate favors the back-arc origin of the WPSP of the plate. Its original configuration would then have been approximately as illustrated in Figure 9a. Given additional reliable paleomagnetic data from other islands, which can be used to determine the past motion of the Palau Kyushu ridge, it should be possible to locate the WPSP precisely.

A second consequence of the clockwise rotation and northward translation of the Philippine Sea Plate is that there must have been considerable convergence between it and Asia. This convergence has been taken up in some combination of subduction and collision events. It is probable that subduction and collision similar to that proposed for more recent history in Mindanao and Halmahera (e.g. Hall, 1988) took place. This predicts that a proto-Philippine trench was active early in the Tertiary and that subducted slabs known to lie beneath the eastern part of the Celebes Sea, west of Talaud should be former Philippine Sea Plate. Moreover, the motion of the Philippine Sea Plate ensures that at least some of this subduction has been oblique, with characteristic deformation of the upper plate as proposed by Fitch (1972), Beck (1980) and Karig (1983). Hence a plate boundary configuration, similar to that shown in Figure 9b, seems likely in Early Tertiary.

By about 30 Ma., the paleomagnetic data suggest that Zambales was on a plate, or microplate rotating counterclockwise and moving northwards (Figure 9c). it seems likely that Zambales was at this time in the sinistral shear generated by the Philippine Sea Plate. Deformation in such a region has been described by two types of models, discussed recently by McKenzie and Jackson (1983, 1986). The first derives from the ideas of Freund (1974) and involves the attachment of the intervening blocks to the two plates between which the shear arises. In an informative mechanical analogue, the intervening blocks are represented by slats pinned to larger slats generating the shear. The second, which is a continuously distributed deformation model, derives from the hydrodynamic relationship between vorticity and instantaneous velocity vectors of fluid flow. Again this has been demonstrated by a floating marker mechanical analogue. It is not clear which model applies immediately to the Philippines. Sometimes the continuous deformation model will be appropriate. At other times, blocks may interact with the boundaries of the shear zone, as in the rotating slats model.

By 18 Ma., the collision with the Palawan microcontinent had begun and Zambales and other parts of Luzon rotated around it, requiring decoupling from the Philippine Sea Plate. Finally during Late Miocene, or Early Pliocene, the northernmost Philippines became associated again with the Philippine Sea Plate and experienced slight clockwise rotation. This motion is not observed in the Central Philippines (McCabe et al., 1986), so that a plate boundary, or a region of diffuse deformation is required between the two regions. This boundary is presumably moving northwards as a part of the interaction between the Manila Trench and the East Luzon trench (Figure 9d).

It appears that the Philippines lie at present in a sinistral shear zone between the Asian land mass and the Philippine Sea Plate. Paleomagnetic evidence suggest that the clockwise rotation of the Philippine Sea Plate has persisted throughout much of the Tertiary. Meanwhile Zambales and possibly substantial parts of a proto-Philippine arc were moving northwards on a counterclockwise rotating plate to the west. The interactions of these plates have largely determined the present configuration of the Philippines.

References.

Aoki, Yutaka, 1972, A post depositional magnetization mechanism of sediments and its application to geophysical problems--paleosecular variation in Japan [Ph.D thesis]: University of Tokyo, 101 p.

Beck, M.E., 1980, Paleomagnetic record of plate margin tectonic processes along the western edge of North America: *Journal of Geophysical Research*, v. **85**, p. 7115-7131.

Fitch, T.J., 1972, Plate convergence, transcurrent faults, and internal deformation adjacent to southeast Asia and the western Pacific: *Journal of Geophysical Research*, v. **77**, p. 4432-4460.

Fuller, M., Duncan, G., Green, G., Lin, J.L., McCabe, R., Toney, K., and Williams, I., 1980, Paleomagnetism of Truk Islands, eastern Carolines and of Saipan, Marianas, in Hayes, D.E., ed., The tectonic and geologic evolution of southeast Asian seas and islands: *American Geophysical Union monograph* 23, p. 235-246.

Fuller, M., Cisowski, S., Hart, M., Haston, R., and Schmidtke, E., 1988, NRM:IRMs demagnetization plots; an aid to the interpretation of natural remanent magnetization: *Geophys. Res. Lett.*, vol. **15**, no. 5, p. 518-521.

Fuller M., Haston R., and Almasco, J., 1988, Paleomagnetism of the Zambales ophiolite, Luzon, northern Philippines: *Tectonics*, in press.

Haston, R., Fuller, M., and Schmidtke, E., 1988, Paleomagnetic results from Palau, West Caroline Islands: A constraint on Philippine Sea plate motion. *Geology*, in press.

Hall, R., 1987, Plate boundary evolution in the Halmahera region, Indonesia, *Tectonophysic*, v. **102**, p. 85-104.

Hawkins, J., and Evans, C., 1983, Geology of the Zambales Range, Luzon, Philippine Islands, Ophiolite derived from an Island Arc-Back Arc pair, in Hayes, D.E., ed., *The tectonic and geologic evolution of southeast Asian seas and islands*, part 2: American Geophysical Union monograph 27, p. 95-123.

Hilde, T.W.C., and Lee, C.S., 1984, Origin and evolution of the West Philippine Basin: A new interpretation: *Tectonophysics*, v. **102**, p. 85-104.

Hilde, T.W.C., Uyeda, S., and Kroenke, L., 1977, Evolution of the western Pacific and its margin: *Tectonophysics*, v. **38**, p. 145-165.

Holloway, N.H., 1982, North Palawan Block Philippines - Its relation to Asian Mainland and Role in the Evolution of south China Sea: *AAPG, Bull.*, v. **66**, no. 9, p. 1355-1383.

Hsu, I-chi, 1971, Magnetic Properties of Igneous rocks in the Northern Philippines, Ph.d. Thesis, University of Wash, 105 p.

Karig, D.E., 1983, Accreted terranes in the northern part of the Philippine Archipelago: *Tectonics*, v. **2**, p. 211-236.

Karig, D.E., Sarewitz, D.R., and Haeck, G.D., 1986, Role of strike-slip faulting in the evolution of allochthonous terranes in the Philippines: *Geology*, v. **14**, p. 852-855.

Keating, B.H., and Helsley, C.E., 1985, Implication of island arc rotations to the studies of marginal terranes. *Journal of Geodynamics*, v. **2**, p. 159-181.

Kodama, K., Keating, B.H., and Helsley, C.E., 1983, Paleomagnetism of the Bonin Islands and its tectonic significance: *Tectonophysics*, v. **95**, p., 25-42.

Larson, E.E., et al., 1975, Paleomagnetism of Miocene volcanic rocks of Guam and the curvature of the southern Mariana island arc: *Geological Society of America Bulletin*, v. **86**, p. 346-350.

Lewis, S.D., Hayes, D.E., and Mrozowski, C.L., 1982, The origin of the West Philippine Basin by inter-arc spreading, in Balce, G.R., and Zanoria, A.S., eds., *Geology and tectonics of the Luzon-Marianas region: Philippine Seatar Commitee Special Publication* Number **1**, p. 31-51.

Louden, K.E., 1976, Magnetic anomalies in the West Philippine Basin, in Sutton, G.H., ed., The geophysics of the Pacific Ocean basin and its margins: *American Geophysical Union monograph* 19, p. 253-276.

Mason, A.C., Corwin, G., Rodgers, C.L., Elmquist, O., Vessel, A.J., and McCracken, R.J., 1956, Military geology of the Palau Islands, Caroline Islands, report, Intelligence Division, Office of the Engineering Headquarters, U.S. Army Forces Far East and Eighth U.S. Army (rear)/U.S. Geological Survey, Washington D.C.

McCabe, R., 1984, Implication of paleomagnetic data on the collision related bending of island arcs: *Tectonics*, v. **3**, p. 409-428.

McCabe, R., and Uyeda, S., 1983, Hypothetical model for the bending of the Mariana Arc, in Hayes, D.E., ed., *The tectonic and geologic evolution of southeast Asian seas and islands*, part 2: American Geophysical Union monograph 27, p. 281-293.

McCabe, R. J., Kikawa, E., Cole, J.T., Malicse, A. J., Baldauf, P.E., Yumul, J., and Almasco, J., 1987, Paleomagnetic results from Luzon and the Central Philippines: *Jour. Geophys. Res.*, v. **92**, p. 555-580.

Meijer, A., Reagan, M., Ellis, H., Shafiquallah, M., Sutter, J., Damon, P., and Kling, S., 1983, Chronology of volcanic events in the eastern Philippine Sea, in Hayes, D.E., ed., *The tectonic and geologic evolution of southeast Asian seas and islands*, part 2: American Geophysical Union monograph 27, p. 349-359.

Molnar, P. , and Tapponier, P., 1975, Cenozoic tectonics of Asia: effects of a continental collision: *Science*, v. **189**, p. 419-426.

Pautot, G., et al., 1986, Spreading direction in the South China Sea: *Nature*, v. **321**, p. 150-153.

Ranken, B., Cardwell, R.K., and Karig, D.E., 1984, Kinematics of the Philippine Sea plate: *Tectonics*, v. **3**, p. 555-575.

Sarewitz, D.R., and Karig, D.E., 1986, Processes of allochthonous terranes, Midoro Island, Philippines: *Tectonics*, v. **5**, p. 525-552.

Shih, T.C., 1980, Marine magnetic anomalies from the western Philippine Sea: Implication for the evolution of marginal basins, in Hayes, D.E., ed., The tectonic and geologic evolution of southeast Asian seas and islands: *American Geophysical Union monograph* **23**, p. 49-76.

Taylor, B., and Hayes, D., 1983, Origin and history of the South China Sea Basin, in Hayes, D.E., ed., *The tectonic and geologic evolution of southeast Asian seas and islands*, part 2: American Geophysical Union monograph 27, p.24-56.

Uyeda, S., and Ben-Avraham, Z., 1972, Origin and development of the Philippine Sea: *Nature*, v. **240**, p. 176-178.

PALAEOMAGNETIC CONSTRAINTS ON THE EARLY HISTORY OF THE MØRE-TRØNDELAG FAULT ZONE, CENTRAL NORWAY

T.H. TORSVIK[1,2], B.A. STURT[2], D.M. RAMSAY[3], A. GRØNLIE[2], D. ROBERTS[2], M.SMETHURST[1], K. ATAKAN[4], R. BØE[2] and H.J. WALDERHAUG[5]

1 *Department of Earth Sciences,*
University of Oxford, Oxford OX1 3PR,
UK

2 *Geological Survey of Norway, P.O. Box 3006,*
N-7002 Trondheim, Norway

3 *Department of Geology,*
University of Dundee, Dundee DD1 4HN,
Scotland

4 *Department of Geology,*
University of Bergen, N-5000 Bergen,
Norway

5 *Department of Geophysics,*
University of Bergen, N-5000 Bergen,
Norway

ABSTRACT. Two principal remanences, SMA and SMB, have been identified from Ordovician and Devonian rocks from Smøla, Central Norway. The proposed youngest, SMB, compares with Late Devonian (Solundian) palaeomagnetic data from Western Norway. SMA, however, is plainly anomalous and it is necessary to infer a tectonic rotation of Smøla to accommodate the palaeomagnetic data. Pre-ORS rocks in the elongate area embracing Smøla, Hitra and Orlandet are characterized by a phase of Mid to Late Ordovician deformation and intrusive activity, and are scarcely affected by the Silurian Scandian orogeny. It is proposed that this "Smøla Terrane" docked against the Western Gneiss Region in Late Devonian time. This collisional docking involved emplacement on a major detachment, and suturing along the Møre-Trøndelag Fault Zone (MTFZ). The nature of this suture was essentially one of a reverse-fault, and the rotation of Smøla is thought to have been accommodated by strike-slip faulting between Smøla and Hitra. The syn- to post-collisional, Late Devonian history of the MTFZ could have been one of sinistral-strike slip, dependent on the orientation of the collisional stress-vector. The MTFZ is an example of a long-lived fault zone, and important phases of post-Devonian reactivation with the formation of a variety of fault-rocks and hydrothermal alteration occurred in Permian, Late Jurassic/Early Cretaceous and Early Tertiary times.

1. Introduction

The Møre-Trøndelag Fault Zone (MTFZ), Central Norway, is commonly depicted as an important Palaeozoic and Mesozoic fault zone, and a variety of geophysical and geological techniques have been employed to date and elucidate the sense of movement and geometry within this zone. Landsat

431

C. Kissel and C. Laj (eds.), Paleomagnetic Rotations and Continental Deformation, 431–457.

432

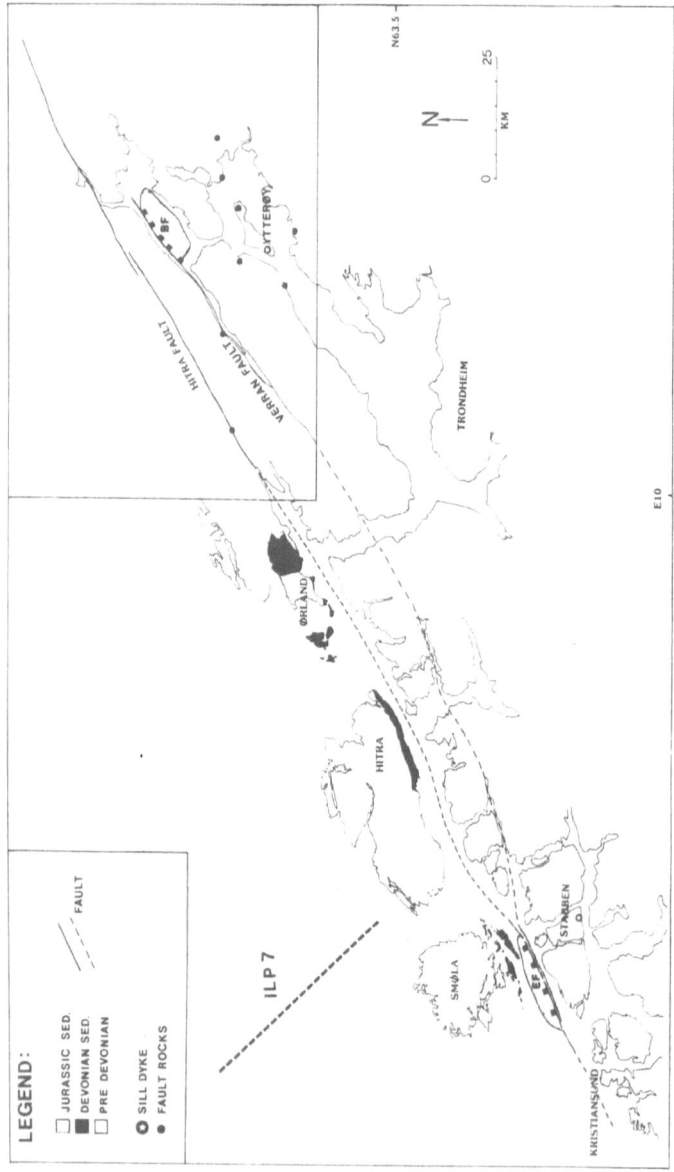

Fig. 1. Sketch map of the Møre-Trøndelag Fault Zone, Central Norway, displaying the two principal tectonic lineaments, i.e. the Hitra-Snasa and Verran Faults, and the distribution of Devonian and Jurassic sediments. The mid-Jurassic basins (EF=Edøyfjord Basin; BF=Beistadfjord Basin) are preserved in half-grabens. The locations of palaeomagnetically tested fault-rocks and dykes/sills (Ytterøy and Stabben) within the MTFZ and the position of Mobil Search Line ILP 7 are indicated.

Multispectral Scanner and Thematic Mapper lineament analysis (Ramberg et al., 1977; Gabrielsen and Ramberg, 1979; Rindstad and Grønlie, 1986), photo- geological studies (Roberts, 1986), seismic reflection profiling (Bøe and Bjerklie, in prep) and field studies have all assisted in mapping out the detailed tectonic lineaments within the MTFZ. In addition, palaeomagnetic studies have offered supplementary data concerning the timing of displacive and/or hydrothermal events (Sturt et al., 1987; Grønlie and Torsvik, in prep).

The MTFZ defines an extensive ENE-WSW fault array and can be traced for nearly 300 km from Nord-Trøndelag southwestward to beyond Smøla (Figure 1), where the two principal tectonic lineaments, i.e. the Hitra-Snasa and Verran Faults converge (Grønlie and Roberts, 1987). The southwestward offshore extension of these lineaments define the southern margin of the Møre Basin (Gabriel sen et al., 1984). The complex MTFZ displays a variety of fault geometries and fault-rock products which point to a protracted history of displacement (Grønlie and Roberts, 1987). Palaeomagnetic dating of carbonate-thorium breccias and associated hydrothermally altered rocks indicates both Late Palaeozoic (Permian) and Mesozoic ages (Sturt et al., 1987; Grønlie and Torsvik, in prep). These breccias, however, follow earlier fault trends, most likely initiated in the Early to Mid Palaeozoic. Secondary, Riedel-like structures point to an early phase of sinistral strike-slip faulting along the Hitra-Snasa Fault (Grønlie and Roberts, 1987), whereas the Mesozoic reactivation history is one of both dextral shear and dip-slip. A northwestw ard downthrow of several hundred metres has been recorded south of Smøla (Bøe and Bjerklie, 1988). These pulses of normal or oblique-slip faulting are post Middle Jurassic in origin, since assumed Middle Jurassic sediments are preserved in a half-graben bounded by the Hitra-Snasa Fault. Middle Jurassic sediments are also preserved in Beistadfjorden (Figure 1) which is bounded on its northwest side by the Verran Fault (Bøe and Bjerklie, in prep). Imbricate fault geometries between the Hitra-Snasa and Verran Faults (Grønlie and Roberts, 1987) have the configuration of a strike-slip duplex (Woodcock and Fisher, 1985), and Grønlie and Roberts (1987) have proposed that the Beistadfjorden Basin formed as an in-line or pull-apart basin associated with dextral movements on the Verran Fault. During the Mesozoic, the Verran also experienced dip-slip displacements (Oftedahl, 1974; Bøe and Bjerklie, 1988), as was the case for the Hitra-Snasa Fault.

The MTFZ can thus be shown to have had a complex history of development, and as with all long-lived fault zones will have a faulting mode dependent on the changes in the orientation of the principal stress vectors. The rapid changes of such stress vectors and the resultant effect on faulting mode has recently been highlighted by Underhill et al. (1988) in a study of Carboniferous faulting in the North of England. It is also well known that rotations of fault blocks occur along complex strike- slip fault zones (Cristie-Blick and Biddle, 1985). In the present case, palaeomagnetic studies along the MTFZ were aimed to test the possibility that block-rotations may have occurred and to establish a time-frame by the study of fault-rocks. In this account we present palaeomagnetic and structural data from the Ordovician and Devonian rocks of Smøla and neighboring islands and consider their possible implications for the Palaeozoic history of the MTFZ.

2. Regional geology sampling and magnetic fabrics

The pre-Devonian rocks of Smøla (Figures 2 and 3) comprise a Lower Ordovician volcano-sedimentary sequence cut by a variety of calc-alkaline Caledonian intrusions and basic dykes (Strand, 1932; Fediuk, 1975; Fediuk and Siedlecki, 1977; Bruton and Bockelie, 1979; Roberts, 1980; Bruton and Harper, 1985; Askvik and Rokoengen, 1985; Gautneb, 1987). A North American fauna has been identified in Arenig-Llanvirn limestones and thus Smøla is commonly portrayed as a

434

Palaeozoic suspect terrane. This terrane is generally considered to have been transported in a SE
direct ion during Ordovician times and accreted onto the Baltic foreland in the Late Silurian, forming
part of the Upper Allochthon in Central Norway (Stephens and Gee, 1985).

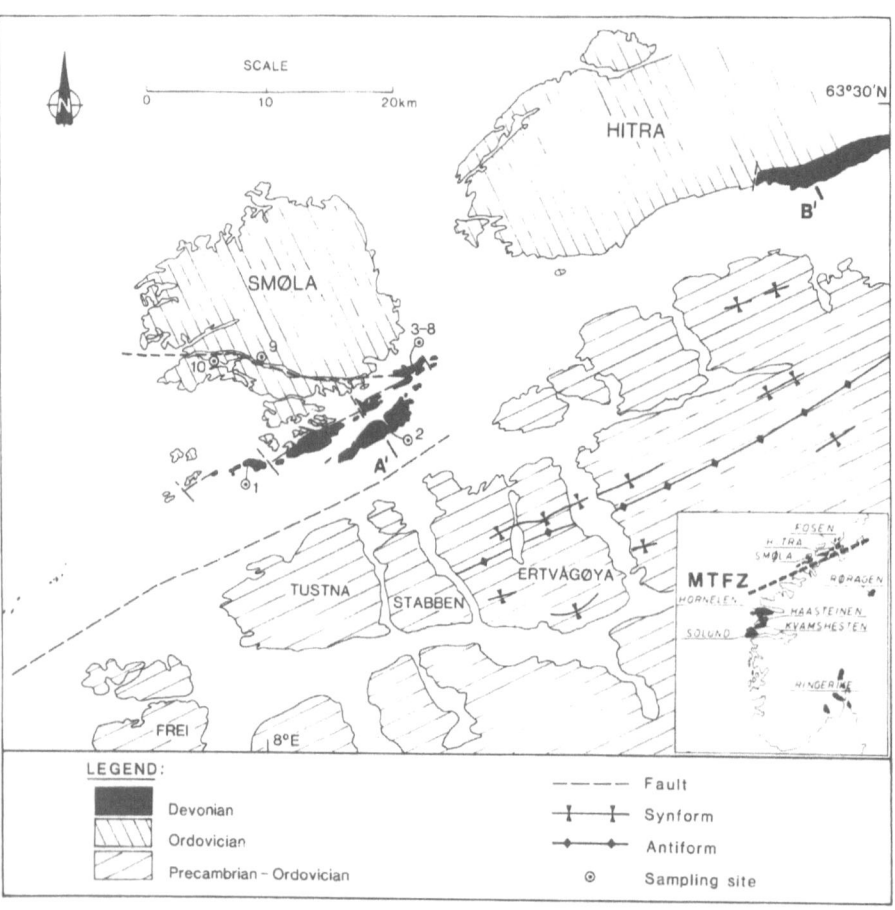

Fig. 2. Geological sketch map of the islands of Smøla and Hitra and the mainland south of
Trondheimsleia (simplified from Askvik and Roekoeng en, 1985). Inset map of southern Norway
show the distribution of Old Red Sandstone deposits in Norway (simplified from Roberts, 1983).
The MTFZ is here indicated as a stippled line (cf. Fig. 1).

Late Silurian to Middle Devonian Old Red Sandstone (hereafter called either ORS or Devonian)
sediments outline a prominent NE-SW trending belt (Figure 2) along the coastal region of Central
Norway. The ORS of SE Smøla (Smøla Group) crops out on a number of small islands, and forms

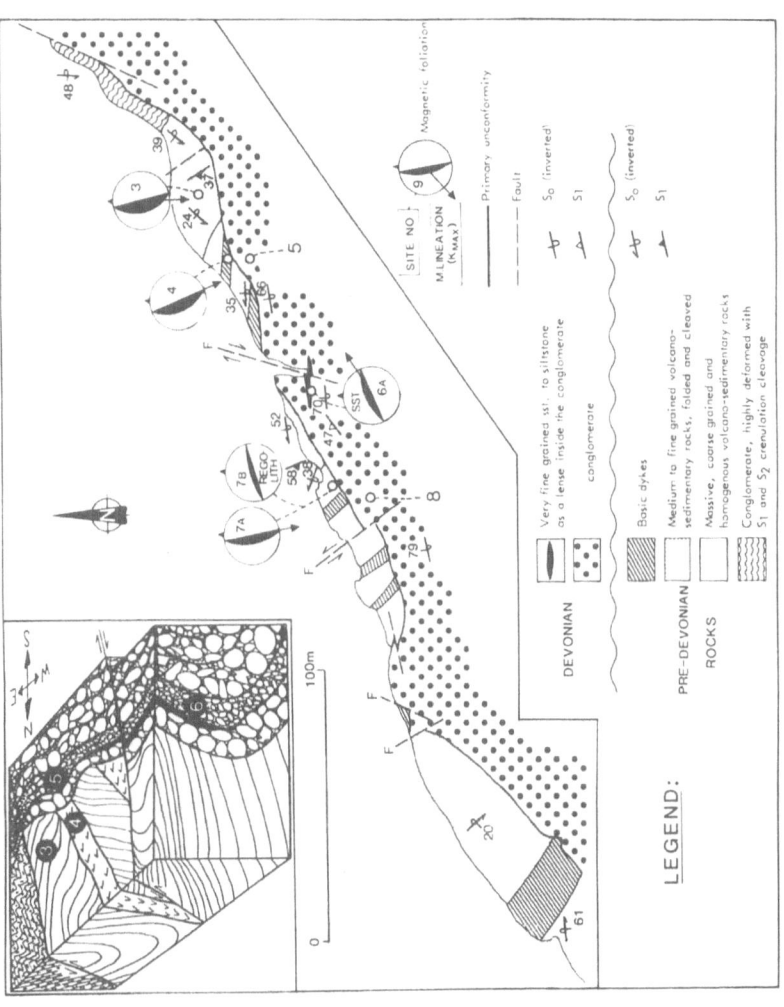

Fig. 3. Geological map of Glassøya, site locations and magnetic fabrics. The magnetic foliation planes and lineations are shown as downward dipping planes and plunges, respectively. Inset map shows a three-dimensional view of the sampling area, embracing sites 3 to 6. Cf. legend and text for details.

a ca. 3.6 km - thick sequence resting unconformably on the pre-Devonian rocks. The NW, and in part the SE margin of the ORS deposits on the nearby island of Hitra (Hitra Group) are similarly defined by a primary unconformity (Siedlecka and Siedlecki, 1972; Bøe, 1986). The Devonian rocks of Smøla are deformed and folded into an overturned asymmetrical syncline, with an axial

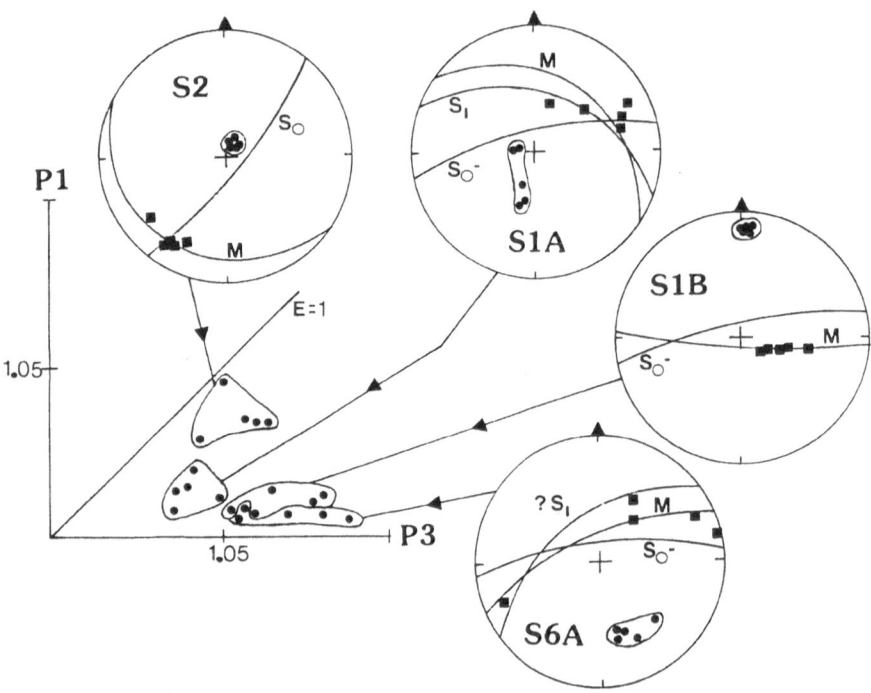

Fig. 4. Details of the magnetic fabric from Boarnøya (sites 1a and b), Edøy (site 2) and Glassøya (site 6; see also Fig. 2.). In the stereoplots the circular and squares symbols represent Kmin and Kmax respectively. Great circles are shown as downard-dipping planes, and are designated M (magnetic foliation; mean value), So (bedding; So : inverted stratigraphy) and S1 (cleavage). Note that sites with distinct intersection (M/So), i.e. sites 2 and 1a, show minimum apparent magnetic oblateness (cf. Flinn plot).

plane dipping at a moderate angle to the NW (Figure 5a and 3). Only the NW limb of this fold-structure is now preserved, and the ORS together with the basal unconformity is inverted along the entire length of the NW margin of the belt.

The ORS deposits of Smøla consist mainly of coarse- to fine- grained conglomerates and sandstones of probable Middle Devonian age. The basal Glassøy Formation (Sites 5a and b, 6a-c, 7B, 8a and b) unconformably overlies Ordovician low-grade metamorphic rocks (Figures 2 and 3). This substrate includes volcanoclastics (sites 3 and 7A), basic dykes (site 3) and conglomerates. Sites 1a and b were sampled on the island of Buarnøya in a zone of fine-grained conglomerate (Middle Buanoya Member). The Glassøy Formation is succeeded by the NW-Kyrhaug Formation which grades upward into the Edøy Formation (Figure 5a). The Edøy Formation (site 2) is a polymict reddish conglomerate with non-overturned dips of 70-75° towards the NW. Palaeomagnetic sampling was confined primarily to a detailed study of the ORS, conglomerate boulders and the immediate substrate from Glassøya (Figure 3 - Table I), but the substrate was also examined at two areas on the main island of Smøla (sites 9 and 10; cf. Figure 2).

On Glassøya, bedding (S_o) and cleavage (S_1) are readily distinguished in the inverted Ordovician volcanoclastics, and locally Ordovician conglomerates (Figure 3) also show a well developed crenulation cleavage (S_2). The anisotropy of magnetic susceptibility (AMS) was studied using a low-field (0.1 mT) induction bridge (KLY-1).The magnetic fabrics in the volcanoclastic sediments and a basic dyke in which no megascopic cleavage is observed are entirely governed by S1.The magnetic foliation (Kmax-Kint) has an intermediate to steep inclination, with a near N-S trend and southerly plunging lineation (Kmax). In the volcanoclastics the degree of anisotropy is in the order of 10 to 15%. The more competent basic dyke, on the other hand, gave a mean anisotropy of 4%. The magnetic fabrics in the basic rocks and dykes from the main island (sites 9 and 10) also yield near N-S and steeply inclined magnetic foliations. Low anisotropies, ca. 3.5%, are recorded, and Kmin compares well with poles to the megascopic foliation (Figure 5c).

The Devonian sandstone-siltstone lens of site 6a is affected by a minor right-lateral offset (Figure 3). The magnetic foliation (Figure 4) defines a planar structure intermediate between So and the plane of a weakly developed megascopic spaced cleavage. Samples from the Devonian regolith define a similarly oriented magnetic foliation (Figure 3), but with no preferred lineation.

The magnetic foliations recorded from sites 1a (Buarnøya) and 2 (Edøy) are clearly post-depositional in origin. At site 2 an axial plane cleavage associated with an intersection type lineation (Figure 4) can be determined and coincides well with the megascopic cleavage.On the other hand, the magnetic foliation from site 1b is almost bedding-parallel, and samples from this site together with site 6a have the most oblate-shaped magnetic ellipsoids (cf. Flinn plot in Fig. 4). The total degree of anisotropy ranges from 4 to 10% in the Devonian rocks.

3. Paleomagnetic and rock-magnetic experiments

The natural remanent magnetization (NRM) was measured on a Molspin spinner and a two-axis cryogenic magnetometer. The demagnetization programme, thermal and alternating field (AF) included 152 specimens and characteristic remanence components were obtained from least square line fitting.

Magnetite and haematite contribute to the bulk magnetic properties from sites 1 and 2 (Figure 7). High-temperature/coercivity blocking components (HB) from sites 1a and b show near E declinations with shallow inclinations, whereas some intermediate blocking (IB) components (T<400°C) from site 1a define a reasonably anti-parallel directional group (Table I). Also at site 2, a considerable part of the NRM is dominated by an IB component (compare Figure 6/S2 and 7). The HB component (T>450-500°C and AF fields>20-30mT) shows declinations near W with shallow downward- pointing inclinations (Figure 6/S2). The IB component is reasona bly separated from the HB component, and is somewhat more NW directed and with steeper inclination (Figure 6/S2 and Table I). The directional differences, however, rarely exceed 10 degrees of arc. IB components are probably resident in magnetite, whereas the HB component is primarily carried by haematite (high coerciv ities and blocking-temperatures exceeding 600°C). Samples from sites 1 and 2 show median destructive (M1/2) AF fields in the 2- 45 mT range (cf. Fig 7).

438

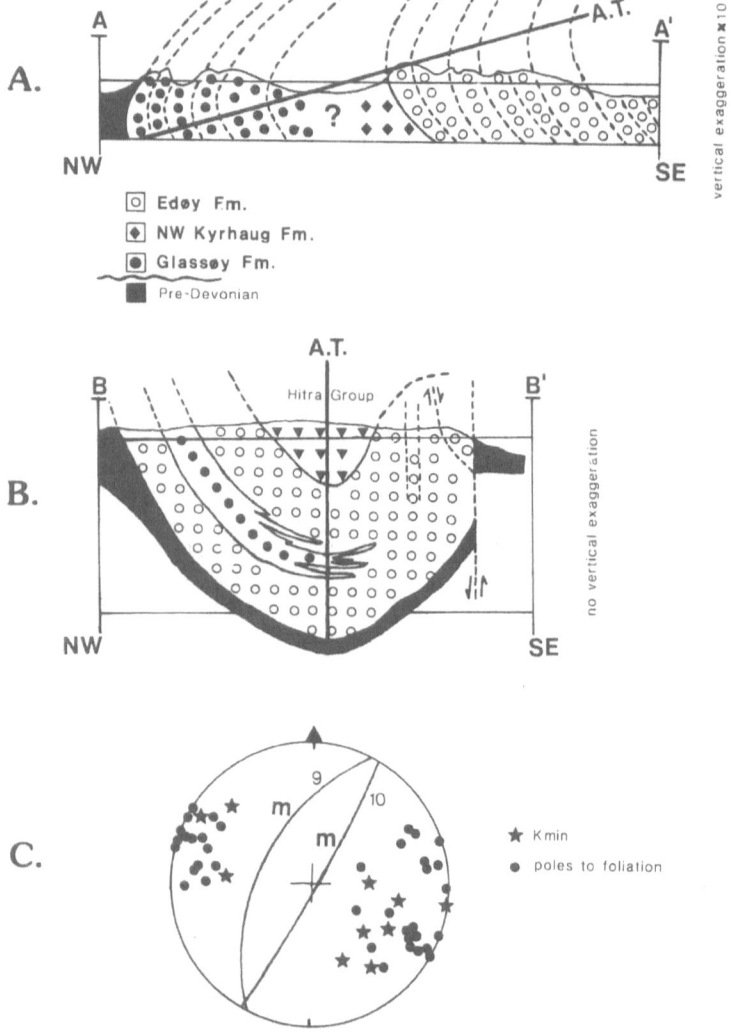

Fig. 5. Vertical section through the ORS of southern Smøla (a) and southwestern Hitra (b). The Hitra profile corresponds to a simplified profile of section A-A'in Bøe (1986). Cf. fig. 2 for scale and profile position.
(c) Comparison of Kmin (sites 9 and 10) and poles to cleavage (Gautneb, 1987) observed from pre-Devonian rocks (diorite/quartz monzodiorite) of southern Smøla.

Table I: Sampling details and site mean statistics.

	DEC	INC	α_{95}	N	Jo	BMP
Devonian rocks						
Buarnøya:						
1A Grey fine-grained CGL	IB:299	+12	26.5	4	7.2	M,H
	HB:123	-1	13.0	10		
1B Grey fine-grained CGL	HB:104	+3	8.5	13	2.9	M,H
Edøya:						
2 Red fine-grained CGL	IB:291	+33	8.5	13	25.1	M,H
	HB:276	+16	8.4	8		
Glassøya:						
6A Red SST-SILT lense	HB:087	-13	3.1	19	9.1	H,M
7B Grey-white regolith	HB:096	-6	10.2	4	8.3	M,H
Devonian conglomerate boulders						
Glassøya:						
5A Basic	IB/HB:182	+16	19.8	4	56.2	M
	HB:091	-16	10.8	6		
5B Basic	HB:195	-45	11.6	3	18.7	M
6B Basic	HB:214	+32	2.7	4	625.8	M
6C Basic	----				7.1	M
8A Volcano-clastic	IB:203	+21	23.7	5	1.3	M
	HB:109	-11	10.7	5		
8B Volcano-clastic	----				0.6	P,M
Ordovician rocks						
Glassøya:						
3 Volcano-clastics	IB:215	+22	15.4	6	0.7	P,M
	HB:120	-22	17.2	6		
4 Basic dyke	IB:197	+17	9.9	5	3.7	P,M
7A Volcano-clastics	HB:132	-13	11.6	5	0.9	P,H
Smøla:						
9 Basaltic rocks	IB:043	+52	17.2	6	5.0	M,P
	HB:096	-22	10.5	3		
10 Basic dyke	LB:027	+51	12.8	7	4.1	M

LB=low blocking; HB=high blocking; DEC=mean declination; INC=mean inclination (dec and inc in insitu co-ordinates; α_{95} = 95 percent confidence circle [61]; N=number of sample directions; Jo=mean NRM intensity; BMP=Bulk magnetic properties; M=magnetite, H=haematite, P=paramagnetic phases.

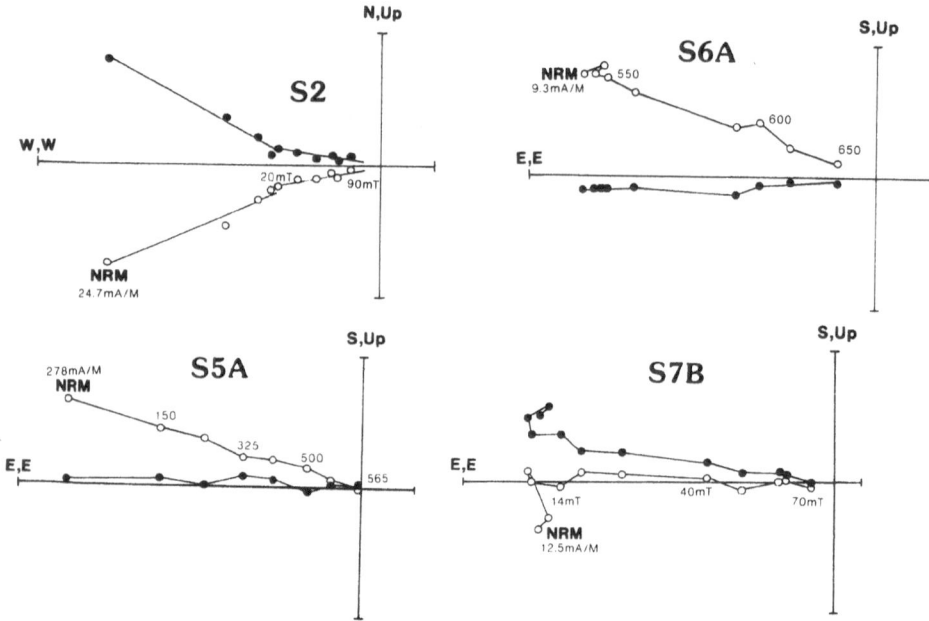

Fig. 6. Vector diagrams showing the characteristic E-W remanence recorded in fine-grained reddish conglomerate (S2), red sandstone/siltsone (S6), a basic conglomerate boulder (S5A) and the Devonian regolith (S7B). In the vector diagrams, open and closed symbols represent points in the vertical and horizontal plane, respectively. The numbers given below or above "NRM" are the natural remanent intensity.

From site 6 (red sandstone/siltstone), near E and shallow upward-pointing single component remanences are recognized (Figure 6/S6A), and haematite is the most important remanence carrier (Figures 6/S6A and 7). $M_{1/2}$ is typically above 90 mT, and thus some AF demagnetized samples were further treated by thermal cleaning. Characteristic remanence components in the Devonian regolith (site 7B) compare with those from site 6a (Figure 6/S7B). A minor LB component, however, proved difficult to isolate.

Four of a total of six sampled conglomerate boulders (TableI, Figure 3) provided sensible palaeomagnetic data, of which three are dominated by a remanence with SSW declinations and downward- pointing inclinations (Figure 8/S5A). The regular E-W component (Figure 6) obtained from 'sandstones' was additionally identified in sites 5a and 8a as a single component magnetization (Figure 6/S5A), or more recurrently defining the HB component if coexisting with the SSW directed remanence.

The magneto-mineralogical properties of the boulders (Table I; Figure 9) are almost entirely governed by magnetite (Tb<580°C; Curie-temperatures around 580°C), and IRM curves repeatedly

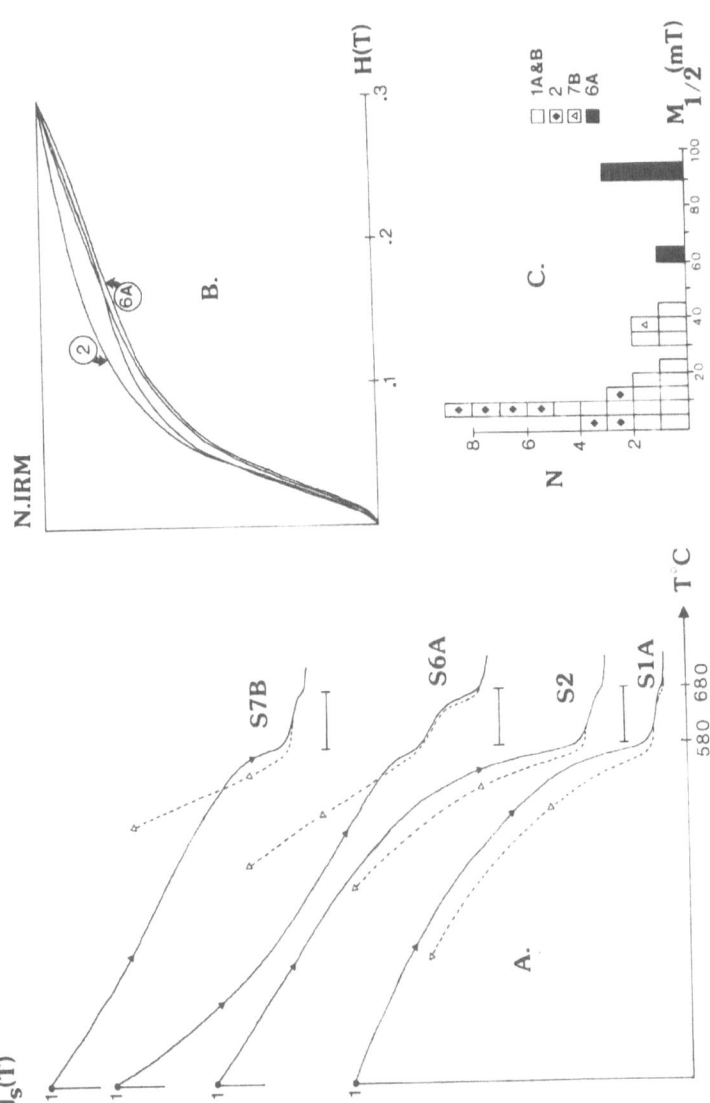

Fig. 7. Thermomagnetic analysis [Js(T)], normalized IRM curves and a histogram showing the frequency of the median destructive field ($M_{1/2}$) as applied during AF demagnetization (Devonian sandstones, fine-grained conglomerate and the regolith).

show saturation in fields of 100-200 mT (compare fig.9/6B). Accessory haematite and/or paramagnetic phases, however, are present, notably in the volcanoclastic boulders.

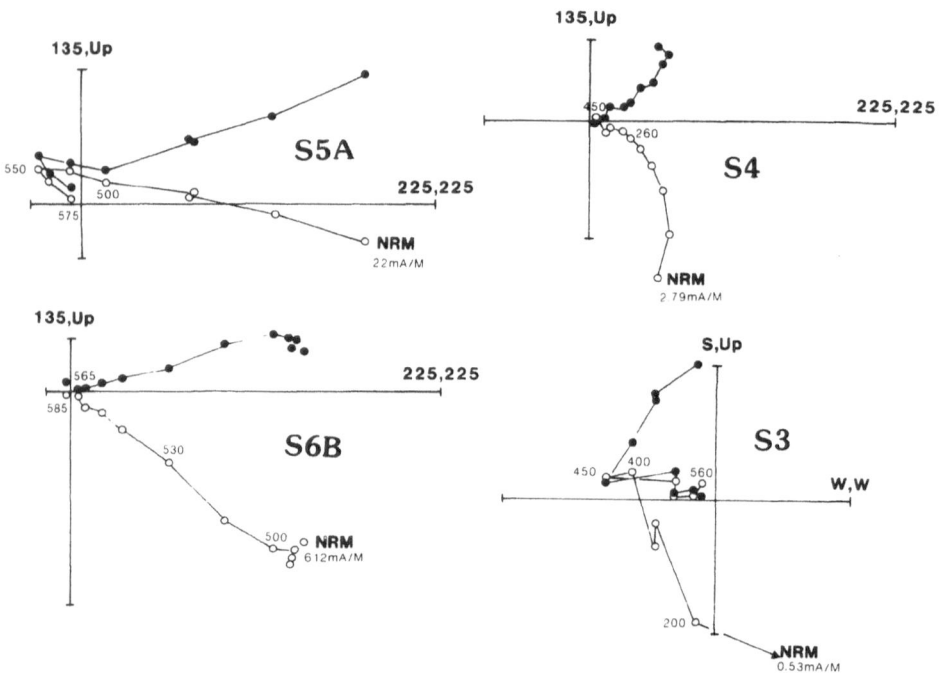

Fig. 8. Further examples of thermal and AF demagnetization. Note that apart from the lower left diagram (S3), the diagrams are shown in a 225,225 projection, thereby optimizing the two-component structure (SMA and SMB interplay) observed from a conglomerate boulder (S5A) compared with two other examples (S4-pre-Devonian and S6B-basic boulder) where the SMB-component is identified above 260 and 500°C. The lower left diagram has a two-component structure, e.g. S5A, but only an expanded part of the diagram is shown (E-W projected) to illustrate the top SMA remane nce in a weakly magnetized volcano-clastic sample (NRM=0.53mA/M).

In the pre-Devonian substrate at Glassøya, site 4 specimens (basic dyke) show a similar SSW remanence to that recorded in the boulders. Incidentally there is a reasonably linear trend toward the origin of the vector-diagrams above 250/350°C (see Figure 8/S4), but in general an HB component is present due to pronounced off-origin trajectory. Directional stability, however, ceased at high temperatures (T>400-500°C), which may relate to viscous effects and magnetomineralogical changes (cf. increase in Js during cooling; Figure 9). Moreover, most specimens (cf. Fig. 8/S4) are strongly biased by a northerly and steeply inclined low-blocking (LB) component (not properly isolated from site 4). At sites 9 and 10 on the main island of Smøla, this steep and northerly magnetization nearly dominates the NRM. As for the Devonian boulders, magnetite is the most important opaque phase.

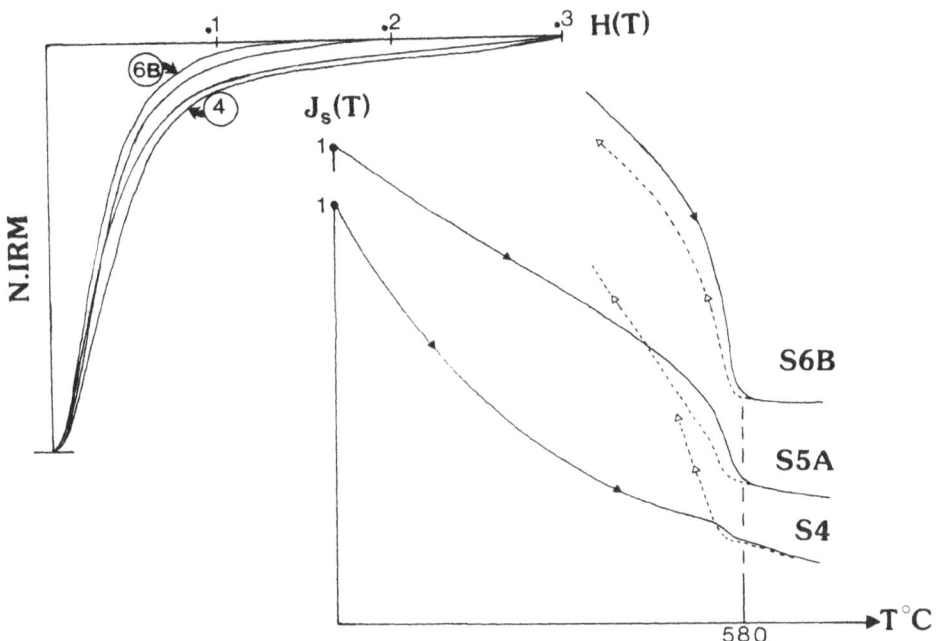

Fig. 9. Normalized IRM curves and thermomagnetic analysis of Devonian boulder specimens (e.g. S6B and S5A) and Ordovician volcanoclastics and dykes (S4). See Fig. 6 for comparison with Devonian "sandstone".

Ordovician volcanoclastics (sites 3 and 7A) gave NRM intensities below 1 mA/m. Nevertheless, it frequently proved possible to demonstrate that these rocks carry two principal magnetizations, i.e. the SSW component superimposed on the E-W component. The latter, however, is generally poorly defined (see fig. 8/S3). Thus, for example sites 5a and 8a (Devonian boulders) and site 3 (Ordovician volcanoclastics) show an almost identical magnetization build-up (compare Figure 8 and Table I). From the pre-Devonian rocks, the E-W remanence was also identified at site 9 (Smøla).

4. Paleomagnetic reference data

In order to determine the magnitude of tectonic rotations (see later), the choice of reference data is of vital importance. Accordingly, Late Devonian to assumed Late Jurassic palaeomagnetic data have been compiled (Table II; Figure 13b and c), and a spherical smoothed spline (cf. Jupp and Kent, 1987) was computed from a total of 22 poles with pre-defined magnetic ages. Apart from palaeomagnetic data from a lamprophyre dyke of possible Permian age from Ytterøy and a Late Carboniferous syenite sill from Stabben (Figure 1) all the included data are located outside the MTFZ. The oldest part of the spline (360 Ma) is based on Late Devonian/Early Carboniferous poles, all derived from rocks (Western/Central Norway) thought to have undergone magnetic overprinting in Late Devonian time (Solundian/Svalbardian event). The spline is most successfully constrained during the Late Carboniferous and Permian (all having reverse polarity) since the palaeomagnetic data are obtained from dykes/sills with good age control. The Permian polar trend, and the suggested westerly movement into the Triassic coincide with most published European APWP's (Torsvik et al., in prep), but the prominent easterly trend of the spline during the Late Jurassic is extraordinary. It is partly an artifact of 'non-smoothing' since the spline <u>ends</u> with an assumed Late Jurassic pole. Late Jurassic palaeomagnetic data from Spitsbergen, however, appear to sustain the suggested easterly polar path during the Jurassic (Halvorsen, 1988). It should be emphasized, however, that all Triassic to Late Jurassic poles from Norway are assumed ages, and they all represent magnetic overprints in physically older rocks and/or have been derived from fault-rocks.

5. Interpretation of remanence data

Two major remanence components (Figure 10 a and b) have been identified from Smøla, denoted SMA (E-W remanence) and SMB (SSW remanence). In addition, a third but less influential low blocking remanence, SMC, has been observed (see Figure 10b). SMA is notably preserved in Devonian 'sandstones'. Two of the site-mean directions are IB components, but their mean directions hardly differ from the HB site-means (Table I). SMA is observed in sandstone/siltstone, fine-grained conglomerates, the regolith (clast size ca. 1 cm), Devonian boulders and the pre-Devonian substrate. This signifies that SMA is of secondary origin. The remanence properties of Devonian sandstone/siltstone (site 6a) are dominated by haematite, whereas the increasing importance of magnetite can be recognized in the fine-grained conglomerates. This reflects the fact that clasts carry a substantial proportion of the 'magnetite' held part of the remanence. SMB is developed in both Devonian boulders and the pre-Devonian substrate, but is unimportant in Devonian 'sandstones'. This we consider as relating to a disparity in magnetic stability, i.e. magneto-mineralogical (haematite/magnetite) and magnetite grain-size contrasts. SMB is carried exclusively by magnetite, and SMA conforms to the upper thermal and coercivity spectra when these remanences are identified at specimen level. SMC is present in a number of specimens, but only firmly established from three sites.

SMA, SMB and SMC all have post-depositional origins. Their precise mode of remanence acquisition is as yet uncertain, though it was probably facilitated through thermo-chemical (TCRM) and/or thermo-viscous (TVRM) processes. There are minor differences in the attitude of bedding between different sampling areas, but stepwise unfolding (pre-Devonian rocks and boulders rotated

445

Table II. Selected late Silurian to early Cretaceous palaeomagnetic poles from Norway, which have been used to construct a spherical smoothed spline (cf. text).

| Code Location (Lat.,Long.) | VGP | | assumed magnetic |
	North	East	AGE
Late Devonian/Early Carboniferous:			
1 Hornelen 1 (61.8,5.3)	12	147	360
2 Hasteinen 1 (61.6,5.4)	16	155	360
3 Kvamshesten 1 (61.4,5.4)	21	144	360
Late Carboniferous/Permian:			
4 Stabben (63.3,8.5)	32	174	290
5 Kvamshesten 2 (61.4,5.4)	32	166	300
6 Ny-Hellesund (58.0,7.8)	38	160	255
7 Arendal 1 (58.4,8.8)	44	161	255
8 Oslo 1 (59.7,10.4)	40	160	260
9 Sunnhordaland 1 (60.0,5.5)	43	162	275
10 Arendal 2 (58.4,8.8)	39	153	255
11 Oslo 2 (58.7,10.4)	45	158	260
12 Hornelen 2 (61.8,5.3)	43	175	260
13 Ytterøy (63.6,11.1)	44	143	255
A2 Atloy 1 (61.3,5.2)	43	144	255
Triassic:			
14 Sunnhordaland 2 (60.0,5.5)	46	117	230
15 Hasteinen 2 (61.6,5.4)	54	111	230
Jurassic/Early Cretaceous:			
16 Hornelen 3 (61.8,5.3)	59	148	175
17 Sotra 1 (60.5,4.0)	66	145	175
18 Sotra 2 (60.5,4.0)	63	182	175
A1 Atloy 2 (61.3,5.2)	61	213	150

1,16-Torsvik et al. (1988); 2,15-Torsvik et al. (1987); 3-Torsvik et al. (1986); 4-Sturt and Torsvik (1987); 5, 12-Smethurst (1987); 6-Halvorsen (1970); 7,10-Halvorsen (1972); 8-Storetvedt et al.(1978); 9,14-Lovlie (1981); 11-Douglass (1987); 13-Torsvik et al. (1988) and this study; A2, A1- Sturt and Torsvik (1988); 17,18-Lovlie and Mitchell (1982).

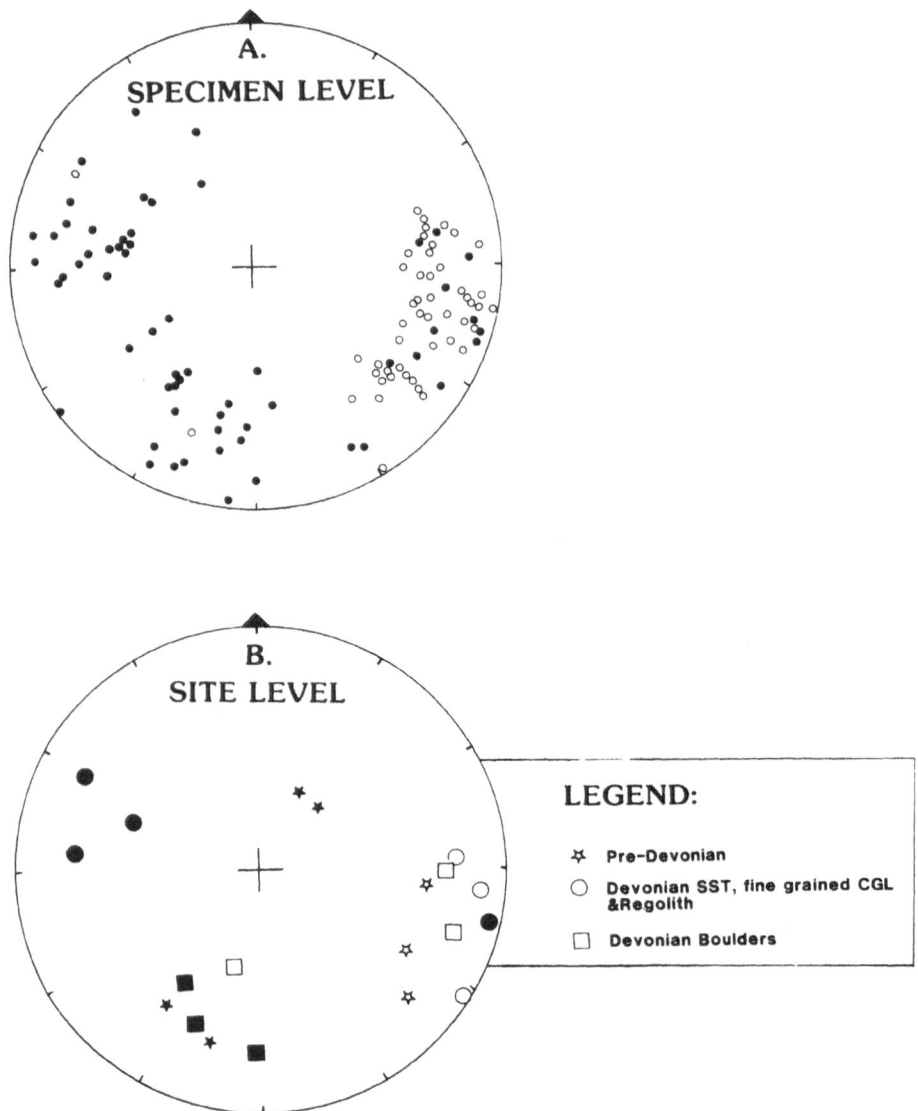

Fig. 10. Stereoplots showing the distribution of characteristic remanence components (in situ) at specimen level (only SMA and SMB remanences) and site level (b).

according to S_0 in the Devonian) points toward a syn- or post- fold origin for SMA (Figures 11 and 12; Table III). The dual-polarity structure of SMA is demolished after tectonic correction (Fig. 11),

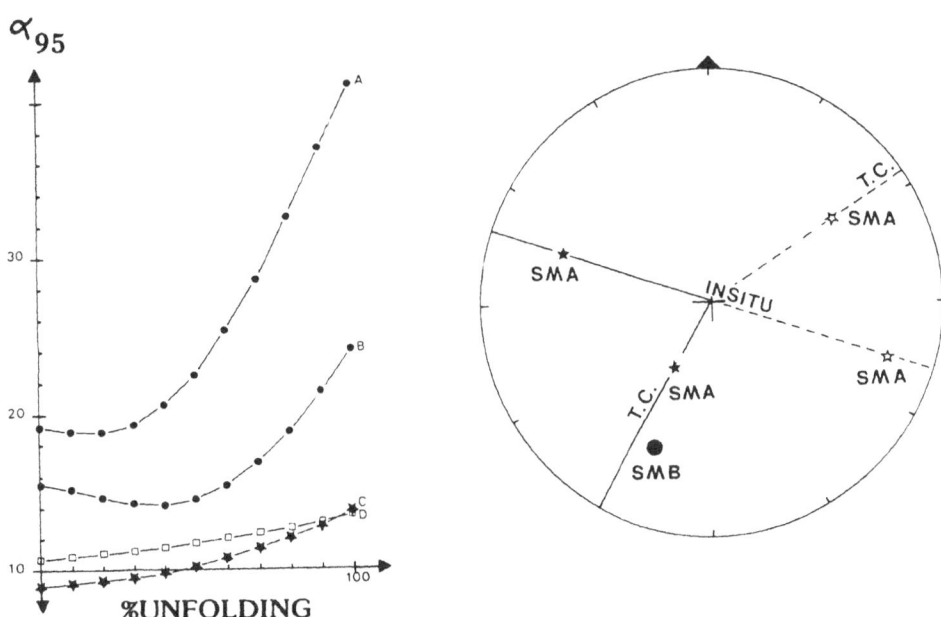

Fig. 11. Stepwise unfolding of the SMA remanence. Curves are as follows:
(a) reverse data, only IB; (b) reverse data, IB+HB; (c) normal data; (d) normal+reverse data. Only (d) is statistically significant at the 95% confidence level. Note that unfolding destroys the dual polarity structure of the SMA component (see stereoplot). In the stereoplot, mean values of SMAr and SMAn (SMA reverse and normal) are shown in-situ and after tectonic correction. For compariosn, the in-situ mean value of the SMB remanence is also shown.

and thus the combination of normal and reverse field directions gave a statistically negative fold-test at the 95% confidence level (Table III). With regard to SMB, a fold test proved inconclusive. A tectonic correction results in a mean direction which is incompatible with any known Palaeozoic direction in Europe. The in situ direction, however, closely compare with Late Devonian/Early Carboniferous palaeomagnetic poles from western Norway and a post-fold magnetization derived from the Devonian rocks from Hitra (cf. Figure 13a and b), and SMB is therefore also considered to be of post-fold origin. The foregoing reasoning involves a circular argument, however, and in this respect an engaging point is that SMA compares better with Late Palaeozoic directions from Norway (and SMB) after tectonic correction. The only case for ignoring the negative fold-test must

448

consider possible strain/deflection of the remanence. It is possible that a considerable angle between an inferred pre-fold remanence and the compressional shortening axis could have produced site-mean directions perpendicular to the shortening axis (remanence deflection). However, we believe that this possibility is unlikely because:
(1) SMA is observed in a variety of different rock types with disparities in magneto-mineralogy, competence and internal layering,
(2) the anisotropy in some of the tested samples is as low as 3 %, and
(3) there is no systematic relationship between the principal axes, remanence directions or within-site scatter.

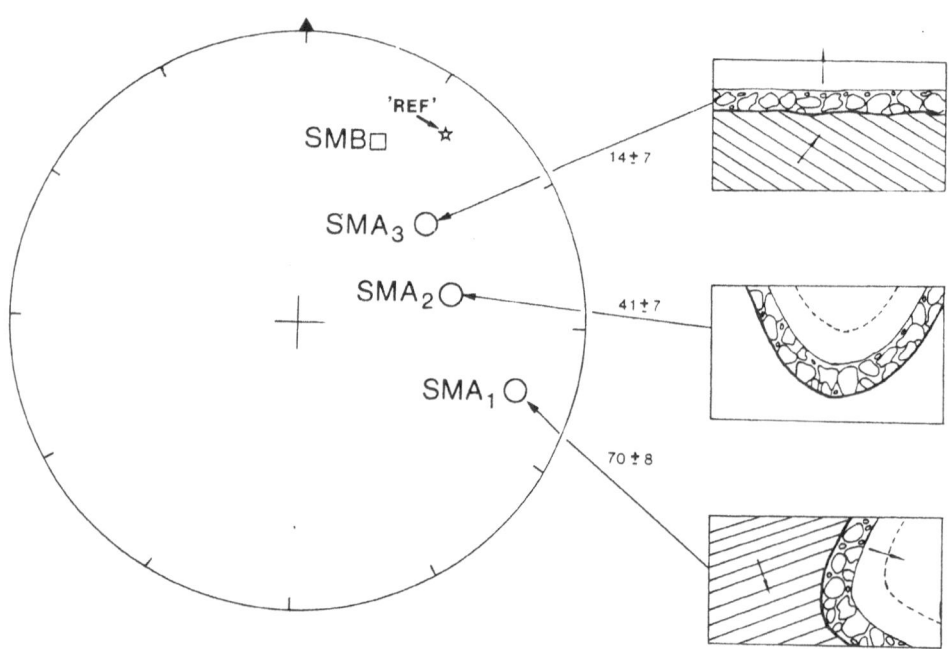

Fig. 12. The SMA and SMB (inverted in diagram) components compared with the reference (expected) Late Devonian field-direction. The SMA is displayed in three manners, i.e. in-situ (SMA1), on the assumption of an original symmetrical structure with vertical axial plane (SMA2) and corrected for bedding, unfolded (SMA3). Numbers represent amount of rotation compared with the expected field-direction. Component SMB is shown in-situ co-ordinated (SMB).

Table III. Fold-tests and final statistics.

	In-Situ					Tectonic correction							
	Dec	Inc	N	α_{95}	k_1	Dec	Inc	α_{95}	k_2	F	DF	SSF	
Group A:													
AN	:108	-10	8	10.8	20.8	056	-25	13.5	13.5	1.54	(14)	NO	
AR (IB, HB)	:289	+21	3	15.6	26.6	211	+54	24.0	11.4	2.89	(4)	NO	
AR (IB)	:295	+23	2	19.2	26.6	210	52	41.1	5.8	4.71	(2)	NO	
AN + AR	:108	-13	11	8.9	22.4	051	-33	13.7	9.5	2.36	(20)	YES	
Group B:	:202	+22	5	10.9	33.0	251	-59	11.5	29.7	1.11	(8)	NO	

Final *in-situ* statistics (including site 9):

Pole

	Dec	Inc	N	α_{95}	k	LAT	LONG	dp	dm
Group A:	107	-14	12	8.4	23.2	N13.9	E265.9	4	9
Group B:	202	+22	5	10.9	33.0	N13.4	E165.8	6	12
Group C:	028	+50	3	9.9	67.3	N53.0	E146.0	9	13

6. Discussion

6.1 LATE PALAEOZOIC/MESOZOIC DEFORMATION IN THE MTFZ

As indicated earlier three secondary remanence components have been identified on Smøla, and the youngest, SMC, is considered to relate to Mesozoic overprinting. Late Palaeozoic and Mesozoic deformation within the MTFZ is evident from the develop ment of fault-rocks (Grønlie and Roberts, 1987), and examination of carbonate-thorium veins and breccias in the inner part of the MTFZ (Figure 1) indicates a complex history with repeated brecciation and hydrothermal alteration (Grønlie and Torsvik, in prep). Three major stages of fault-rock development are indicated (Fig 13a). Palaeomagnetic data indicate that the oldest breccias, including fault-breccias located in the Verran Fault zone, are probably of Permian age. This phase appears to be coeval with alkaline intrusive activity (e.g. a lamprophyre dyke on Ytterøy; Carstens, 1961) and extensional tectonics in the North Sea (Beach et al., 1986). Subsequent reactivation of the MTFZ is indicated during the Jurassic, and fault-rocks and fault-related magnetic overprints (Stabben Sill; Figure 13a) suggest a Mid-Late Jurassic age. This stage may correspond to a phase of dextral strike-slip faulting along the Verran Fault (Grønlie and Roberts, 1987 and in prep), or alternatively post-Mid Jurassic normal dip-slip and subsequent dextral strike-slip in connection with the development of the Edøyfjorden and Beistadfjorden Basins (Bøe and Bjerklie, in prep). The relative pole-position of the SMC component from Smøla also appears to be of Jurassic age (Figure 13a). SMC has been firmly identified from three sites, and it is interesting to note that two of these sites are located on either side of a major fault which displaces both the Devonian and the pre-Devonian rocks (Figure 2; sites 9 and 10).

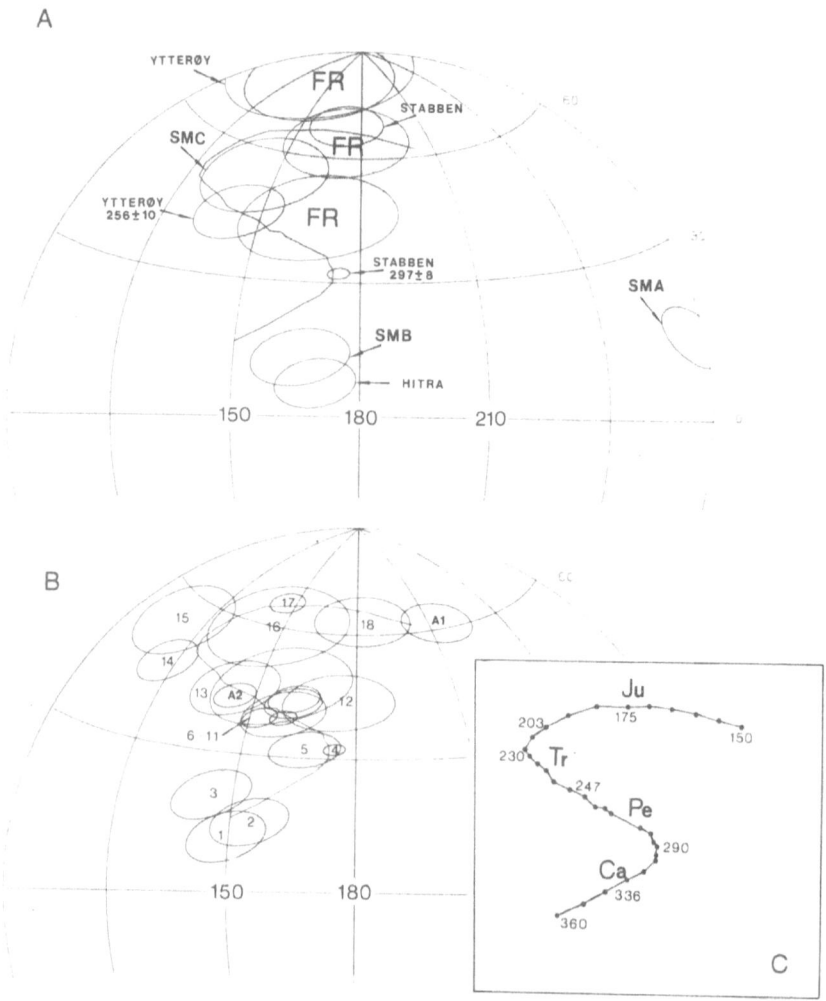

Fig. 13. a) Palaeomagnetic data obtained from rocks within the MTFZ together with a suggested spherical smoothed spline. The included poles are: SMA, SMB and SMC (this study); Ytterøy lamprophyre dyke (Torsvik et al., 1988a); Stabben syenite sill (Sturt and Torsvik, 1987), FR=fault-rocks (Grønlie and Torsvik, in prep); HI=Hitra ORS (Bøe et al., 1988). Cf. Figure 1 for geographic locations. b) Selected Late Devonian to Late Jurassic palaeomagnetic poles from Norway, and a calculated spherical smoothed spline. Cf. Table II and text for details. All poles in (a) and (b) shown as dp/dm semi- confidence ellipses (equal-area projection). c) Details in the spherical smoothed spline. Ca=Carboniferous, Pe=Permian, Tr=Triassic, Ju=Jurassic. Numbers in million years.

Finally, in the course of the tectonic history of the MTFZ, the latest activity yet recorded palaeomagnetically probably occurred in Early Tertiary times. This latter tectonic phase coincides with moderate normal dip- slip and brecciation, and was accompanied by hydrothermal alter ation.

6.2 THE EARLY HISTORY OF THE MTFZ

The SMB (in-situ) component correlates with the Late Devonian syn- to post-fold magnetic overprints observed from other Devonian massifs in western Norway and Hitra (Figure 13a and b).The older SMA component, on the other hand, is clearly anomalous when compared with either Late Devonian or younger palaeomagnetic data from Norway (Figure 13a and b). In order to explain this, it is necessary to infer a clockwise (based on the dual-polarity of SMA) tectonic rotation in the order of $70 \pm 8°$. In this case SMA would be a pre-rotational component and SMB post-rotational. The estimated amount of rotation could be reduced to about 15° by a bedding correction (unfolding), but this is precluded by the negative fold-test.

We may, however, speculate regarding the abrupt change in fold style between Hitra and Smøla. The ORS rocks of Hitra are folded into a near symmetrical syncline with a subvertical axial plane in marked contrast to the overturned asymmetrical nature of the syncline on Smøla with its quite gently NW-dipping axial plane (Figure 5a and b). If we make the assumption that the synclines originally had approximately the same styles, it would require a rotation on a horizontal axis (255/-60) to bring the axial plane on Smøla into the vertical. If this can be justified, the block rotation of Smøla would accordingly be reduced to $41 \pm 7°$ (Fig. 12). Minor adjustments must also be made for the southerly dip of the Jurassic sediments in Edøyfjorden (Figure 1), but this makes a difference of less than 10 degrees. Even when such adjustments are made, a clockwise rotation of Smøla in the order of 30-60° has to be accepted. Such a rotation would also explain the marked difference between the pre-Devonian structural grain of Smøla (N-S) and the neighboring areas (ENE-WNW). Acceptance of the rotation of the Smøla block carries the implication that the trend of the major syncline must have formerly had a NE-SW to NNE-SSW trend. This leads to a consideration that Devonian palaeogeography may possibly have been one of upstanding fault-blocks separated by subsiding depot-centres with differing trends, a pattern which is clearly marked in the Carboniferous of northern England (Johnson, 1967).

Recent studies of the geology of the Smøla-Hitra region have emphasized the anomalous character of not only the Devonian rocks but also their substrate. The marked differences between this pre-Devonian basement and the rocks SE of the MTFZ are compelling. There is now good evidence that the main deformation and low-grade metamorphism of the Arenig-Llanvirn volcano-sedimentary rocks of Smøla occurred during the Middle to Late Ordovician (Sundvoll and Roberts, 1977; Roberts, 1980; Gautneb, 1987; R. Bøe and R. Tucker pers. comm., 1988), in contrast to Mid to Late Silurian Scandian age for the medium-grade metamorphism in nappes and basement rocks on the mainland south of the MTFZ. This led to the suggestion (Roberts, 1987; Sturt and Roberts, 1987) that the rocks of the Smøla-Hitra-Orlandet tract formed a distinctive terrane, which these authors named the Smøla Terrane; the implication being that this separete terrane unit docked against the Western Gneiss Region sometime in the period from Late Silurian to Late Devonian times along the approximate site of the MTFZ. We will now consider the possible nature of this terrane amalgamation. A model for the docking of the Smøla Terrane requires either:

(1) An emplacement of the terrane along the approximate line of the MTFZ by a strike-slip mode, or
(2) emplacement along a major detachment at depth with suturing along the approximate line of the MTFZ.

A combination of these mechanisms may also be applicable.

Although possible strike-slip motions on the MTFZ in Devonian times have been proposed in the literature (Steel et al. 1985; Grønlie and Roberts, 1987), there is in fact no direct evidence for such a postulate. Nor is there any direct evidence for the type of detachment suggested above, though a certain amount of evidence is available, particularly recent information from Mobil Search Lines (C. Hurich; pers. comm., 1988).

On the non-migrated section of profile ILP7 (Figure 1; 20km NW of Smøla-Hitra), a narrow zone of reflectors with 15° apparent dips toward NW lies at a depth of approximately 10 km. Assuming a continuation with similar apparent dips beneath Smøla- Hitra, this zone of reflectors can be predicted to lie at about 5km beneath the southern coast of Hitra.The possibility can thus be envisaged that this may represent the lower boundary of a major thrust nappe. Above the narrow zone of reflectors a number of other reflectors can be observed, dipping in the same direction.

A zone of strong reflectors in the lower and middle crust (with general northerly apparent dips in the order of 30°) comes up to a depth of 15 km off Stadtlandet (ILP 11b; N-S line) just N of the trace of the MTFZ line. To the north this zone descends to Moho depths. It is possible that this zone of reflectors links with the MTFZ, as they are not apparent south of its trace, even though the observed faults along the line of the MTFZ at high crustal levels are vertical to steeply inclined structures. It is reasonable to assume that the marked reflection zone represents an older crustal fracture, i.e. earlier than the obvious Mesozoic structural signature. A similar zone of very strong dipping reflectors is seen on profile ILP7 at a depth of 18 km (20 km offshore Hitra) descending to Moho depths. The zones in the two profiles show almost identical signatures and would appear to correlate, and may represent a deep-seated decollement consistent with a collision along the line of the MTFZ.The trace of the MTFZ offshore also coincides with a linear high-gravity belt (>50-60 milligals) between Shetland and Møre (Grønlie and Ramberg, 1970; Bott and Watts, 1971; Talwani and Eldholm, 1972; Vogt, 1986).

The Mobil Search data, together with local evidence of southerly directed thrusting in the ORS on Hitra (Bøe et al., 1988), makes an attractive combination in a model which also allows for the recorded rotation of the Smøla block. If we assume that a detachment is present beneath Smøla-Hitra, and with southerly directed translation of the hanging-wall, this would demand that the nature of the proposed MTFZ-suture was essentially that of a reverse fault. At the same time, to allow for a clockwise rotation of Smøla this would involve a sinistral strike-slip fault or lateral ramp separating Hitra from Smøla and penetrating to the assumed thrust detachment surface at ca. 5km depth (Figure 14). The proposed model thus requires:

(a) That Devonian sedimentation north of the MTFZ occurred in a palaeogeographic setting of fault-blocks and troughs allowing for an initial difference in trend of the Hitra and Smøla ORS basins;

(b) The presence of a detachment surface or surfaces at depth along which the Smøla Terrane was translated until docking with the Western Gneiss Region occurred along the approximate line of the MTFZ;

(c) The presence of a sinistral strike-slip fault between Hitra and Smøla which penetrated to the proposed thrust/ detachment. This would allow for the independent clockwise rotation of Smøla and adjacent islands;

(d) Depending on the principal stress vector during the time of Late Devonian folding and docking, the MTFZ could have taken up a strike-slip or oblique-slip motion, i.e. if the principal stress vector was not orthogonal to the MTFZ. Assuming southerly thrusting, then sinistral shear is likely.

We realize that this model is at best speculative, but it does satisfy a number of geological and geophysical constraints and recorded features outlined in the text.

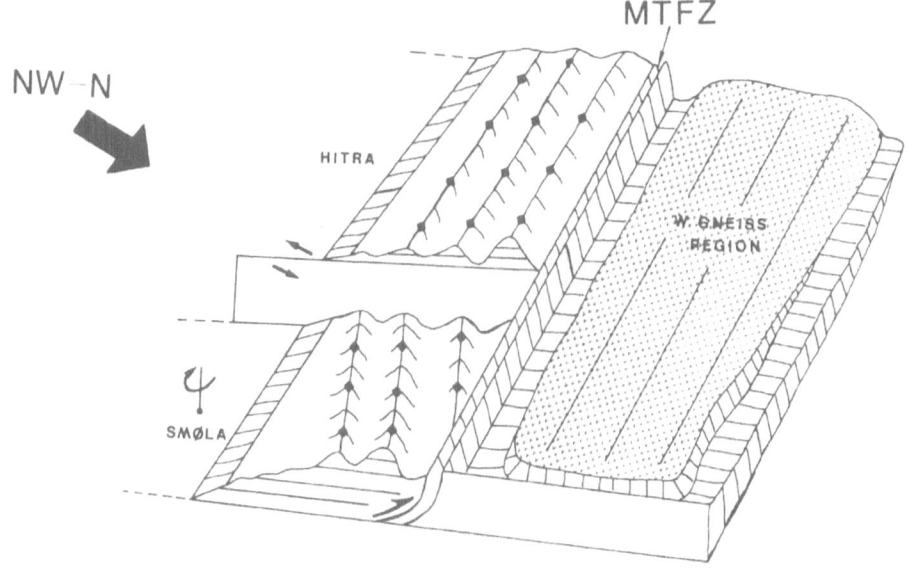

Fig. 14. Schematic cartoon illustrating the suggested collision of the Smøla Terrane with the Western Gneiss Region, with reverse or high-angle oblique-slip faulting approximately along the site of the MTFZ. The rotation of Smøla is considered to be accommodated by sinistral strike- slip faulting between Smøla and Hitra. This fault must terminate at depth against a low-angle thrust/detachment surface. See text for explanations.

South of the MTFZ the main Devonian massifs were folded and faulted during Late Devonian (Solundian) crustal shortening (approximately N-S) to produce a prominent E-W fold belt (Sturt, 1981; Roberts, 1983; Torsvik et al., 1986; 1987; 1988b). It is of interest to note the progressive change in the axial orientation of folds in the Western Gneiss Region north of Hornelen from E-W to parallel with the MTFZ (cf. Figures 2 and 15). This probably implies that a major fault structure already existed along the site of the MTFZ prior to the suggested Late Devonian docking. Thus, it is quite likely that the collision-vector for the Smøla Terrane was not orthogonal to the MTFZ, i.e. if it was close to N-S a syn- to post-collisional sinistral shear displacement along the MTFZ would have been initiated, as proposed by Sturt et al. (1987) and Grønlie and Roberts (1987).

When the rocks of West Shetland are restored for minimum movements on the Walls Boundary Fault (Flinn, 1979), i.e. ca. 50 km sinistral displacement, and allowing for ca. 100% E-W extension of the Mesozoic Viking Graben, the Devonian rocks of W. Shetland will be placed approximately opposite the West Norwegian Devonian massifs (Figure 15). The east-west trend of the West Shetland Devonian would thus appear to be a continuation of the E-W fold belt of the Norwegian mainland. In many ways the Devonian geology of the West Shetland epitomizes the problem that we have addressed. Not only does it contain part of a once continuous fold-belt, but the metamorphic grades (anchizone to lower greenschist facies) are also compatible. One of the

454

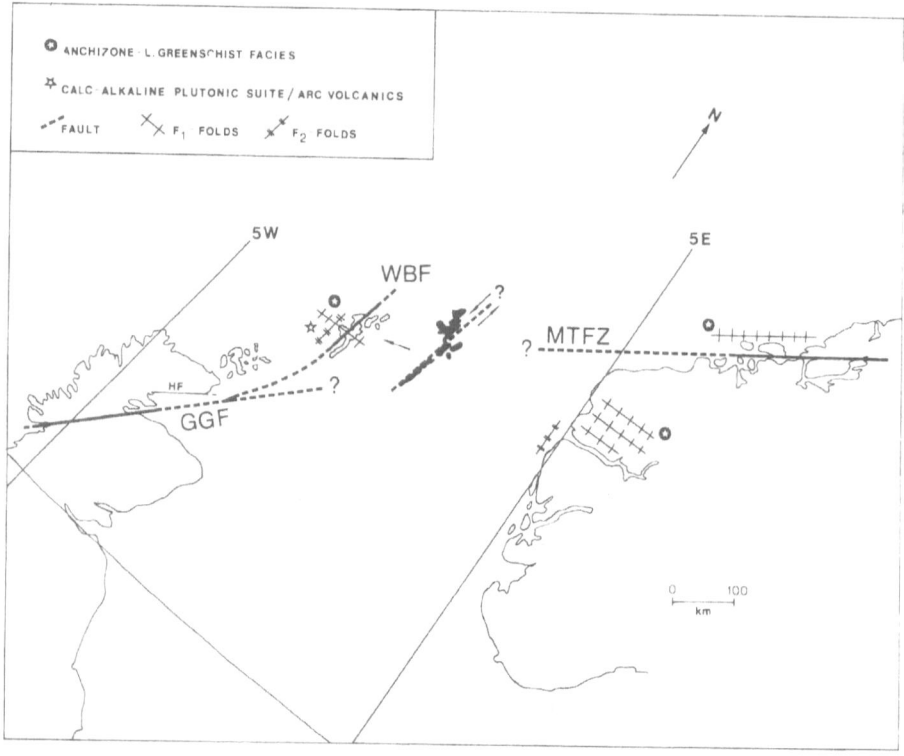

Fig. 15. Fold-trends in the deformed Devonian rocks of Western and Central Norway and West Shetland. West Shetland is restored for ca. 50 km. sinistral movement along the Walls Boundary Fault, and ca. 100% E-W extension in the northern North Sea. GGF=Great Glen Fault, HF=Helmsdale Fault.

striking features of the West Shetland ORS sequence is the presence of a thick pile of Lower-Mid Devonian calc-alkaline lavas which show a mature island-arc geochemical signature (Thirlwall, 1981).The sequence is further cut by a calc-alkaline batholithic complex. Both the volcanic and the plutonic rocks are elements which are normally associated with subduction.This raises an interesting question, namely the possibility that the final closure of part of the Iapetus Ocean did not occur until Mid-Late Devonian times.

Acknowledgements

The Norwegian Research Council for the Humanities and Sciences (NAVF), the Geological Survey of Norway (NGU) and Statoil are thanked for financial support. R. Van Der Voo is thanked for comments on an early version of the manuscript. Norwegian Lithosphere Contribution (37).

References

Askvik, H. and Rokoengen, K., 1985. Bed rock map KRISTIANSUND -M. 1.250.000, *Nor. Geol. Unders.*

Beach, A., Bird, T. and Gibbs, A., 1987. Extensional tectonics and crustal structure; deep seismic reflection data from the northern North Sea Viking Graben. In M.P. Coward, J.F. Dewey and P.L. Hancock (Eds.) Continental Extensional Tectonics, *Geol. Soc. Lond. Special Pub.*, **28**, 467-476.

Bott, M.H.P. and Watts, A.B., 1971. Deep structure of the continental margin adjacent to the British Isles. In The Geology of the East Atlantic continental margin, *Inst. of Geol. Sci.*,, **2**, 70/14, 89-109.

Bruton, D.L. and Bockelie, J.F., 1979. The Ordovician sedimentary sequence of Smøla, West central Norway. *Nor. Geol. Unders.*, **348**, 21-32.

Bruton, D.L. and Harper, A.T., 1985. Early Ordovician (Arenig-Llanvirn) fauna from oceanic islands in the Appalachian-Caledonian orogen. In:D.G. Gee and B.A. Sturt (Eds.) *The Caledonian Orogen and related areas*, John Wiley and Sons Ltd, 359-368.

Bøe, R., 1986. Some sedimentological and structural studies of the Old Red Sandstone Hitra Group, Hitra, Sor-Trøndelag, *Cand. Scient. Thesis*, Univ. of Bergen, Dept. of Geology, Norway, 238 pp.

Bøe, R. and Bjerklie, K. On the occurence of Mesozoic sedimentary rocks in Edøyfjorden and Beistadfjorden, Central Norway: Implications for the structural history of the Møre-Trøndelag-Fault-Zone (in prep).

Bøe, R., Atakan, K. and Sturt, B.A., 1988. The style of deformation of he Devonian rocks of Hitra and Smøla (in progess).

Carstens, H., 1961. A post-Caledonian ultrabasic lamprophyre dyke on the island of Ytterøy in the *Nor. Geol. Tidsskr.*, **47**, 10-21.

Cristie-Blick, N. and Biddle K., 1985. Deformation and basin formation along strike-slip faults. *Soc. Econ. Pal. Min.*, 2-34.

Douglass, D.N. and Kent, D.V., 1986. Multicomponent magnetization of the Upper Silurian-Lower Devonian Ringerike Sandstone, adjacent dykes and Permian Lavas, Oslo, Norway. *Eos*, 67, 267.

Fediuk, F., 1975. Smøla, Berggrunnsgeologisk kart 1321, M:1:50000, Nor. Geol. Unders.

Fediuk, F. and Siedlecki, S., 1977. Beskrivelse til det berggrunns-geologiske kart 1321. *Nor. Geol. Unders.*, **330**, 1-23.

Fisher, R.A., 1953. Dispersion on a sphere. *Proc. R. Soc. Lond.Ser.A.*, **217**, 295-305.

Flinn, D., 1979. Transcurrent faults and associated cataclasis in Shetland. *J. Geol. Soc. Lond.*, **133**, 231-248.

Gabrielsen, R.H. and Ramberg, I.B., 1979. Fracture patterns in Norway from Landsat imagery: results and potential use. Proc. Noregian Sea Symposium, Troms. *Nor. Petr. Soc.*, 1-28.

Gabrielsen, R.H., Ferseth, R., Hamar, G. and Ronnevik, H., 1984. Nomenclature of the main structural features on the Norwegian Continental Shelf north of the 62nd parallel. Petroleum Geology of the North European Margin, *Nor. Petr. Soc.*,Graham and Trotman, 41-60.

Gautneb, H., 1987. Hoy-K dioritter og assosierte bergarter pa sydlige Smøla Arkipel, More og Romsdal, Cand. *Scient thesis*, Univ. of Bergen, Dept. of Geology, Norway, 265 pp.

Grønlie, A. and Roberts, D., 1987. Dextral strike-slip duplexes of Mesozoic age along the Hitra-Snasa and Verran Faults, Møre-Trøndelag Fault Zone, Central Norway. *NGU report*, 87.139, 24pp.

Grønlie, R. and Ramberg, I.B., 1970. Gravity indications of Deep sedimentary Basins below the Norwegian Continental Shelf and the Voring Plateau. *Nor. Geol. Tidsskr.*, 50, 357-391.

Halvorsen, E., 1970. Palaeomagnetism and the age of the younger diabases in the Ny-Hellesund areas, S. Norway. *Nor. Geol. Tidskr.*, 50, 157-166.

Halvorsen, E., 1972. On the palaeomagnetism of the Arendal diabases. *Nor. Geol. Tidskr.*, 52, 217-228.

Halvorsen, E., 1988. Palaeomagnetism of Upper Jurassic/Lower Cretaceous Dolerites from Hinlopenstretet, Svalbard; Implications for the Mesozoic Apparent Polar Wander Path for Western Europe (in progress).

Johnson, G.A.L., 1967. Basement control of Carboniferous sedimentation in Northern England. *Proc. Yorkshire Geol. Soc.*, 36, 175-194.

Jupp, P.E. and Kent, J.T., 1987. Fitting smooth paths to spherical data. *Applied Statistics*, 36, 14-46.

Lovlie, R., 1981. Palaeomagnetism of coast-parallel alkaline dykes from western Norway; ages of magmatism and evidence for crustal uplift and collapse. *Geophys. J. R. astr. Soc.*, 66, 417-426.

Lovlie, R. and Mitchell, J.G., 1982. Complete remagnetization of some Permian dykes from western Norway induced during burial/uplift. *Phys. Earth. Planet. Inter.*, 30, 415-421.

Oftedahl, C., 1975. Middle Jurassic graben tectonics in mid-Norway, *Proc. Jur. North Sea Symp.*, Stavanger.

Ramberg, I., Gabrielsen, R.H., Larsen, B. and Solli, A., 1977. Analysis of fracture patterns in southern Norway. *Geol. en Mijnbouw*, 56, 295-310.

Rindstad, B. and Grønlie, A., 1986. Landsat TM-data used in the mapping of large-scale geological structures in coastal areas of Trøndelag, Central Norway. *NGU Bull.*, 407, 1-12.

Roberts, D., 1980. Petrochemistry and palaeogeographic setting of the Ordovician volcanic rocks of Smøla, Central Norway. *Nor. Geol. Unders.*, 359, 43-60.

Roberts, D., 1983. Devonian tectonic deformation in the Norwegian Caledonides and its regional perspectives. *Nor. Geol. Unders.*, 380, 85-96.

Roberts, D., 1986. Structural-photogeological and general tectonic features of the Fosen-Namsos Western Gneiss Region of Central Norway. *Nor. Geol. Unders.*, 407, 13-25.

Roberts, D., 1987. Terranes in the Scandinavian Caledonides. Manus in prep. (*IGCP Project* 233).

Smethurst, M., 1987. Old Red Sandstone palaeomagnetism of Ireland and Norway and palaeogeography of the North Atlantic region. *Ph.D. Thesis*, University of Leeds.

Siedlecka, A. and Siedlecki, S., 1972. A contribution to the geology of the Downtonian sedimentary rocks from Hitra. *Nor. Geol. Unders.*, 275, 1-28.

Steel, R. J., Siedlecka, A. and Roberts, D., 1985. The Old Red Sandstone Basins of Norway and their deformation: A review. In D.G. Gee and B.A. Sturt (eds.) *The Caledonide Orogen Scandinavia and related area*, John Wiley and Son Ltd., 293-315.

457

Stephens, M.B. and Gee, D.G., 1985. A tectonic model for the evolution of the eugeoclinal terranes in the central Scandinavian Caledonides. In D.G. Gee and B.A. Sturt (eds.) *The Caledonide Orogen Scandinavia and related area*, John Wiley and Son Ltd., 953-977.

Storetvedt, K.M., Pedersen, S., Lovlie, R. and Halvorsen, E., 1978. Palaeomagnetism in the Oslo rift. In J.B. Ramberg and E.R. Neumann (Eds.) *Tectonics and Geophysics of Continental Rifts*, D. Reidel Publ. Co., Holland, 289-296.

Strand, T., 1932. A lower Ordovician fauna from the Smøla Island, *Norway. Nor. Geol. Tids.*, **11**, 356-366.

Sturt, B.A., 1983. Late Caledonian and possible Variscan stages in the orogenic evolution of the Scandinavian Caledonides (abstract), in:The Caledonide Orogen - *IGCP Project 27*, Morocco and Palaeozoic Orogeneses, Symposium de Rabat.

Sturt, B.A. and Torsvik, T.H., 1987. A late Carboniferous palaeomagnetic pole recorded from a syenite sill, Stabben, Central Norway. *Physics Earth. Planet. Intr.*, **49**, 350-359.

Sturt, B.A. and Torsvik, T.H., 1988. Palaeomagnetism and dating of fault movements (submitted to *Geology*).

Sturt, B.A. and Roberts, D., 1987. Terrane linkage in the final stages of the Caledonian orogeny in Scandinavia. Extended abstract *IGCP Project 233* symposium, Nouakchott, Mauritania, 193-196.

Sturt, B.A., Torsvik, T.H. and Grønlie, A., 1987. Dating of Hydrother-alteration-zones and breccias in the Trondheimsfjord region: The Møre Trøndelag Fault Zone - A preliminary report. *NGU report*, **87**.138, 11 .

Sundvoll, B. and Roberts, D., 1977. Fremgangsmate pa datering og geokjemi av eruptivbergarter pa Smøla og Hitra. *Nor. Geol. Unders* (Unpublished report).

Talwani, M. and Eldholm, O., 1972. The continental margin off Norway: A geophysical study. *Geol. Soc. Am. Bull.*, **83**, 3575-3606.

Thirlwall, M.F., 1981. Implications for Caledonian plate tectonic models of chemical data from volcanic rocks of the British Old Red Sandstone. *J. Geol. Soc. Lond.*, **138**, 123-138.

Torsvik, T.H., Sturt, B.A., Ramsay, D.M. and Vetti, V., 1987. The Tectono-Magnetic signature of the Old Red Sandstone and Pre-Devonian strata in the Hasteinen area, Western Norway, and implications for the later stages of the Caledonian Orogeny. *Tectonics*, **6**, 3305-322.

Torsvik, T.H., Sturt, B.A., Grønlie, A. and Ramsay D.M., 1988a. Palaeomagnetic data bearing on the age of the Ytterøy Dyke, Central Norway. *Phys. Earth Planet. Inter.*, (in press).

Torsvik, T.H., Sturt, B.A., Ramsay, D.M. , Kisch, H.J. and Bering, D., 1986. The tectonic implications of Solundian (Upper Devonian) magnetization of the Devonian rocks of Kvamshesten, western Norway. *Earth Planet. Sci. Lett.*, **80**, 337-347.

Torsvik, T.H., Sturt, B.A., Ramsay, D.M., Bering, D. and Fluge, P., 1988b. Palaeomagnetism, magnetic Fabrics and the structural style of the Hornelen Old Red Sandstone, Western Norway. *J. Geol. Soc. Lond.* (in press)

Torsvik, T.H., Smethurst, M., Briden, J.C. and Sturt, B.A., in prep. *A review of Palaeozoic Palaeomagnetic data from Europe.*

Underhill, J.R., Gayer, R.A., Woodcock, N.H., Donnelly, R., Jolley, E.J. and Simpson, I.G., 1988. The Dent Fault System, northern England - reinterpreted as a major oblique-slip fault zone. *J. Geol. Soc. Lond.*, **145**, 303-316.

Vogt, P.R., 1986. Geophysical and geochemical signatures and plate tectonics. In B.G. Hurdle (Ed.) *The Nordic Seas*. Springer-Verlag, 413-662.

Vogt, T., 1928. Den norske fjellkjedens revolusjonsshistorie. *Nor. Geol. Unders.*, **122**, 97-115.

Woodcock, N.H. and Fischer, M., 1985. Strike-slip duplexes. *Struct. Geol.*, **8**, 725-735.

PALEOMAGNETICALLY OBSERVED ROTATIONS ALONG THE HIKURANGI MARGIN OF NEW ZEALAND

R.I.WALCOTT
Research School of Earth Sciences,
Victoria University of Wellington,
PO Box 600,
Wellington,
New Zealand.

ABSTRACT. The Hikurangi Margin is the active subduction complex on the east coast of the North Island and the northern South Island of New Zealand that lies between, and structurally links, the Tonga-Kermadec subduction zone with the Alpine Fault. The current deformation of the Hikurangi margin is well determined from seismicity and geodetic strain observations. Extension in the north perpendicular to the plate boundary zone changes to shortening in the south so that present-day rotation of the margin as a whole is indicated: such rotation is observed paleomagnetically. The amounts and rates of rotation vary with time and location but on average the margin has rotated at about 4 degrees/Ma for at least the last 10Ma. Although the rotation is not uniform over the margin as a whole, it is so over discrete parts of it. These parts are termed domains and are of the order of 100kms in lateral dimension; they appear to have rotated in a coherent, uniform manner. The overall rotation of the margin is considerd to record a progressive change in orientation of the underlying subducted plate and such a rotation is inferred from plate tectonic reconstructions. Two domains are rotating rapidly today, in excess of 5 degrees/Ma and behind them the back-arc region is itself deforming in one case by spreading and the other by simple shear on a set of major transcurrent faults. In the middle of the margin is a domain that apparently has not rotated in the last 2Ma and it is bounded in the back-arc region by a block that has been relatively stable over the period. The driving mechanism of rotation is the rotation of the underlying slab, itself a natural outcome of the particular configuration of plate boundaries of the region.

1. Introduction

Between East Cape and Kaikoura on the east coast of New Zealand (Figure 1) lies the actively deforming, frontal wedge of the Hikurangi margin, the southernmost part of that long subduction zone extending southward from the Tonga Islands.The subduction zone terminates abruptly at Kaikoura where continental rocks lie on both sides of the plate boundary which then continues southward as a transform along the Alpine Fault and the Macquarie Ridge to the triple junction of the Australian, Antarctic and Pacific Plates.

For the last ten years paleomagnetic studies have been made along the Hikurangi margin to determine the amount and timing of rotation about vertical axes of the rocks of the region. Rotations were first described from there by Walcott, Christoffel and Mumme (1981) where it was suggested that 8Ma old sedimentary rocks within the margin had rotated 20 to 30 degrees clockwise compared to rocks of the same age lying on the Pacific Plate. Since that time numerous further results have been reported (Walcott and Mumme,1983; Mumme and Walcott, 1985; Wright and

C. Kissel and C. Laj (eds.), Paleomagnetic Rotations and Continental Deformation, 459–471.

Walcott, 1986; Mumme, Lamb and Walcott, in prep). This paper summarises the observations available to the present. Its principal aim, however, is to investigate the mechanism of the rotation. We show that fundamentally the mechanism is a simple one nvolving the rotation of the underlying subducted plate which itself is a natural consequence of the evolution of the subduction zone. We show, however, that the amount and nature of the deformation is strongly dependent on the modes of deformation possible in the crustal rocks at the boundaries of the rotating block, ie. on the Australian side of the frontal wedge.

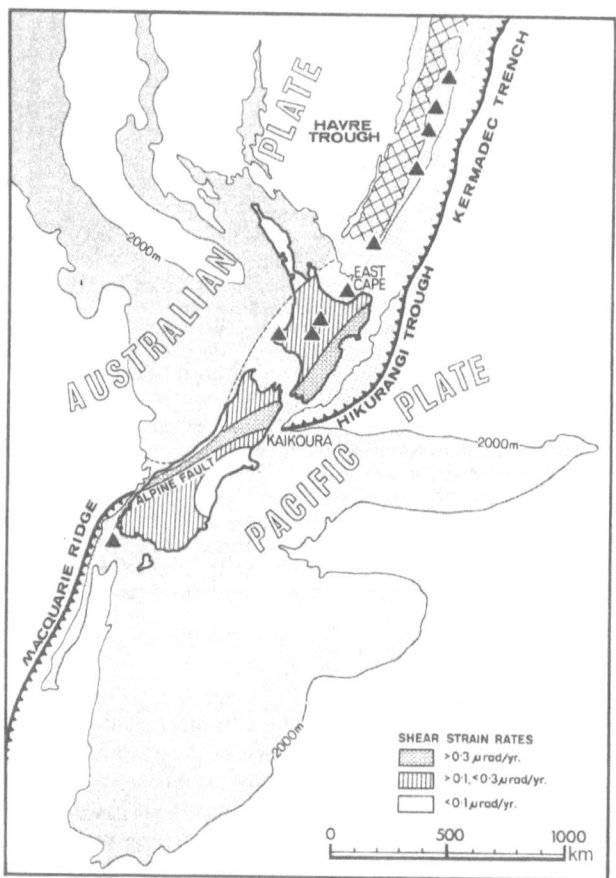

Figure 1. The tectonic setting of the Hikurangi Margin: it lies at the southern end of the Tonga-Kermadec subduction zone between East Cape and Kaikoura. The deformation rates shown on land are inferred from geodetic data. The 2000m isobath is taken to define the limits of continental andoceanic lithosphere. Subducted oceanic lithosphere, based on the distribution of intermediate and deep seismicity, is stippled. Active andesitic volcanoes are shown by solid triangles.

2. Plate reconstructions

2.1. CENOZOIC PLATE MOTIONS IN THE S.W. PACIFIC

The relative motions of the Pacific, Antarctic and Australian Plates are well determined by sea-floor spreading along the Antarctic Plate boundary (Stock and Molnar, 1982). These three plates have existed since at least Anomaly 6 and probably Anomaly 18 time, shortly after a very widespread reorganisation of plate motions occurred. It was at this time that the plate boundary is presumed to have propagated through New Zealand separating the Australian and the Pacific plates which had been one plate since the end of spreading in the Tasman sea at anomaly 25 time. In Figure 2, the relative position of the plates is shown based on the data of Stock and Molnar (1982). The plates are identified by present day coastal outlines, the 2000m isobath and present-day latitude and longitude grid. Those parts of the present coast along the Hikurangi margin that were involved in the deformation within the plate boundary zone are omitted and this region is shown simply by a rectangular gridded area. Note that these diagrams are drawn with the Pacific Plate fixed and the Australian Plate moves (rotates) relative to it. This is a convenient format since the declination

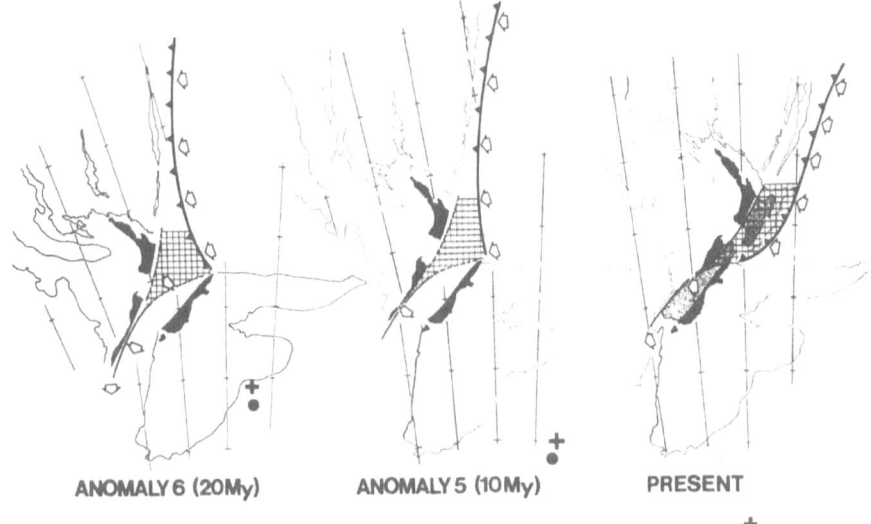

ANOMALY 6 (20My)   ANOMALY 5 (10My)   PRESENT

Figure 2. The relative position of the Australian and Pacific Plates marked by the present day latitude and longitude grid with present day coastlines and 2000m isobaths, based on the sea-floor spreading data of Stock and Molnar (1982). The position of the plate boundary marked by the subduction zone to the north and the transform to the south of New Zealand are as discussed in the text. The gridded area in the plate boundary zone is the Hikurangi margin; its area remains much the same excluding the recent extension in the Central Volcanic Region of the North Island. The Pacific plate is held fixed in this diagram; note that the Australian plate has rotated about 20 degrees since anomaly 6 time.

of the earth's magnetic field has remained meridional with respect to the Pacific plate but has rotated relative to the Australian Plate. The amount of rotation based on the Apparent Polar Wander Path of (Idnurm, 1985) is 20 degrees in the last 20Ma and this is identical to the relative rotation of the two plates from reconstruction based on sea-floor spreading data over the same period.

2.2. THE PLATE BOUNDARY ZONE THROUGH NEW ZEALAND

The location of the plate boundary in the Cenozoic is constrained to a narrow zone in the last 20 Ma by the nature of the deformation of Tertiary rocks in continental areas and by the ages of sea-floor in oceanic areas.

North of New Zealand the South Fiji Basin is of Oligocene age and was likely formed by back-arc spreading so that the subduction of the time (and immediately later periods) lay to the east along the Colville Ridge which was presumably (it has not been shown to be) the active arc in Miocene times. The Havre Trough is floored by Pliocene to Recent oceanic crust and is actively today being formed by back-arc spreading. The Kermadec Arc evidently separated from the Colville Ridge and has been the active arc of the region since the begining of the Pliocene, around 5 Ma ago. It has been suggested by a number of people that an additional subduction zone existed during the Tertiary to the west of this principal one: a subduction zone facing west under the Three Kings Rise is one such suggestion. There do not appear to be any kinematic reasons why such a feature need exist and the reasons put forward for the existence are not compelling. Nevertheless we do not know the age of, for example, the Three Kings Rise nor the New Caledonia Basin and it is always possible that a somewhat more complicated system of plates may have existed in the past. However, it is unlikely that a more complex model is needed for the last 20 Ma than that at of the present.

South of New Zealand the plate boundary during the last 20 Ma is tightly constrained to lie along the Macquarie Ridge, west of the Oligocene sea-floor of the Emerald Basin (Weissel, Hayes and Herron, 1978), and east of the older sea floor of the south Tasman Sea. The plate boundary in the South Island is bounded on the west by the Alpine Fault and on the east by the foothills of the Southern Alps, some 100 km further to the east; beyond these bounds Late Tertiary rocks are little deformed, but within these bounds they are strongly deformed.

The Hikurangi margin lies between the relatively well constrained positions of the northern and southern parts of the plate boundary. It is transitional, linking the southern transform with the northern subduction, and it does this by a large amount of internal deformation and rotation.

3. The Hikurangi margin

3.1. PRESENT-DAY TECTONICS

A map generalising strain rates estimated from repeated geodetic observations over about the last 50 years is given in Figure 3. The method, details of collection and analysis of these data have been extensively published (Walcott, 1978a,b; 1984, 1987) and will not be repeated here. Figure 3a shows the pure shear component and Figure 3b the simple shear component of deformation related to axes parallel and perpendicular to the regional NE/SW structure of the region. Positive pure shear represents shortening and negative pure shear extension perpendicular to the trend of the plate boundary. Positive simple shear represents dextral shear parallel to the plate boundary. The rates are given as microstrain/year.

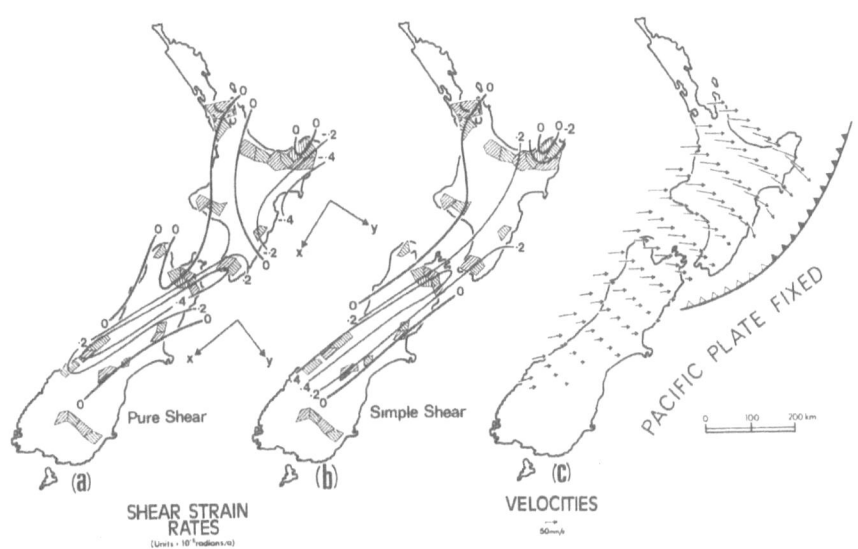

Figure 3. (a) and (b). Generalised maps of the shear strain rates estimated from repeated geodetic data, (a) is referred to here as PureShear and involves shortening or extension parallel to the Y-axis, (b) is Simple Shear and involves dextral shear parallel to the X-axis. (c) shows velocities determined from the shear strain rates by integration holding the Pacific Plate fixed.

The northern and southern parts of the Hikurangi margin differ markedly in their deformation. In the north the deformation is dominated by extension. In the Central Volcanic Region, normal faulting on faults dipping west and striking parallel to the plate boundary is observed. Immediately to the east, north-south striking dextral strike-slip faults accommodate the strain and further east again on the coast extensively developed normal faults can be observed. It is likely that these latter faults are rather superficial and sole out at depths of 10 km or less (Walcott, 1987). Beneath that, sedimentary rocks are inferred to be accumulating by an underplating mechanism, thickening the crust and inducing the surface extension. The extension of the frontal wedge is therefore likely to be superficial and is not a measure of the deformation of the Hikurangi margin, including the subducted plate, as a whole. The extension behind the wedge however contributes a southeasterly component of velocity to the wedge so that in the north where the Central Volcanic Region is opening fastest the wedge is encroaching upon the Pacific plate faster than further south along the margin. This is clearly shown in the velocity map of

In the south, the strain rates indicate shortening perpendicular to the plate boundary zone and integrating the strain across the full width of the zone shows that all of the relative plate motion is accommodated in the crustal rocks of the zone. No slip is therefore kinematically required on the subduction thrust and it is said to be locked (Walcott, 1978b). The relative plate motion in the northern part of the South Island has been examined in detail by Bibby (1982) who has shown that the relative velocity between the Australian and Pacific plates estimated from geodetic data is the same, within error estimates, as that inferred from sea-floor spreading data given by Chase (1978) or Minster and Jordan (1978). Moreover, he has shown the the motion is partitioned with the strike slip motion, ie that parallel to the plate boundary, taken up behind the frontal wedge and shortening taken up in the frontal wedge itself, as it is in the southern part of the North Island too (Walcott, 1978b).

3.2. ROTATION OF THE HIKURANGI MARGIN.

With the northern part of the frontal wedge of the Hikurangi margin having a southeasterly directed velocity and the southern being locked to the Pacific Plate the whole margin must be rotating relative to that plate. East Cape lies at a distance of about 500 km from the Pacific Plate just south of Kaikoura so that a velocity at East Cape of 60 mm/a would correspond to a rotation rate of 6 degrees /Ma relative to the Pacific Plate. It is however very difficult to estimate the rate of

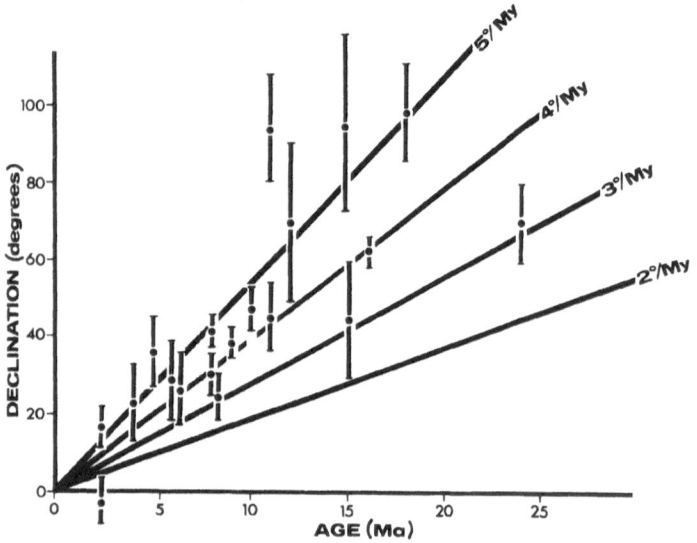

Figure 4. Declination of the primary remanent magnetisation in sediments from the Wairoa and Raukumara Domains of the Hikurangi margin.The declination increases with age in general, at a rate of about 4 degrees/Ma.

rotation of the frontal wedge as a whole since data is available only on the land area and there may be substantial strain in the offshore regions. Since, also, it is the velocity of the trench axis that is important in determining the rotation of the subducted plate and there may well be shortening in the frontal part of the wedge the present day velocity at the northern end of the Hikurangi trough is likely to be less than this value. A rotation rate of 4 degrees/Ma appears to be an appropriate value although rates as high 6 and 7 degrees/Ma are possible on the exisiting data.

The rotation must be somewhat greater today than earlier in the Late Tertiary because of the additional plate velocity imparted by the opening of the Havre Trough and Central Volcanic Region. This allows us to infer with some confidence that the orientation of the margin at 10 and 20 Ma was north/south and northwest/southeast respectively, a direction appropriate both for the relative plate motion given by the estimated instantaneous pole (Figure 2) and the linking of the two more precisely determined parts of the plate boundary.

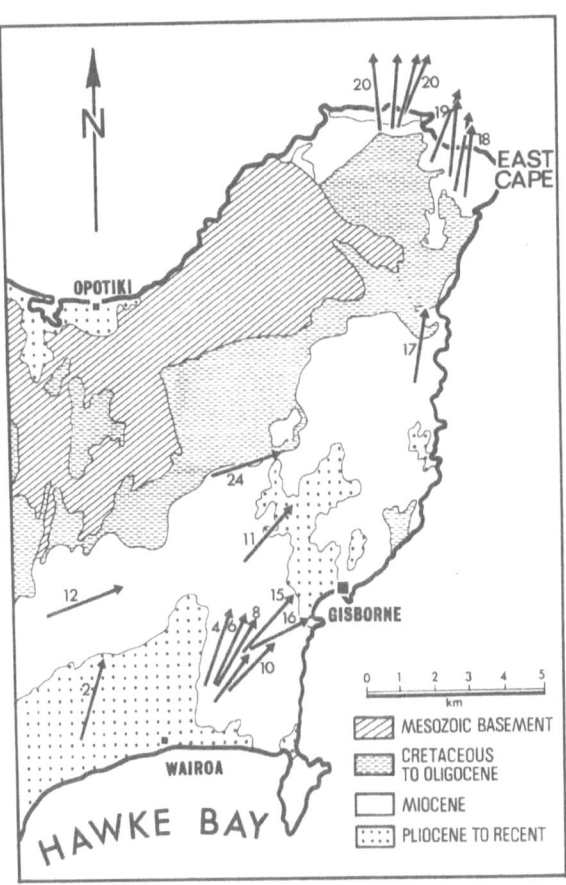

Figure 5. Directions of the declination of primary remanent magnetisation in the Raukumara Peninsular. The boundary separating the rotated Wairoa Domain from the unrotated Raukumara Domains is a zone some 50 kms wide northeast of Gisborne and it may cut across the peninsuala to Opotiki. Numbers adjacent to the arrows showing declination are the ages of the rocks estimated either from paleontology or magnetic-stratigraphy.

Table I. Summary of paleomagnetic data: Hijurangi Margin

Location	Thick	Bedding	Num	Age	D	I	k	α95	Ref
A. Raukumara Domain									
1. HK 1	52m	083/31	25	18±2	173	38	18.4	6.7	WM
2. HK 2	15m	088/37	14	"	005	-55	29.8	8.0	"
3 HK 3	86m	081/27	24	"	019	-65	16.1	7.1	"
4 HK 4	200m	083/26-37	98	20±2	022	-42	11.8	4.1	MLW
5 TA 2	26m	166/18	19	"	203	25	12.9	8.9	"
6 TA 3	10m	196/28	27	18±1	201	36	29.6	5.0	"
7 TA 4	40m	206/19	33	"	009	-45	18.5	5.7	"
8 TA 6	200m	280/40-70	42	19±1	004	-42	14.3	5.7	"
9 TA 7	21m	344/31	28	18±1	010	-47	30.9	4.8	"
10 MA 1	20m	025/28	35	17±1	189	44	35.0	4.0	"
B. Waira Domain									
1 MK 1	46m	349/17	27	16±1	063	-63	25.1	5.4	WM
2 MK 2	29m	357/17	9	15±3	225	50	20.8	12.5	"
3 MK 3	6m	055/20	6	"	095	-62	14.2	15.2	"
4 MK 4	50m	003/26	15	4±1	022	-52	8.1	8.9	"
5 MK 5	"	006/27	43	5±1	041	-42	5.2	9.4	"
6 MK 6	"	011/25	30	6±1	029	-52	9.4	8.3	"
7 MK 7	100m	013/24	10	10±1	047	-57	76.9	3.0	WW
8 MK 8	"	013/25	110	9±1	040	-62	174.4	2.0	"
9 MK 9	"	019/25	70	8±1	041	-61	110.1	3.2	"
10 MK 10	"	019/25	100	6±1	206	54	8.4	8.4	"
11 WH 1	15m	010/10	31	2.3±.1	197	57	65.1	3.1	"
12 HA1	95m	018/17	22	11±4	045	-44	15.9	7.2	WM
13 HA 2	60m	048/19	· 10	"	094	-72	12.8	9.5	"
14 HA 3	40m	244/44	15	"	209	37	7.2	9.9	"
15 WK 1			16	12±2	250	67	6.4	13.9	"
16			40	24±1	070	-50	6.4	8.8	MW
C. Wairarapa Domain									
1. MP 1			48	2.2±.2	176	52	28.5	3.8	L
2. HI 1	100m	070/16	21	8±2	030	-63	82.3	3.4	WM
D. North Marlborough Domain									
1 BR 1			58	5±1	216	60	6.0		R
2 CC 1	300m	190/30-46	12	8±2	024	-61	56.2	5.4	WM
3 DM 1	80m	029/57	46	18±2	279	59	10.5	6.4	"

Refs.L= Lamb (in prep); MLW= Mumme, Lamb and Walcott (in prep); MW = Mumme and Walcott (1985); R= Roberts(1986); WCM= Walcott, Christoffel and Mumme (1981); WM = Walcott and Mumme (1982); WW= Wright and Walcott (1986).

The Hikurangi margin, and the subduction zone are believed to have terminated at Kaikoura for about the last 20 Ma. Evidence of strong deformation first appeared in the Tertiary rocks of the Hikurangi margin of that age: near Kaikoura a very thick extensively developed coarse conglomerate and breccia appeared around 18 Ma ago and in the north similar tectonic indicators may be a little older at about 20 to 22 Ma ago. So on initiation the subduction zone may have propagated southward but it had already reached its limit, at the continental rocks of the Chatham Rise on the Pacific Plate, by about the beginning of the period discussed here.

4. Paleomagnetic studies on the Hikurangi margin

The presently available data on the Hikurangi margin and adjacent areas are listed in Table I and plotted as a function of age in Figure 4. In this figure only the data from the Wairoa, Wairarapa and Marlborough Domains are shown, and these show a general increase in declination with age at around 4 degrees/ Ma, but the scatter is considerable arising partly from errors in the data and partly from a variation in the rotation in different parts of the margin. The best determined area is the northern part of the margin where 26 different sites have given useful paleomagnetic data. These data are shown in Figure 5 and plotted as a function of age in Figure 6. In the vicinity of East Cape the available data show that the region has rotated relative to the Pacific Plate at about the same rate as the Australian Plate and therefore there is no differential or tectonic rotation. The name Raukumara Domain is given to this non-rotated block and it is believed to form the southern-most limit of the Kermadec structural arc.

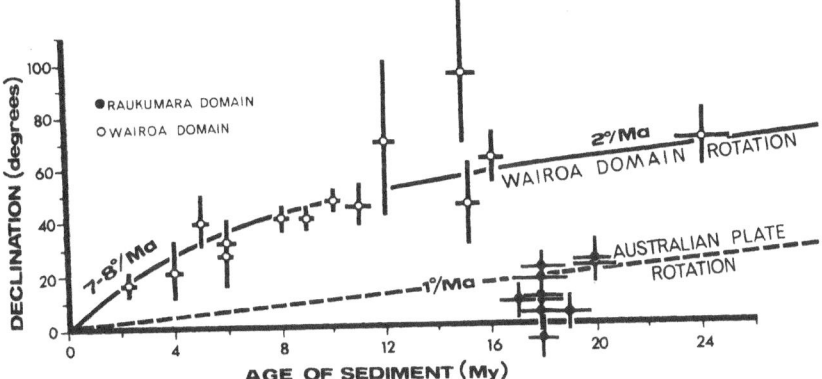

Figure 6. Declination plotted against age for the Wairoa and Raukumara Domains. The Australian Plate rotation relative to the Pacific Plate is about 1 degree/Ma. The rotation of the Wairoa Domain prior to about 12Ma ago is not well constrained.

To the south east all available data indicate a very substantial rotation at a rate that has increased toward the present. A rotation of 80 degrees is measured for the oldest rocks so far sampled. The present day rate of rotation is about 7 degrees/ Ma. The data are coherent and the region which is given the name Wairoa Domain appears to have rotated as a block. The structure of the Domain is dominated by the Wairoa Syncline a fore-arc basin with gently dipping limbs.

Although the available geodetic data show the same regional extension as further north the Wairoa Syncline does not appear to be as extensively faulted as the southern part of the Raukumara Domain but in these rocks of commonly poorly bedded, massive grey mudstones and siltstone, faults are not easy to identify.

Only two locations in the Wairarapa Domain (Figure 7) have so far given stable magnetisations. In rocks about 8 Ma old a declination of some 30 degrees is obtained and in rocks about 2 Ma old no rotation is observed. This indicates a rather different history for the Wairarapa Domain to the those domains north and south of it.

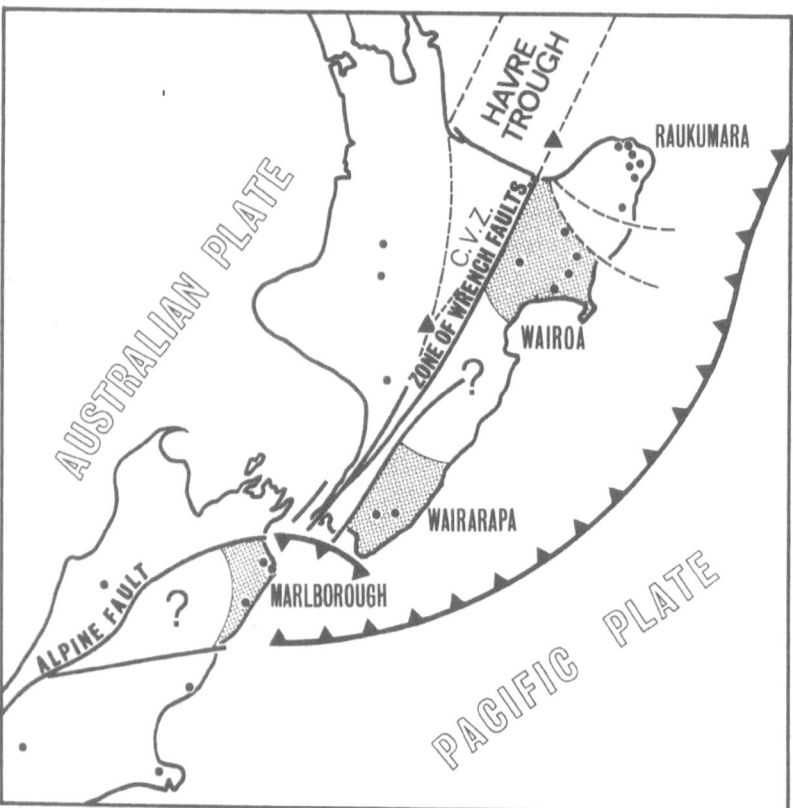

Figure 7. The tectonic domains on the Hikurangi Margin. The western boundary of the Hikurangi margin is taken to be the zone of wrench faulting the extends northward from Cook Strait to Opotiki in the North Island and the southernmost of the Marlborough faults in northeastern South Island.

In the Marlborough Domain very substantial rotations have been observed but these are insufficient in number to determine the history of rotation with any confidence. It is evidently rotating fast today judging from the very high strain rates given by the geodetic data. It appears to have rotated about 9 degrees within the last 18Ma, but whether as a domain or as separate blocks within the domain is still not determined.

In both the Pacific and the Australian Plates the observed paleomagnetic declinations for the last 20 My are small with the declination near zero on the Pacific Plate and, for the Australian Plate the declination amounts to about 1 degree for each million years of age as referred to above. The central part of the North Island probably experienced considerable compressive deformation in the Oligocene, with substantial rotations, but it had become stable by the Miocene (Mumme and Walcott, 1983).

5. Mechanics of rotation

5.1. THE MARGIN AS A WHOLE

The mechanics of rotating the margin as a whole at around 4 degrees/Ma as suggested by the data of Figure 4 is most easily understood to be the result of the the rotation of the subducted plate underneath the margin. This rotates because it links the subduction zone to the north with the transform system to the south; one end is anchored to the Pacific Plate and the other is rotating back into the plate because of the subduction. The frontal wedge therefore rotates passively on the subjacent plate. For one part of the subducted plate to rotate faster than another and to have a different trend than an adjacent part requires there to be a fault or a tear in the subducted plate separating the two parts. We can therefore explain the difference in tectonic behaviour of the Raukumara and Wairoa Domains as due to a tear in the subducted plate approximately along the line between the two domains with rotation of the subducted plate south of, but none north of, the tear.

A number of tears have been suggested to exist in the subducted Pacific Plate at the Hikurangi margin. These include the Cook Strait region between North and South Islands where Robinson (1986) has identified a discontinuity between that part of the subducted plate under the North Island and that part under the Strait and South Island with a step of some 7 km, up to the northeast. It has also been suggested that there are marked discontinuities in both trend and subducted plate geometry in the region just north of East Cape. On the basis of this suggestion for the mechanism of rotation there is yet another tear in the subducted plate between the Wairoa and Raukumara Domains. It would have to be a major one probably lasting the whole life of the subduction zone since its initiation (Figure 2).

According to this model the rate of subduction in terms of actual slip on the subduction thrust decreases markedly southward from East Cape to Kaikoura. In the Marlborough region of northeastern South Island the Hikurangi margin is not so much an active subduction zone as a thrust zone where continental rocks are being thrust over the subducted plate. In this region structures in the basement of the Pacific Plate can be traced continuously into the plate boundary zone where they swing round to the northeast above the subducted plate. The reality of rotation as a tectonic phenomenon in this part of the plate boundary is thus made very clear when it is recognised that continental rocks of the Pacific plate overly oceanic lithosphere of the same plate.

5.2. ROTATION OF THE WAIROA DOMAIN

For rotation of any part of the frontal wedge to occur at all requires the hinterland to deform in such a way as to permit the rotation. The Wairoa domain is rotating because in its hinterland the Australian Plate is being ripped apart by the back-arc basin of the Central Volcanic Region which is opening faster to the north, adjacent to the Bay of Plenty coastline, than to the south around Taupo and at Ruapehu.

Another contribution to the rotation is the southward movement of the domain along the curved strike-slip faults that lie to the west. It is rather likely however that the two features, angular opening in the Central Volcanic Region and displacement on the strike slip faults, are mechanically linked and are not independent. The two are kinematic consequences of the rotation of the Hikurangi margin, and it is not obvious which is cause and which is effect. But rotation of the subducted plate must be one of the more fundamental aspects of the system.

5.3. ROTATION OF THE WAIRARAPA DOMAIN

The Wairarapa Domain lying between Hawke Bay and Cook Strait does not appear to have rotated at all in the last 2 Ma or so but has rotated some 30 degrees in the last 10 Ma. Certainly the region behind the domain in the axial ranges of the North Island and the area further to the west shows no structural features that would permit an eastward extension behind the north of the domain or a compression behind the south, which is what would be needed for rotation to occur. So in this case we believe that even though the underlying plate is rotating steadily at 4 degrees/Ma the overlying frontal wedge is prevented from following by the constraints imposed by its hinterland.

The rotation earlier in the Tertiary implies that the hinterland did at that time deform in the required manner. The region west of the axial ranges is the Pliocene- Quaternary Wanganui Basin with around 5 km of sediments. The structure of the basement is essentially unknown and it is possible that in the Late Miocene it may well have been cut by a series of strike-slip faults such as those of the Marlborough region today and these would permit the required deformation of the region.

5.4. ROTATION OF THE MARLBOROUGH DOMAIN

The rotation of the Marlborough Domain is discussed by Lamb (this volume). He proposes a mechanism by which the movement on the major transcurrent faults that cut the hinterland of the region is transferred into rotation of the margin.

6. Conclusions

There are geological, plate tectonic, and kinematic reasons for concluding that the subduction zone extending southward from Tonga to the New Zealand region has been active for at least the last 20 Ma and throughout that time it ended against the continental rocks of the Chatham Rise. The 600 km of the subduction zone north of the Chatham Rise form a mechanical link between the transform of the Alpine Fault and the subduction to the north. Rotation of this link -the Hikurangi margin- is an inevitable consequence of continued plate movement since one end is fixed and the other continuously moves back into the Pacific Plate. This is suggested to be the fundamental mechanism by which a variety of structural features of the region are produced including the rotations. But this overall, regional rotation is strongly modulated by what is mechanically permitted by the plate behind the margin. Where this region can deform it will do so to permit the rotation and at the present day it does this by oblique strike-slip faulting as in Marlborough or by differential opening as in the Central Volcanic Region. There may also be other modes of such deformation possible.

References

Bibby, H.M., 1981. Geodetically determined strain across the southern end of the Tonga-Kermadec-Hikurangi subduction zone. *Geophys. J. R. Astr. Soc.*, **66**, 513-533

Chase, C.G., 1978. Plate Kinematics: the Americas, East Africa and the rest of the world. *Earth Planet. Sci. Lett.*, **37**, 353-368

Minster, J.B. and Jordan, T.H., 1978. Present day plate motions. *J. Geophys. Res.*, **83**, 5331-5354

Mumme, T and Walcott, R.I. 1985. Paleomagnetic studies at Geophysics Division 1980 to 1985. Dept of Scientific and Industrial Research, *Geophysics Division Report* no. **204**. 62 pages.

Roberts, A.P., 1986. Paleomagnetic Study of the Blind River Section, Marlborough, New Zealand. B.Sc(Hons) *Project in Geophysics*, Victoria University of Wellington, Wellington, New Zealand. 101pp.

Robinson, R. 1986. Seismicity, structure and tectonics of the Wellington region, New Zealand. Geophys. *J. R. Astr. Soc*, **87**, 379-409

Stern, T.A. 1985. A back-arc basin formed within continental lithosphere: the Central Volcanic Region of New Zealand. *Tectonophysics*, **112**, 385-409

Stock, J. and Molnar, P. 1982. Uncertainties in the relative position of the Australia, Antarctica, Lord Howe and Pacific Plates since the Late Cretaceous. *J. Geophys. Res.*, **87**, 4679-4714.

Walcott, R.I., Christoffel, D.A. and Mumme, T.M., 1981. Bending within the axial tectonic belt of New Zealand in the last 9Ma from paleomagnetic data. *Earth Planet. Sci. Lett.*, **52**, 427-434

Walcott, R.I. and Mumme, T.M., 1982. Paleomagnetic study of Tertiary sedimentary rocks from the East Coast of the Division *Report no.189*. 44 pages.

Walcott, R.I. 1978a. Present-day tectonics and Late Cenozoic evolution of New Zealand. Geophys. *Jour. Roy. Atron. Soc.*, **79**, 613-633

Walcott, R.I., 1978b. Geodetic strain and large earthquakes in the axial tectonic belt of New Zealand. *J. Geophys. Res.*, **83**, 4419-4429

Walcott, R.I. 1987. Geodetic strain and the deformational history of the North Island of New Zealand during the late Cainozoic. *Phil. Trans. R. Soc. London*, **A 321**, 163-181.

Weissel, J.K., Hayes, D.E. and Herron, E.M., 1976. Plate tectonic synthesis: the relative motion between the Australian, New Zealand and Antarctic continental fragments since the Late Cretaceous. *Mar. Geol.*, **25**, 231-277

Wright, I.C. and Walcott, R.I., 1986. Large tectonic rotation of part of New Zealand in the last 5 Ma. *Earth Planet. Sci. Lett.*, **80**, 348-352.

ROTATIONS ABOUT VERTICAL AXES IN PART OF THE NEW ZEALAND PLATE-BOUNDARY ZONE, THEORY AND OBSERVATION

S. LAMB
Roxford
Hertingfordbury
Hertford
Herts, SG14 2LF
U.K.

ABSTRACT. Palaeomagnetism gives information on rigid body rotations. A simple model of rigid bodies floating on the surface of a slowly deforming continuous medium suggests that the relation between rigid body rotation and deformation is complex. Rotations about vertical axes of crustal blocks in part of the New Zealand plate-boundary zone may be controlled by both the dimensions and orientations of the crustal blocks and the 3-D geometry of the plate-boundary zone.

1. Introduction

Palaeomagnetism can be used to determine rigid body rotations of crustal blocks in a deforming plate-boundary zone. In particular, it is sometimes possible to determine rotations about vertical axes relative to the margins of the deforming zone. However, it may not be easy to relate these rotations to the deformation within the deforming zone, determined from a knowledge of the relative plate motion across the plate-boundary zone, or from seismicity (Jacksonand McKenzie,1988), repeated triangulation studies, and structural mapping. Thus, rigid body rotations about vertical axes were often unsuspected prior to palaeomagnetic analysis. It is likely that such rotations are a general feature of deformation in continental crust. However, the fundamental controls on the behaviour of crustal blocks are not clear. McKenzie and Jackson (1983) suggested that such rotations are a consequence of the vertical vorticity of the underlying and continuously deforming lithosphere, and the rotation rates will equal half this vorticity. Block models based on specific patterns offaulting in the brittle crust often predict a more complex pattern of rotations (Garfunkel 1974, Ron et al.,1985). However, such patterns of faulting may not be an accurate description of the actual deformation in a plate-boundary zone.

If it is assumed that rigid body rotations about vertical axes in the brittle crust are controlled by he deformation in the underlying and continuously deforming lithosphere, as suggested by Mckenzie and Jackson (1983), then the behaviour of isolated rigid floats on the surface of a slowly deforming continuous medium may provide insight into the expected pattern of rotations (floating block model). Of course, this will only be a useful model for the behaviour of crustal blocks if the interference between neighbouring blocks in a real zone of deformation is a second order effect. This may be the case if the horizontal dimensions of crustal blocks are small compared to the width of the deforming zone and are at east as great as their thickness.

C. Kissel and C. Laj (eds.), Paleomagnetic Rotations and Continental Deformation, 473–488.
© *1989 by Kluwer Academic Publishers.*

2. Floating block model

2.1. SAND-BOX MODEL

Markers on the surface of a model zone of deformation, such as a mud/sand mixture, illustrate the difference between the well understood passive marker rotation and rigid body rotation (Figure1 a,b). Circles and lines painted on the surface of the mud/sand mixture form passive markers, while rigid floats (corks) rest on the surface. With deformation, as a consequence of oblique convergence against a barrier, the circles become ellipses and the painted lines rotate with respect to the margins of the deforming zone (Figure 1b). The floats also rotate about vertical axes because of the coupling with the deforming mud/sand mixture. However, their rotation is not the same as the marker lines (Figure1b). In addition, if the floats were not circular but elliptical in cross-section, then their rotations would vary with ellipticity and initial orientation (Lamb, 1987). Vorticity in the deforming medium is not a requirement for rigid body rotation.

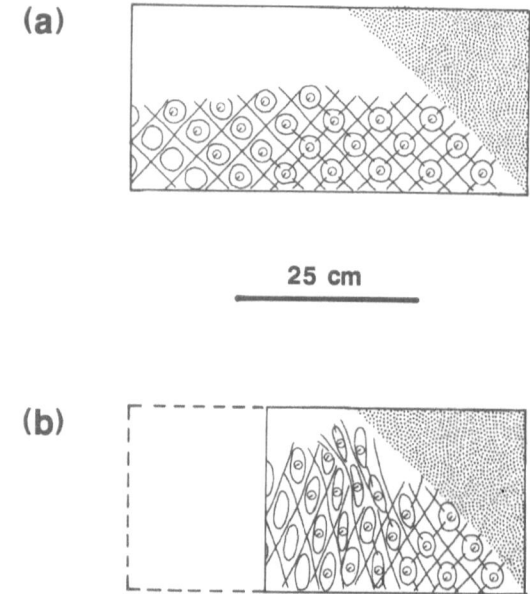

Figure 1. Sketches of plan views of a sand-box model of deformation, using a mud/sand mixture which is squashed against a rigid barrier. (a) Model before deformation: lines and circles are painted on the surface of a mud/sand mixture (passive markers), while circular corks in the centres of the circles form rigid markers. Triangular ornamented region on the right hand side of each sketch forms a rigid barrier. (b) After deformation: circles become ellipses, lines and corks rotate.

2.2. ROTATION PARAMETERS

One can define various parameters which describe an ellipsoidal float within a deforming continuous medium (with horizontal principal axes a and b; Figure 2). The aspect ratio (k) is the ratio of the shortest to longest dimension (b/a), while the orientation can be specified by the angle ϕ between the long axis and the margins of the deforming zone. If $\phi = \pm 90°$ then the long axis is at right angles to the margins of the deforming zone, while if $\phi = 0$, then the long axis is parallel to the margins. If there are no velocity gradients along the length (x axis) of the deforming zone, then the deformation in the deforming zone can be specified by θ which is the angle between the margin and the relative velocity of the two margins. If $\theta = 0$, then the deformation is dextral simple shear, and if $\theta = -90°$, then the deformation is normal compression. In these cases, the vorticity (W) of deformation is defined as $\partial u / \partial y = W$, and u, v are the components of velocity, along the x and y axes respectively, of a point with coordinates x, y (Figure 2).

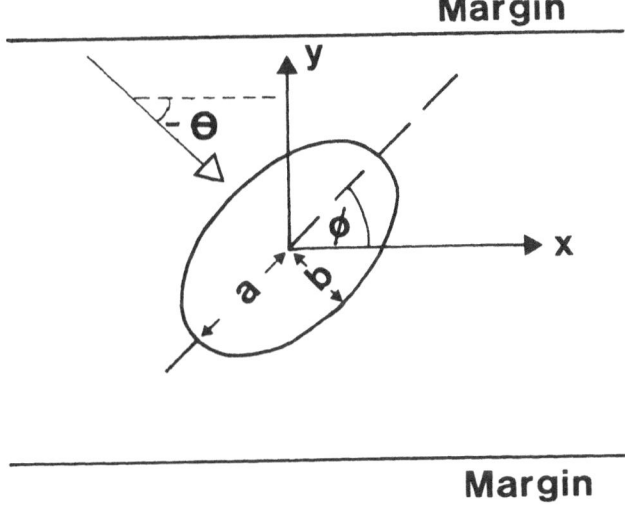

Figure 2. Diagram illustrating various parameters defined in text. The fixed axes have an origin in the centre of the elliptical float: x, y horizontal; z vertical. The orientation of the ellipse (principal dimensions a and b) is defined by ϕ which is the angle between the long axis of the ellipse and the x axis. θ is the angle between the relative velocity vector and the margins (x axis). Negative θ (as shown) implies dextral shear and compression.

2.3. ROTATION HISTORIES

Rotation histories of single isolated rigid floats on the surface of a slowly deforming continuous medium can be determined experimentally and theoretically (Jeffery 1923, Lamb, 1987). Figure 3 shows graphs of the orientation (ϕ) of elliptical floats with various aspect ratios (k), plotted against time. These have been calculated assuming that the deforming medium in contact with the float is at rest relative to the float and is deforming slowly in aspecified manner at a great distance from the float (Jeffery,1923, Lamb,1987). For instance, elliptical floats on the surface of a zone deforming by dextral simple shear ($\theta = 0$; Figure 3a), if they start with their long axes at right angles to shear zone margins, will initially change orientation rapidly, rotating clockwise. However, as they become subparallel to the shear zone margins, they will spend a long time in a limited range of orientations before eventually rotating rapidly towards right angles with the margins again. However circular floats will always rotate at a constant rate which equals half the vorticity of the velocity field ($W/2$) of the deforming medium.

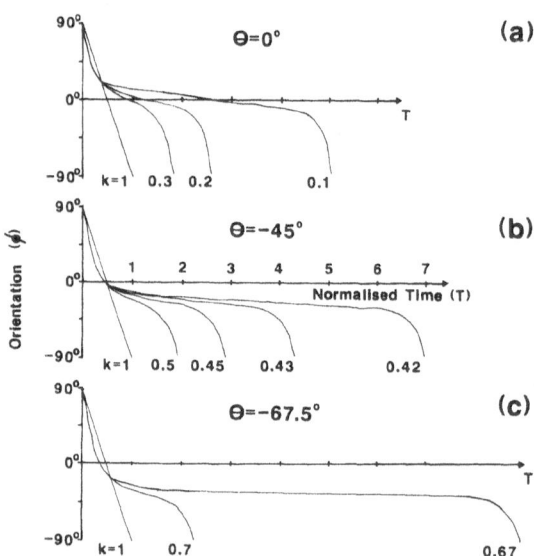

Figure 3. Rotation histories of elliptical floats which have sufficiently large aspect ratios that they rotate continuously. The orientation of floats, initially orientated with their long axes at right angles to the deforming zone, are plotted against time (t, normalised time = $\pi W t/2$) for various values of k and the θ.

In deformation with a component of pure shear or normal compression (Figures 3 b,c; $\theta = -45°$, $-67.5°$) the rotation histories are similar to those in simple shear except that the float will only rotate if its aspect ratio exceeds a critical value. For a certain range of orientations, sufficiently elongate floats will either remain stationary or rotate in the opposite direction to the overall sense of shear (Figure 4).

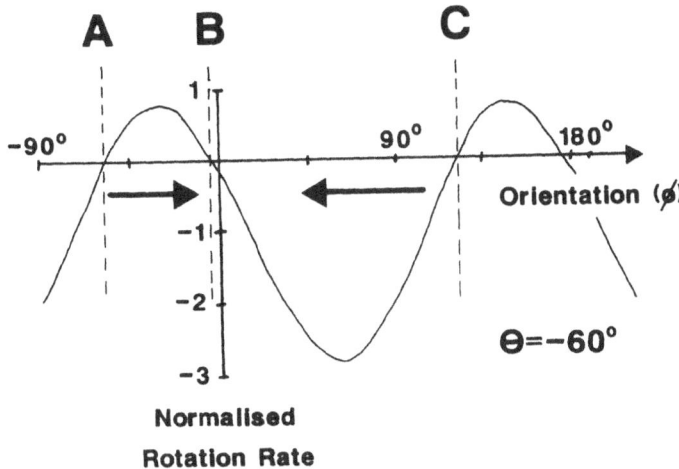

Figure 4. Diagram illustrating two possible rotation histories, depending on the initial orientation, for an elongate float with a sufficiently small spect ratio, floating on a zone deforming with components of simple shear and normal compression. Plot shows the variation in instantaneous rotation rate (r, normalised rotation rate = $2r/W$ with orientation. If the initial orientation is between A and B, then the float will rotate anticlockwise towards B (given the particular deformation illustrated). If the initial orientation is between B and C, then the float will rotate clockwise towards B, and if the initial orientation is at A or B, then the float will remaistationary.

2.4. IMPLICATIONS FOR CRUSTAL BEHAVIOUR

The floating block model suggests a surprisingly complex pattern of rigid body rotations for floats with different aspect ratios and initial orientations. Thus, brittle crustal blocks in a real deforming zone, if they are floating on a continuously deforming underlying lithosphere (McKenzie and Jackson, 1983, Lamb,1987) and the interference between neighbouringblocks can be ignored, might be expected to exhibit similar complexities in their pattern of rotations. One should be prepared for rigid body rotations about vertical axes in any zone of deformation, which can only be fully understood in terms of a structural analysis at the scale of the individual rotating crustal blocks. However, the floating block model suggests that if deformation is dominated by simple shear and crustal blocks are not markedly elongate and have rotated through angles between 20° and 90°, then half the vertical vorticity ($W/2$) of the underlying lithosphere will be a good estimate of the average rotation rate about vertical axes for the majority of crustal blocks. In addition, there may be a tendancy in any real zone of deformation for the interference between neighbouring crustal blocks, and also the discrepancy between the velocity field of the surface brittle crust and the underlying continuously deforming lithosphere, to be minimised. This may be achieved if elongate crustal blocks break up so that individual blocks are small and equidimensional, rotating at a rate given by half the vertical vorticity of the underlying lithosphere.

2.5. CHANGES IN THE VERTICAL BOUNDARY CONDITIONS

Experiments with brittle material floating on a highly viscous medium (Brun *et al.*, this volume) suggest that the horizontal spacing of the dominant block boundaries is roughly equal to the brittle layer thickness. Thus, the characteristic horizontal dimensions of crustal blocks in a wide zone of deformation through continental crust would be expected to be 10-30 km and equal to the thickness of the brittle crustal layer. Much smaller crustal blocks in narrow intense zones of deformation would be expected to be floating at a depth of a few kilometres on a zone of more distributed deformation.

In many convergent plate-boundary zones the deforming continental crust rests directly on the subducted plate, forming a crustal wedge generally less than 30 km thick. In such cases, the crust cannot be said to be floating on a continuously deforming medium and hence the floating block model may not be applicable to large (>10km across) crustal blocks. Here, the shear traction on the subducted slab interface, as well as changes in the geometry of the subducted slab (Walcott, this volume), will have an important effect on the behaviour of crustal blocks.

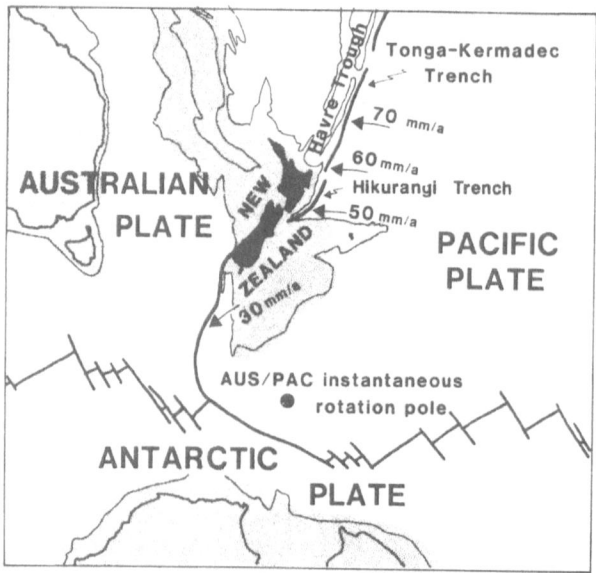

Figure 5. Map showing the boundaries between the Australian, Pacific, and Antartic plates in the New Zealand region. Areas of continental crust are stippled. The Australian/Pacific instantaneous pole of relative plate motion (Chase1978), and some velocities for the Pacific plate relative to the Australian plate, are also shown.

3. New Zealand plate-boundary zone

Rotations about vertical axes have been observed in part of the New Zealand plate-boundary zone between the obliquely convergent Pacific and Australian plates (Figure 5, Walcott *et al.* 1981, Walcott and Mumme, 1983, Mumme and Walcott, 1985, Wright and Walcott,1986, Roberts,1986, Mumme *et al.*, in review, Lamb, in review, Walcott, this volume). Here, the Pacific plate is being subducted beneath the Australian plate at ca. 50 mm/a in the Hikurangi subduction zone. The crust above the subducted plate forms a zone of deformation up to 200 km wide. Figure 6 shows an estimate of the smoothed finite deformation during the last 4 Ma of an originally orthogonal latitude/longitude grid (Lamb, in review). Clockwise rigid body rotations about vertical axes relative to the Pacific plate, determined or inferred from palaeomagnetism, are shown in brackets.

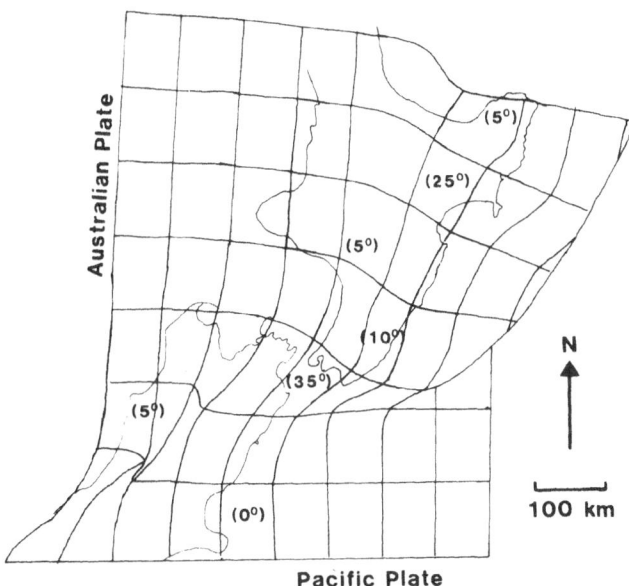

Figure 6. Diagram showing an estimate of the smoothed out deformation of a 1°0 longitide/latitude grid system during the last 4 Ma (Lamb in review). Grid was orthogonal at 4 Ma. Observed or inferred palaeomagnetic clockwise rotations about vertical axes relative to the Pacific plate are shown in brackets.

Marlborough fault system. A wide zone of dextral transpression, referred to as the Marlborough fault system, occurs at the southern end of the Hikurangi subduction zone, forming essentially a transform linking the trench with the Alpine fault (Figure 7). The deformation is dominated by major active faults with a dominant dextral strike-slip component of motion. Palaeomagnetic sampling in Miocene and Pliocene sedimentary rocks has shown that crustal blocks have locally rotated clockwise about a vertical axis ca. 100° in the last18 Ma relative to the Pacific plate (locality 18 in Figure 7, Mumme & Walcott,1985). Elsewhere a clockwise rotation of ca.35° has been observed in Pliocene sediments (locality 16 in Figures 7 and 11a, Roberts,1986). However, only part of the crust in the Marlborough fault system is underlain by the seismically active subducted Pacific plate at a depth less than 30 km (Ansell and Adams 1986, Figure 7), and therefore the Marlborough fault system can be divided into two domains. The part which rests on the seismically active subducted slab is referred to as the northern Marlborough domain, while that

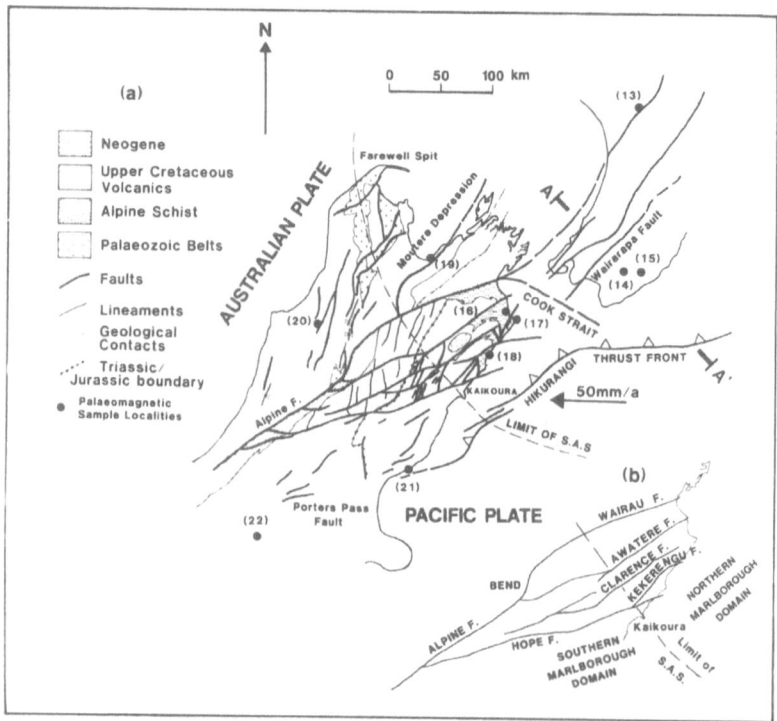

Figure 7. (a) Geological map (after New Zealand Geological Survey 1972, 1:1,000 000 geological maps) of the plate-boundary zone near the northern end of the South Island, showing the major structural features. Note the position of the southern limit of the underlying seismically active subducted Pacific plate (SAS, Ansell and Adams,1986) and the numerous north trending faults and lineaments to the southwest of this. Palaeomagnetic sample sites are also shown. (b) Map showing the principal faults in the Marlborough fault system, and the division into the northern and southern Marlborough domains.

part which is away from the subducted slab is referred to as the southern Marlborough domain (Figure 7b).

4. Northern Marlborough domain

4.1. LARGE CRUSTAL BLOCKS

Within the crust resting on the seismically active subducted slab at a depth less than 30 km, much of the regions between the major strike-slip faults are essentially underformed except for a regional tilt, which is generally less than 25° (Lamb, in review). Thus, crustal blocks have dimensions up to 20 km x 80 km across. An integration of the short term deformation, determined from repeated triangulation studies (Bibby,1981) and fault-slip data, gives estimates of the

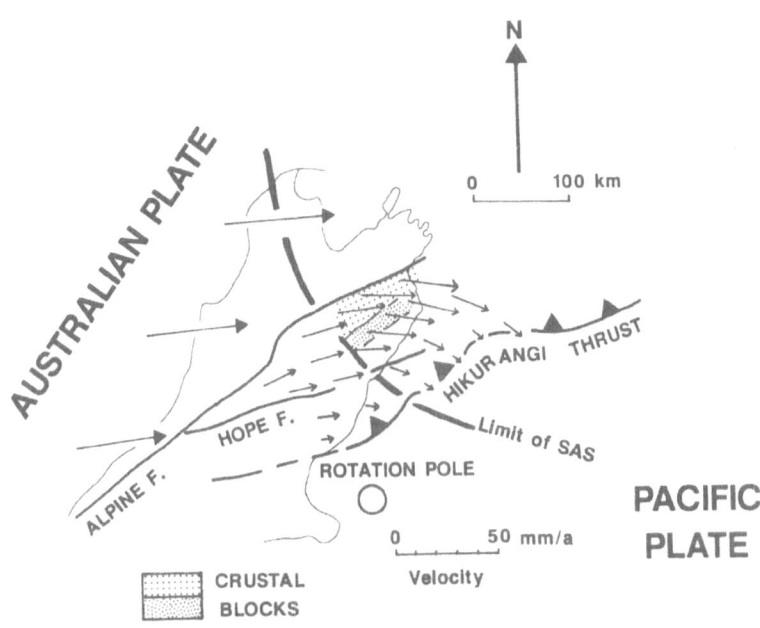

Figure 8. Arrows show estimates of the horizontal instantaneous velocities of the crust in the Marlborough fault system, relative to the Pacific plate. Large arrows show the velocities of the Australian plate relative to the Pacific plate, calculated from the instantaneous rotation pole and rate (Chase,1978). Other velocities were estimated from an integration of the available short term deformation (Bibby 1981, Lamb in review). The southern limit of the seismically active subducted slab (SAS, Ansell and Adams,1986) is shown by the heavy dashed line. Velocity pattern suggests that large crustal blocks resting on the subducted slab are rotating clockwise at 8.5° ± 5.5°/Ma about poles located about 100 km south of the region.

velocities of the large crustal blocks relative to the Pacific plate (Figure 8). These suggest that these blocks are rotating clockwise about poles located about 100 km south of the region at 8.5° ± 5.5° /Ma (Lamb, in review). This rotation rate cannot be directly compared with palaeomagnetic observations as sample sites are in different crustal blocks.For instance, a ca. 35° clockwise rotation of Pliocene sediments (locality 16 in Figures 7 and 11a, Roberts,1986) is representative of a crustal block at the northeastern end of the Awatere and Clarence faults (Lamb and Bibby, in review). The rotation of this block has taken up dextral shear on the Clarence fault, accommodated by north-south compression along an east trending fault zone (Figures 9a and 11a), increasing towards the east. Thus, the rotation of the large crustal blocks further to the

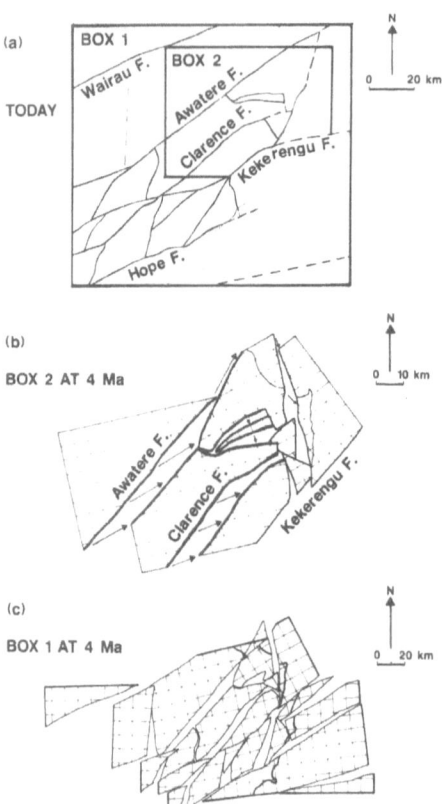

Figure 9. Cut-out reconstructions of part of the Marlborough fault system (northern Marlborough domain) at 4 Ma. (a) Map of major structures in northern Marlborough domain. (b) Reconstruction of box 2 in (a); heavy lines are block boundaries which have been reconstructed with some confidence, other boundaries are speculative. Gaps represent shortening during the last 4 Ma. (c) Highly speculative reconstruction of box 1 in (a).

southwest is less, but may still be compatible with the inferred short term rate (ca. 20° in 4 Ma, Figure 9b).

The large block rotations in the crust resting on the subducted slab can be understood in terms of the commonly observed partitioning of deformation in obliquely convergent subduction zones (Fitch, 1972; Walcott,1984, Lamb and Vella,1987). The component of plate motion parallel to the trench is commonly taken up on a series of strike-slip faults at the back of the crustal wedge resting on the subducted slab, while the component of plate motion normal to the trench is taken up by a fold and thrust belt in the frontal parts of the crustal wedge. If a large crustal block straddles the plate-boundary zone, then this partitioning can be accommodated by the pivoting of the crustal block about an axis near the trench (Figure10).

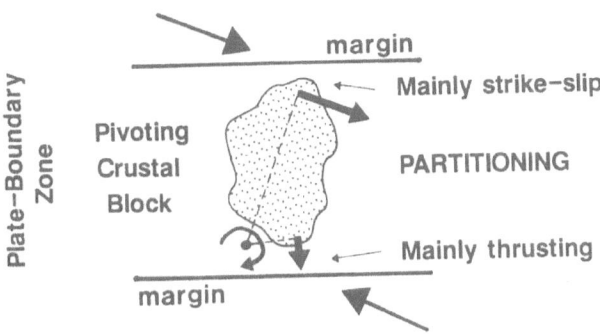

Figure 10. Diagram illustrating how rotation of a large crustal block, straddling the plate-boundary zone, is equivalent to partitioning of the deformation into normal compression in the frontal parts and strike-slip at the back of the crustal wedge overlying the subducted slab.

4.2. SMALL CRUSTAL BLOCKS

In detail the deformation in the crust resting on the subducted slab is complex. Localised intense zones of deformation, which formed in the last 4 Ma, contain crustal blocks less than 2 km across. For instance, the region adjacent to the active dextral strike-slip Kekerengu fault forms a zone up to 10 km wide of intense faulting and folding (Figure11, Lamb in review, Lamb and Bibby, in review). Palaeomagnetic sampling from one crustal block (locality 18 in Figure11b) suggests a clockwise rotation about a vertical axis relative to the Pacific plate in excess of 100° for 18 Ma sedimentary rocks. Part of this rotation may record the progressive swing of fold axial traces into parallelism with the Kekerengu fault (Figure11b, Lamb and Bibby, in review).

The behaviour of small crustal blocks in the intense zone of deformation may approximate to that of crustal blocks floating on an underlying distributed zone of deformation at a depth of a few kilometres, described by the floating block model. In this case, if the short term deformation determined from repeated triangulation studies (Bibby,1981) represents that of the underlying distributed deformation, and there are no velocity gradients parallel to the general structural trend,

484

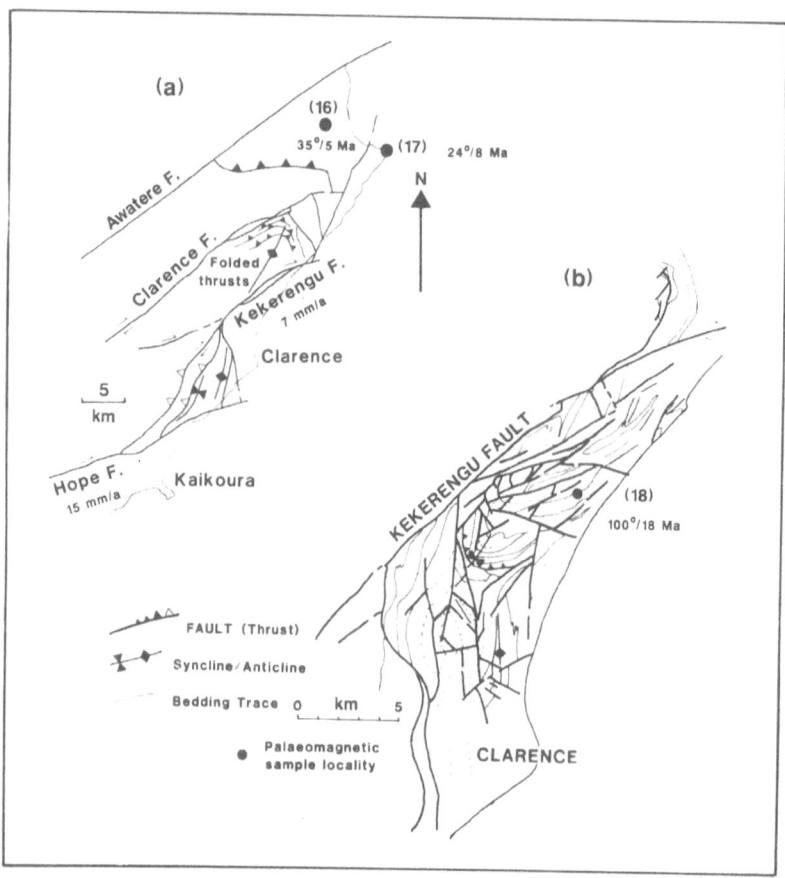

Figure 11. (a) Structural map of part of the northern Marlborough domain (see Figure 7) showing major structures and palaeomagnetic sample sites. (b) Detailed structural map of the region adjacent to the Kekerengu fault (Lamb, in review). This is an area of structural complexity in which crustal blocks are generally less than 2 km across. Palaeomagnetic sample site is shown, which suggests a clockwise rotation in excess of 100° for18 Ma sediments.

then it is possible to place constraints on the expected rotation rates. Figure12 shows the calculated variation in clockwise instantaneous rotation rate for a traverse across the Marlborough fault system, based on the geodetic deformation and floating block model. Thus, maximum clockwise rotations would be expected near the centre of the Marlborough fault system, with maximum possible instantaneous rotation rates in excess of 30°/Ma. These high rotation rates would be for markedly elongate crustal blocks (aspect ratio $k \ll 1$) in suitable orientations, and they would not be expected to be maintained for long (less than a million years). Equidimensional blocks ($k = 1$) would show a similar variation in rotation rates across the northern Marlborough domain, though with instantaneous rotation rates less than 16°/Ma. These rotation rates would be closer to the long term average rotation rates. Palaeomagnetically observed clockwise rotations, assuming all the rotation occurred in the last 4 Ma during the development of the intense zones of deformation, have a range which is similar to the calculated rotation rates (Figure12). Thus, the floating block model may be a reasonable approximation to the behaviour of the sampled crustal blocks, despite the many differences between the model and a real zone of deformation.

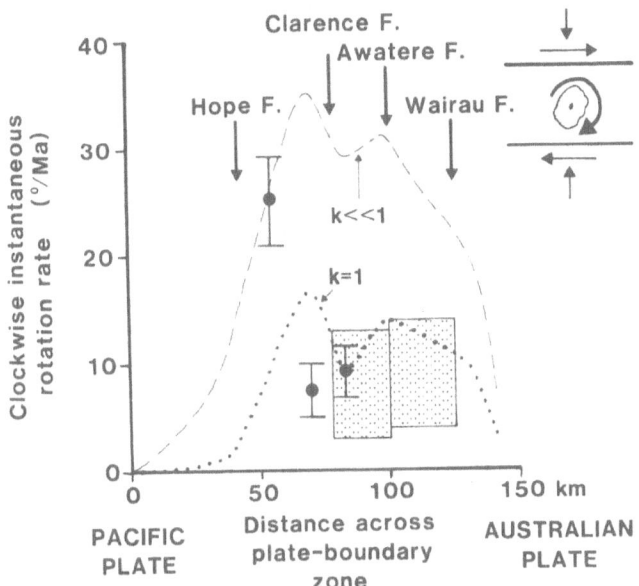

Figure 12. Graph showing the variation in instantaneous rotation rate, calculated for small crustal blocks with different shapes (equidimensional, k=1; elongate, $k \ll 1$) using the floating block model and the geodetic deformation for an approximately SE trending traverse across the northern part of the Marlborough fault system. It is assumed that there is no velocity gradient parallel to the general structural trend. Also shown are the average palaeomagnetically determined rotation rates (solid dots with 95% confidence limits), assuming all rotation occurred in the last 4 Ma. Dotted boxes show estimates of rotation rates of large crustal blocks resting on the subducted slab, deduced from the variation in the short term velocities along their lengths (Figure 8).

5. Southern Marlborough domain

The abundance of north trending faults and lineaments in the crust which is not underlain by the seismically subducted slab (Figure 7a), suggests that here crustal blocks are generally less than 10 km across. The short term deformation suggests that the crust is deforming by distributed dextral shear with a shear direction close to that of the relative plate velocity vector (Figure 8).

486

6. Development of Marlborough domains

Deformation in a trellis system linked to a large rigid block (Figure 13) produces a similar velocity field to the inferred short term velocity field in the Marlborough domains (Figure 8). Dextral shear across the trellis results in rotation of the cross-bars and pivoting of the large rigid block. This model implies that the rotation rate of crustal blocks is equal to the vorticity of deformation as a consequence of motion on the trellis system (McKenzie and Jackson 1983, 1986). If crustal blocks are not pinned, then their rotation rate will be different to that required by the trellis system, though the trellis system may still produce the same velocity field as that in the underlying continuously deforming lithosphere (McKenzie and Jackson,1983, 1986).

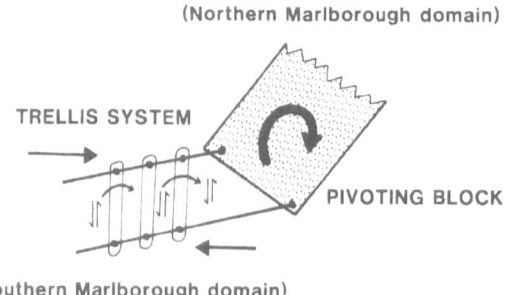

Figure 13. Diagram illustrating possible rotation kinematics of the Marlborough fault system, as suggested by the velocity pattern in Fig. 8.

The trellis system may describe the motion of small crustal blocks between the major faults in the southern Marlborough domain, while the pivoting block may describe the motion of the large blocks in the northern Marlborough domain. Thus, depending on the slip rates on the major faults in the northern Marlborough domain, the rotation of these large blocks may accommodate the shear across the whole of the southern Marlborough domain, or that part which is not taken up by slip on the major faults in the southern Marlborough domain. Similarly, displacement on the Clarence fault (shear in the trellis system) is taken up by rotation of the crustal block at the northern end of the Clarence fault (Figures 9a and 11a).

The model suggests that the angular relation between the faults in the southern and northern Marlborough domains has not remained constant, but has decreased through time. Thus, early in the evolution of the system, the faults in the northern Marlborough domain may have been markedly oblique to those in the southern Marlborough domain. The difference in behaviour between the two domains must be a consequence of the presence of the underlying subducted slab, which essentially changes the boundary conditions, as well as affecting the thermal structure of the overlying crust. Walcott (1987, this volume) has suggested that the overall pattern of rotations in the northern part of the New Zealand plate-boundary zone records the progressive clockwise rotation of both the leading edge of the overlying crustal wedge and the trend of the underlying subducted slab. However, the deforming crust overlying the subducted slab is not a rigid body. Thus, the leading edge of the deforming crustal wedge should be treated as a passive marker line which may follow the change in trend of the underlying subducted slab. However, palaeomagnetic observations record rigid body rotations which need not be the same as the passive marker rotation.

7. Conclusion

The previous analysis suggests that palaeomagnetically determined rotations within a deforming zone can only be fully understood within the context of a 3-D structural analysis of the plate-boundary zone at the scale of the individual crustal blocks. The dimensions and orientations of crustal blocks, as well as their relative position in the plate-boundary zone, appear to exert a control on the rotation of the blocks. Thus, the behaviour of large crustal blocks resting on the subducted slab appear to be related to the plate boundary conditions. The rotation of small crustal blocks may be more closely related to the local deformation. However, the rotation rates in either case, determined from palaeomagnetic measurements, are the same order of magnitude as those predicted by the floating block model. The overall effect of the deformation is the progressive rotation of the Australian plate margin, as described by Walcott (1987, this volume).

Acknowledgements

This work was carried during the tenure of a New Zealand University Grants Committee post-doctoral fellowship at Victoria University, Wellington, New Zealand. Harold Wellman and Dick Walcott have been an endless source of encouragement and ideas. However all misconceptions are entirely my own.

References

Ansell, J., Adams, D., 1986. Unfolding the Wadati-Benioff zone in the Kermadec-New Zealand region. *Phys. Earth Planet. Int.* , **44,** 274-280

Bibby, H.M., 1981. Geodetically determined strain across the southern end of the Tonga-Kermadec-Hikurangi subduction zone. *Geophys. J. R. Astr. Soc.* , **66,** 513-533.

Chase, C.G., 1978. Plate Kinematics: the Americas, East Africa and the rest of the world. *Earth planet. Sci. Lett.,* **37,** 353-368.

Fitch, T.J., 1972. Plate convergence, transcurrent faults and internal deformation adjacent to southeast Asia and the western Pacific. *J. Geophys. Res.,* **77,** 4432.

Garfunkel, Z.V.I.,1974. Model for the late Cenozoic tectonic history of the Mojave desert, California, and its relation to the adjacent regions. *Geol. Soc. Am. Bull.,* **85,** 1931-1944.

Jackson, J.A., McKenzie, D.P., 1988. The relationship between plate motions and seismic moment tensors, and the rates of active deformation in the Mediterranean and Middle East. *Geophys. J.,* **93,** 45-73.

Jeffery, G.B., 1923. The motion of ellipsoidal particles immersed in a viscous fluid. *Proc. R. Soc. London Ser. A,* **102,** 161-179.

Lamb, S.H., 1987. A model for tectonic rotations about a vertical axis. *Earth planet. Sci. Lett.,* **84,** 75-86.

Lamb, S.H., Vella, P., 1987. The last million years of deformation in part of the New Zealand plate-boundary zone. *J. Struct. Geol.*, **9**, p877.

Lamb, S.H., in review. Tectonic rotations about vertical axes during the last 4 Ma in part of the New Zealand plate boundary zone. *J. Struct. Geol.*

Lamb, S.H., Bibby, H.M., in review. The last 25 Ma of deformation with a component of horizontal simple shear in part of the New Zealand plate-boundary zone. *J. Struct. Geol.*

McKenzie, D.P., Jackson, J.A., 1983. The relationship between strain rates, crustal thickening, palaeomagnetism, finite strain, and fault movements within a deforming zone. *Earth Planet. Sci. Lett.*, **65**, 182-202.

McKenzie, D.P., Jackson, J.A., 1986. A block model of distributed deformation by faulting. *J. Geol. Soc. London*, **143**, 349-353.

Mumme, T.C., Walcott, R.I., 1985. Palaeomagnetic studies at Geophysics Division 1980-1983. Department of Scientific and Industrial Research, *Geophysics Division Report* **204**, 62pp.

Mumme, T.C., Lamb, S.H., Walcott, R.I., in review. Palaeomagnetic studies of the Upper Tertiary rocks from the Te Araroa to Tokamaru Bay region, East Coast, North Island, New Zealand. *J. Royal Society of New Zealand.*

New Zealand Geological Survey, 1972. 1:1,000,000 Geological map series. *Department of Industrial and Scientific Research*, Wellington.

Roberts, A.P., 1986. Palaeomagnetic study of the Blind River Section, Marlborough, South Island, New Zealand. *Unpubl. B.Sc. (Hons) dissertation*, Victoria University, Wellington. 101pp.

Ron, H., Freund, R., Garfunkel, Z., Nur, A. 1984. Block rotation by strike-slip faulting: structural and palaeomagnetic evidence. *J. Geophys. Res.*, **89**, 6256-6270.

Walcott, R.I., 1984. The kinematics of the plate-boundary zone through New Zealand: a comparison of short and long-term deformation. *Geophys. J.R. Astron. Soc.*, **79**, 613 633.

Walcott, R.I., 1987. Geodetic strain and the deformational history of New Zealand during the late Cainozoic. *Phil. Trans. R. Soc. Lond. A*, **321**, 163-181.

Walcott, R.I., Christoffel, D.A., Mumme, T.C. 1981. bending within the axial tectonic belt of New Zealand in the last 9 Ma from palaeomagnetic data. *Earth planet. Sci. Lett.*, **52**, 427-434.

Walcott, R.I., Mumme, T.C., 1982. Palaeomagnetic study of the Tertiary sedimentary rocks from the East Coast of the North Island, New Zealand. Department of Scientific and Industrial Research, *Geophysics Division Report*, **189**, 44pp.

Wright, I.G., Walcott, R.I., 1986. Large tectonic rotation of part of New Zealand in the last 5 Ma. *Earth planet. Sci.Lett.*, **80**, 348-352.

PALEOMAGNETIC ROTATIONS IN THE COASTAL AREAS OF ECUADOR AND NORTHERN PERU

C. LAJ[1], P. MITOUARD[1,2], P. ROPERCH[3], C. KISSEL[1], T. MOURIER[4], F. MEGARD[5]

1 - Centre des Faibles Radioactivités,
Laboratoire Mixte CNRS-CEA,
91198 Gif-sur-Yvette Cedex, France.

2- Institut Français d'Etudes Andines,
Casilla 18-1217,
Lima 18, Peru.

3 - ORSTOM,
Laboratoire de Géophysique Interne,
Université de Rennes, 35042 Rennes, France.

4 - Laboratoire de Géologie Historique,
Université Paris XI, 91405 Orsay, France.

5 - Centre Géologique et Géophysique,
USTL, 34000 Montpellier, France.

ABSTRACT. We report a paleomagnetic study of about 850 cores from over 80 sites sampled in Paleozoic to Tertiary volcanic, plutonic and sedimentary formations in western Ecuador and Northern Peru. Most of the sampled lithologies carry a stable primary remanent magnetization whose direction is significantly different from that of coeval formations of stable South America. In western Ecuador the results are consistent with the progressive disappearance of a marginal basin accompanied by a clockwise rotation of about 70°. In Northern Peru four sites from Late Carboniferous (Pennsylvanian) formations from the Amotape-Tahuin range in the Piura province show a 110° clockwise rotation and yield evidence for a northward displacement. The results from the Cretaceous to Paleogene formation from the same region indicate a clockwise rotation ranging from 90° for the lowermost units to 25° for the uppermost ones. When considered together with previous geological studies, these data are consistent with the hypothesis of the accretion of an Amotape-Tahuin continental terrane to the Peruvian margin in Neocomian times. The accretion was followed by in-situ rotation, suggesting a dextral shear regime. Preliminary results obtained here from Cretaceous and Paleogene formations in the Central Andes and published results from other authors indicate that this pattern of clokwise rotation changes to anticlockwise south of the major Huancabamba deflection (4° S). This rotational pattern could be related to a general shear regime, dextral north of the deflection and sinistral south of it, or to a recently proposed mechanism involving along-strike variations in the amount of late Cenozoic shortening.

1. Introduction.

During the last decade a large number of paleomagnetic studies have shown that the western active margin of North America is a mosaic of allochtonous accreted terranes of widely differing sizes, some of which have undergone large latitudinal transport. The accretion and subsequent coastwise

489

C. Kissel and C. Laj (eds.), Paleomagnetic Rotations and Continental Deformation, 489–511.
© *1989 by Kluwer Academic Publishers.*

490

translation and rotation of these terranes are a major characteristic of the North American orogenic belt (Cox, 1957; Beck, 1976, 1980, Beck et al., 1980; Simpson and Cox, 1977; Wells, this volume).

In contrast, the role and extent of microblock collisions in the building and shaping of the Andean Cordillera is still an open problem, and one which may have different answers in the three different major segments of the Cordillera documented by the geological observations (Figure 1).

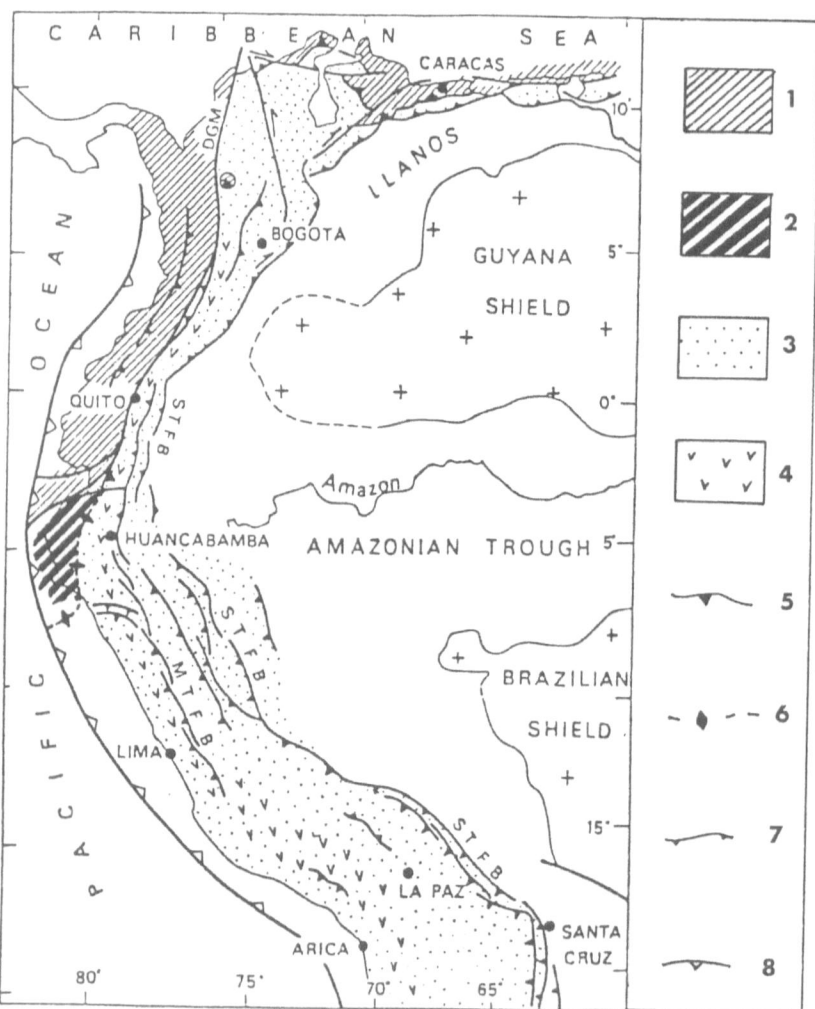

Fig. 1. Main geological characteristics of northwestern South America (modified from Mégard, 1987): 1 - Accreted oceanic terranes (Northern Andes). 2 - Coastal area of the Huancabamba Andes. 3 - Integral Peruvian Andes (Central Andes). 4 - Volcanic and plutonic Cenozoic belt. 5 - Sutures. 6 - Presumed suture. 7 - Main intracontinental overthrusts. 8 - Actual trench.

North of 3° S latitude, the Andes of Colombia and Ecuador appear to be a cordilleran orogen related to the accretion of oceanic crust, as evidenced by ophiolitic sutures included in the belt and by recent geological studies (Feininger,1987; Feininger and Bristow, 1980; McCourt et al., 1984; Aspden and McCourt, 1986; Mégard, 1987). Ophiolites are also present in the Magellan Andes of Chile and Argentina in the extreme south. The paleomagnetic data from this segment have been interpreted in terms of oroclinal bending (Dalziel et al., 1973), but it has been recently pointed out that block rotation in a distributed sinistral shear could account for the results as well (Beck, 1988).

The Central Andes, in which no ophiolitic suture has yet been recognized, have generally been considered as a genuine marginal orogen formed exclusively by subduction since the Early Jurassic (James, 1971; Audebaud et al., 1973; Mégard, 1978). The paleomagnetic results from this zone indicate counterclockwise rotations in Peru and northernmost Chile and clockwise rotations farther south in Chile, and have also generally been interpreted in terms of an oroclinal bending (the Arica deflection) (Palmer et al., 1980; Heki et al., 1984; May and Butler, 1985). More recently Beck (1988) has proposed an alternative explanation in terms of in-situ block rotations in response to shear (sinistral to the north and dextral to the south of the Arica bend).

Clearly, although the existing paleomagnetic results already allow some tectonic speculations, many more are needed before a pattern can emerge with unambiguous tectonic implications, such as for the North American Cordillera. In this article we report on the paleomagnetic studies conducted in the last two years on Paleozoic to Tertiary volcanic, plutonic and sedimentary formations in southern Ecuador and northern Peru. Part of this work has already been published (Roperch et al., 1987; Mourier et al., 1988a).

2. Choice of sites and sampling methods.

After the initial choice of the sampling areas, the structural position and the freshness of the outcrops were of course the main criteria for selecting the sampled sites. Careful attention was paid to the tectonic setting of the volcanic sites since it is necessary to restore the formations to their original horizontal position for the correct interpretation of the paleomagnetic results. In some cases interbedded sediments yielded precise bedding corections. In others less reliable criteria were used, such as the attitude of pillow-lava paleoslopes or columnar jointing in massive flows and sills. In these last cases only sites with tilts of the order of 15-20° were selected to avoid generating substantial declination errors when correcting the remanent magnetization vector back to its pre-tilt position (McDonald, 1980).

In the sedimentary formations, very fresh outcrops located in main structures and with clear bedding criteria were selected. Also, only sites with very fine grained lithologies were chosen because their sedimentation rate is usually low enough to average out completely the secular variation of the geomagnetic field and also because they are low energy deposits. The final choice of the sedimentary formations was made using a portable spinner magnetometer directly in the field. This allowed to select the highest magnetizations among geologically equivalent outcrops. Sites with magnetization lower than the noise level of this instrument ($\sim 5 \times 10^{-4}$ A/m) were however sampled in regions where it proved impossible to locate better magnetized outcrops.

Only the sites sampled in the granodioritic intrusions were devoid of bedding plane indications and we have assumed that these post-tectonic intrusions have not been significantly tilted.

The cores were obtained using either a gasoline powered drill for the volcanic formations or an electric drill powered by a portable generator for the sedimentary ones. The corer is a water cooled

25-mm barrel with a sintered diamond cutting edge. At each site a minimum of 8-10 cores were drilled and each core was oriented with both a magnetic and a sun compass.

3. The sampled regions

3.1 WESTERN ECUADOR.

Ecuador lies immediately adjacent to a NNE - SSW trending submarine trench along which the oceanic Nazca plate is being consumed beneath the western margin of South America. The margin

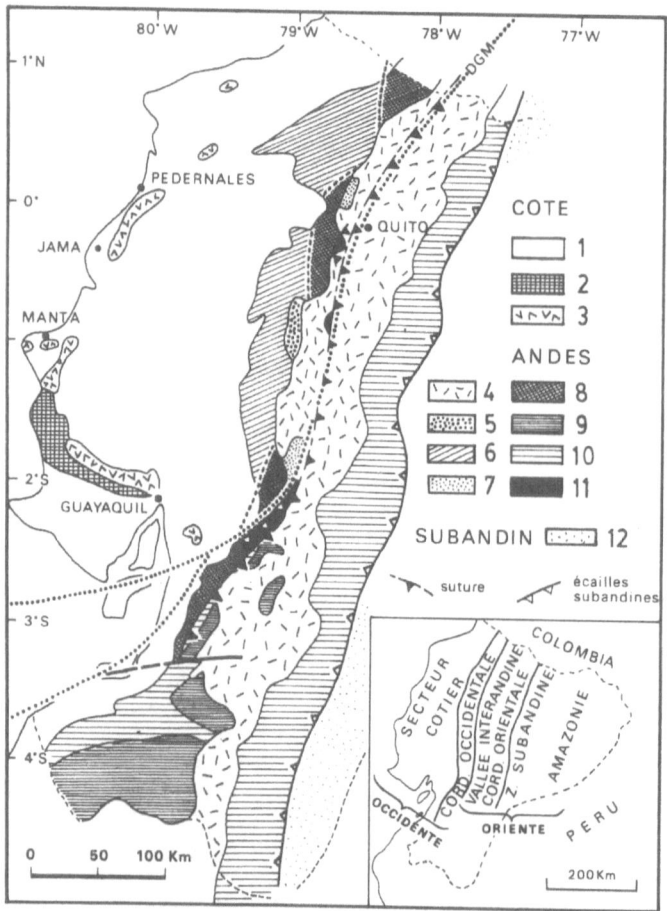

Figure 2. Schematic geological map of Ecuador based on the geological maps of the Servicio Nac. Geol. Min. 1969 (from Mégard, 1987). Sampling was conducted in the Piñon formation (label 3) near Guayaquil, Manta and Pedernales, and in the volcanic Macuchi arc (label 6).

consists of the low elevation Costa, a 100-150 km wide zone between the coast line and the toe of the Andes. The latter consists of the parallel submeridian Western and Eastern Cordilleras, separated by the graben-like inter-Andean valley. An intracontinental volcanic arc of Cenozoic age, which includes active volcanoes, forms the highest part of both Cordilleras (Figure 2).

The Costa is underlain by a coherent basement of mid-Cretaceous age, consisting of pillow basalts, dolerite sills and dykes and scarce gabbros and peridotites. These rocks are called the Piñon formation and are believed to be part of a basement igneous complex of oceanic origin extending over western Ecuador responsible for the prominent positive (Bouger) gravity anomaly measured over the Costa (Feininger and Seguin, 1983). The lithologies of the Piñon formation are oceanic tholeites geochemically similar to the MORBs from the Galapagos rise and also to the tholeites from western Colombia (Lebrat et al., 1987). Basaltic flows of latest Cretaceous age are interspersed in the series in the Manta area (Goossens and Rose, 1973). They display distinct transitional MORB geochemical characteristics and are informally called the San Lorenzo formation. Similar basalts are associated to the Cayo formation farther north near Pedernales (Faucher et al., 1971; Henderson, 1979).

The paleomagnetic sampling was carried out in three geographically separate areas of the Costa over a distance exceeding 200 km. Essentially the Piñon and San Lorenzo formations were sampled, because in most localities they have been subjected only to minor warping and block faulting. A few sites were also sampled in the volcanic rocks of the Macuchi arc. A total of over 350 cores were obtained from 34 sites.

3.2 THE HUANCABAMBA ANDES OF SOUTHERN ECUADOR AND NORTHERN PERU.

The Huancabamba Andes extend between 3°S and 8°S over southern Ecuador and northern Peru, and represent the connexion between the Northern and the Central Andes (Figures 1 and 3). One of the major features of this segment is the change of the Andean trend, from N 20W in the northern part of the Central Andes to N 20E in the Northern Andes. This bend is known as the Huancabamba deflexion.

In the Coastal area of the Huancabamba Andes, the pre-Mesozoic basement outcrops mainly in the large Amotape-Tahuin range, where it consists of polyphase metamorphic rocks of Precambrian and early Paleozoic age, unconformably overlain by conformable to disconformable Devonian to Permian marine and continental sedimentary series. There is a complete hiatus of post-Permian to pre-Albian deposits. Albian carbonates unconformably overlie the pre-Mesozoic basement and are in turn overlain by flysch series of late Cretaceous age. Both these and the Albian carbonates grade eastward into mixed volcanic and partly volcanoclastic sedimentary rocks which fill the Lancones synclinorium.

A different evolution is observed in the Olmos Paleozoic massif situated farther east, where the basement is unconformably overlain by a conformable to disconformable platform sequence, similar to that observed in the remainder of northern Peru, consisting of carbonates of Triassic to Liassic age, volcanic and/or clastic rocks of middle to late Jurassic age, deltaic siliciclastic rocks of Neocomian age and carbonates of Albian to Santonian age. No pre-Devonian deformational and metamorphic event is observed in the different Paleozoic massifs of northern Peru and southern Ecuador which are also characterized by late Devonian-early Carboniferous Eohercynian folding not observed in the Amotape-Tahuin massif (Mourier et al., 1988b).

A paleogeographic reconstruction of the Huancabamba Andes shows that two different volcanic arcs were successively active during the Mesozoic (Mourier et al., 1988b). A middle to

494

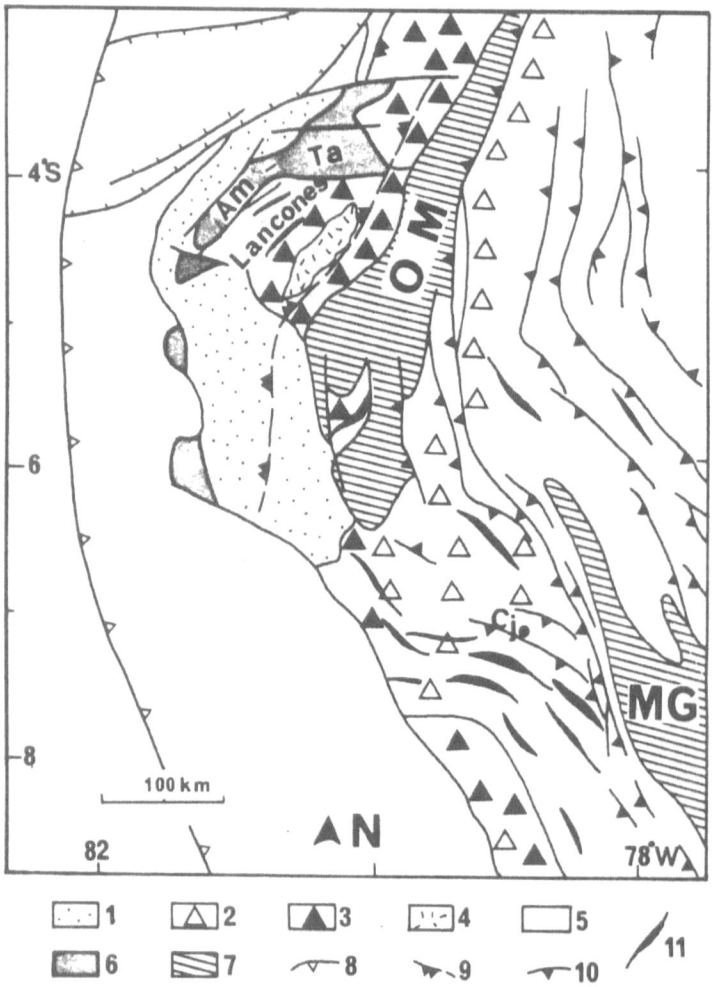

Figure 3. Main geologic features of the Huancabamba Andes and of the northern part of the Central Andes. Also shown the "S" shaped Cajamarca deflection. 1 - Tertiary basins. 2 - Jurassic volcanic arc. 3 - Cretaceaous volcanic arc. 4 - Pre-Albian basic complex. 5 - Meso-Cenozoic undifferenciated series. 6 - Paleozoic units of the Amotape-Tahuin coastal range (Am-Ta). 7 - Precambrian-Paleozoic basement (OM: Olmos massif; MG: Maranon geanticline). 8 - Peru trench. 9 - Hypothetical suture. 10 - Main thrusts. 11 - Axes of major folds. Cj: Cajamarca.

late Jurassic and earliest Cretaceous volcanic arc, situated east of the Olmos massif,was suddenly replaced in the early Cretaceous by a younger arc situated about 150 km to the west of the Olmos massif, much closer to the present trench. Paleontological and radiometric data (Feininger and Bristow, 1980; Canfield et al., 1982) indicate that the arc jump occurred between the Valanginian and the Aptian about 130 Ma (Mourier et al., 1988b)

The Cretaceous volcanics are mainly located in the Lancones synclinorium, between the Amotape-Tahuin range and the Olmos massif (Figure 3). They consist of a basal pre-Albian undated basic complex represented by pillow-lava flows intercalated with hyaloclastic breccias and scarce volcanoclastic strata, intruded by dykes and sills of basaltic to andesitic composition. This basal complex is unconformably overlain by the Albian to Senonian volcanic arc and volcanoclastic series of the Lancones synclinorium. These in turn are intruded by granodioritic plutons not precisely dated but probably of post-Paleocene and pre-Oligocene age (Beaufils, 1978).

A total of more than 350 cores were obtained from 32 sites in the different Cretaceous volcanic units and post-Paleocene intrusions of the synclinorium Four sites were also obtained in the tilted but undeformed Carboniferous siltstones of the Amotape range.

3.3. THE CAJAMARCA REGION.

Between 7° and 8° S one of major structural features of the Andean Cordillera is the Cajamarca deflection characterized by a progressive change in the strike of regional scale upright folds over 200 km from the Cordillera to the coastal area. The eastern and western limbs of structure trend N160 south of 7°30 along the Cordillera and north of 7°S along the coastal andean foothill respectively, and are thus parallel to the main structural, geomorphologic and magmatic feature of the Central Andes. The scatter of the local fold axes about this general trend is much larger at the western than at the eastern termination. On the other hand, the trend of the main structures is very closely E-W in the Cajamarca transect along 7°S (Figure 3).

The formations present in this region consist dominantly of a well exposed Mesozoic to Lower Tertiary sedimentary sequence and Cenozoic volcanic units. Large bodies of acid plutonic rocks (Andean batholite) also outcrop along the coastal Andean foothills.

Since the studies of Benavides (1956), the stratigraphy of the Cretaceous sedimentary sequence of the Cajamarca area is quite well established and used as a reference in Peruvian geology. It consists of Neocomian deltaic sandstones and Aptian to Santonian platform carbonates which are subdivised into 11 different well-dated formations with specific facies and characteristic sedimentary sequences (Benavides, 1956; Janjou, 1981; Jaillard, 1986; Cordova, 1986). Most of the sampling has been done on the black mudstones and marls of the Middle Albian Pariatambo formation and also occasionally in other Cretaceous formations (Chulec formation of Lower Albian and Cajamarca formation of Turonian age). Three sites are located in the post tectonic granodioritic intrusions dated at 43 Ma (Stewart et al., 1979) along the Jequetepeque river in the coastal andean foothills and some sites were sampled in the Cenozoic volcanic units which uncomformably overlie EW to NW-SE trending structures on the Llama and Cajamarca road section.

Although the sampling has covered the entire deflection, in this paper we only consider the results obtained from the andine-trending eastern and western terminations of the deflexion because the results from the E-W transect clearly reflect local thin-skinned tectonic phenomena. The complete paleomagnetic results will be discussed elsewhere (Mitouard et al., to be published).

4. Results

4.1. GENERAL MAGNETIC PROPERTIES.

Measurements of remanent magnetization were made using either a spinner magnetometer or a LETI 3-axis cryogenic magnetometer, depending on the intensity of the samples. For the sedimentary units the isothermal remanent magnetisation (IRM) acquisition in fields up to 1.6 T and subsequent AF demagnetization of the saturation IRM (SIRM) show that in most cases the IRM saturates in fields of 0.15 - 0.2 T and that the median AF destructive field of the SIRM is always lower than 3 mT. Measurements of the NRM and Curie balance experiments indicate that the pillows from the Piñon formation have magnetic properties quite similar to those expected from submarine basalts, i.e. high magnetization intensity and low Curie point. These pillow have the highest magnetization intensities ranging from 0.5 to 15 A/m, while the pillows from the Lancones substratum are on the contrary characterized by higher Curie points and significant lower magnetizations. The lowest NRMs, of the order of 2×10^{-4} A/m are observed in the sedimentary paleozoic units from the Amotape-Tahuin range. In the majority of the lithologies the results are consistent with a magnetic mineralogy dominated by magnetite.

Both thermal and AF stepwise demagnetizations were used. In a general way both methods isolated the same stable component of magnetization for samples of the same core, however the AF method sometimes failed to decrease the remanent intensity to values lower than 10-15% of the NRM. The thermal method, on the contrary, always yielded very consistent results and was used routinely.

All samples were stepwise demagnetized with 12-15 steps between room temperature and the limit of reproducible results, which ranges between 450° and 580 °C. At each step the low field magnetic susceptibility was measured for volcanic and sedimentary samples. The maximum changes observed over the entire temperature interval were of the order of a factor of three. This indicates that the magnetic carriers have not changed drastically during the thermal treatment.

4.3. PALEOMAGNETIC RESULTS.

In general the rocks yielded reliable paleomagnetic results. Stable and consistent primary paleomagnetic directions were isolated after heating at 200-250 °C, sometimes less, and were precisely determined using orthogonal diagrams. Most of the sites are characterized by well grouped paleomagnetic directions. As an exception some sites are characterized by an unreasonably high within-site scatter, although single samples present perfectly linear demagnetization diagrams. We have chosen to reject these highly scattered sites (K<10). The results obtained from the different regions are described separately below:

4.3.1. Western Ecuador. The sites sampled in the Macuchi arc gave very scattered results probably reflecting a complex tectonics with very strongly tilted bedding planes, and are for the moment not interpretable. On the other hand all the sites from western Ecuador gave quite satisfactory results, except for three which were rejected because of large within-site scatter. Diametrically opposite normal and reversed polarities are recorded in all the sampled areas and their mean directions are clearly different from the present axial dipole field and the direction of coeval sites from stable South America (Table I and Figure 4). However the two polarities cannot be unambiguously distinguished because of the low inclinations recorded by these formations. The most likely interpretation is that the directions in the north-easterly quadrant are normal polarities and that the Piñon basement has been rotated clockwise by an angle of about 75°. The rather high

Table I: Paleomagnetic results from western Ecuador

Site	N	in Situ		after B.C.		K	α_{95}
		Dec	Inc	Dec	Inc		
AN20	14	67	-24	71	5	14	10.9
AN21	12	76	-25	80	5	22	9.9
AN22	9	62	-11	64	3	86	5.6
AN23	12	44	12	39	10	120	3.0
AN24	11	265	10	265	-5	18	11.1
AN25	9	269	9	269	-11	11	17.3
AN26	8	262	12	263	-3	82	6.1
AN27	10	60	-17	60	-17	382	2.3
AN28	10	58	-31	90	-27	25	9.9
AN29	6	52	-27	83	-29	297	3.3
AN30	11	76	-17	82	-8	24	9.6
AN31	12	75	-23	74	7	10	12.7
AN33	8	52	-18	49	-48	25	8.8
AN34	10	68	2	69	-27	10	15.9
AN35	7	54	14	54	14	20	11.0
AN36	16	66	-26	59	-10	399	1.8
AN40	8	239	31	234	25	58	7.4
AN41	9	226	31	222	24	174	3.9
EQ05	10	90	-41	90	-21	410	2.4
EQ06	10	110	-42	106	-23	21	10.8
EQ07	10	43	-44	65	7	485	2.2
EQ08	11	12	-73	64	-18	17	11.3
EQ09	10	80	-62	80	2	35	8.8
EQ11	11	286	5	286	5	33	8.0
EQ16	13	261	6	254	26	10	14.1

Mean Direction: N = 23/25   D = 75°   I = -9°   K = 14 $\alpha_{95} = 7.7°$

scatter in declination among the different sites might result from the record of a differential rotation between sites, the age of which is not well constrained (Roperch et al., 1987).

4.3.2. *The Amotape-Tahuin range.* The results obtained from the Pennsylvanian formations in the Amotape-Tahuin range are reported in Table II. Although all the samples have very weak magnetizations, the four studied sites are characterized by reasonably grouped paleomagnetic directions (N=4/4, D=257, I=50.3, k=19.2, α_{95}=15.9, after bedding correction). Because the sites have similar bedding plane attitudes it was unfortunately not possible to perform a fold test. However, the very high value of the inclination observed before bedding correction (~75°) can hardly be related to an overprint of the remanent magnetization since the deposition of the sediments. For this reason we interpret the result as a primary, clockwise rotated reverse paleomagnetic direction.

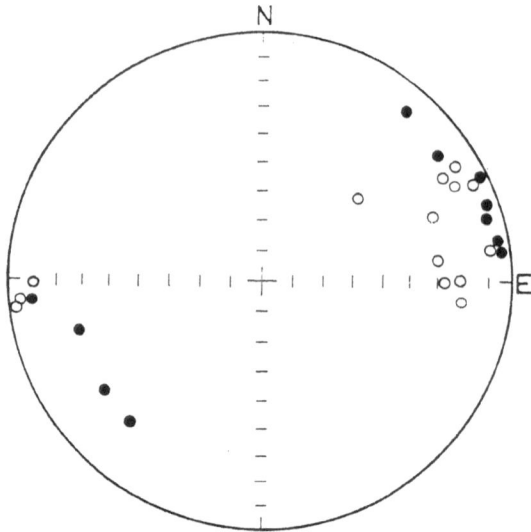

Figure 4. Equal area projection of the mean paleomagnetic directions obtained from the stable sites in western Ecuador after bedding correction. Open circle: upper hemisphere; full circles: lower hemisphere.

Table II : Paleomagnetic results from the Pensylvanian formations of the Amotape-Tahuin range

Sites	n	Before B.C.		After B.C.		K	α_{95}
		Decl.	Incl.	Decl.	Incl.		
AND86 76A	14	277.0	80.0	237.0	58.7	32	6.6
AND86 76B	9	284.0	70.0	273.5	41.0	18	10.8
AND86 12B	10	332.0	72.0	249.5	66.6	21	9.7
AND86 19	8	273.0	63.6	259.0	32.2	50	7.0
		Mean value (after BC):					
N=4/4	D=257°		I=50.3°		K=19.2		α_{95} =15.9°

4.3.3. The Lancones synclinorium. The results obtained from 26 reliable sites in the Lancones synclinorium are reported in Table III and represented on the stereograms of Figure 5. These results show that the mean paleomagnetic direction (N=12/12, D= 90.3, I= -12.3, K= 21.8, α_{95} = 8.6) obtained after bedding correction from the sites sampled in the pre-Albian basement is significantly different from that obtained for Albian to Senonian volcanics above the unconformity (N= 8/9 , D= 58.9 , I= -16.7, K= 36.3 , α_{95} = 8.2 after bedding correction). Although no fold test could be performed, because some of the sites are virtually horizontal and the others only slightly tilted, we regard all these directions as primary. Finally a third distinct paleomagnetic direction (N= 6/6; D= 25.0; I= -19.0; K= 48; α_{95} = 6.9) was obtained from the post-Paleocene intrusive formations. No bedding correction was possible in this cases. We think however that the measured rotation is not a spurious effect related to an undetected tilt (McDonald, 1980). Indeed the tilts, if any, are certainly very small and a spurious effect would hardly result in the observed tightly grouped directions from the different sites and an inclination value consistent with that obtained from sites with clear bedding plane attitude.

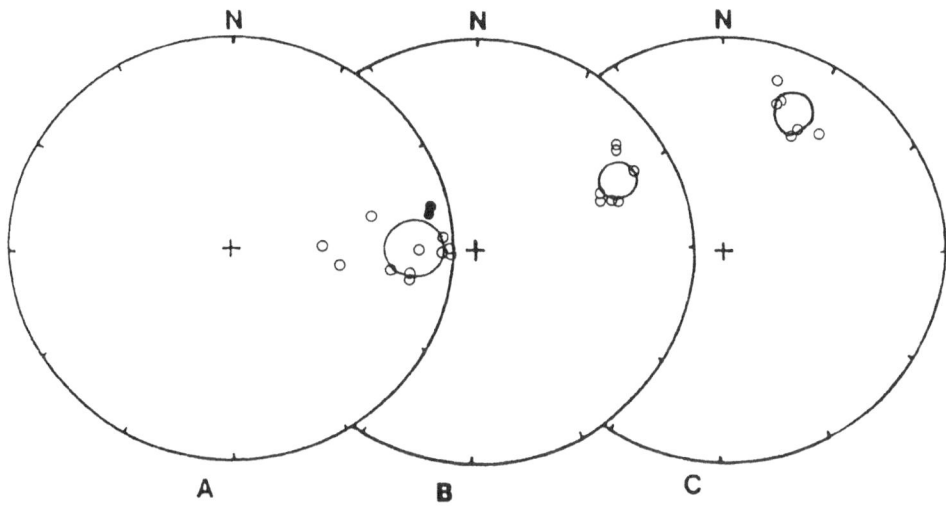

Fig. 5. Stereographic projection of the mean directions of paleomagnetic vectors for the Lancones synclinorium formations. Open circles correspond to upper-hemisphere projections. Full circles correspond to lower-hemisphere projections. Big circles correspond to the 95% confidence area about the mean. A - Basic substratum of the Lancones synclinorium. B - Volcanic formations of the Lancones synclinorium. C - Post-Cretaceous granodioritic pluton of the Lomas area.

4.3.4. The Cajamarca region. As mentioned above we only consider here the results from the andine trending eastern and western terminations. At the eastern termination all the results concern Albian (Pariatambo and Chulec) sedimentary formations, while at the western termination stable directions were obtained from the Pariatambo formation and from volcanic and intrusive sites of

Table III : Paleomagnetic results from the Substratum and the volcanic formations of the Lancones Basin.

Sites	n	Before B.C.		After B.C.		K	α_{95}
		Decl.	Incl.	Decl.	Incl.		
Substratum							
AND86 78	10	97.4	0.8	98.8	-11.3	28	8.4
AND86 79	10	98.4	1.7	99.7	-11.5	13	13.7
AND86 80	7	89.7	5.2	90.2	-9.0	12	15.3
AND86 81	9	93.6	-23.0	89.7	-44.3	25	10.4
AND86 82	9	100.0	-36.2	100.0	-36.2	37	7.5
AND86 83	10	79.3	-12.0	78.0	-23.0	43	6.8
AND86 84	13	85.5	-4.7	86.5	-3.2	51	5.0
AND86 85	10	78.0	0.8	79.0	5.2	115	4.0
AND86 86	10	91.3	-0.2	91.3	-0.2	20	9.8
AND86 87	9	91.3	-2.6	91.3	-2.6	48	6.7
AND86 88	5	81.6	5.8	81.6	5.8	216	5.2
AND86 89	5	99.2	-17.5	99.2	-17.5	76	8.8
Mean value (after BC):							
N= 12/12	D=90.3°		I=-12.3°		K=21.8°		α_{95} =8.6°
Volcanic formations							
AND86 90	13	49.7	-18.9	51.0	-10.2	268	2.4
AND86 91	7	49.5	-26.5	24.8	-35.5	43	8.0
AND86 92	11	52.0	-11.0	52.0	-11.0	19	9.0
AND86 93	9	50.4	-29.3	62.9	-24.1	28	8.0
AND86 94	7	62.2	-24.4	69.2	-19.5	27	10.2
AND86 95	11	56.0	-22.5	61.0	-10.1	143	3.0
AND86 96	7	71.6	-16.0	67.9	-21.0	108	5.1
AND86 97	10	66.0	-25.0	66.0	-25.0	98	4.5
Mean value (after BC):							
N=7/8	D=61.3°		I=-17.4°		K=70.0		α_{95} =6.3°
Intrusive formations							
AND86 73	12	37.7	-19.1	37.7	-19.1	526	1.0
PE 87.09	9	29.5	-25.8	29.5	-25.8	76	5.3
PE87.11	10	30.5	-22.8	30.5	-22.8	73.5	5.1
PE87.18	12	16.3	-9.9	16.3	-9.9	208	2.7
PE87.15	7	18	-18	18	-18	105	5.0
PE87.16	8	19	-17	19	-17	26	9.7
Mean value (uncorrected for B.C.)							
N=6/6	D=25.0°		I=-18.9°		K=48.1		α_{95}=6.9°

post-Paleocene age. Both sedimentary formations are precisely dated biostratigraphically, either as early Albian (Chulec formation) or middle Albian (Pariatambo formation) (Benavides, 1956). These periods belong to the Cretaceous long normal chron, so that, contrarily to what is observed, only normal polarity magnetizations should be recovered. The occurrence of reverse polarities is a puzzling problem which we do not fully understand. It could be a consequence of a very unprobable undetected reverse polarity chron during this period, or an indication that the time of acquisition of the measured stable remanent magnetization significantly postdates the deposition. However a positive fold test performed on many sites of the Albian formations establishes that the magnetization predates the folding, and is thus older than the late Cretaceous.

Table IV. Paleomagnetic results from the Cajamarca fold belt.

Site	N	in Situ		after B.C.		K	α_{95}
		Dec	Inc	Dec	Inc		
Eastern termination (Albian sites):							
AN 02	10	120.0	22.2	165.0	26.8	99	4.4
AN 03	7	316.2	-30.4	348.7	-10.7	259	3.3
AN 61	13	5.2	-17.8	350.6	-25.8	163	3.0
AN 63	6	15.4	-4.6	351.5	-31.4	212	3.8
AN 64	6	317.0	-40.8	348.2	-26.4	23	12.0
AN 65	13	307.8	-19.4	336.5	-23.0	87	4.2
Mean Direction:							
N = 7/7		D = 347.2		I = -25.1	K =98.4		α_{95} = 5.5
Western termination (Albian sites):							
PE87.28	8	70.6	32.6	64.7	4.2	22	11.0
PE87.29	11	336.0	-16.2	336.0	-16.2	364	2.2
PE87.31	7	136.9	37.3	113.2	32.6	166	4.0
PE87.32	10	164.8	28.3	139	27.5	141	4.1
PE88.57	10	333	-23	316	-43	31	7.9
PE88.59	9	331	-32.7	331	-32.7	98	4.9
PE88.73	10	349.4	-9.5	349.3	-13.6	73	5.2
PE88.76	8	22.4	-21.6	22.4	-21.6	50	5.9
Mean Direction:							
N = 6/8		D = 325°		I = -28.7°	K = 16		α_{95} = 14.3°
Western termination (Post paleocene volcanics and intrusions):							
PE87.30	8	9.7	-2.6	9.7	-2.6	485	2.2
PE87.37	11	319.5	-44.8	318.3	-13.0	23	8.7
AN 54	9	329.0	-16.7	329.0	-16.7	29	8.7
AN 55	9	341.3	-11.2	341.3	-11.2	50	6.6
AN 56	6	346.3	-11.0	342.3	-11.0	317	3.2
Mean Direction:							
N=4/5		D=333°		I=-13.2°	K=51		α_{95} =9.8°

As reported in Table IV, the paleomagnetic directions recovered from the eastern termination (N=7/7, D=347.2 , I= -25.0 , K= 94 , α_{95} = 5.5) are slighltly westerly. At the western termination the declinations are rotated anticlockwise with respect to the eastern termination and the rotation is similar for the Albian formation (N= 6/8, D= 325 , I= -28.7 , K= 16, α_{95} = 14.3 and for the plutonic and volcanic formation (N= 4/5 D= 333, I= -13.2, K=51, α_{95} = 9.8) with a difference in the inclination value.

5. Tectonic implications of the data.

To interpret the data in terms of tectonic evolution, the paleomagnetic directions recovered from the sampled lithologies must be compared with those from coeval formations of stable South America (SOAM). Unfortunately, the apparent polar wandering path (APWP) for SOAM is quite ill-defined, for the entire period from Carboniferous to Paleocene-Eocene considered here, so that appreciable uncertainties exist for a precise geodynamical interpretation of the results.

For the interpretation of the results from the Amotape Tahuin Range, where the sampled sediments have been assigned biostratigraphycally to the Pennsylvanian stage, the only available pole is the 300 Ma given by Irving and Irving (1982).

For the Cretaceous, on the other hand, different reference poles are available. The APWP given by Irving and Irving (1982) is characterized by a highly elongate pattern with a total spread of over 30°, which, if real, would imply that the pole swung back and forth several times during the Cretaceous. The reality of this streak has however been questioned by Ernesto [1986] on the basis of paleomagnetic results from the Mesozoic volcanic rocks from the Serra Geral Formation in Brasil. Also, Beck (1988) has recently analyzed possible explanations for the streak (genuine apparent polar wander, intra-cratonal deformation or a geomagnetic explanation) and concluded that they do not readily account for the elongated pattern. Consequently, he has assumed a single Cretaceous pole for SOAM (85.5 S; 73.0 E) not significantly different from the Early Cretaceous pole of Ernesto.

Because the paleomagnetic results reported here have been obtained from formations of diffferent ages within the Cretaceous, the assumption of a unique pole for this entire period is certainly a rather crude approximation. For this reason in Table V the results are analyzed in terms of concordance/discordance using both the pole given by Beck and the pole of appropriate age from the APWP for South America given by Irving and Irving. The discrepancies in the obtained results give an immediate measure of the uncertainties resulting from the lack of a well established APWP for South America. The rotation (R) and the flattening (F) parameters calculated by the equations given by Beck (1980) and modified by Demarest (1983) are used to obtain the tectonic significance of the data.

The mean paleomagnetic direction recovered from the Piñon formation in western Ecuador differs from the expected one by a large clockwise rotation of the order of 72 ± 8°. Both reference poles suggest that some northward motion has occurred with respect to SOAM, while only a very slight difference in the absolute latitude is inferred from a comparison with the geocentered dipole field predictions.

The Amotape Tahuin Range appears to have undergone a large clockwise rotation (108 ± 21°) and a significant flattening (25.8 ± 9°) implying a northward transport of some 17° in latitude with respect to stable South America. The evidence for the northward transport does not depend on the particular choice of the 300 Ma pole: the 290 or the 280 Ma poles would have given the same results. However we are well aware that the limited number of studied sites and the large uncertainties of the Irving and Irving APWP for South America do not allow to establish the allochtonous character of the Amotape Tahuin block on purely paleomagnetic arguments. In the interpretation given below, previous geological arguments play a major role. Refinements of APWP for South America and additional sampling are certainly needed for a more accurate description of the geodynamical evolution of the Amotape Tahuin block.

Table V. Concordance/Discordance statistics.

		Beck	I & I (110)	I & I (80)	Dipole field
Western Ecuador	R	77± 8.2	----	76 ± 18.7	75± 7.8
	F	18 ± 8.8	----	7 ± 34.8	9 ± 7.7
Lancones Basin					
Substratum	R	92.5 ± 9	93.5± 12.6	----	90.3 ± 8.8
	F	13 ± 9.6	8.6 ± 20	----	4.2 ± 8.6
Volcanic formations	R	63.5 ± 7	----	62.9 ± 18.2	61.3 ± 6.6
	F	18.3 ± 7.6	----	7.1 ± 34.5	9.4 ± 6.3
Intrusive formations	R	----	----	----	25 ±7
	F	----	----	----	11 ±7
Cajamarca region					
Eastern termination	R	-10.5 ± 6.4	-9.5 ± 10.8	----	-12.8 ± 6.0
(Albian)	F	20 ± 7	15.3 ± 18.8	----	11.3 ± 5.5
Western termination	R	- 32.8 ± 16.4	-31.5 ± 18.6	----	-35 ± 16.3
(Albian)	F	23.5 ± 15	18.9 ± 23	----	14.9 ± 14
Western termination	R	----	----	----	-27 ± 10.2
(Paleocene volcanics	F	----	----	----	-0.6 ± 9.8
and intrusive)					

The Cretaceous data from the Lancones basin indicate that for both the pre-Albian basement and the volcanic units, the flattening parameters are characterized by large errors so that they are barely significantly different from zero when the 110 and 80 Ma poles of Irving and Irving are used respectively. On the other hand, the use of Beck's pole indicates a northward transport with respect to SOAM. The evidence of latitudinal movement of these units with respect to the stable craton thus depends on the particular choice of the reference pole, and is consequently not really compelling. A comparison with the centered dipole field predictions suggests a small, barely significant northward drift of the order of 3-4°.

The rotation parameters of the units from the Lancones basin are on the contrary significantly different from zero and from each other: the pre-Albian basement of the Lancones synclinorium has undergone a large (92 ± 9°) clockwise rotation significantly different from the (63.5 ± 7°) clockwise rotation of the volcanic formations situated above the unconformity which in turn significantly differs from the $(25 \pm 7°)$ clockwise rotation of the post-Paleocene intrusions (in this last case we have referred the data to the geocentered dipole because the APWP of Irving and Irving is really ill defined for this period).

In the Cajamarca region, whatever the reference pole, the slightly westerly declinations observed at the eastern termination indicate that a small anticlockwise rotation of the order of 10° might have occurred but the rotation parameter is quite ill-defined. As mentioned above, we have also obtained results from the E-W transect, which will not be discussed here. Indeed the large anticlockwise rotations measured in this transect do not result from the phenomena studied here but arise largely from the local tectonics responsible for the arcuate shape of the Cajamarca deflection (Mitouard et al., to be published; Mourier, to be published). In this respect, the large anticlockwise rotations obtained in this region by Heki et al. (1984) from the Pariatambo and Chulec formations cannot, in our opinion, be interpreted directly in terms of rotation of the entire northern Peru. At the western termination where the general trend of the structures is andine, the paleomagnetic directions from the Pariatambo formation appear to be significantly rotated anticlockwise by an angle of the order of $(35 \pm 16°)$. The amount of rotation of the younger intrusive and volcanic formations, again calculated using the centered dipole field, is of the same order of magnitude $(27 \pm 10°)$. Thus the coastal area of the Cajamarca region apppears to have rotated significantly more than the internal zone, and this result is independent of any particular choice of the reference pole because coeval formations are compared. This result needs to be confirmed along other transects.

For both the Pariatambo and the Chulec formations in the Cajamarca region the value of the inclination is consistent with a paleolatitude some 8-9° south of the present one, i.e. the northward motion has been slightly greater than for the Lancones basin. Moreover the concordance/discordance analysis yields values of the F parameter slightly different from zero when Irving and Irving APWP is used and significantly different from zero when Beck's reference pole is used. This would imply a northward displacement of the order of a few hundreds of kilometers of the sampled regions with respect to the stable craton.

6. Discussion

In summary the main conclusions of the paleomagnetic study described above are: a) the observed inclination values suggest that since the Cretaceous western Ecuador has not moved significantly from the present latitudinal position. Parts of the coastal areas of Peru on the contrary appear to have undergone some latitudinal drift during this period. b) Since the Cretaceous very large rotations have taken place all along the western coast from 1°S to 8°S, with a greater amplitude to the north of the Huancabamba deflection. c) The pattern of rotations is consistently clockwise north of the Huancabamba deflection and anticlockwise south of it.

Although insufficient in both geographical and temporal extension, these results already allow some conclusions about the evolution of the western margin of South America when considered together with the previous geological observations.

In western Ecuador, for instance, considering the absence of major faults in the Coastal area, the results from the Piñon formation are best explained by a large scale block rotation than by local rotations of smaller blocks in a strike-slip environment as proposed in similar settings. Furthermore the oceanic affinity of the Piñon basement suggests an allochtonous origin, while the value of the inclination shows that a geographical gap might have existed between the Coastal block and the continent. Our interpretation (Roperch et al., 1987), shown schematically in Figure 6, involves the progressive consumption of a marginal oceanic area leading to the collision of the

505

Macuchi arc from south to north. Although evidence for island arc rotations before collision has been reported (Keating and Helsley, 1985) we think that the paleomagnetic data are better explained by a rotation of the Coastal block during or after collision in a dextral shear regime. Feininger and Bristow (1980) have however proposed a different solution which is also consistent with the paleomagnetic data.

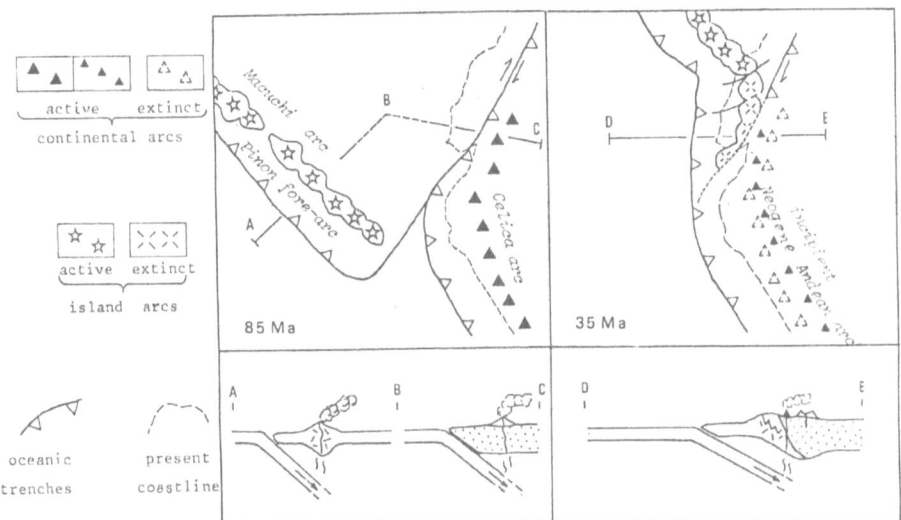

Figure 6. Oversimplified restored map and cross sections for the proposed plate regime for southern Colombia, Ecuador and northern Peru in Senonian and latest Eocene to Earliest Oligocene times.

The paleomagnetic results from the Paleozoic and Cretaceous units of Northwestern Peru are consistent with a previous hypothesis (Mourier et al., 1988b) of the accretion of an Amotape-Tahuin allochtonous continental terrane to the active Peruvian margin in the Early Cretaceous, after a moderate northward displacement of this block with respect to the stable craton. As a direct consequence of this accretion, we assume that the late Jurassic-Early Cretaceous subduction zone died out and was replaced by a late Cretaceous subduction zone situated ~ 150 Km more to the west, to which the Albian to late Cretaceous Lancones arc is associated. In our interpretation (Mourier et al., 1988a,b) the lowermost basic units of the Lancones synclinorium are considered to be part of an island arc caught in the suture zone between the Amoptape-Tahuin block and continental SOAM. The difference in the rotation parameters between the basic complex and the overlying volcanic units suggests that about 25-30° of the total rotation of the Lancones area are related to pre- or syn-accretion processes, as has also been documented in other studies elsewhere (Beck, 1988; Keating and Helsley, 1985; Wells and Coe, 1985).

As no large northward drift is suggested from the paleomagnetic data for the Cretaceous Lancones units, most of the 60° rotation of these units appears to have occurred in situ. Coherent block rotations about a vertical axis have been documented by paleomagnetic studies in many areas

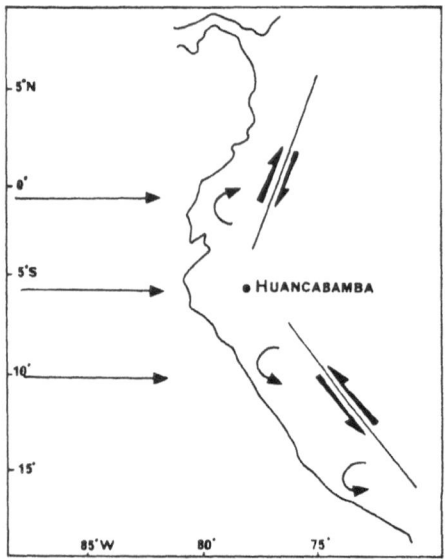

Figure 7. The E-W motion of the Nazca plate in the last 28 m.y. results in dextral oblique convergence north of Huancabamba, sinistral to the South. In turn oblique convergence may cause opposed block rotations (see also Beck, 1988).

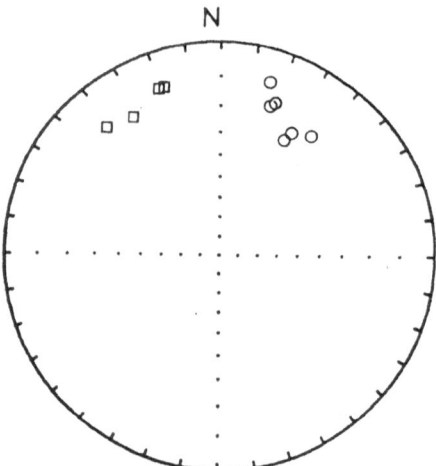

Figure 8. Equal area projection of the mean paleomagnetic directions determined in post-Paleocene intrusions north (circles) and south (squares) of the Huancabamba deflection.

of tectonic activity (Beck, 1976; Beck et al., 1980; Wells and Coe, 1985; McKenzie and Jackson, 1986; Luyendyk et al., 1980; Garfunkel, 1974; Kissel et al., this volume) and are a widespread characteristic of lithospheric deformation in response to shear. The results obtained here thus suggest that a dextral shear regime must also exist in this zone between the Nazca plate, southwestern Ecuador, the Amotape-Tahuin block and stable South America. However, one would expect such a shear to be associated with strike-slip faulting, but none has yet been documented in the field. So far, the geological evidences of Cretaceous and post-Cretaceous active faulting along the Amotape range have been considered as purely extensional features, although strongly dipping fold axes in the faulted edges of the Amotape-Tahuin Paleozoic range could be related to strike-slip faulting. However, evidence for right lateral strike-slip faulting active in the Oligo-Miocene exists slightly to the north of the studied region, in Southern Ecuador although the amplitude of the movements appears rather limited (Noblet et al., 1988).

The results from the Cajamarca region are more delicate to interpret. A northward transport of the Albian formations is inferred from a comparison with the dipole field predictions and the concordance/discordance analysis yields values of the F parameter significantly different from zero. This is a notable difference with most of the paleomagnetic data available for the Central and Southern Andes which all document significant rotations without any compelling evidence for large latitudinal displacement. A northward drift of a few degrees of the entire craton is not inconsistent with the present global reconstructions (Pilger, 1983). Wrench faulting along the Marañon Thrust and Fold Belt of northern Peru have been documented for the Neogene (Janjou, 1981; Mourier, in preparation) but the northern termination of this structural feature around 6°S is not consistent with large scale displacements exceeding a few tens of kilometers. It should be stressed that while the values of the F parameters depend on the choice of the reference pole, the shortening between the Lancones region and the Cajamarca region is inferred by data obtained from coeval formations and is thus independent of any particular choice of the reference pole. It is thus clear that a better understanding of the possible latitudinal displacement of the different sections of the coastal area will only be achieved when both more data from these regions and from the stable craton will be available.

The interpretation of the anticlockwise rotations obtained by our group and of those obtained by several authors (Heki et al., 1983; Kono et al., 1985; May and Butler, 1985) in other regions of the Central Andes is also not straightforward. Some investigators, notably Kono et al. (1985) argue that the results support the idea of oroclinal bending, although the complete straightening of the Arica bend requires far greater rotations than the observed ones, so that in this hypothesis the embayement of western South America appears as an inherited structure enhanced but not created by Mesozoic or Cenozoic shortening (Isacks, 1988). As an alternative explanation Beck (1988) has suggested that distributed shear caused by oblique subduction could be a more likely explanation for the rotated Chilean and Peruvian poles. For the region studied here, for instance, dextral and sinistral shear respectively north and south of the Huancabamba deflection could arise as a result of the angular relation between the trend of the margin and the direction of the convergence, which has been closely E-W in the last 28 m.y. (Figure 7) (Pilger, 1983; Pardo Casas and Molnar, 1987). The results from post-paleocene intrusions north or south of the Huancabamba defection, for instance, can easily be interpreted in terms of shear (Figure 8). Also the westward increase in counterclockwise rotation from the main chain to the coast observed in the Cajamarca region, if confirmed in other transects, is an indication for the existence of sinistral shear in the coastal zone (Wells and Heller, 1988). The hypothesis of sinistral shear is not in contradiction with the northward displacement, because the onset of the shear regime, linked to the direction of convergence in the last 28 m.y., may largely postdate the northward displacement, as suggested by a comparison between the F values for the Almbian and the Post Paleocene

508

formations in the Cajamarca region. On the other hand, there is no clear evidence of large scale left handed strike-slip faulting over the entire region affected by the rotation, as would be expected from distributed shear.

The paleomagnetic results which are presently available are also consistent with a recent completely different suggestion by Isacks (1988) that the concave bend of western South America, the "Bolivian orocline", may have been enhanced by an along-strike variation in the amount of late Cenozoic shortening. Indeed calculations using two versions of this model yield values of rotations quite consistent with the observed ones, although a problem may exist because counterclockwise rotations are found south of the presumed point where the rotations should reverse sense according to the model.

7. Conclusions.

The paleomagnetic results from the Coastal areas of South America reported here yield new evidence for possible oceanic and continental terrane accretion in western Ecuador and northern Peru and for in-situ post-accretion clockwise rotations, suggesting that the geodynamical evolution of northern Peru is more closely related to the processes observed in the Northern Andes than to those classically assumed for the Peruvian Andes. As discussed above, they also infer the existence of strike-slip faulting in the Huancabamba Andes which is presently poorly documented in the field. South of the Huancabamba deflection counterclockwise rotations are recorded, in agreement with results from other authors. The results are however not unambiguously interpretable and are consistent with at least two main and largely diverging models. Additional investigation of the changes in rotation along transects from the coast inward may clarify this problem. Also, the possible existence of northward transport of the coastal area needs to be further scrutinized before any definite interpretation can be made.

Clearly, and in spite of their imitations, the present results show that careful additional paleomagnetic studies may contribute to a better understanding of the lithospheric proccesses occuring along the Andean Cordillera.

Acknowledgements.

We wish to thank Yves Saint-Geours, Director of the Institut Français d'Etudes Andines (IFEA), in Lima for solving all kinds of major and minor problems for us. The Director of the Instituto Geofisico del Perù kindly provided the necessary permits. C. Noblet, A. Farfan Medrano, F. Monge, S. Benitez and E. Navarrete helped in the field work. The financial support has been given by the CEA, the CNRS, the ORSTOM and the INSU ASP Blocs et Collisions. This is contribution 971 from the CFR.

References.

Aspden, J.A. and McCourt W.J. , 1986, Mesozoic oceanic terrane in the Central Andes of Colombia, *Geology*, **14**, 415.
Audebaud, E., Capdevila, R., Dalmayrac, B., Debelmas, J., Laubacher, G., Lefevre, C., Marocco, R., Martinez, C., Mattauer, M., Mégard, F.,Paredes, J. and Tomas, T., 1973, Les

traits géologiques essentiels des andes centrales (Pérou-Bolivie), *Rev. Geog.Phys. Geol. Dyn.* **15**, 73.

Beaufils, G., 1978, Recherche de concentrations metalliques sulfurées et cartographie géologique dans la région de Tambo Grande. Unpublished report n° 007/78, BRGM, Lima.

Beck, M.E.,Jr., 1976, Discordant paleomagnetic pole positions as evidence of regional shear in the western Cordillera of North America, *American Journal of Science*, **276**, 694.

Beck, M.E.Jr., 1980, Paleomagnetic record of plate-margin tectonic processes along the western edge of North America, J. *Geophys. Res.*, **85**, 7115.

Beck, M.E., Jr, Cox,A. and Jones, D.L., 1980, Mesozoic and Cenozoic microplate tectonics of Western North America, *Geology*, **8**, 455.

Beck, M.E., Jr., 1988, Analysis of Late-Jurassic-Recent Paleomagnetic Data from Active Plate Margins of South America, *J. South American Earth Sci.*, **1**, 39-52.

Benavides, V., 1956, Cretaceous systems in northern Peru, *Bull. Am. Mus. nat. Hist.*, **108**, 252-494.

Cox, A., 1957, Remanent magnetism of Lower to Middle Eocene basalt flows from Oregon, *Nature*, **179**, 685.

Dalziel, I.W.D., Kligfield, R., Lowrie, W., and Opdyke, N.D., 1973, Paleomagnetic data from the southernmost Andes and the Antarctandes, in I*mplications of continental drift to the Earth Sciences*, D.H. Tarling and S.K. Runcom eds, Academis Press, London, Vol.1, p.37.

Demarest, H.H. Jr., 1983, Error analysis for the determination of tectonic rotation from paleomagnetic data, *J. Geophys. Res.* **88**, 4321.

Ernesto, M., Paleomagnetismo da formaçao Serra Geral: contribuiçao ao estudo do processo de abertura do Atlantico sul , 1985, PhD. Thesis, Sao Paulo, Bresil, 154 pp.

Faucher, B. Vernet, R., Bizon, J.J., Grekoff, N., Lys, M., and Sigal, J., 1971, Sedimentary formations in Ecuador. A stratigraphic and sedimentological survey. BEICIP, Paris.

Feininger, T., 1987, Allochtonous terranes in the Andes of Ecuador and Northwestern Peru. *Canadian J. Earth Sci.*, **24**, 266.

Feininger, T. and Bristow, C.R., 1980, Cretaceous and Paleogene geologic history of coastal Ecuador., *Geol. Rdsch.*, **69**, 849.

Feininger, T., and Seguin, M.K., 1983, Simple Bouger gravity anomaly field and the inferred crustal structure of continental Ecuador, 1983, *Geology*, **11**, 40-44.

Goossens, P.J. and Rose, W.I., 1973, Chemical composition and age determination of tholeitic rocks in the Basic Igneous Complex Ecuador, *Bull. Geol. Soc. Am.*, **84**, 1043-1052.

Garfunkel, Z., 1974, Model for the Cenozoic history of the Mojave Desert, California, and for its relation to adjacent regions, *Geol. Soc. Am. Bull.*, **85**, 1931.

Heki, K., Hamano, Y., Konoshita, H., Taira, A. and Kono, M., 1984, Paleomagnetic study of Cretaceous rocks of Peru, South America: Evidence for rotation of the Andes, *Tectonophysics*, **108**, 267.

Henderson, W.G., 1979, Cretaceous and Eocene volcanic arc activity in the Andes of northern Ecuador, *J. Geol. Soc. London*, **136**, 367-378.

Irving, E. and Irving , G.A., 1982, Apparent polar wander paths Carboniferous through Cenozoic and the assembly of Gondwana, *Geophysical Surveys* **5**, 141-188.

Isacks, B., 1988, Uplift of the Central Andean Plateau and Bending of the Bolivian Orocline, *J. Geophys. Res.*, **93**, 3211-3231.

James, D.E.,, 1971 Plate tectonic model for the evolution of the Central Andes, *Geol. Soc. Am. Bull.*, **82**, 3325.

Jaillard, E. (1988), Sedimentary evolution of an active margin during Upper Cretaceous times: the North Peruvian margin from Upper Aptian to Senonian, *Geol. Rundsch.*, in press.

510

Janjou, D., 1981, Données géologiques pour un modèle d'évolution des Andes nord-péruviennes, *Mem. Sci. Terre, Univ. P. et M. Curie*, Paris, 81-35, 170p.

Keating, B.H. and Helsley, C.E., 1985, Implications of island arc rotations to the studies of marginal terranes, *J. Geodynamics*, **2**, 159.

Kissel, C., Laj, C. and Mazaud, A., 1986, First paleomagnetic results from Neogene formations in Evia, Skyros and the Volos region and the deformation of Central Aegea, *Geophys. Res. Lett.* **13**, 1446.

Kono, M., Heki, K. and Hamano, Y., 1985, Paleomagnetic study of the Central Andes: counterclockwise rotation of the Peruvian Block, *J. Geodynamics*, **2**, 193.

Lebrat, M., Mégard, F., Dupuy, C. and Dostal, J., 1987, Geochemistry of the Cretaceous volcanic rocks of Ecuador: geodynamic implications., *Bull. Geol. Soc. Am.*, in press.

Luyendyk, B.P., Kamerlingh, M.J. and Terres, R.R., 1980, Geometric model for Neogene crustal rotations in Southern California, *Geol. Soc. Am. Bull.*, **91,**,211.

May, S.R. and Butler, R.F., 1985, Paleomagnetism of the Puente Piedra Formation, central Peru, *Earth Planet. Sci. Lett.*, **72**, 205.

McCourt, W.J., Aspden, J.A. and Brook, M., 1984, New geological and geochronological data from the Colombian Andes. Continental growth by multiple accretion, *J.Geol. Soc. London* **141**, 831.

McDonald, W.D., 1980, Net tectonic rotation, apparent tectonic rotation, and the structural tilt correction in paleomagnetic studies, *J. Geophys. Res.*, **85**, 3659-3670.

McKenzie, D. and Jackson, J.A., 1986, A block model of distributed deformation by faulting, *J. Geol. Soc. London*, **143**, 349..

Mégard, F., 1987, Cordilleran Andes and marginal Andes: a review of andean geology north of the Arica elbow (18° S), in Circum Pacific orogenic belts and evolution of the Pacific Ocean Basin, J.W.H Monger and J. Francheteau eds., *Ameri. Geophys. Union, Geodynamics series*, **18**, 71.

Mégard, F., 1978, Etude géologique des Andes du Pérou Central, *Mem. ORSTOM*, **86**, Paris.

Mourier, T., Laj, C., Mégard, F., Roperch, P., Mitouard, P. and Farfan Medrano, A., 1988a, An accreted continental terrane in northwestern Peru, *Earth. Plan. Scie. Lett.*, **88**, 182-192.

Mourier, T., Mégard, F., Pardo, A. and Reyes, L., 1988b, L'évolution mesozoïque des Andes de Huancabamba (3° -8°S) et l'hypothèse de l'accrétion du microcontinent Amotape-Tahuin. *Bull. Soc. Geol. Fr.*, **8**, 69.

Noblet, C., Lavenu, A. and Schneider F., 1988, to be published.

Palmer, H.C., Hayatsu, A. and MacDonald, W.D., 1980, The Middle Jurassic Camaraca Formation, Arica, Chile: paleomagnetism, K-Ar dating and tectonic implications, *Geophys. J.*, **62**, 155.

Pardo Casas, E. and Molnar, P., 1987, Relative motion of the Nazca (Farallon) and South America Plate since late Cretaceous time, *Tectonic*, **6**, 233-248.

Pilger, R.H., Jr., 1983, Kinematics of the South American subduction zone from global plate reconstructions. In: *Geodynamics of the eastern Pacific region, Caribbean and Scotia arcs*, R. Cabre ed., Geodynamics Series, Am. Geophys. Union, p 113.

Roperch, P., Mégard, F., Laj, C., Mourier, T., Clube, T., and Noblet, C., 1987, Rotated oceanic blocks in Western Ecuador, *Geophys. Res. Lett.*, **14**, 558.

Simpson, R.W. and Cox, A., 1977, Paleomagnetic evidence for tectonic rotation of the Oregon Coast Range, *Geology*, **5**, 585.

Wells, R.E. and Coe, R.S., 1985, Paleomagnetism and Geology of Eocene Volcanic Rocks of Southwest Washington, Implications for Mechanisms of Tectonic Rotation, *J. Geophys. Res.* **90**, 1925-1947.

Wells, R.E. and Heller, P.L., 1988, The relative contribution of accretion, shear, and extension to Cenozoic tectonic rotation in the Pacific Northwest, *Geol. Soc. Am. Bull.*, **100,** 325-338.

LIST of PARTICIPANTS

S. ALLERTON
School of Environmental Sciences,
University of East Anglia,
Norwich NR4 7TJ,
U.K.
(Currently:
Centre des Faibles Radioactivités,
Laboratoire mixte CEA-CNRS,
Av. de la Terrasse, B.P.1,
91198 Gif-sur-Yvette Cedex).

M. BECK
Dept. of Geology,
Western Washington University,
Bellingham, Wa 98225,
U.S.A.

J.-P. BRUN
Institut de Géologie,
Université de Rennes,
35042 Rennes Cedex,
France.

P. CHOUKROUNE
Institut de Géologie,
Université de Rennes,
35042 Rennes Cedex,
France.

R. COE
Earth Sciences,
University of California,
Santa Cruz, CA 95064,
USA

P. CROSS
Department of surveying,
University of Newcastle,
Old Brewery Building Haymarket,
Newcastle upon tyne, NE1 7RU,
U.K.

P. DAGLEY
University of Liverpool,
Dept. of Geological Sciences,
Jane Herdman Laboratories,
Brownlow street, P.O. Box 147,
Liverpool L69 3BX,
U.K.

P. DAVY
Institut de Géologie,
Université de Rennes,
35042 Rennes Cedex,
France.

P. C. ENGLAND
Department of Earth Sciences,
University of Oxford,
Parks Road, Oxford OX1 3PR,
U.K.

M. FULLER
Dept. of Geological Sciences,
University of California,
Santa Barbara, CA 93106,
U.S.A.

R. FUNICIELLO
Dipartimento Scienze della Terra,
Università di Roma,
Piazzale Aldo Moro, 5,
00100 Roma,
Italy.

Z. GARFUNKEL
Department of Geology,
Hebrew University of Jerusalem,
Jerusalem,
Israel.

513

C. GASPARINI
Istituto Nazionale di Geofisica
Via di Villa Ricoti,
00161 Roma,
Italy

E. GEISS
Deutsches Geodätisches
Forschungsinstitut,
Marstallplatz 8,
D-800 München 22,
R.F.A.

N. GORUR
Istanbul Technical University,
Jeoloji Bölümü,
Maçka-Istanbul,
Turkey.

D. HATZFELD
Lab. de Géophysique Interne et
Tectonophysique,
IRIGM, BP 68,
38402 St Martin d'Hères Cédex,
France.

J. JACKSON
Dept. of Earth Sciences,
University of Cambridge,
Bullard Laboratories
Madingley Road,
Cambridge CB3 0EZ,
U.K.

C. KISSEL
Centre des Faibles Radioactivités,
Laboratoire mixte CEA/CNRS
Avenue de la Terrasse, B.P.1
91198 Gif-sur-Yvette Cedex,
France.

C. LAJ
Centre des Faibles Radioactivités,
Laboratoire mixte CEA/CNRS
Avenue de la Terrasse, B.P.1
91198 Gif-sur-Yvette Cedex,
France.

S. H. LAMB
33 Famshawe Street,
Hertford S.G.143 AT,
U.K.

B. LUYENDYK
Dept. of Geological Sciences,
University of California,
Santa Barbara, CA 93106,
U.S.A.

F. MALTEZOU
Public Petroleum Company
Kifissias 199,
151 24 Maroussi, Athens,
Greece.

E. McCLELLAND
Dept. of Earth Sciences,
University of Oxford,
Parks Road, Oxford, OX1 3PR,
U.K.

D. McKENZIE
University of Cambridge,
Bullard Laboratories,
Madingley Road,
Cambridge CB3 0EZ,
U.K.

J. L. MERCIER
Géologie Dynamique interne,
Université Paris XI,
91405 Orsay Cedex,
France.

P. MOLNAR
Dept. of Earth, Atmospheric and
Planetary Sciences,
MIT Cambridge, Mass 02139
U.S.A

C. NICHOLSON
Institute for crustal studies,
University of California,
Santa Barbara, CA 93106,
U.S.A.

A. NUR
Department of Geophysics,
Stanford University,
Stanford CA 94305-2215,
U.S.A

M. L. OSETE
D. Fisica de la Tierra, Astronomia
y Astrofisica,
Universidad Complutense,
Madrid 28040,
Spain.

G. A. PAPADOPOULOS
Section of Seismicity and
Earthquake Resistant Structures,
Ministery of environment and
Public works,
17 Louisis Riankour Street,
11523 Athens,
Greece.

S. PAPAMARINOPOULOS
Department of Geology
University of Patra
261 10 Patra
Greece.

S. PAVLIDES
University of Thessaloniki,
Geological Department 351-1,
54006 Thessaloniki,
Greece.

F. SALVINI
Università di Pisa,
25 Via Santa Maria,
Pisa,
Italy.

M. SARIBUDAK
Istanbul Technical University,
Jeofizik Bölümü,
Maçka-Istanbul,
Turkey.

A. M. C. SENGOR
Istanbul Technical University,
Jeoloji Bölümü,
Maçka-Istanbul,
Turkey.

K. SIMEAKIS
IGME,
70 Messoghion Street,
Athens 608,
Greece.

D. B. STONE
University of Alaska,
Geophysical Institute,
Fairbanks, AK 99775,
U.S.A.

T. H. TORSVIK
Geological Survey of Norway,
P.O. Box 3006,
N-7001 Trondheim,
Norway.

M. TOZZI
Dipartimento Scienze della Terra,
Università di Roma,
Piazzale Aldo Moro, 5,
00100 Roma,
Italy.

K. VEROSUB
Dept. of Geology,
University of California,
Davis, CA 95616,
U.S.A.

R. I. WALCOTT
Geology Department,
Victoria University of Wellington,
Private Bag, Wellington,
New Zealand.

R. E. WELLS
U. S. Geological Survey,
345 Middlefield Road,
MS 75, Menlo Park, CA 94025,
U.S.A.

R. WESTAWAY
University of Liverpool,
Dept. of Geological Sciences,
Jane Herdman Laboratories,
Brownlow street, P.O. Box 147,
Liverpool L69 3BX,
U.K.